LONDON MATHEMATICAL SOCIETY LECTURE NOTE

Managing Editor: Professor M. Reid, Mathematics Institute,
University of Warwick, Coventry CV4 7AL, United Kingdom

The titles below are available from booksellers, or from Cambridge University Pres

Theory of *p*-adic Distributions:
Linear and Nonlinear Models

S. ALBEVERIO

Institut für Angewandte Mathematik, Universität Bonn,
Endenicherallee 60, D-53115 Bonn, Germany;
H. C. M., I.Z.K.S., SFB 611; BiBoS.

A. YU. KHRENNIKOV

International Center for Mathematical Modeling in Physics
and Cognitive Sciences MSI, Växjö University,
SE-351 95, Växjö, Sweden.

V. M. SHELKOVICH

Department of Mathematics, St. Petersburg State Architecture and
Civil Engineering University, 2-ja Krasnoarmeiskaya 4,
190005, St. Petersburg, Russia.

CAMBRIDGE
UNIVERSITY PRESS

CAMBRIDGE UNIVERSITY PRESS
Cambridge, New York, Melbourne, Madrid, Cape Town,
Singapore, São Paulo, Delhi, Tokyo, Mexico City

Cambridge University Press
The Edinburgh Building, Cambridge CB2 8RU, UK

Published in the United States of America by
Cambridge University Press, New York

www.cambridge.org
Information on this title: www.cambridge.org/9780521148566

First published 2010

A catalogue record for this publication is available from the British Library

ISBN 978-0-521-14856-6 Paperback

Cambridge University Press has no responsibility for the persistence or
accuracy of URLs for external or third-party internet websites referred to in
this publication, and does not guarantee that any content on such websites is,
or will remain, accurate or appropriate. Information regarding prices, travel
timetables, and other factual information given in this work is correct at
the time of first printing but Cambridge University Press does not guarantee
the accuracy of such information thereafter.

To Solvejg, Irina, Evgenia

Contents

Preface

For a few hundred years theoretical physics has been developed on the basis of real and, later, complex numbers. This mathematical model of physical reality survived even in the process of the transition from classical to quantum physics – complex numbers became more important than real, but not essentially more so than in the Fourier analysis which was already being used, e.g., in classical electrodynamics and acoustics. However, in the last 20 years the field of p-adic numbers $\mathbb{Q}_p$ (as well as its algebraic extensions, including the field of complex p-adic numbers $\mathbb{C}_p$) has been intensively used in theoretical and mathematical physics (see [1]–[3], [10]–[15], [24], [35], [38], [43], [44], [55], [64], [65], [77]–[79], [89], [94], [115]–[122], [123], [124], [133], [157], [158], [174], [198], [241]–[246] and the references therein). Thus, notwithstanding the fact that p-adic numbers were only discovered by K. Hensel around the end of the nineteenth century, the theory of p-adic numbers has already penetrated intensively into several areas of mathematics and its applications.

The starting point of applications of p-adic numbers to theoretical physics was an attempt to solve one of the most exciting problems of modern physics, namely, to combine consistently quantum mechanics and gravity, thus to create a theory of quantum gravity. In spite of the considerable success of some models, there is still no satisfactory general theory. In the physical literature one could even find discussions that one should not exclude the possibility that either quantum mechanics or gravitation theory is wrong, and radical new ideas are needed to solve the problems of quantum gravity. Physical models with p-adic space appeared as an attempt to provide a less radical approach to such problems (see [39], [40], [41], [77], [79]). The main idea behind p-adic theoretic physics (at least in the first period of its development) was that troubles with consistency of quantum mechanics and gravity are induced by the use of an infinitely divisible real continuum as the basic mathematical model of the underlying physical space-time. One should then look for new mathematical models of space-time and of structures defined on it. The fields of p-adic numbers provide an excellent new possibility. In the literature on applications of p-adic numbers in physics numerous arguments have been put forward in favor of the use of p-adic numbers. For example, in cosmology and string theory, speculations have been formulated according to which space-time at the so-called Planck scale (of the order of 10^{-23} cm) should be disordered and the Archimedean axiom might be violated for measurements which would be performed on such scales (see [115], [116], [245], [246]). Since the fields of p-adic numbers are disordered as well as non-Archimedean it seems appropriate to exploit such fields in this sense. We also mention the following argument [116], [241]. If one starts with the field of rational numbers[1] $\mathbb{Q}$ and wants to obtain a complete field with K valuation ("absolute value"), then, as a consequence of the famous Ostrovsky theorem (Theorem 1.3.2 in Chapter 1), there are only two possibilities for K: either K is the field of real numbers or K is one of the fields of p-adic numbers $\mathbb{Q}_p$.

p-adic based models can be considered as having given an important new contribution to string theory, gravity and cosmology. However, the most important consequence of this p-adic physical activity was that it induced a kind of "Carnot cycle": physics $\rightarrow$ mathematics $\rightarrow$ physics... Physical applications induce developments of new mathematical theories which in their turn induce new physical applications (which in their turn induce new mathematical theories and so on). p-adic string theory, gravity and cosmology stimulated development, and new applications of p-adic Fourier analysis, the theory of p-adic distributions and pseudo-differential equations, the theory of self-adjoint operators in $\mathcal{L}^2(\mathbb{Q}_p)$, as well as, e.g., the theory of stochastic differential equations over the field of p-adic numbers, Feynman path integration over p-adics, the theory of p-adic valued probabilities and dynamical systems (see, e.g., the monographs [36], [115], [130], [157], [162], [241] and the papers [11], [13], [32]–[34], [129]–[130], [249]–[251]). In their turn these mathematical theories provide new possibilities for physical applications of p-adic mathematics – e.g. in the theory of disordered systems (spin glasses) ([43], [44], [162], [164], [199]). Applications were, however, not only restricted to physics. There were also proposed p-adic models in psychology, cognitive and social sciences, and, e.g., in biology, image analysis (see, e.g., [116], [117]).

---

[1] We recall that in all real physical experiments measurements have only a finite precision. Therefore the result of any measurement is always a rational number, see [241], [116] for discussions.

Those new applications stimulated the development of new branches of p-adic analysis, in particular, the theory of p-adic wavelets ([21], [131], [132], [142], [143], [144], [146], [147], [148], [162], [163], [164], [165], [225]). The latter theory provides new possibilities for the study of p-adic pseudo-differential equations, surprisingly not only linear but also nonlinear ones ([23], [143], [162], [166]). The wavelet theory also gives a proper technique to solve the p-adic Schrödinger equation with point interactions ([31], [169]). And again the possibility to obtain wavelet-like solutions of the p-adic Schrödinger equation implied interesting physical consequences [133].

Further applications demonstrated that one should consider more general non-Archimedean models (e.g., for spin glasses, psychology, neurophysiology) and develop a corresponding analysis on general ultrametric spaces ([131], [132], [135]–[138], [162]).

Of course, one should never forget that mathematics has its own self-stimulating forces in the development of new concepts and formalisms. Concerning p-adic analysis we should mention the development of harmonic analysis on general locally compact groups as well as its applications inside mathematics ([203], [105]). However, many problems are intrinsically of a p-adic nature and they could not be formulated or even if formulated they could not be studied in such details without the p-adic framework. We remark that a part of p-adic analysis was also developed intensively without reference to physics; see, e.g., Gouvêa [98], Katok [112], Koblitz [152], Robert [204], Schikhof [214], Taibleson [230]. Nevertheless, many fundamental developments in p-adic analysis, such as [115]–[118], [157], [241], were indeed stimulated by physics and other applications.

When one speaks about p-adic analysis and applications, one should always mention which kind of p-adic model is under consideration. There are two main frameworks: either concerned with the relevant maps going from $\mathbb{Q}_p$ into $\mathbb{C}$, where $\mathbb{C}$ is the field of complex numbers, or from $\mathbb{Q}_p$ into $\mathbb{Q}_p$. The first stage of development of p-adic mathematical physics for the model with $\mathbb{C}$-valued maps as well as the corresponding analysis was presented in the excellent book by Vladimirov, Vololvich, and Zelenov [241] which became like a kind of "Bible" for physicists using p-adic models as well as for mathematicians looking for new problems connected with p-adics. Let us also mention the book by Kochubei [157], which is also basically concerned with applications, especially covering probabilistic aspects. Physical, cognitive and psychological models using $\mathbb{Q}_p$-valued maps as well as the corresponding analysis were presented in a series of monographs by one of the authors of this book [115]–[117]. The latter models are more extensively represented (at the book level) than the models based on maps from $\mathbb{Q}_p$ into $\mathbb{C}$.

We think the time has come to present the new mathematical theories which were developed in connection with p-adic mathematical physics in a monograph which would, on the one hand, update the book [241] by presenting and using new recent results and, on the other hand, open new possibilities for the development of p-adic analysis for maps from $\mathbb{Q}_p$ into $\mathbb{C}$. Our book should serve for such a purpose. In this book a wide-ranging survey of many important topics in p-adic analysis is presented. Moreover, it also gives a self-contained presentation of p-adic analysis itself.

The main attention is given to the development of p-adic distribution theory, including nonlinear models, Fourier and wavelet analysis, the theory of pseudo-differential operators and equations, Tauberian theorems, and asymptotic methods.

The book consists of fourteen chapters and four appendixes. Every chapter begins with a brief summary of its contents and a brief survey of related results. Chapters 1–5 cover standard material to introduce the reader to the theory of p-adic numbers, p-adic functions and p-adic distributions. These chapters are based first of all on the books by Vladimirov, Vololvich, Zelenov [241], Taibleson [230], and Katok [112], as well as on those by Koblitz [152], Gouvêa [98], Robert [204], and Schikhof [214]. The next nine chapters are more advanced and are based on the authors' original results (see References). These chapters represent new developments in p-adic harmonic analysis, in the p-adic theory of generalized functions, in the p-adic theory of wavelets, and in p-adic pseudo-differential operators and equations. In each chapter we compare p-adic results with the corresponding results in the real case.

In Chapter 6, we develop the theory of p-adic *associated homogeneous distributions* and *quasi associated homogeneous distributions*. The results of this chapter are closely connected with the theory of real *quasi associated homogeneous distributions*, which is presented in Appendix A and gives the solution of an important problem. In Chapter 7, we introduce the p-adic Lizorkin spaces of test functions and distributions and study their properties. These spaces are invariant under pseudo-differential operators (in particular, fractional operators), and, consequently, they provide "natural" domains of definition for them (see Chapters 9, 10). We would like to recall that in the real case this type of space was introduced by P. I. Lizorkin [180]–[182]. In the real setting such spaces play a key role in some problems related to fractional operators and are used in applications [210], [211], [206]. In Chapter 8, we develop a p-adic *wavelet theory*, as well as a p-adic *multiresolution analysis* and construct wavelet bases. The wavelet theory plays a key role in applications of p-adic analysis and gives a new powerful technique for solving p-adic problems (see Chapters 9–11). In Chapter 9 one class of multidimensional pseudo-differential

operators is studied. This class includes fractional operators. In this chapter
the spectral theory of pseudo-differential operators is developed. We derive
a criterion for multidimensional p-adic pseudo-differential operators to have
multidimensional wavelets (constructed in Chapter 8) and characteristic functions of balls as eigenfunctions. In particular, the multidimensional wavelets
are eigenfunctions of the Taibleson fractional operator. In Chapter 10, linear
and nonlinear p-adic evolutionary pseudo-differential equations are studied.
To solve the Cauchy problems we develop the *"variable separation method"*
which is an analog of the classical Fourier method. This method is based on the
fact that the wavelets constructed in Chapter 8 are eigenfunctions of pseudo-
differential operators constructed in Chapter 9. In Chapter 11 we continue
the investigation of p-adic Schrödinger-type operators with point interactions
started by A. N. Kochubei [156], and study $D^{\alpha} + V_Y$, where D^{α} ($\alpha > 0$) is
the operator of fractional differentiation (which was studied in Chapter 9)
and V_Y is a singular potential containing the Dirac delta functions. In Chapter 12, we extend the notion of regular variation introduced by J. Karamata
to the p-adic case and prove multidimensional *Tauberian type theorems* for
distributions. In Chapter 13, we study the asymptotic behavior of p-adic singular Fourier integrals. All constructed asymptotic formulas have the *stabilization* property (in contrast to the asymptotic formulas in the real setting).
Theorems which give asymptotic expansions of singular Fourier integrals are
related with the Abelian type theorems which were proved in Chapter 12.
In Chapter 14, we develop the nonlinear theory of distributions (generalized
functions). We construct the algebraic technique which allows us to solve both
linear and nonlinear singular problems of the p-adic analysis related with the
theory of distributions. Notwithstanding the fact that the real case problems
related with the multiplication of distributions were studied in many books and
papers (see [57], [66], [68], [86], [97], [108], [176]–[179], [194]–[197], [218]–
[222]), p-adic analogs of above mentioned problems have not been studied
so far.

In our opinion, our book could serve as a basic course on p-adics and its
applications, as well as for graduate courses such as: "the theory of p-adic
distributions", "p-adic harmonic analysis", "the theory of p-adic wavelets",
"the theory of p-adic pseudo-differential operators".

This research was supported in part by the grant of Växjö University: "Mathematical Modelling". The first and the third authors (S. A. and V. S.) were also
supported in part by DFG Project 436 RUS 113/809 and DFG Project 436 RUS
113/951. The third author (V. S.) was also supported in part by Grants 05-01-
04002-NNIOa and 09-01-00162 of Russian Foundation for Basic Research.

We wish to thank Ya. I. Belopolskaya, R. L. Benedetto, B. G. Dragovich, Yu. N. Drozzinov, A. Escassut, W. Karwowski, A. V. Kosyak, S. V. Kozyrev, S. Kuzhel, Hassan A. Marić, G. Parisi, V. I. Polischook, M. A. Skopina, V. S. Vladimirov, I. V. Volovich, B. I. Zavialov, V. V. Zharinov, K. Yasuda for fruitful discussions. We also are greatly indebted to Evgenia Sorokina for careful reading of this book and elimination of misprints.

Sergio Albeverio, Andrei Khrennikov, Vladimir Shelkovich

Bonn–Växjö–St. Petersburg–Moscow

1

p-adic numbers

1.1. Introduction

In this chapter some basic facts on the field of *p*-adic numbers $\mathbb{Q}_p$ are presented. Here we follow some sections from the books [47], [96], [98], [152], [241], and especially from the textbook [112]. Section 1.9.3 follows [254] and [241, 1.6.]. Section 1.9.4 is based on [163] and [162, 2.4.].

1.2. Archimedean and non-Archimedean normed fields

Denote by $\mathbb{N}$, $\mathbb{Q}$, $\mathbb{Z}$, $\mathbb{R}$, $\mathbb{R}_+$, $\mathbb{C}$ the sets of positive integers, rational numbers, integers, real numbers, nonnegative real numbers and complex numbers, respectively, and set $\mathbb{N}_0 = 0 \cup \mathbb{N}$.

Definition 1.1. Let X be a nonempty set. A *distance* or *metric* on X is a map $d : X \times X \to \mathbb{R}_+$ such that for all $x, y, z \in X$ we have
(1) $d(x, y) = 0 \Leftrightarrow x = y$;
(2) $d(x, y) = d(y, x)$;
(3) $d(x, y) \leq d(x, z) + d(z, y)$ (*triangle inequality*).
A set together with a metric is called a *metric space*.

A metric d is called *non-Archimedean* (or *ultra-metric*) if it satisfies the additional condition
(3′) $d(x, y) \leq \max(d(x, z), d(z, y))$ (*strong triangle inequality*).
The corresponding metric space is called an *ultrametric space*.

Since $\max(d(x, z), d(z, y))$ does not exceed the sum $d(x, z) + d(z, y)$, condition (3′), the *strong triangle inequality*, implies condition (3), the *triangle inequality*.

The same set X can give rise to many different metric spaces (X, d).

1

Definition 1.2. (i) Let X be a metric space with respect to metric d. A sequence $\{x_n : x_n \in X\}$ is called a *Cauchy sequence* if for any $\varepsilon > 0$ there exists a number $N(\varepsilon)$ such that $d(x_m, x_n) < \varepsilon$ for all $m, n > N(\varepsilon)$, i.e.,

$$\lim_{m,n \to \infty} d(x_m, x_n) = 0.$$

(ii) If any Cauchy sequence $\{x_n\}$ in the metric space X has a limit in X, then X is called a *complete metric space*.

(iii) A subset $M \subset X$ is called *dense* in X if every open ball $U_r^-(a) = \{x \in X : d(x, a) < r\}$ around every element $a \in X$ contains an element of M, i.e., for every $a \in X$ and every $r > 0$ we have $U_r^-(a) \cap M \neq \emptyset$.

If X is not a complete metric space, it can be completed by using an explicit construction of the completion. The proof of this important theorem can be found in many textbooks on functional analysis. This proof consists of an *explicit construction of the completion* $\widehat{X}$ and the metric $\widehat{d}$ on it.

In Section 1.4, we will give the complete proof of this theorem in the particular case of metric spaces called *normed fields*.

Definition 1.3. We say that two metrics d_1 and d_2 on a metric space X are *equivalent* if a Cauchy sequence with respect to d_1 is also a Cauchy sequence with respect to d_2, and vice versa.

The sets X we will be dealing with will mostly be fields. Recall that a field F is a set together with two operations $+$ and $\cdot$ such that F is a commutative group under $+$ and $F^\times = F \setminus \{0\}$ is a commutative group under $\cdot$, and the distributive law holds. $F^\times$ is called the *multiplicative group of the field*.

Denote by $\mathbb{Z}(F)$ the ring generated in F by its unity element. If F has zero characteristic, Char $F = 0$, i.e., for any $n = 1, 2, \ldots$

$$n \cdot 1 = \underbrace{1 + \cdots + 1}_{n \text{ times}} \in F, \quad n \cdot 1 \neq 0,$$

then $\mathbb{Z}(F)$ is isomorphic to the ring of integers $\mathbb{Z}$. Therefore in this case we can consider $\mathbb{Z}$ as a subring of the field F. We shall denote the element $n \cdot 1$ by the same symbol n as the corresponding natural number. In what follows we consider only normed rings F which have *zero characteristic*.

Definition 1.4. Let F be a field. A *norm* on F is a map $\| \cdot \| : F \to \mathbb{R}_+$ such that for all $x, y \in F$ we have

(1) $\|x\| = 0 \Leftrightarrow x = 0$;

(2) $\|xy\| = \|x\|\,\|y\|$[1];

(3) $\|x + y\| \leq \|x\| + \|y\|$ (*triangle inequality*).

A norm $\|\cdot\|$ is called *non-Archimedean* if it satisfies the additional condition

(3′) $\|x + y\| \leq \max(\|x\|, \|y\|)$ (*strong triangle inequality*).

The norm $\|\cdot\|$ is called *trivial* if $\|0\| = 0$ and $\|x\| = 1$ for all $x \neq 0$.

Using a norm $\|\cdot\|$, one can introduce a metric $d(x, y) = \|x - y\|$ which is induced by this norm. In this case we can regard the field F as a metric space. It is easy to see that a metric induced by a non-Archimedean norm is also non-Archimedean.

Definition 1.5. We say that two norms $\|x\|_1$ and $\|x\|_2$ on a normed field F are *equivalent* ($\|x\|_1 \sim \|x\|_2$) if they induce equivalent metrics.

Theorem 1.2.1. ([47, Ch. I,3.], [98, 3.1.]) *Let* $\|x\|_1 \sim \|x\|_2$.

(i) *If* $\|x\|_1$ *is trivial then* $\|x\|_2$ *is also trivial.*

(ii) $\|x\|_1 < 1$ *if and only if* $\|x\|_2 < 1$; $\|x\|_1 > 1$ *if and only if* $\|x\|_2 > 1$; $\|x\|_1 = 1$ *if and only if* $\|x\|_2 = 1$.

Theorem 1.2.2. *Let* $\|x\|_1$ *and* $\|x\|_2$ *be two norms on a field* F. *Then* $\|x\|_1 \sim \|x\|_2$ *if and only if there exists a positive real* α *such that*

$$\|x\|_2 = \|x\|_1^{\alpha}, \quad \forall x \in F. \tag{1.2.1}$$

Proof. Let $\|x\|_1 \sim \|x\|_2$. If $\|x\|_1$ is trivial, then, according to Theorem 1.2.1, $\|x\|_2$ is also trivial. Consequently, (1.2.1) is satisfied for any α.

Suppose that $\|x\|_1$ is non-trivial. In this case there exists an element $a \in F$ such that $\|a\|_1 \neq 1$. If necessary, one can replace a by a^{-1}, and consequently, one can assume that $\|a\|_1 < 1$. Let us define

$$\alpha = \frac{\log \|a\|_2}{\log \|a\|_1}.$$

According to Theorem 1.2.1, $\|a\|_1 < 1$ implies $\|a\|_2 < 1$. Thus $\alpha > 0$.

Now we will show that α satisfies (1.2.1). Choosing $x \in F$ such that $\|x\|_1 < 1$, we consider the set

$$S = \{r = \frac{m}{n} : m, n \in \mathbb{N}, \|x\|_1^r < \|a\|_1\}. \tag{1.2.2}$$

For any $r \in S$ we have $\|x\|_1^m < \|a\|_1^n$, i.e., $\|\frac{x^m}{a}x^n\|_1 < 1$. In view of Theorem 1.2.1, we conclude that $\|\frac{x^m}{a}x^n\|_2 < 1$, i.e., $\|x\|_2^m < \|a\|_2^n$ and $\|x\|_2^r < \|a\|_2$.

---

[1] In general, instead of this axiom the following one is used: $\|xy\| \leq \|x\|\,\|y\|$. But in this book to define a norm on the field we will use axiom (2).

Using the same arguments, one can see that

$$S = \{r = \frac{m}{n} : m, n \in \mathbb{N}, \|x\|_2^r < \|a\|_2\}. \tag{1.2.3}$$

The conditions (1.2.2) and (1.2.3) can be rewritten as

$$r > \frac{\log \|a\|_1}{\log \|x\|_1}, \qquad r > \frac{\log \|a\|_2}{\log \|x\|_2}. \tag{1.2.4}$$

The inequalities (1.2.4) imply that

$$\frac{\log \|a\|_1}{\log \|x\|_1} = \frac{\log \|a\|_2}{\log \|x\|_2}.$$

Otherwise there would be a rational r between these two numbers and only one of the conditions in (1.2.4) would be satisfied. Consequently,

$$\frac{\log \|x\|_2}{\log \|x\|_1} = \frac{\log \|a\|_2}{\log \|a\|_1} = \alpha,$$

i.e., (1.2.1) holds.

The cases $\|x\|_1 > 1$ and $\|x\|_1 = 1$ follow from Theorem 1.2.1. $\qquad \square$

Proposition 1.2.3. *The norm* $\|x\| = |x|^\alpha$, $\alpha > 0$, *is a norm on* $\mathbb{Q}$ *if and only if* $\alpha \leq 1$. *In this case it is equivalent to the norm* $|x|$.

Proof. Let $\alpha \leq 1$. It is clear that the first two properties of the norm from Definition 1.4 hold. Let us examine the third property (the triangle inequality). Suppose that $|y| \leq |x|$. Then

$$|x + y|^\alpha \leq \left(|x| + |y|\right)^\alpha = |x|^\alpha \left(1 + \frac{|y|}{|x|}\right)^\alpha$$

$$\leq |x|^\alpha \left(1 + \frac{|y|}{|x|}\right) \leq |x|^\alpha \left(1 + \frac{|y|^\alpha}{|x|^\alpha}\right) \leq |x|^\alpha + |y|^\alpha.$$

If $\alpha > 1$, the triangle inequality does not hold. For example, $|1 + 1|^\alpha = 2^\alpha > |1|^\alpha + |1|^\alpha = 2$. $\qquad \square$

Now we give a criterion for a norm to be non-Archimedean.

Theorem 1.2.4. *The norm* $\| \cdot \|$ *is non-Archimedean if and only if* $\|n\| \leq 1$ *for any* $n \in \mathbb{Z}$.

Proof. We prove this proposition by induction. Let the norm $\| \cdot \|$ be non-Archimedean. It is clear that $\|1\| = 1 \leq 1$. Let us assume that $\|k\| = 1 \leq 1$ for all $k = 1, 2, \ldots, k - 1$. Next, it follows from our assumption that $\|n\| = \|1 + (n - 1)\| \leq \max(\|1\|, \|(n - 1)\|) = 1$. Thus, according to the induction axiom, we have $\|n\| \leq 1$ for any $n \in \mathbb{N}$. Since $\| - n\| = \|n\|$, we conclude that $\|n\| \leq 1$ for any $n \in \mathbb{Z}$.

Conversely, assume that $\|n\| \leq 1$ for any $n \in \mathbb{Z}$. Since the binomial coefficients $C_n^k = \frac{n!}{k!(n-k)!}$, $k \leq n$, are integers, we have $\left\| C_n^k \right\| \leq 1$. Thus

$$\|x + y\|^n = \|(x + y)^n\| = \left\| \sum_{k=0}^{n} C_n^k x^k y^{n-k} \right\| \leq \sum_{k=0}^{n} \left\| C_n^k \right\| \|x\|^k \|y\|^{n-k}$$

$$\leq \sum_{k=0}^{n} \|x\|^k \|y\|^{n-k} \leq (n+1)\big(\max(\|x\|, \|y\|) \big)^n.$$

Hence, for every integer n we have $\|x + y\| \leq \sqrt[n]{n+1} \max(\|x\|, \|y\|)$. Letting n go to ∞, we obtain $\|x + y\| \leq \max(\|x\|, \|y\|)$, i.e., the norm $\| \cdot \|$ is non-Archimedean. $\qquad\square$

Corollary 1.2.5. *The norm* $\| \cdot \|$ *is non-Archimedean if and only if* $\sup\{\|n\| : n \in \mathbb{Z}\} = 1$.

By Theorem 1.2.4 one can observe the difference between non-Archimedean and Archimedean norms. According to this theorem, a norm $\| \cdot \|$ is Archimedean if and only if it satisfies the *Archimedean property (axiom)*: *for any* $x, y \in F$, $x \neq 0$, *there exists* $n \in \mathbb{N}$ *such that* $\|nx\| \geq \|y\|$.

Indeed, if $\|y\| > \|x\|$, then the Archimedean property implies the existence of $n \in \mathbb{N}$ such that $\|n\| > \|y\|/\|x\| > 1$, i.e., the norm is Archimedean. Conversely, if a norm $\| \cdot \|$ is Archimedean, there exists $n \in \mathbb{N}$ such that $\|n\| > 1$. Hence, $\|n\|^k \to \infty$, as $k \to \infty$. Consequently, for any $x, y \in F$, $x \neq 0$, there exists k such that $\|n^k\| > \|y\|/\|x\|$. Thus the Archimedean property $\|n^k x\| > \|y\|$ holds.

If the norm is non-Archimedean, for any $n \in \mathbb{Z}$ we have $\|nx\| \leq \|x\|$.

The non-Archimedean property of a norm has some strange consequences.

Proposition 1.2.6. *If a field* F *is non-Archimedean, then for* $x, y \in F$

$$\|x\| \neq \|y\| \implies \|x + y\| = \max\big(\|x\|, \|y\| \big).$$

Thus, any triangle in an ultra-metric space is isosceles and the length of its base does not exceed the lengths of the sides.

Proof. Exchanging x and y if necessary, we may suppose that $\|x\| > \|y\|$. By the strong triangle inequality $(3')$ in Definition 1.4,

$$\|x + y\| \leq \max(\|x\|, \|y\|) = \|x\|.$$

On the other hand,

$$\|x\| = \|(x + y) - y\| \le \max(\|x + y\|, \|y\|).$$

Since $\|x\| > \|y\|$, we must have $\|x\| \le \|x + y\|$. Consequently, $\|x\| = \|x + y\|$. □

Thus for a non-Archimedean field $\|x \pm y\| \le \max(\|x\|, \|y\|)$, and in the case where $\|x\| \ne \|y\|$, the above inequality becomes an equality.

Proposition 1.2.7. *If the field F is non-Archimedean, then any point of an open ball $U_r^-(a) = \{x \in F : \|x - a\| < r\}$ is its center, i.e., if $b \in U_r^-(a)$, then $U_r^-(b) = U_r^-(a)$. The same statement is true for a closed ball $U_r(a) = \{x \in F : \|x - a\| \le r\}$.*

Proof.

$$x \in U_r^-(a) \Rightarrow \|x - a\| < r \Rightarrow \|x - b\| = \|(x - a) + (a - b)\|$$
$$\le \max(\|x - a\|, \|a - b\|) < r \Rightarrow x \in U_r^-(b).$$

The reverse implication is proved in the same way.

Replacing $<$ with $\le$, we prove the statement for a closed ball $U_r(a)$. □

1.3. Metrics and norms on the field of rational numbers

1.3.1. *p*-adic norm

We know that there exists a norm of the field $\mathbb{Q}$: the ordinary absolute value $|\cdot|$. This norm induces the ordinary Euclidean metric $d(x, y) = |x - y|$. The question arises: are there any other norms on $\mathbb{Q}$? It turns out that *there are other norms*. Below we describe all such norms.

Definition 1.6. Let p be a prime number. We define the *p-adic order* $\mathrm{ord}_p(x)$ *of a rational number* $x \in \mathbb{Q}$ by the following definition:

(i) If $x \in \mathbb{Z}$, then $\mathrm{ord}_p(x)$ is equal to the highest power of p which divides x.

(ii) If $x = a/b$, where $a, b \in \mathbb{Z}$, then $\mathrm{ord}_p(x) = \mathrm{ord}_p(a) - \mathrm{ord}_p(b)$. The p-adic order of $x \in \mathbb{Q}$ is also called the *p-adic additive valuation* and denoted as $v_p(x)$.

(iii) We set $\mathrm{ord}_p(0) = +\infty$.

The reason to set $\mathrm{ord}_p(0) = +\infty$ is that we can divide 0 by p^n for each $n \in \mathbb{N}$.

It is clear that for all $x, y \in \mathbb{Q}$:

$$\begin{aligned} \mathrm{ord}_p(xy) &= \mathrm{ord}_p(x) + \mathrm{ord}_p(y), \\ \mathrm{ord}_p(x+y) &\geq \min\left(\mathrm{ord}_p(x), \mathrm{ord}_p(y)\right). \end{aligned} \tag{1.3.1}$$

Now we define a map $| \cdot |_p : \mathbb{Q} \to \mathbb{R}_+$ as follows:

$$|x|_p = \begin{cases} p^{-\mathrm{ord}_p(x)}, & \text{if} \quad x \neq 0, \\ 0, & \text{if} \quad x = 0. \end{cases} \tag{1.3.2}$$

We point out that the definition $|0|_p = 0$ follows from Definition 1.6 (iii).

It is clear that the function $| \cdot |_p$ can take only a discrete set of values $\{p^\gamma : \gamma \in \mathbb{Z}\}$.

Note that if $x, y \in \mathbb{N}$, then $x \equiv b \pmod{p^n}$ if and only if $|x - y|_p \leq p^{-n}$. By the notation $x \equiv b \pmod{z}$ we mean that z divides $x - y$.

Theorem 1.3.1. *The map $| \cdot |_p$ is an non-Archimedean norm on the field of rational numbers $\mathbb{Q}$, i.e., it satisfies the axioms (1), (2), (3′) from Definition 1.4.*

Proof. It is clear that axiom (1) holds. Because $\mathrm{ord}_p(xy) = \mathrm{ord}_p x + \mathrm{ord}_p y$, axiom (2) also holds.

Let us verify axiom (3′). If $x = 0$ or $y = 0$, or $x + y = 0$, property (3′) is trivial, so we assume that x, y, $x + y$ are nonzero. Let $x = a/b$, $y = c/d$ be written in lowest terms. Then we have $x + y = (ad + bc)/bd$ and

$$\begin{aligned} \mathrm{ord}_p(x+y) &= \mathrm{ord}_p(ad + bc) + \mathrm{ord}_p(bd) \\ &\geq \min\left(\mathrm{ord}_p(ad), \mathrm{ord}_p(bc)\right) - \mathrm{ord}_p(b) - \mathrm{ord}_p(d) \\ &= \min\left(\mathrm{ord}_p(a) + \mathrm{ord}_p(d), \mathrm{ord}_p(b) + \mathrm{ord}_p(c)\right) - \mathrm{ord}_p(b) \\ &\quad - \mathrm{ord}_p(d) \\ &= \min\left(\mathrm{ord}_p(a) - \mathrm{ord}_p(b), \mathrm{ord}_p(c) - \mathrm{ord}_p(d)\right) \\ &= \min\left(\mathrm{ord}_p(x), \mathrm{ord}_p(y)\right). \end{aligned}$$

Consequently,

$$\begin{aligned} |x + y|_p = p^{-\mathrm{ord}_p(x+y)} &\leq \max\left(p^{-\mathrm{ord}_p(x)}, p^{-\mathrm{ord}_p(y)}\right) \\ &= \max\left(|x|_p, |y|_p\right) \leq |x|_p + |y|_p. \end{aligned}$$

$\square$

Remark 1.1. If in definition (1.3.2) instead of a prime number p we use an arbitrary integer number $m > 1$, then axiom (2) from Definition 1.4 may fail. For example, let $m = 4$. Then $|2|_4 = 1$, but $|2 \cdot 2|_4 = \frac{1}{4} \neq |2|_4 |2|_4 = 1$.

Remark 1.2. According to Proposition 1.2.6, if $|x|_p \neq |y|_p$ then

$$|x + y|_p = \max\left(|x|_p, |y|_p\right).$$

For $p = 2$, if $|x|_2 = |y|_2$ then

$$|x + y|_2 \leq \frac{1}{2}|x|_p.$$

The latter inequality follows from definition (1.3.2).

It is clear that $|p^n|_p \to 0$ as $n \to \infty$, so that *high* powers of p are *small* with respect to the p-adic norm (1.3.2).

In view of Theorem 1.2.4, for any $n \in \mathbb{Z}$ we have $|nx|_p \leq |x|_p$.

Remark 1.3. The norms $|\cdot|_{p_1}$ and $|\cdot|_{p_2}$ are not equivalent if p_1 and p_2 are different primes. Indeed, for the sequence $x_n = (p_1/p_2)^n$ we have $|x_n|_{p_1} \to 0$, but $|x_n|_{p_1} \to \infty$, as $n \to \infty$.

Thus we observe that the field $\mathbb{Q}$ admits the p-adic norms $|\cdot|_p$ for each prime p, as well the ordinary norm (ordinary absolute value) $|\cdot|$. The last norm sometimes is denoted by $|\cdot|_\infty$ for $p = \infty$, and $p = \infty$ sometimes is called the infinite prime.

1.3.2. The Ostrovski theorem

Now we prove that *there are no other norms on $\mathbb{Q}$ except for $|\cdot|_p$ and $|\cdot|$.*

Theorem 1.3.2. *(Ostrovski's theorem) Any* non-trivial norm $\|\cdot\|$ *on the field* $\mathbb{Q}$ *is equivalent either to the real norm $|\cdot|$ or to one of the p-adic norms $|\cdot|_p$.*

Proof. 1. Suppose that $\|\cdot\|$ is Archimedean, i.e., there exists $n \in \mathbb{N}$ such that $\|n\| > 1$. Let n_0 be the least such n. Of course we can find some positive real number α so that $\|n_0\| = n_0^\alpha$.

Let us write any positive integer n in base n_0, i.e., in the form

$$n = a_0 + a_1 n_0 + \cdots + a_s n_0^s, \tag{1.3.3}$$

where $0 \leq a_i \leq n_0 - 1$, $i = 0, 1, \ldots, s$, $a_s \neq 0$. Then

$$\|n\| \leq \|a_0\| + \|a_1 n_0\| + \cdots + \|a_s n_0^s\| = \|a_0\| + \|a_1\| \|n_0^\alpha + \cdots + \|a_s\| n_0^{s\alpha}.$$

Since we chose n_0 to be the smallest integer whose norm was greater than 1, we know that $\|a_i\| \leq 1$. Hence

$$\|n\| \leq 1 + n_0^\alpha + \cdots + n_0^{s\alpha} = n_0^{s\alpha}\left(1 + n_0^{-\alpha} + \cdots + n_0^{-s\alpha}\right)$$

$$\leq n_0^{s\alpha} \sum_{i=0}^{\infty} n_0^{-i\alpha} = n_0^{s\alpha} \frac{n_0^\alpha}{n_0^\alpha - 1}.$$

Since $n \geq n_0^s$, the latter inequality implies

$$\|n\| \leq Cn_0^{s\alpha} \leq Cn^\alpha,$$

where $C = \frac{n_0^\alpha}{n_0^\alpha - 1}$ does not depend on n. Substituting an integer of the form n^N in the above inequality instead of n, we have

$$\|n^N\| \leq Cn^{N\alpha} \Leftrightarrow \|n\| \leq \sqrt[N]{C} n^\alpha.$$

Letting $N \to \infty$ for n fixed, we obtain

$$\|n\| \leq n^\alpha. \tag{1.3.4}$$

Now we prove the opposite inequality. It follows from (1.3.3) that $n_0^s \leq n < n_0^{s+1}$, and, consequently,

$$n_0^{(s+1)\alpha} = \|n_0^{s+1}\| = \|n + n_0^{s+1} - n\| \leq \|n\| + \|n_0^{s+1} - n\|.$$

Thus

$$\|n\| \geq \|n_0^{s+1}\| - \|n_0^{s+1} - n\| \geq n_0^{(s+1)\alpha} - (n_0^{s+1} - n)^\alpha,$$

because according to (1.3.4), we have $\|n_0^{s+1} - n\| \leq (n_0^{s+1} - n)^\alpha$. Now since $n_0^s \leq n < n_0^{s+1}$, it follows that

$$\|n\| \geq n_0^{(s+1)\alpha} - (n_0^{s+1} - n)^\alpha = n_0^{(s+1)\alpha}\left(1 - \left(1 - \frac{1}{n_0}\right)^\alpha\right)$$

$$= C'n_0^{(s+1)\alpha} \geq C'n_0^\alpha,$$

and once again $C' = 1 - \left(1 - \frac{1}{n_0}\right)^\alpha$ does not depend on n and is positive. As before, we now use the latter inequality for n^N, take the Nth root, and let $N \to \infty$, obtaining

$$\|n\| \geq n^\alpha. \tag{1.3.5}$$

From (1.3.4) and (1.3.5), we deduce that $\|n\| = n^\alpha$ for all $n \in \mathbb{N}$. Using property (2) of the norm, one can see that $\|x\| = |x|^\alpha$ for all $x \in \mathbb{Q}$. In view of Theorem 1.2.2, we can conclude that such a norm is equivalent to the ordinary absolute value $|\cdot| \equiv |\cdot|_\infty$.

2. Now suppose that $\|\cdot\|$ is non-Archimedean. Then according to Theorem 1.2.4, we have $\|n\| \leq 1$ for every $n \in \mathbb{N}$. Since $\|\cdot\|$ is non-trivial, there exists a smallest $n_0 \in \mathbb{N}$ such that $\|n_0\| < 1$. First, we observe that n_0 must be a prime number. Indeed, if $n_0 = n_1 n_2$, where n_1 and n_2 are smaller than n_0, then by our choice for n_0 we would have $\|n_1\| = \|n_2\| = 1$, and so $\|n_0\| = \|n_1\|\|n_2\| = 1$. Thus n_0 is a prime number. Denote this prime number by p.

Next, we will prove that if $n \in \mathbb{Z}$ is not divisible by p, then $\|n\| = 1$. If we divide n by p we will have a remainder, so that we can write $n = rp + s$, where $0 < s < p$. By the minimality of p, $\|s\| = 1$. We also have $\|rp\| < 1$, because $\|p\| < 1$ and $\|r\| \leq 1$ (since r is an integer). Consequently,

$$\|n - s\| = \|rp\| < \|s\| = 1,$$

and by Proposition 1.2.6, $\|n\| = \|s\| = 1$. Finally, given $n \in \mathbb{Z}$, one can write $n = p^v n'$, where p does not divide n'. Hence

$$\|n\| = \|p^v n'\| = \|p\|^v,$$

where $\rho = \|p\| < 1$. Then $\rho = (1/p)^\alpha$ for some positive real α. Therefore

$$\|n\| = (1/p)^{v\alpha} = \|n\|_p^\alpha.$$

Now, using property (2) of the norm, it is easy to see that the same formula holds for any nonzero rational number x in place of n. In view of Theorem 1.2.2, we have $|\cdot| \equiv |\cdot|_p$.

Thus the theorem is proved. ☐

Proposition 1.3.3. *For any $x \in \mathbb{Q}^\times = \mathbb{Q} \setminus \{0\}$ the following relation holds:*

$$\prod_{2 \leq p \leq \infty} |x|_p = 1$$

Proof. Expanding a rational number x by prime factors $x = \epsilon p_1^{\alpha_1} p_2^{\alpha_2} \cdots p_n^{\alpha_n}$, where $\epsilon = \pm 1$, p_j are different prime numbers, $\alpha_j \in \mathbb{Z}$, and using definition (1.3.2) of p-adic norm and the ordinary absolute value, we obtain

$$|x|_{p_j} = p_j^{-\alpha_j}; \qquad |x|_p = 1, \ p \neq p_j; \qquad |x|_\infty = p_1^{-\alpha_1} p_2^{\alpha_2} \cdots p_n^{\alpha_n}.$$

These facts imply our statement. ☐

The formula from Proposition 1.3.3 establishes the relation between all non-trivial norms of $\mathbb{Q}$ [2].

1.4. Construction of the completion of a normed field

Starting from an arbitrary normed field $(F, \|\cdot\|)$ not necessary complete with respect to its norm $\|\cdot\|$, we construct the field $\widehat{F}$ containing F, and supply it with a norm (induced by the norm $\|\cdot\|$ of F) in such a way that the field

---

[2] Formulas of this type are called *"adelic products"*. For details on adelic formulas, see for example [233]–[235], [244].

$\widehat{F}$ will be a *complete* normed field. Below we give the proof of this statement from [112, 1.3].

1. In the completion procedure the main role will be played by the Cauchy sequences (see Definition 1.2 (i)). Let us denote by $\{F\}$ the collection of all Cauchy sequences in $(F, \| \cdot \|)$. It is easy to see that if $\{a_n\}$, $\{b_n\}$ are Cauchy sequences then $\{a_n \pm b_n\}$, $\{a_n b_n\}$ are also Cauchy sequences, i.e., Cauchy sequences can be added, subtracted and multiplied (see the well-known standard results in analysis). Consequently, $\{F\}$ constitutes a commutative ring with identity elements under addition and multiplication $\widehat{0} = \{0, 0, 0, \dots\}$ and $\widehat{1} = \{1, 1, 1, \dots\}$, respectively. Since for every $a \in F$ the *constant sequence*

$$\widehat{a} = \{a, a, a, \dots\}$$

belongs to $\{F\}$, the set $\{F\}$ contains a subring isomorphic to F.

2. Let us introduce the set P of all infinitesimal sequences from $\{F\}$, i.e., all Cauchy sequences $\{a_n\}$ such that $\lim_{n \to \infty} \|a_n \cdot \| = 0$. According to the well-known standard results in analysis, if $\{a_n\}$, $\{b_n\} \in P$ then $\{a_n \pm b_n\} \in P$. Next, if $\{a_n\}$ is a Cauchy sequence then it is a bounded sequence: according to Definition 1.2 (i), $\|a_n \cdot \| \leq \|a_{N(\varepsilon)+1} \cdot \| + (\varepsilon)$ for all $n > N(\varepsilon)$. Thus, if $\{a_n\} \in P$ and $\{b_n\} \in F$ (and consequently it is bounded) then $\{a_n \cdot b_n\} \in P$. Therefore, P is an *ideal* in $\{F\}$, i.e., a subring in $\{F\}$ such that for all $p \in P$ and $a \in \{F\}$ we have $ap \in P$.

Let

$$\widehat{F} \stackrel{def}{=} \{F\}/P.$$

Elements of $\widehat{F}$ are equivalence classes of Cauchy sequences in $(F, \| \cdot \|)$. Here two Cauchy sequences are equivalent if their difference is an infinitesimal sequence, i.e., this difference belongs to P. We stress that for different $a \in F$ constant sequences $\widehat{a} = \{a, a, a, \dots\}$ belong to different equivalence classes in $\widehat{F}$. We shall denote by $A = (\{a_n\})$ the equivalence classes of the Cauchy sequences $\{a_n\}$, so $A = (\{a_n\}) \in \widehat{F}$. We will identify $a \in F$ with $(\widehat{a}) \in \widehat{F}$ and will thus consider F as a subset of $\widehat{F}$.

Theorem 1.4.1. *The set $\widehat{F}$ is a field.*

Proof. We introduce on $\widehat{F}$ the operations of addition and multiplication in the following way: if $\{a_n\} \in A$, $\{b_n\} \in B$, then $A + B \stackrel{def}{=} (\{a_n + b_n\})$, $AB \stackrel{def}{=} (\{a_n b_n\})$. It is easy to see that these operations do not depend on the choice of representatives in the equivalence classes and $\widehat{F}$ constitutes a commutative ring with identity elements $(\widehat{0})$ and $(\widehat{1})$ under addition and multiplication, respectively.

Now we prove that $\widehat{F}$ is a field. Let A be an equivalence class in $\widehat{F}$ different from the zero class $\widehat{(0)} = P$, and let $\{a_n\} \in A$. Since $\{a_n\} \in A$ is not an infinitesimal sequence, it is easy to see that there exist a positive number c and a positive integer N such that

$$\|a_n\| > c \qquad \forall n \geq N.$$

Now we consider a new sequence $\{a_n^*\}$, where

$$a_n^* = \begin{cases} 0, & \text{if } 1 \leq n \leq N-1, \\ 1/a_n, & \text{if } n \geq N. \end{cases}$$

The sequence $\{a_n^*\}$ is a Cauchy sequence. Indeed, if $m, n \geq N$, then

$$0 \leq \|a_m^* - a_n^*\| = \left\| \frac{1}{a_m} - \frac{1}{a_n} \right\| = \frac{\|a_m - a_n\|}{\|a_m\| \cdot \|a_n\|} \leq c^{-2} \|a_m - a_n\|.$$

Since $\{a_n\}$ is a Cauchy sequence, the latter inequality implies that $\{a_n^*\}$ is also a Cauchy sequence. Let us denote the equivalence class of the Cauchy sequence $\{a_n^*\}$ by A^{-1}. Then

$$\{a_n\}\{a_n^*\} = \{\underbrace{0, \ldots, 0}_{N-1}, 1, 1, 1, \ldots\},$$

where the Cauchy sequence on the right differs from $\widehat{1}$ by the infinitesimal sequence

$$\{\underbrace{-1, \ldots, -1}_{N-1}, 0, 0, 0, \ldots\}.$$

Thus $AA^{-1} = \widehat{(1)}$.   □

Now we extend the norm $\| \cdot \|$ from F to the field $\widehat{F}$ in the following way. If $\{a_n\}$ is any Cauchy sequence in $A \in \widehat{F}$, we set

$$\|A\| \overset{def}{=} \lim_{n \to \infty} \|a_n\|. \tag{1.4.1}$$

This norm is well defined. Firstly, since $|\|a_n\| - \|a_m\|| \leq \|a_n - a_m\|$, we have that $\{\|a_n\|\}$ is a Cauchy sequence of real numbers with respect to the absolute value. Due to the fact that $\mathbb{R}$ is complete, the limit $\|A\|$ exists. Secondly, this limit does not depend on the choice of the Cauchy sequence $\{a_n\}$ in A. Let $\{a_n'\}$ be a second Cauchy sequence in A. Then $0 \leq \lim_{n \to \infty} |\|a_n\| - \|a_n'\|| \leq \lim_{n \to \infty} \|a_n - a_n'\| = 0$. Thus $\lim_{n \to \infty} \|a_n\| = \lim_{n \to \infty} \|a_n'\|$.

Proposition 1.4.2. *The function* (1.4.1) *is a norm on* $\widehat{F}$.

Proof. Let us verify the first three properties from Definition 1.4.

(1) If $A = (\widehat{0})$, then $\{a_n\}$ is a null sequence, and, consequently, $\|A\| = 0$. If $A \neq (\widehat{0})$ and $A = (\{a_n\})$, then there exist positive numbers c and N such that for all $n \geq N$ we have $\|a_n\| \geq c > 0$. Hence $\|A\| > 0$.

(2) Now let $A = (\{a_n\})$, $B = (\{b_n\})$. Using the properties of real limits, we obtain

$$\|AB\| = \lim_{n \to \infty} \|a_n b_n\| = \lim_{n \to \infty} \|a_n\| \|b_n\|$$

$$= \lim_{n \to \infty} \|a_n\| \lim_{n \to \infty} \|b_n\| = \|A\| \|B\|.$$

(3) Similarly, we obtain $\|A + B\| \leq \|A\| + \|B\|$. $\square$

Theorem 1.4.3. *The field $\widehat{F}$ is complete with respect to the norm $\|\cdot\|$, and F is a dense subfield of $\widehat{F}$.*

Proof. We first prove the second part. Let $A \in \widehat{F}$, and let $\{a_m\}$ be a Cauchy sequence in F representing A. For each fixed positive integer n, we consider the constant sequence $\widehat{a}_n$. Then the sequence $\{a_m - a_n\}_{m=1}^{\infty}$ represents the element $A - (\widehat{a}_n)$, and since $\{a_m\}$ is a Cauchy sequence, we can write

$$\lim_{n \to \infty} \|A - (\widehat{a}_n)\| = \lim_{n,m \to \infty} \|a_m - a_n\| = 0. \tag{1.4.2}$$

Thus F is dense in $\widehat{F}$.

Now we suppose that $\{A_n\} = \{A_1, A_2, \dots\}$ is a Cauchy sequence in $\widehat{F}$. By the density of F in $\widehat{F}$, for any A_n there exists an element $a_n \in F$ such that

$$\|A_n - (\widehat{a}_n)\| < \frac{1}{n} \tag{1.4.3}$$

Therefore $\{A_n - (\widehat{a}_n)\}$ is a null sequence, and, consequently, it is a Cauchy sequence in $\widehat{F}$. We have

$$\{(\widehat{a}_n)\} = \{A_n\} - \{A_n - (\widehat{a}_n)\};$$

hence $\{(\widehat{a}_n)\}$ is a Cauchy sequence in $\widehat{F}$. Since all elements of $\{(\widehat{a}_n)\}$ belong to F, $\{a_n\}$ itself is a Cauchy sequence in F. Let us denote the equivalence class of $\{a_n\}$ by A (in our notation, $(\{a_n\}) = A$). It follows from (1.4.2) and (1.4.3) that $\{A - (\widehat{a}_n)\}$ and $\{A_n - (\widehat{a}_n)\}$ are null sequences in $\widehat{F}$, and hence their difference

$$\{A - A_n\} = A - \{(\widehat{a}_n)\} - \{A_n - (\widehat{a}_n)\}$$

is a null sequence in $\widehat{F}$. This implies that

$$\lim_{n \to \infty} \|A - A_n\| = 0,$$

i.e., $A = \lim_{n \to \infty} A_n$. $\square$

Using standard results in analysis, one can prove the following statement.

Proposition 1.4.4. *The operations on* $\widehat{F}$ *are extended from F by continuity,*
i.e., if

$$A = \lim_{n \to \infty} (\widehat{a}_n), \qquad B = \lim_{n \to \infty} (\widehat{b}_n),$$

then

$$A + B = \lim_{n \to \infty} (\widehat{a}_n + \widehat{b}_n), \qquad A \cdot B = \lim_{n \to \infty} (\widehat{a}_n \cdot \widehat{b}_n).$$

1.5. Construction of the field of *p*-adic numbers $\mathbb{Q}_p$

It is well known that the field of rational numbers $\mathbb{Q}$ *is not complete with*
respect to any of its non-trivial norms [98, Lemma 3.2.3.]. Here all *non-trivial*
norms are given by Ostrovski's theorem 1.3.2. The field of real numbers $\mathbb{R}$ is
the completion of $\mathbb{Q}$ with respect to the real norm $|\cdot|$. Now we introduce the
field $\mathbb{Q}_p$ *of p-adic numbers* as the completion of the field $\mathbb{Q}$ with respect to the
p-adic norm $|\cdot|_p$ of (1.3.2). Thus the space $\mathbb{Q}_p$ is ultrametric. In view of the
Ostrovski theorem there are two "universes": the real universe and the *p*-adic
one.

We construct $\mathbb{Q}_p$ by the completion procedure of Section 1.4. The elements
of $\mathbb{Q}_p$ are equivalence classes of Cauchy sequences in $\mathbb{Q}$ with respect to the
p-adic norm. According to Section 1.4, $\mathbb{Q}$ can be identified with the subfield
of $\mathbb{Q}_p$ consisting of equivalence classes of constant Cauchy sequences.

Let $x \in \mathbb{Q}_p$, and let $\{x_n\}$ be a Cauchy sequence of rational numbers representing a. Then by definition (1.4.1)

$$|x|_p \stackrel{def}{=} \lim_{n \to \infty} |x_n|_p. \tag{1.5.1}$$

If $|x|_p \neq 0$, then the sequence of norms $\{x_n\}$ stabilizes (in the sense of $|x_n|_p =$
$|x|_p$ for sufficiently large n). This fact is also a consequence of the *strong*
triangle inequality. Indeed, since $|x_n - x|_p < |x|_p$ for sufficiently large n,
according to the strong triangle inequality (see Theorem 1.3.1 and Definition 1.4
(3′)), for sufficiently large n we have

$$|x_n|_p = |(x_n - x) + x|_p = \max(|x_n - x|_p, |x|_p) = |x|_p. \tag{1.5.2}$$

Thus the *p*-adic absolute value is extended to $\mathbb{Q}_p$ and we have

$$\{|x|_p : x \in \mathbb{Q}_p\} = \{|x|_p : x \in \mathbb{Q}\} = \{p^\gamma : \gamma \in \mathbb{Z}\} \cup \{0\}.$$

In this sense the behavior of *p*-adic numbers is quite different from the behavior
of real numbers. When we extend $\mathbb{Q}$ to $\mathbb{R}$, the Euclidean absolute value takes
all nonnegative real values.

According to Proposition 1.4.2, the norm $|\cdot|_p$ on $\mathbb{Q}_p$, which is defined by (1.5.1), has the following properties. If $x, y \in \mathbb{Q}_p$, then

(1) $|x|_p \geq 0$, $|x|_p = 0 \Leftrightarrow x = 0$;

(2) $|xy|_p = |x|_p |y|_p$;

(3) $|x + y|_p \leq \max(|x|_p, |y|_p)$;

Moreover, if $|x|_p \neq |y|_p$, then

(3') $|x + y|_p = \max(|x|_p, |y|_p)$.

Thus the norm (1.5.1) on $\mathbb{Q}_p$ satisfies the *strong triangle inequality* (3), i.e., is non-Archimedean. For any $n \in \mathbb{N}$ we have $|nx|_p \leq |x|_p$.

Now we can extend the p-adic valuation (order) from $\mathbb{Q}$ to $\mathbb{Q}_p$: for any $x \in \mathbb{Q}_p \setminus \{0\}$ we set

$$v_p(x) = \mathrm{ord}_p(x) = -\log_p |x|_p, \quad \text{and} \quad v_p(0) = \mathrm{ord}_p(0) = \infty.$$

It is clear that properties (1.3.1) also hold if we replace $\mathbb{Q}$ by $\mathbb{Q}_p$.

1.6. Canonical expansion of p-adic numbers

Let us consider the series

$$\frac{d_{-m}}{p^m} + \frac{d_{-m+1}}{p^{m-1}} + \cdots + d_0 + d_1 p + d_2 p^2 + \cdots, \qquad (1.6.1)$$

where $0 \leq d_i \leq p - 1, i \geq -m, d_{-m}i \neq 0$. The partial sums $S_n = \sum_{i=-m}^{n} d_i p^i$ of the series (1.6.1) form a Cauchy sequence with respect to the $|\cdot|_p$ norm. Indeed, for any $\varepsilon > 0$ one can chose $N(\varepsilon)$ such that $p^{-N(\varepsilon)} < \varepsilon$ and for $k > n > N(\varepsilon)$ we have

$$\left| \sum_{i=-m}^{k} d_i p^i - \sum_{i=-m}^{n} d_i p^i \right|_p = \left| \sum_{i=n+1}^{k} d_i p^i \right|_p \leq \max_{n+1 \leq i \leq k} |d_i p^i|_p \leq p^{-N(\varepsilon)} < \varepsilon.$$

Thus, any series (1.6.1) represents an element of the field $\mathbb{Q}_p$.

Conversely, any equivalence class of Cauchy sequences in $\mathbb{Q}$ (which defines an element of $\mathbb{Q}_p$) contains a unique *canonical* representative Cauchy sequence, which is the sequence of partial sums of a series of the form (1.6.1). To construct this *canonical* representative, we need the following lemma.

Lemma 1.6.1. *Let $x \in \mathbb{Q}$ and $|x|_p \leq 1$. Then for any positive integer i there exists an integer $\alpha \in \mathbb{Z}$ such that $|\alpha - x|_p \leq p^{-i}$. Moreover, the integer α can be chosen in the set $\{0, 1, 2, \ldots, p^i - 1\}$ and it is unique if it is chosen in this range.*

Proof. Let $x = a/b$ be an irreducible fraction. Since $|x|_p \leq 1$, it follows that p does not divide b and, consequently, b and p^i are relatively prime. Thus one can find two integers m and n such that $mb + np^i = 1$. Let $\alpha = am$. Then

$$
\begin{aligned}
|\alpha - x|_p &= |am - a/b|_p = |a/b|_p \, |mb - 1|_p \\
&\leq |mb - 1|_p = |np^i|_p = |n|_p \, p^{-i} \leq p^{-i}.
\end{aligned}
$$

Finally, according to the strong triangle inequality, one can add a multiple of p^i to the integer α to get an integer α' such that $|\alpha' - x|_p \leq p^{-i}$, where α' is between 0 and p^i. Thus the lemma is proved.   □

Theorem 1.6.2. *Every equivalent class $a \in \mathbb{Q}_p$ satisfying $|a|_p \leq 1$ has exactly one representative Cauchy sequence $\{a_i\}$ such that*
(1) $a_i \in \mathbb{Z}$, $0 \leq a_i < p^i$ for $i = 1, 2, \ldots$;
(2) $a_i \equiv a_{i+1} \pmod{p^i}$ for $i = 1, 2, \ldots$.

Proof. Let $\{b_i\}$ be the Cauchy sequence representing a. We shall construct an equivalent Cauchy sequence $\{a_i\}$ satisfying (1) and (2).

For every $j = 1, 2, \ldots$ we denote by $N(j)$ a non-negative integer such that $|b_i - b_{i'}|_p \leq p^{-j}$, for all $i, i' \geq N(j)$. Let us note that we may take the sequence $N(j)$ to be strictly increasing with respect to j, so $N(j) \geq j$. Since $|b_i|_p \to |a|_p \leq 1$, as $i \to \infty$, truncating several initial terms if necessary, we may assume that $|b_i|_p \leq 1$ for all i.

According to Lemma 1.6.1, one can find integers a_j such that $0 \leq a_j < p^j$ and

$$
\left| a_j - b_{N(j)} \right|_p \leq p^{-j}.
$$

Let us show that $a_j \equiv a_{j+1} \pmod{p^j}$ and $(a_j) \sim (b_j)$. According to the following inequality

$$
\begin{aligned}
\left| a_{j+1} - a_j \right|_p &= \left| (a_{j+1} - b_{N(j+1)}) + (b_{N(j+1)} - b_{N(j)}) + (a_j - b_{N(j)}) \right|_p \\
&\leq \max\left(\left| a_{j+1} - b_{N(j+1)} \right|_p, \left| b_{N(j+1)} - b_{N(j)} \right|_p, \left| a_j - b_{N(j)} \right|_p \right) \\
&\leq \max\left(p^{-j-1}, p^{-j}, p^{-j} \right) = p^{-j}
\end{aligned}
$$

the first assertion holds. To prove the second assertion, let us take any j. Then for $i \geq N(j)$ we have

$$
\begin{aligned}
\left| a_i - b_i \right|_p &= \left| (a_i - a_j + (a_j - b_{N(j)}) - (b_i - b_{N(j)}) \right|_p \\
&\leq \max\left(\left| a_i - a_j \right|_p, \left| a_j - b_{N(j)} \right|_p, \left| b_i - b_{N(j)} \right|_p \right) \\
&\leq \max\left(p^{-j}, p^{-j}, p^{-j} \right) = p^{-j}.
\end{aligned}
$$

Hence, as $i \to \infty$, we have $|a_i - b_i|_p \to 0$.

Now we prove the uniqueness. Let $\{a_i'\}$ be a different sequence satisfying the requirements of the theorem such that $a_{i_0}' \neq a_{i_0}$ for some i_0. In this case we have $a_{i_0}' \not\equiv a_{i_0} \pmod{p^{i_0}}$, since both a_{i_0}' and a_{i_0} are between 0 and p^{i_0}. Consequently, according to the condition (2), for $i > i_0$

$$a_i \equiv a_{i_0} \not\equiv a_{i_0}' \equiv a_i' \pmod{p^{i_0}},$$

i.e., $a_i \not\equiv a_i' \pmod{p^{i_0}}$. Consequently,

$$|a_i - a_i'|_p \geq p^{-i_0}, \qquad \forall\, i \geq i_0,$$

and $(a_i) \not\sim (a_i)$. □

If $a \in \mathbb{Q}_p$ and $|a|_p \leq 1$, then from Theorem 1.6.2, one can write all the terms a_i of the representative sequence $\{a_i\}$ in the form

$$a_i = d_0 + d_1 p + d_2 p^2 + \cdots + d_{i-1} p^{i-1},$$

where $d_k \in \{0, 1, \ldots, p-1\}$ for all $0 \leq k \leq i$. Condition (2) of Theorem 1.6.2 means that

$$a_{i+1} = d_0 + d_1 p + d_2 p^2 + \cdots + d_{i-1} p^{i-1} + d_i p^i,$$

where the p-adic digits $d_0, d_1, \ldots, d_{i-1}$ are the same as for a_i. Thus a is represented as the convergent series

$$a = \sum_{k=0}^{\infty} d_k p^k.$$

If $a \in \mathbb{Q}_p$ and $|a|_p > 1$, then we can multiply a by a power $p^m = |a|_p$. Thus the p-adic number $a' = ap^m$ satisfies $|a'|_p = 1$, and, according to the above reasoning,

$$a = \sum_{k=-m}^{\infty} d_k p^k,$$

where $d_k \in 0, 1, \ldots, p-1$ for all $k \leq -m$, $d_{-m} \neq 0$, and the series converges.

Thus, any p-adic number $x \in \mathbb{Q}_p$, $x \neq 0$, can be uniquely represented in the following *canonical form*

$$x = \sum_{k=\gamma}^{\infty} d_k p^k = p^{\gamma} \sum_{j=0}^{\infty} x_j p^j, \qquad (1.6.2)$$

where $\gamma = \gamma(x) \in \mathbb{Z}$, $x_j = d_{j+\gamma}$, $x_j = 0, 1, \ldots, p-1$, $x_0 \neq 0$, $j = 0, 1, \ldots$. Since $|p^{\gamma} x_j p^j|_p = p^{-\gamma-j}$, $j = 0, 1, 2, \ldots$, the series (1.6.2) converges in the p-adic norm $|\cdot|_p$. The above series can be treated as a number written in base

p which extends infinitely far to the left or has infinitely many p-adic digits before the point and finitely many after:

$$x = \ldots d_k \ldots d_2 d_1 d_0 . d_{-1} d_{-2} \ldots d_\gamma. \tag{1.6.3}$$

By definition (1.5.1), the norm $|x|_p$ is a limit of the sequence of p-adic norms $|S_n|_p$ of the partial sums $S_n = p^\gamma \sum_{j=0}^n x_j p^j$ of the series (1.6.2). Since $|S_n|_p = p^{-\gamma}$, we have $|x|_p = p^{-\gamma}$.

The order defined for rational numbers by Definition 1.6 can be extended to all p-adic numbers as: if $x \in \mathbb{Q}_p$ then $\mathrm{ord}(x) = -\gamma$ for $x \in \mathbb{Q}_p \setminus \{0\}$ and $\mathrm{ord}(0) = -\infty$.

Example 1.6.1. For every prime p we have the following expansion of -1,

$$-1 = (p-1) + (p-1)p + (p-1)p^2 + \cdots,$$

since $1 + (p-1) + (p-1)p + (p-1)p^2 + \cdots = 0$.

Example 1.6.2. In $\mathbb{Q}_2$, the rational number $1/3$ has the expansion

$$1/3 = 1 + 1 \cdot 2 + 0 \cdot 2^2 + 1 \cdot 2^3 + 0 \cdot 2^4 + \cdots.$$

By means of representation (1.6.2), the *fractional part* $\{x\}_p$ of a p-adic number $x \in Q_p$ is defined as the following

$$\{x\}_p \overset{def}{=} \begin{cases} 0, & \text{if } \gamma(x) \geq 0 \quad \text{or} \quad x = 0, \\ p^\gamma \sum_{r=0}^{|\gamma|-1} x_r p^r, & \text{if } \gamma(x) < 0. \end{cases} \tag{1.6.4}$$

It is easy to prove that if $\gamma(x) < 0$ then

$$p^\gamma \leq \{x\}_p \leq 1 - p^\gamma.$$

The *integer part* $[x]_p$ of a p-adic number $x \in Q_p$ is defined as

$$[x]_p \overset{def}{=} \begin{cases} x, & \text{if } \gamma(x) \geq 0 \quad \text{or} \quad x = 0, \\ p^\gamma \sum_{r=|\gamma|}^{\infty} x_r p^r, & \text{if } \gamma(x) < 0. \end{cases}$$

A sum $z = x + y$ of two p-adic numbers in the canonical form (1.6.2)

$$x = \sum_{k=\alpha}^{\infty} a_k p^k, \qquad y = \sum_{k=\beta}^{\infty} b_k p^k$$

is represented in the canonical form

$$z = \sum_{k=\gamma}^{\infty} c_k p^k, \qquad \gamma = \min(\alpha, \beta),$$

where c_k is defined by the relation

$$\sum_{k=\alpha}^{n} a_k p^k + \sum_{k=\beta}^{n} b_k p^k \equiv \sum_{k=\gamma}^{n} c_k p^k \pmod{p^{n+1}}, \quad \forall n \geq \gamma.$$

The field $\mathbb{Q}_p$ is a commutative and associative group with respect to addition. The set $\mathbb{Q}_p^\times = \mathbb{Q}_p \setminus \{0\}$ is the *multiplicative group of the field* $\mathbb{Q}_p$.

The field $\mathbb{Q}_p$ is a completion of the rational numbers $\mathbb{Q}$ and contains $\mathbb{Q}$ as a dense subset. The following statement gives a description of rational numbers as p-adic numbers.

Proposition 1.6.3. ([112, 1.6.]) *The canonical expansion* (1.6.3) *of a p-adic number represents a rational number if and only if it is eventually periodic to the left.*

1.7. The ring of p-adic integers $\mathbb{Z}_p$

The p-adic numbers $\mathbb{Z}_p = \{x \in \mathbb{Q}_p : |x|_p \leq 1\}$ (i.e., such that in (1.6.2) $\gamma(x) \geq 0$ or $\{x\}_p = 0$) are called *p-adic integers*. In view of (1.6.2), $\mathbb{Z}_p$ consists of p-adic numbers

$$x = \sum_{k=0}^{\infty} x_k p^k. \tag{1.7.1}$$

Proposition 1.7.1. $\mathbb{Z}_p$ *is a subring of the ring* $\mathbb{Q}_p$.

Proof. Let $x, y \in \mathbb{Z}_p$. Then $|x + y|_p \leq \max(|x|_p, |y|_p) \leq 1$ and $|xy|_p = |x|_p |y|_p \leq 1$, i.e., $x + y, xy \in \mathbb{Z}_p$. Further $|1|_p = 1$, that is $1 \in \mathbb{Z}_p$. For the additive inverse $-x$ we have $|-x|_p = |-1|_p |x|_p \leq 1$, i.e., $-x \in \mathbb{Z}_p$. The proposition is thus proved. $\qquad\square$

Example 1.6.1 implies that $1 - p$ is invertible in $\mathbb{Z}_p$ for multiplication with the inverse given by $\sum_{k=0}^{\infty} p^k$. Since

$$p \cdot \sum_{k=0}^{\infty} x_k p^k = x_0 p + x_1 p^2 + \cdots \neq 1 + 0p + 0p^2 + \cdots,$$

the prime p is not invertible in $\mathbb{Z}_p$ for multiplication, $\frac{1}{p} \notin \mathbb{Z}_p$.

Proposition 1.7.2. ([112, 1.5.], [204, 1.5.]) *A p-adic integer $x \in \mathbb{Z}_p$ has a multiplicative invertible element in $\mathbb{Z}_p$ if and only if in* (1.7.1) $x_0 \neq 0$.

Thus the group of invertible elements in $\mathbb{Z}_p$ constitutes the set

$$\mathbb{Z}_p^\times = \left\{ x \in \mathbb{Z}_p : |x|_p = 1 \right\} = \left\{ x \in \mathbb{Z}_p : x = \sum_{k=0}^{\infty} x_k p^k, \quad x_0 \neq 0 \right\}.$$

It is a *multiplicative group of the ring* $\mathbb{Z}_p$. For all $x \in \mathbb{Z}_p^\times$ we have $|x|_p = 1$. Integers $x \in \mathbb{Z}_p$ such that $|x|_p = 1$ are called *units* in $\mathbb{Q}_p$.

Recall that an ideal I of the ring R is called *maximal* if it is not strictly contained in any other proper ideal of the ring. A *principal right (left) ideal I of the ring R* is the ideal such that $I = rR$ $(I = Rr)$ for some $r \in R$.

The ring of integers $x \in \mathbb{Z}_p$ such that $|x|_p < 1$ (i.e., such that in (1.7.1) $x_0 = 0$ or $|x|_p \leq 1/p$) forms a unique *maximal ideal of the ring* $\mathbb{Z}_p$; this ideal has the form $p\mathbb{Z}_p = \mathbb{Z}_p \setminus \mathbb{Z}_p^\times$. The *residue field* $\mathbb{Z}_p / p\mathbb{Z}_p$ consists of p elements.

Proposition 1.7.3. ([112, 1.8.], [204, 1.6.]) *The ring* $\mathbb{Z}_p$ *is a* principal ideal domain. *More precisely, all its ideals are the principal ideals* $\{0\}$ *and* $p^k \mathbb{Z}_p$ *for all* $k \in \mathbb{N}$.

We have

$$\mathbb{Z}_p \supset p\mathbb{Z}_p \cdots \supset p^k \mathbb{Z}_p \supset \cdots \supset \bigcap_{k \geq 0} p^k \mathbb{Z}_p = \{0\}.$$

Any rational integer number is also a p-adic integer. Indeed, a rational integer $m \geq 0$ can be represented as a finite sum: $m = \sum_{k \geq 0} m_k p^k$, $m_k = 0, 1, \ldots, p-1, k \geq 0$. Taking into account Example 1.6.1, a negative rational integer $-m$ can be represented as

$$-m = (-1)m = \sum_{j=0}^{\infty} (p-1)p^j \cdot \sum_{k \geq 0} m_k p^k.$$

The right-hand side of the above relation can be rewritten in the form (1.7.1).

The set of positive integers $\mathbb{N} = \{1, 2, \ldots\}$ is dense in $\mathbb{Z}_p$. Indeed, let $x = \sum_{k=0}^{\infty} x_k p^k \in \mathbb{Z}_p$. For any $n \in \mathbb{N}$ we set $x_{(n)} = \sum_{k=0}^{n} x_k p^k$. Then $x_{(n)} \in \mathbb{N}$ and $|x - x_{(n)}|_p \leq p^{-n}$, and, consequently, our statement holds.

However, there are p-adic integers among rational fractions. Indeed, in view of Example 1.6.1,

$$\frac{1}{1-p} = \sum_{k=0}^{\infty} p^k \in \mathbb{Z}_p.$$

1.8. Non-Archimedean topology of the field $\mathbb{Q}_p$

It is possible to introduce in $\mathbb{Q}_p$ the metric $\rho_p(x, y) = |x - y|_p$, $x, y \in$ $\mathbb{Q}_p$. Then $\mathbb{Q}_p$ becomes a complete metric space. Since the norm $|\cdot|_p$ is non-Archimedean, the corresponding metric ρ_p satisfies the strong triangle inequality:

$$\rho_p(x, y) \leq \max\left(\rho_p(x, z), \rho_p(z, y)\right), \qquad x, y, z \in \mathbb{Q}_p.$$

This kind of metric is called an *ultrametric*.

Since the p-adic norm has a discrete set of values $\{p^\gamma : \gamma \in \mathbb{Z}\} \cup \{0\}$, we need only consider balls of radiuses $r = p^\gamma$, where $\gamma \in \mathbb{Z}$. Denote by $B_\gamma(a) \equiv$ $U_{p^\gamma}(a) = \{x : |x - a|_p \leq p^\gamma\}$ the closed ball of radius p^γ with the center at a point $a \in \mathbb{Q}_p$, by $B_\gamma^-(a) \equiv U_{p^\gamma}^-(a) = \{x : |x - a|_p < p^\gamma\}$ the corresponding open ball, and by $S_\gamma(a) = \{x : |x - a|_p = p^\gamma\}$ the sphere of radius p^γ with the center at a point $a \in \mathbb{Q}_p$, $\gamma \in \mathbb{Z}$. We set $B_\gamma(0) = B_\gamma$ and $S_\gamma(0) = S_\gamma$.

Since $\mathbb{Q}_p$ is an *ultrametric* space, $B_\gamma(a)$ is an *abelian* group.

It is clear that $S_\gamma(a) = \{x : |x - a|_p = p^\gamma\} = B_\gamma(a) \setminus B_{\gamma-1}(a)$, $B_\gamma(a) \subset$ $B_{\gamma'}(a)$ if $\gamma < \gamma'$, and

$$B_{\gamma-1}(a) = \{x : |x - a|_p < p^\gamma\}, \qquad B_\gamma(a) = \bigcup_{\gamma' \leq \gamma} S_{\gamma'}(a),$$

$$\bigcup_\gamma B_\gamma(a) = \bigcup_\gamma S_\gamma(a) = \mathbb{Q}_p, \qquad \bigcap_\gamma B_\gamma(a) = \{a\}.$$

Let us remark that $B_0 = \mathbb{Z}_p$ is the set of p-adic integers, $S_0 = \mathbb{Z}_p^\times$ is the group of invertible elements (multiplicative group) in $\mathbb{Z}_p$, and $B_{-1} = p\mathbb{Z}_p$ is a unique maximal ideal of the ring $\mathbb{Z}_p$.

Theorem 1.8.1. (1) *Every point of the ball $B_\gamma(a)$ is its center.*

(2) *The ball $B_\gamma(a)$ and the sphere $S_\gamma(a)$ are both open and closed sets in* $\mathbb{Q}_p$.

(3) *Two balls in $\mathbb{Q}_p$ have nonempty intersection if and only if one is contained in the other, i.e.,*

$$B_\gamma(a) \cap B_{\gamma'}(b) \neq \emptyset \iff B_\gamma(a) \subset B_{\gamma'}(b) \quad or \quad B_{\gamma'}(b) \subset B_\gamma(a).$$

Thus two balls in $\mathbb{Q}_p$ are either disjoint or one is contained in the other.

Proof. (1) This statement follows from Proposition 1.2.7.

(2) It is well known that in every metric space the ball $B_\gamma(a)$ is closed. Now we prove that $B_\gamma(a)$ is open. According to the proof of Proposition 1.2.7, if

$b \in B_\gamma(a)$ then $B_\gamma(b) \subset B_\gamma(a)$ (it is much more than we need to prove that $B_\gamma(a)$ is open).

It is well known that in every metric space the sphere $S_\gamma(a)$ is closed. Now we prove that $S_\gamma(a)$ is open. Let $x \in S_\gamma(a)$ and $\gamma' < \gamma$. Let us prove that $B_{\gamma'}^-(x) \subset S_\gamma(a)$. Indeed, if $y \in B_{\gamma'}^-(x)$ then $|x - y|_p < |x - a|_p = p^\gamma$, and according to property (3′) of the p-adic norm (see Section 1.5), we have $|y - a|_p < |x - a|_p = p^\gamma$, i.e., $y \in S_\gamma(a)$.

(3) If one ball is contained in the other, they have a nonempty intersection. To prove the converse, we assume that $\gamma \leq \gamma'$ and $y \in B_\gamma(a) \cap B_{\gamma'}(b)$. Then according to statement (1) of this theorem, $B_\gamma(a) = B_\gamma(y)$ and $B_{\gamma'}(b) = B_{\gamma'}(y)$. Because $B_\gamma(y) \subset B_{\gamma'}(y)$, we have the required inclusion.   □

Sets which are open and closed at the same time are called *clopen*. We use the word *clopen* as an abbreviation for *closed and open*.

It follows from statement (2) of Theorem 1.8.1 that the sphere $S_\gamma(a)$ *is not a boundary of the open ball* $B_\gamma^-(a)$. Moreover, statement (2) of Theorem 1.8.1 implies that the ball $B_\gamma^-(a)$ *has no boundary*. And, of course, the closed ball

$$B_\gamma(a) = \{x : |x - a|_p \leq p^\gamma\}$$
$$= \{x : |x - a|_p < p^{\gamma+1}\} = B_{\gamma+1}^-(a)$$

is not the closure of the open ball $B_\gamma^-(a)$.

Proposition 1.8.2. *The set of all balls in* $\mathbb{Q}_p$ *is countable.*

Proof. Let us write the center of the ball $B_\gamma(a)$ as $a = \tilde{a} + p^{-\gamma}\xi$, where $\tilde{a} = \sum_{k=-m}^{-\gamma-1} a_k p^k$ is a rational number, $\xi \in \mathbb{Z}_p$. Then

$$B_\gamma(a) = \{x : |x - a|_p \leq p^\gamma\} = \{x : |p^\gamma(x - a)|_p \leq 1\}$$
$$= \{x : |p^\gamma(x - \tilde{a}) + \xi|_p \leq 1\} = \{x : |p^\gamma(x - \tilde{a})|_p \leq 1\} = B_\gamma(\tilde{a}).$$

Thus each ball can be characterized by a pair of numbers $(\tilde{a}, \gamma)$ where both numbers belong to countable sets. Therefore the set of all pairs $(\tilde{a}, \gamma)$ is also countable and so is the set of all balls in $\mathbb{Q}_p$.   □

Statement (3) of Theorem 1.8.1 and Proposition 1.8.2 imply the following statement.

Proposition 1.8.3. *Every open set in* $\mathbb{Q}_p$ *is a union at most of a countable set of disjoint balls.*

Let us recall some well-known definitions.

Definition 1.7. (*a*) A topological space X is called *disconnected* if it can be represented as a union of two disjoint nonempty open subsets. Equivalently, it can be represented as a union of two disjoint nonempty closed subsets.

    (*b*) If (*a*) does not hold then X is called a *connected* topological space.

    (*c*) A subset A of X is called *connected* if it is a connected space in the induced topology.

    (*d*) A topological space X is called *totally disconnected* if the only connected subsets of X are the empty set and the points $\{a\}$, $a \in X$. In other words, the connected component of any point coincides with this point.

Theorem 1.8.4. *The space* $\mathbb{Q}_p$ *is totally disconnected.*

Proof. According to Theorem 1.8.1, for each $a \in \mathbb{Q}_p$ and $\gamma \in \mathbb{N}$ the set

$$B_{-\gamma}(a) = \{x \in \mathbb{Q}_p : |x - a|_p \le p^{-\gamma}\}$$
$$= \{x \in \mathbb{Q}_p : |x - a|_p < p^{-\gamma+1}\} = B^-_{-\gamma+1}(a)$$

is an open and closed neighborhood of the point a. Suppose that $a \in A$ so that $A \ne \{a\}$. Then there is $\gamma \in \mathbb{N}$ such that $B_{-\gamma}(a) \cap A \ne A$. Thus,

$$A = \left(B_{-\gamma}(a) \cap A\right) \cup \left((\mathbb{Q}_p \setminus B_{-\gamma}(a)) \cap A\right),$$

where both $B_{-\gamma}(a)$ and its complement $\mathbb{Q}_p \setminus B_{-\gamma}(a)$ are open and non-empty sets. This implies that A is not connected.   □

Definition 1.8. (*a*) A topological space X (or a subset K of X) is called *compact* if each of its open covers contains a finite subcover.

    (*b*) If (*a*) does not hold then X (or K) is called *non-compact*.

    (*c*) A topological space X is called *locally compact* if every point of X has a compact neighborhood.

Theorem 1.8.5. ([241, 1.3.]) *A set* $K \subset \mathbb{Q}_p$ *is compact in* $\mathbb{Q}_p$ *if and only if it is closed and bounded in* $\mathbb{Q}_p$.

Proof. The necessity of the condition is obvious. Let us prove the sufficiency. Since $\mathbb{Q}_p$ is a complete metric space, it is sufficient to prove the countable compactness of any bounded closed (infinite) set K. That is, we need to prove that every infinite set $M \subset K$ contains at least one limit point. Let $x \in M$. Since M is bounded, then $|x|_p = p^{-\gamma(x)} \le C$, i.e., $\gamma(x)$ is bounded below.

    We consider two cases. (1) If $\gamma(x)$ is not bounded above on M then there exists a sequence $\{x_n \in M, n \to \infty\}$ such that $\gamma(x_n) \to \infty$ as $n \to \infty$. It means that $|x_n|_p = p^{-\gamma(x_n)} \to 0$ as $n \to \infty$. Thus $x_n \to 0$ in $\mathbb{Q}_p$ as $n \to \infty$, and, consequently, $0 \in K$.

(2) If $\gamma(x)$ is bounded above on M then there exists γ_0 such that M contains an infinite set of points of the form

$$p^{\gamma_0}(x_0 + x_1 p + x_2 p^2 + \cdots), \quad x_j = 0, 1, \ldots, p - 1, \quad x_0 \neq 0.$$

Since x_0 takes only $p - 1$ values then there exists an integer $a_0 \in \{1, 2, \ldots, p - 1\}$ such that M contains an infinite set of points of the form $p^{\gamma_0}(a_0 + x_1 p + x_2 p^2 + \cdots)$, where a_0 is fixed. If we continue this process, we obtain a sequence $\{a_j \in \{0, 1, 2, \ldots, p - 1\}, a_0 \neq 0 : j = 0, 1, 2, \ldots\}$. The desired limit point is $p^{\gamma_0}(a_0 + a_1 p + a_2 p^2 + \cdots) \in K$.   □

Corollary 1.8.6. ([112, 2.1.], [241, 1.3.])

(1) *Every ball $B_\gamma(a)$ and every sphere $S_\gamma(a)$ is compact. In particular, the set $\mathbb{Z}_p$ is compact.*

(2) *The space $\mathbb{Q}_p$ is locally compact.*

(3) *In the space $\mathbb{Q}_p$ the Heine–Borel lemma is valid: for every infinite covering of a compact K it is possible to choose a finite covering of K.*

Proposition 1.8.7. ([241, 1.3.]) *Every compact in $\mathbb{Q}_p$ can be covered by a finite number of disjoint balls of a fixed radius.*

Proposition 1.8.8. ([241, I.3, Examples 1.]) *A sphere S_γ is represented by the union of $(p - 1)p^{\gamma-\gamma'-1}$ disjoint balls $B_{\gamma'}(a)$, $\gamma' < \gamma$:*

$$S_\gamma = \bigcup_a B_{\gamma'}(a), \quad \gamma' < \gamma \tag{1.8.1}$$

where

$$a = p^{-\gamma}\left(a_0 + a_1 p + \cdots + a_{\gamma-\gamma'-1} p^{\gamma-\gamma'-1}\right), \tag{1.8.2}$$

$0 \le a_j \le p - 1$, $a_0 \neq 0$. *In particular,*

$$S_\gamma = \cup_{a_0=1}^{p-1} B_{\gamma-1}(p^{-\gamma} a_0) \tag{1.8.3}$$

and

$$S_0 = \{|x|_p = 1\} = \cup_{r=1}^{p-1} B_{-1}(r),$$

where $B_{-1}(r) = \{x \in Z_p : x_0 = r\} = r + p\mathbb{Z}_p$, $r = 1, \ldots, p - 1$.

Proof. Any point $x = p^{-\gamma}(x_0 + x_1 p + \cdots) \in S_\gamma$ can be uniquely represented in the form $x = a + x'$, where a is given by (1.8.2) and $x' \in B_{\gamma'}$. Therefore

$$S_\gamma = \bigcup_a \{a + B_{\gamma'}\} = \bigcup_a B_{\gamma'}(a)$$

The number of balls $B_{\gamma'}(a)$ is $(p-1)p^{\gamma-\gamma'-1}$ and in view of Theorem 1.8.1 (3) these balls are disjoint. $\square$

We call the covering given by (1.8.1) and (1.8.2) the *canonical covering* of the sphere S_γ.

Proposition 1.8.9. *The ball B_γ is represented by the union of $p^{\gamma-\gamma'}$ disjoint balls $B_{\gamma'}(a)$, $\gamma' < \gamma$:*

$$B_\gamma = B_{\gamma'} \bigcup \bigcup_a B_{\gamma'}(a), \qquad (1.8.4)$$

where $a = 0$ and $a = a_{-r}p^{-r} + a_{-r+1}p^{-r+1} + \cdots + a_{-\gamma'-1}p^{-\gamma'-1}$ are the centers of the balls $B_{\gamma'}(a)$, $0 \le a_j \le p-1$, $j = -r, -r+1, \ldots, -\gamma'-1$, $a_{-r} \ne 0$, $r = \gamma'+1, \gamma'+2, \ldots, \gamma-1, \gamma$. In particular, the ball B_0 is represented by the union of p disjoint balls

$$B_0 = B_{-1} \cup \cup_{r=1}^{p-1} B_{-1}(r), \qquad (1.8.5)$$

where $B_{-1}(r) = \{x \in Z_p : x_0 = r\} = r + pZ_p$, $r = 1, \ldots, p-1$; $B_{-1} = \{|x|_p \le p^{-1}\} = pZ_p$.

Proof. The proof follows from Proposition 1.8.8 if we take into account that

$$B_\gamma = B_{\gamma'} \bigcup \bigcup_{r=\gamma'+1}^{\gamma} S_r,$$

where the sets $B_{\gamma'}$ and $S_r, r = \gamma'+1, \gamma'+2, \ldots, \gamma$, are disjoint. The number of balls is

$$1 + (p-1) \sum_{r=\gamma'+1}^{\gamma} p^{r-\gamma'-1} = 1 + (p-1)\frac{1 - p^{\gamma-\gamma'}}{1-p} = p^{\gamma-\gamma'}.$$

$\square$

We call the covering (1.8.4) the *canonical covering* of the ball B_γ.

1.9. $\mathbb{Q}_p$ in connection with $\mathbb{R}$

1.9.1. Cantor-like sets

Now we introduce Cantor-like sets to use them later as a geometric model for p-adic numbers.

Let us recall the Cantor set construction. Let us define

$$C_0 = I = [0, 1], \qquad C_1 = \left[0, \frac{1}{3}\right] \cup \left[\frac{2}{3}, 1\right],$$

$$C_2 = \left[0, \frac{1}{9}\right] \cup \left[\frac{2}{9}, \frac{1}{3}\right] \cup \left[\frac{2}{3}, \frac{7}{9}\right] \cup \left[\frac{8}{9}, 1\right], \ldots.$$

Here each set C_n is the union of 2^n closed intervals, each of length 3^{-n}, and $C_0 \supset C_1 \supset C_2 \supset \cdots$. By definition, the *Cantor set* is

$$C \stackrel{def}{=} \bigcap_{n=0}^{\infty} C_n.$$

It contains all points in the interval I that are not deleted at any step in this infinite process. Since each $C_n \neq \emptyset$ is closed, $C \neq \emptyset$ is a closed subset of the unit interval.

The Cantor set C has measure zero in the sense that the sum of the lengths of the removed intervals is equal to 1:

$$\frac{1}{3} + \frac{2}{9} + \frac{4}{27} + \cdots + \frac{2^{n-1}}{3^n} + \cdots = 1.$$

The Hausdorff dimension of the Cantor set is equal to log 2/log 3.

Let us represent real numbers in the unit interval $I = [0, 1]$ as infinite fractions in base 3. Then the following statement holds.

Proposition 1.9.1. ([112, 2.2.]) *The Cantor set C consists of all points of I which can be represented in base 3 by the digits 0 and 2.*

Proof. Let us represent real numbers in $I = [0, 1]$ as infinite fractions in base 3. Notice that all endpoints of the intervals constituting C_n have two expansions in base 3, e.g.,

$$\frac{1}{3} = 1 \cdot 3^{-1} + 0 \cdot 3^{-2} + \cdots + 0 \cdot 3^{-k} + \cdots$$

$$= 0 \cdot 3^{-1} + 2 \cdot 3^{-2} + \cdots + 2 \cdot 3^{-k} + \cdots = 0.0222\ldots;$$

$$\frac{2}{9} = 0 \cdot 3^{-1} + 2 \cdot 3^{-2} + 0 \cdot 3^{-3} + \cdots + 0 \cdot 3^{-k} + \cdots$$

$$= 0 \cdot 3^{-1} + 1 \cdot 3^{-2} + 2 \cdot 3^{-3} + \cdots + 2 \cdot 3^{-k} + \cdots = 0.0122\ldots.$$

By choosing the representations of the endpoints which contain no 1's, we see that every point of the Cantor set C can be expressed by an expansion in base 3 containing only the digits 0 and 2, since the remaining points of C have a unique representation in base 3, and by construction, this representation contains no 1's.

Conversely, every point of $I = [0, 1]$ expressed by an expansion in base 3 containing only the digits 0 and 2 belongs to the intersection of the sets C_n. Indeed, if the first digit is 0, it belongs to $[0, 0.0222\ldots]$; if the first digit is 2, it belongs to $[0.2, 0.222\ldots]$; the second digit determines whether it belongs to the left (0) or right (2) interval of C_2, etc.

The above reasoning implies our statement.   □

Proposition 1.9.2. ([112, 2.2.])

(1) *The Cantor set C is uncountable.*

(2) *The Cantor set C is perfect, i.e., it is closed and does not contain isolated points.*

Now we introduce the Cantor-like set $C^{(p)}$, which is the set of all real numbers in $I = [0, 1]$ whose expansions in base $2p - 1$ have only even digits. $C^{(p)}$ can be obtained by a procedure similar to that of obtaining the Cantor set C. In order to obtain $C_1^{(p)}$, we divide I into $2p - 1$ equal subintervals and remove every second open interval. The set $C_1^{(p)}$ consists of all points in I that can be written in base $2p - 1$ as $0.a_1 a_2 a_3 \ldots$ with a_1 even. Repeating the same procedure with every subinterval of $C_1^{(p)}$, we obtain a closed set $C_2^{(p)}$ of points with the first two digits a_1 and a_2 even, and so on. The intersection

$$C^{(p)} \overset{def}{=} \bigcap_{n=0}^{\infty} C_n^{(p)}$$

is the Cantor-like set. This set is uncountable and perfect.

1.9.2. $\mathbb{Z}_p$ and the Cantor-like sets

Now we consider the Cantor-like sets as a geometric model for p-adic integers.

Theorem 1.9.3. ([112, 2.2.]) *The set of dyadic integers $\mathbb{Z}_2$ with the dyadic norm $|\cdot|_2$ is homeomorphic to the Cantor set C with the Euclidean absolute value $|\cdot|$. The homeomorphism can be realized by the map $\psi_{(2)} : \mathbb{Z}_2 \to C$, which is given as*

$$\psi_{(2)} : \sum_{k=0}^{\infty} a_k 2^k \to \sum_{k=0}^{\infty} \frac{2a_k}{3^{k+1}}.$$

Proof. Let us prove that $\psi_{(2)}$ is a homeomorphism. First, we observe that due to the uniqueness of the representation of elements in both $\mathbb{Z}_2$ and C, $\psi_{(2)}$ is a bijection.

If $x_1, x_2 \in C$ such that $|x_1 - x_2| < 3^{-n}$, then x_1 and x_2 belong to the same or adjacent interval in the partition of I into subintervals of length 3^{-n}. But since closed intervals constituting C_n do not have common endpoints, x_1 and x_2 belong to the same component $I_n \subset C_n$, and hence their first n digits coincide.

Conversely, if the first n digits of $x_1, x_2 \in C$ coincide, they belong to the same component $I_n \subset C_n$. On the other hand, the first n digits of two 2-adic numbers $y_1, y_2 \in \mathbb{Z}_2$ coincide if and only if $|y_1 - y_2|_2 < 2^{-n}$. Both 2^{-n} and 3^{-n} tend to 0 as $n \to \infty$, and we conclude that $\psi_{(2)}$ and $\psi_{(2)}^{-1}$ are both continuous. $\square$

Now we consider $\mathbb{Z}_p$. Repeating the proof of Theorem 1.9.3 almost word for word, we obtain the following assertion.

Theorem 1.9.4. ([112, 2.2.]) *The map* $\psi_{(p)} : \mathbb{Z}_p \to C^{(p)}$, *which is given as*

$$\psi_{(p)} : \sum_{k=0}^{\infty} a_k p^k \to \sum_{k=0}^{\infty} \frac{2a_k}{(2p-1)^{k+1}},$$

is a homeomorphism.

1.9.3. $\mathbb{Q}_p$ and the Cantor-like sets

Following [241, 1.6.] and [254], we will construct a homeomorphism ϕ from the set of p-adic numbers $\mathbb{Q}_p$ to some Cantor-like subset $K = \phi(\mathbb{Q}_p)$ of real numbers $\mathbb{R}$. Let us introduce the map $\phi : \mathbb{Q}_p \to \mathbb{R}_+$ by the formula

$$\phi(x) = |x|_p \sum_{k=0}^{\infty} x_k p^{-2k}, \qquad (1.9.1)$$

where $x_k = 0, 1, \ldots, p-1, x_0 \neq 0$, are determined by the canonical representation (1.6.2) of the p-adic number x.

We set the *order in* $\mathbb{Q}_p$ as follows: $x < y$, if either $|x|_p < |y|_p$, or when $|x|_p = |y|_p$ and there exists an integer $j \geq 0$ such that $x_0 = y_0, x_1 = y_1, \ldots, x_{j-1} = y_{j-1}, x_j < y_j$. It is clear that the transitive axiom is fulfilled: if $x < y$ and $y < z$ then $x < z$. Thus $\mathbb{Q}_p$ is a *completely ordered set*: $0 < x$ for any $x \in \mathbb{Q}_p^*$.

Lemma 1.9.5. $\phi(x) > \phi(y)$ *if and only if* $x > y$.

Proof. 1. Let $x > y$ and $|x|_p > |y|_p \neq 0$, i.e., $|x|_p \geq p|y|_p$. According to (1.9.1), we have

$$\phi(y) = |y|_p \sum_{k=0}^{\infty} y_k p^{-2k} \leq (p-1)|y|_p \sum_{k=0}^{\infty} p^{-2k} = |y|_p \frac{p^2}{p+1},$$

and

$$\phi(x) - \phi(y) \leq |x|_p - |y|_p \frac{p^2}{p+1} \leq p|y|_p \left(1 - \frac{p}{p+1}\right) = |y|_p \frac{p}{p+1} > 0.$$

Now let $x > y$ and $|x|_p = |y|_p$, $x_0 = y_0$, $x_1 = y_1, \ldots, x_{j-1} = y_{j-1}, x_j > y_j$ for some $j \geq 0$. Then from (1.9.1),

$$\phi(x) \geq |x|_p \sum_{k=0}^{j} x_k p^{-2k},$$

$$\phi(y) \leq |y|_p \sum_{k=0}^{j} y_k p^{-2k} + |y|_p \frac{p-1}{1-p^{-2}} p^{-2(j+1)}$$

$$= |x|_p \sum_{k=0}^{j-1} x_k p^{-2k} + |x|_p y_j p^{-2j} + |x|_p \frac{1}{p+1} p^{-2j},$$

and

$$\phi(x) - \phi(y) \leq |x|_p p^{-2j} \left(x_j - y_j - \frac{1}{p+1}\right) > 0.$$

2. Conversely, if $\phi(x) - \phi(y) > 0$ then either $x = y$ or $x < y$ is impossible, as was proved above. $\square$

It follows from Lemma 1.9.5 that the map ϕ is one-to-one.

Lemma 1.9.6. *The function ϕ is continuous.*

Proof. The proof follows from the estimate

$$\left|\phi(x) - \phi(y)\right| \leq p|x - y|_p, \qquad x, y \in \mathbb{Q}_p,$$

which can be proved by using arguments from Lemma 1.9.5. $\square$

These lemmas imply the following statement.

Corollary 1.9.7. ([241, 1.6.]) *The map (1.9.1) is a homeomorphism from $\mathbb{Q}_p$ to some subset $K = \phi(\mathbb{Q}_p) \subset \mathbb{R}$.*

Theorem 1.9.8. ([241, 1.6.]) *The set $K = \phi(\mathbb{Q}_p)$ is a countable union of disjoint perfect nowhere dense sets of the Lebesgue measure zero.*

Proof. Since $\mathbb{Q}_p = \cup_{\gamma \in \mathbb{Z}} S_\gamma$, we have

$$K = \bigcup_{\gamma \in \mathbb{Z}} K_\gamma, \qquad K_\gamma = \phi(S_\gamma), \qquad K_\gamma \cap K_{\gamma'} = \emptyset, \qquad \gamma \neq \gamma'.$$

We will study the structure of the set K_0, since the structure of remaining sets K_γ is similar. In view of (1.9.1), $1 \le \phi(x) < p$ for $x \in S_0$, and, consequently, $K \subset [1, p)$.

Let us consider the following system of disjoint intervals in $[1, p)$:

$$I_n^a = \left(\sum_{j=0}^{n} a_j p^{-2j} + \frac{p^{-2n}}{p+1}, \ \sum_{j=0}^{n} a_j p^{-2j} + p^{-2n} \right), \quad n = 0, 1, 2, \ldots$$

$a = (a_0, a_1, \ldots, a_n)$, $a_j = 0, 1, \ldots, p-1$, $j = 0, 1, 2, \ldots, n$, $a_0 \ne 0$, $a_n \ne p-1$. Set

$$I_n = \bigcup_a I_n^a, \qquad I = \bigcup_{n=0}^{\infty} I_n.$$

The Lebesgue measure of the set I is equal to

$$\mu(I) = \sum_a \sum_{n=0}^{\infty} \mu(I_n^a)$$

$$= (p-1)\left(1 - \frac{1}{p+1}\right) + \sum_{n=1}^{\infty} p^{n-1} p^{-2n} p^{n-1}\left(1 - \frac{1}{p+1}\right)$$

$$= \frac{p(p-1)}{p+1} + \frac{(p-1)^2}{p+1} \sum_{n=1}^{\infty} p^{-n} = p - 1. \tag{1.9.2}$$

By using Lemma 1.9.5, one can prove that there is no $x \in S_0$ and $n = 0, 1, 2, \ldots$ such that

$$\sum_{j=0}^{n} a_j p^{-2j} + \frac{p^{-2n}}{p+1} < \phi(x) < \sum_{j=0}^{n} a_j p^{-2j} + p^{-2n},$$

i.e., $I \cap K_0 = \emptyset$. Now using the lemma on embedded segments, we observe that $K_0 \cup I = [1, p)$. This fact and (1.9.2) imply that K_0 has Lebesgue measure zero. The procedure of construction of the set $K_0 = [1, p) \setminus I$ is similar to that of obtaining the Cantor perfect set C on the segment $[0, 1]$. Therefore the proofs of the remaining properties of K_0 are similar to the proofs of corresponding properties of the Cantor set C. $\qquad\square$

1.9.4. $\mathbb{Q}_p$ and the Monna map

Let us consider the map $\rho : \mathbb{Q}_p \to \mathbb{R}_+$ given by the formula

$$\rho : \sum_{j=\gamma}^{\infty} x_j p^j \to \sum_{j=\gamma}^{\infty} x_j p^{-j-1}, \quad x_j = 0, 1, \ldots, p-1, \quad \gamma \in \mathbb{Z}. \tag{1.9.3}$$

This map was considered by Monna [188][3]. This is a surjective map but it is not one-to-one. The map ρ is continuous. Moreover, the following lemma holds.

Lemma 1.9.9. ([163], [162, 2.4.]) *The map ρ satisfies Hölder's inequality:*

$$|\rho(x) - \rho(y)| \le |x - y|_p, \quad x, y \in \mathbb{Q}_p. \qquad (1.9.4)$$

Proof. Let us consider two p-adic numbers $x = \sum_{j=\alpha}^{\infty} x_j p^j$, $y = \sum_{j=\beta}^{\infty} y_j p^j$. Without loss of generality we assume that $\alpha \le \beta$. Then

$$\rho(x) - \rho(y) = \sum_{j=\alpha}^{\beta-1} x_j p^{-j-1} + \sum_{j=\beta}^{\infty} (x_j - y_j) p^{-j-1}.$$

If $\alpha < \beta$ then

$$|\rho(x) - \rho(y)| \le (p-1) \sum_{j=\alpha}^{\infty} p^{-j-1} = p^{-\alpha} = |x - y|_p.$$

If $\alpha = \beta$ then $|x - y|_p = p^{-\gamma}$, where $\gamma \ge \alpha$. In this case we have

$$\rho(x) - \rho(y) = \sum_{j=\gamma}^{\infty} (x_j - y_j) p^{-j-1},$$

and

$$|\rho(x) - \rho(y)| \le (p-1) \sum_{j=\gamma}^{\infty} p^{-j-1} = p^{-\gamma} = |x - y|_p.$$

The lemma is thus proved. $\qquad\qquad\qquad\qquad\qquad\qquad\qquad\qquad\qquad$ $\square$

The restriction of the map ρ on $\mathbb{Q}_p/\mathbb{Z}_p$ is the one-to-one map

$$\rho : \mathbb{Q}_p/\mathbb{Z}_p \to \mathbb{N}_0 = \{0\} \cup \mathbb{N}.$$

Here the set $\mathbb{Q}_p/\mathbb{Z}_p$ consists of elements of the form $a = \sum_{j=\gamma}^{-1} a_j p^j$.

Lemma 1.9.10. ([163], [162, 2.4.]) *The map ρ has the following properties:*

$$\rho(p^\gamma x) = p^{-\gamma} \rho(x), \quad x \in \mathbb{Q}_p; \qquad (1.9.5)$$

$$\rho(a + x) = \rho(a) + \rho(x), \quad a \in \mathbb{Q}_p/\mathbb{Z}_p, \quad x \in \mathbb{Z}_p; \qquad (1.9.6)$$

---

[3] In Section 8.2, the Monna map establishes a relation between the real and p-adic Haar wavelets.

in particular

$$\rho(a - 1) = \rho(a) + 1. \tag{1.9.7}$$

For $a \in \mathbb{Q}_p/\mathbb{Z}_p$, $x \in \mathbb{Q}_p$, and $k \leq 0$

$$\chi_{[0, \, p^k]}\big(\rho(x - a)\big) = \chi_{[0, \, p^k]}\big(\rho(x) - \rho(a)\big), \tag{1.9.8}$$

where χ_A is a characteristic function of the set A.

Proof. Properties (1.9.5) and (1.9.6) are obvious. Let us prove (1.9.7). If $a \in \mathbb{Q}_p/\mathbb{Z}_p$, we have $a = \sum_{j=\gamma}^{-1} a_j p^j$ and (see Example 1.6.1)

$$a - 1 = \sum_{j=\gamma}^{-1} a_j p^j + \sum_{j=0}^{\infty}(p - 1)p^j.$$

Then

$$\rho(a - 1) = \rho\Big(\sum_{j=\gamma}^{-1} a_j p^j + \sum_{j=0}^{\infty}(p - 1)p^j\Big)$$

$$= \sum_{j=\gamma}^{-1} a_j p^{-j-1} + \sum_{j=0}^{\infty}(p - 1)p^{-j-1} = \rho(a) + 1.$$

Formula (1.9.6) implies (1.9.8). □

Lemma 1.9.11. ([163], [162, 2.4.]) *For the map ρ and for $a \in \mathbb{Q}_p/\mathbb{Z}_p$ and $m, k \in \mathbb{Z}$ we have*

$$\rho : p^m a + p^k \mathbb{Z}_p \to p^{-m}\rho(a) + [0, \, p^k], \tag{1.9.9}$$

$$\rho : \mathbb{Q}_p \setminus \{p^m a + p^k \mathbb{Z}_p\} \to \mathbb{R}_+ \setminus \{p^{-m}\rho(a) + [0, \, p^k]\}, \tag{1.9.10}$$

up to a finite set of points.

Proof. Let us prove (1.9.9). For simplicity we consider the case $m, k = 0$. Let $x \in \{a + \mathbb{Z}_p\}$. Setting in (1.9.4) $y = a$ and $y = a - 1$, and taking into account (1.9.7), we observe that $|\rho(x) - \rho(a)| \leq |x - a|_p \leq 1$ and $|\rho(x) - \rho(a) - 1| \leq |x - a + 1|_p \leq 1$. These inequalities imply that $\rho(x) \in [\rho(a), \rho(a) + 1]$. A general case (1.9.9) is proved in a similar way. Thus the set $p^m a + p^k \mathbb{Z}_p$ is mapped into the segment $p^{-m}\rho(a) + [0, \, p^k]$.

If $x \notin \{a + \mathbb{Z}_p\}$, $a \in \mathbb{Q}_p/\mathbb{Z}_p$, then $x \in \{a' + \mathbb{Z}_p\}$ for some $a' \in \mathbb{Q}_p/\mathbb{Z}_p$, $a' \neq a$. And at the same time $|\rho(a) - \rho(a')| \geq 1$. If $|\rho(a) - \rho(a')| > 1$ then in view of (1.9.3), $\rho(\{a + \mathbb{Z}_p\}) \cap \rho(\{a' + \mathbb{Z}_p\}) = \emptyset$. If $|\rho(a) - \rho(a')| = 1$ then the sets $\rho(\{a + \mathbb{Z}_p\})$ and $\rho(\{a' + \mathbb{Z}_p\})$ intersect only at one point. Repeating our reasoning for the case $m, k \in \mathbb{Z}$, we can prove that the set $\mathbb{Q}_p \setminus \{p^m a + p^k \mathbb{Z}_p\}$ is mapped outside of the segment $p^{-m} \rho(a) + [0, \, p^k]$ in $\mathbb{R}_+$ up to endpoints of this segment. $\qquad\qquad\square$

Lemma 1.9.12. *([163], [162, 2.4.]) The map ρ transforms the Haar measure on $\mathbb{Q}_p$ to the Lebesgue measure on $\mathbb{R}_+$: if $X \subset \mathbb{Q}_p$ is measurable then*

$$\mu(X) = l(\rho(X)),$$

where $\mu(X)$ is the Haar measure of the set X and $l(\rho(X))$ is the Lebesgue measure of the set $\rho(X) \subset \mathbb{R}_+$.

Proof. According to Lemma 1.9.11, balls in $\mathbb{Q}_p$ are mapped onto closed intervals in $\mathbb{R}_+$, and their measures are preserved. The map ρ is surjective, and in view of Lemma 1.9.11, nonintersecting balls are mapped onto intervals whose intersection is either empty or has measure zero.

The lemma is thus proved. $\qquad\qquad\square$

1.10. The space $\mathbb{Q}_p^n$

The space $\mathbb{Q}_p^n = \mathbb{Q}_p \times \cdots \times \mathbb{Q}_p$ consists of points $x = (x_1, \ldots, x_n)$, where $x_j \in \mathbb{Q}_p$, $j = 1, 2 \ldots, n$, $n \geq 2$. The p-adic norm on $\mathbb{Q}_p^n$ is

$$|x|_p = \max_{1 \leq j \leq n} |x_j|_p, \quad x \in \mathbb{Q}_p^n. \tag{1.10.1}$$

This norm is non-Archimedean. Indeed, for any $x, y \in \mathbb{Q}_p^n$

$$\begin{aligned} |x + y|_p &= \max_{1 \leq j \leq n} |x_j + x_j|_p \leq \max_{1 \leq j \leq n} \max \left(|x_j|_p, |y_j|_p \right) \\ &= \max \left(\max_{1 \leq j \leq n} |x_j|_p, \max_{1 \leq j \leq n} |y_j|_p \right) = \max \left(|x|_p, |y|_p \right) \end{aligned}$$

The space $\mathbb{Q}_p^n$ is a complete metric locally compact and totally disconnected space.

The scalar product of vectors $x, y \in \mathbb{Q}_p^n$ is defined as

$$x \cdot y = \sum_{j=1}^{n} x_j y_j.$$

We have

$$|x \cdot y|_p \leq |x|_p |y|_p, \quad x, y \in \mathbb{Q}_p.$$

Denote by $B_\gamma^n(a) = \{x : |x - a|_p \leq p^\gamma\}$ the ball of radius p^γ with the center at a point $a = (a_1, \ldots, a_n) \in \mathbb{Q}_p^n$ and by $S_\gamma^n(a) = \{x : |x - a|_p = p^\gamma\} =$

$B_\gamma^n(a) \setminus B_{\gamma-1}^n(a)$ the corresponding sphere, $\gamma \in \mathbb{Z}$. We set $B_\gamma^n(0) = B_\gamma^n$, and $S_\gamma^n(0) = S_\gamma^n$. For $n = 1$ we will omit the upper index 1 in our notations.

It easy to see that

$$B_\gamma^n(a) = B_\gamma(a_1) \times \cdots \times B_\gamma(a_n), \qquad (1.10.2)$$

where $B_\gamma(a_j) = \{x_j : |x_j - a_j|_p \le p^\gamma\}$ is a ball of radius p^γ with the center at a point $a_j \in \mathbb{Q}_p$, $j = 1, 2 \ldots, n$.

Many one-dimensional theorems hold for the multidimensional case; for example, Ostrovski's Theorem 1.3.2, Theorem 1.8.1, Corollary 1.8.6, etc.

2

p-adic functions

2.1. Introduction

In Section 2.2 of this chapter we consider basic convergence properties of sequences and series in $\mathbb{Q}_p$. Here we follow some sections from [112], [152], [230, I], [241]. In Section 2.3, which substantially follows Section III of the book [241], we present some results on additive and multiplicative characters of the field $\mathbb{Q}_p$.

2.2. *p*-adic power series

2.2.1. *p*-adic sequences

Proposition 2.2.1. *If*

$$\lim_{n \to \infty} x_n = x, \quad x_n, x \in \mathbb{Q}_p, \quad |x|_p \neq 0,$$

then the sequence of norms $\{|x_n|_p : n \in \mathbb{N}\}$ must stabilize for sufficiently large n, i.e., there exists N such that

$$|x_n|_p = |x|_p \quad \forall n \geq N.$$

Proof. It is clear that there exists N such that $|x_n - x|_p < |x|_p$ for all $n \geq N$. Therefore, according to the *strong triangle inequality* (see Section 1.5 (3)), for all $n \geq N$ we have

$$|x_n|_p = |(x_n - x) + x|_p = \max(|x_n - x|_p, |x|_p) = |x|_p.$$

$\square$

According to Section 1.8, $\mathbb{Q}_p$ is a complete metric space, and, consequently, every Cauchy sequence converges. Cauchy sequences are characterized as follows.

Theorem 2.2.2. *A sequence $\{a_n\}$ in $\mathbb{Q}_p$ is a Cauchy sequence, and therefore convergent, if and only if it satisfies*

$$\lim_{n \to \infty} |a_{n+1} - a_n|_p = 0. \tag{2.2.1}$$

Proof. If $\{a_n\}$ is a Cauchy sequence then $\lim_{n \to \infty, m \to \infty} |a_m - a_n|_p = 0$. Setting $m = n + 1$, we obtain (2.2.1).

Conversely, let us assume that (2.2.1) holds. This means that for any $\varepsilon > 0$ there exists a positive integer $N(\varepsilon)$ such that for any $n > N(\varepsilon)$ we have

$$|a_{n+1} - a_n|_p < \varepsilon.$$

Now take any $m > n > N(\varepsilon)$ and due to the strong triangle inequality we obtain

$$|a_m - a_n|_p = |a_m - a_{m-1} + a_{m-1} - a_{m-2} + \cdots - a_n|_p$$
$$\leq \max(|a_m - a_{m-1}|_p, \ldots, |a_{n+1} - a_n|_p) < \varepsilon.$$

$\square$

2.2.2. *p*-adic series

Now let us consider a numerical series

$$\sum_{j=1}^{\infty} a_j, \qquad a_j \in \mathbb{Q}_p. \tag{2.2.2}$$

We say that this series converges if the sequence of its partial sums $S_n = \sum_{j=1}^{n} a_j$ converges in $\mathbb{Q}_p$, and it converges absolutely if the series $\sum_{j=1}^{\infty} |a_j|_p$ converges in $\mathbb{R}$.

Proposition 2.2.3. *A series (2.2.2) converges in $\mathbb{Q}_p$ if and only if* $\lim_{n \to \infty} a_n = 0$, *in which case*

$$\left| \sum_{j=1}^{\infty} a_j \right|_p \leq \max_j |a_j|_p.$$

Proof. The series converges if and only if the sequence of partial sums $S_n = \sum_{j=1}^{n} a_j$ converges. Since $a_n = S_{n+1} - S_n$, from Theorem 2.2.2 the series converges if and only if a_n tends to 0.

Now assume that $\sum_{j=1}^{\infty} a_j$ converges and its sum is equal to S. If $S = \sum_{j=1}^{\infty} a_j = 0$, then there is nothing to prove. If not, by Proposition 2.2.1 we have $|S|_p = |S_N|_p$ for large enough N. Taking into account that $a_n \to 0$ and the strong triangle inequality, we obtain

$$|S_N|_p = \left|\sum_{j=1}^{N} a_j\right| \leq \max_{1 \leq j \leq N} |a_j|_p = \max_j |a_j|_p.$$

Thus the proposition is proved. ☐

It follows from Proposition 2.2.3 that if the series (2.2.2) converges then its sum does not depend on the reordering.

Proposition 2.2.4. *If the series $\sum_{j=1}^{\infty} |a_j|_p$ converges in $\mathbb{R}$, then $\sum_{j=1}^{\infty} a_j$ converges in $\mathbb{Q}_p$.*

Proof. Since the series $\sum_{j=1}^{\infty} |a_j|_p$ converges, the sequence of its partial sums is a Cauchy sequence, i.e., for any $\varepsilon > 0$ there exists an integer $N(\varepsilon)$ such that, for all n, m such that $m > n > N$, we have

$$\sum_{j=n+1}^{m} |a_j|_p < \varepsilon.$$

By the triangle inequality,

$$|S_m - S_n|_p = \left|\sum_{j=n+1}^{m} a_j\right|_p < \sum_{j=n+1}^{m} |a_j|_p < \varepsilon$$

for all $m > n > N$. This implies that $\{S_n\}$ is Cauchy and thus the series $\sum_{j=1}^{\infty} a_j$ converges in $\mathbb{Q}_p$. ☐

A formal power series is an expression of the form

$$f(x) = \sum_{n=0}^{\infty} a_n x^n, \tag{2.2.3}$$

where $a_n \in \mathbb{Q}_p$ and $x \in \mathbb{Q}_p$.

Let us consider the power series (2.2.3). We already know that it converges if and only if $|a_n x^n|_p \to 0$. Thus if $r \geq 0$ denotes a real number such that $|a_n|_p r^n \to 0$, then the series $\sum_{n=0}^{\infty} a_n x^n$ converges at least for $|x|_p \leq r$.

The radius of convergence of a power series $f(x) = \sum_{n=0}^{\infty} a_n x^n$ with coefficients in $\mathbb{Q}_p$ is defined by

$$r(f) = \sup\{r \geq 0 : |a_n|_p r^n \to 0\} \qquad (0 \leq r \leq \infty).$$

Just as in the Archimedean case (power series over $\mathbb{R}$ or $\mathbb{C}$), the radius of convergence can be computed by means of Hadamard's formula:

$$r(f) = \frac{1}{\overline{\lim_{n \to \infty}} |a_n|_p^1 / n}. \tag{2.2.4}$$

Proposition 2.2.5. ([112, 3.2.], [152, IV], [241, 2.1.]) *Let* $0 < r(f) < \infty$. *Then a power series* (2.2.3) *converges if* $|x|_p < r(f)$ *and diverges if* $|x|_p > r(f)$.

Proof. 1. Let $|x|_p < r(f)$. Setting $|x|_p = (1 - \varepsilon)r(f)$, we have

$$|a_n x^n|_p = \left(r(f)|a_n|_p^1 / n \right)^n (1 - \varepsilon)^n.$$

Since there are only finitely many n for which $|a_n|_p^1 / n > \frac{1}{r(f) - \varepsilon r/2}$, we have

$$\lim_{n \to \infty} |a_n x^n|_p \leq \lim_{n \to \infty} \left(\frac{(1 - \varepsilon)r(f)}{(1 - \varepsilon/2)r(f)} \right)^n = \lim_{n \to \infty} \left(\frac{(1 - \varepsilon)}{(1 - \varepsilon/2)} \right)^n = 0.$$

Thus the general term of the series (2.2.3) tends to 0, and the series converges.

2. Analogously, if $|x|_p > r(f)$, setting $|x|_p = (1 + \varepsilon)r(f)$, we have

$$|a_n x^n|_p = \left(r(f)|a_n|_p^1 / n \right)^n (1 + \varepsilon)^n.$$

Since there are infinitely many n for which $|a_n|_p^1 / n > \frac{1}{r(f) + \varepsilon r/2}$, we have

$$\lim_{n \to \infty} |a_n x^n|_p \geq \lim_{n \to \infty} \left(\frac{(1 + \varepsilon)r(f)}{(1 + \varepsilon/2)r(f)} \right)^n = \lim_{n \to \infty} \left(\frac{(1 + \varepsilon)}{(1 + \varepsilon/2)} \right)^n \neq 0.$$

Thus, the series diverges. $\square$

What happens on the "boundary" $|x|_p = r(f)$? In the Archimedean case ($\mathbb{R}$ or $\mathbb{C}$) the behavior on the boundary of the interval or disc of convergence may be quite complicated. In the non-Archimedean case there is a single answer for all points $|x|_p = r(f)$. This is because a series converges if and only if its terms approach zero, if and only if $|a_n|_p |x^n|_p \to 0$, and this depends only on the norm $|x|_p$ and not on particular values of x with a given norm.

For the non-Archimedean case we have the following.

Proposition 2.2.6. ([112, 3.2.], [241, 2.1.]) *The domain of convergence of power series* (2.2.3) *is a ball* $D = \{x \in \mathbb{Q}_p : |x|_p \leq R(f)\}$ *for some* $R(f) \in \{p^\gamma, \gamma \in \mathbb{Z}\} \cup \{0\} \cup \{\infty\}$, *and the series converges uniformly on* D.

Proof. Let the radius of convergence of $f(x)$ be r. We know that the series converges in the open ball $\{|x|_p < r(f)\}$. As for the points of the "boundary" $|x|_p = r(f)$, the necessary and sufficient condition for convergence is $|a_n x_n|_p \to 0$ as $n \to \infty$, which depends only on the norm $|x|_p$, not on the specific value of x. Thus, either for all points with $|x|_p = r(f)$ the series converges, in which case $R(f) = r(f)$, or for all points with $|x|_p = r$ the series diverges, in which case $R(f) = p^{-1}r(f)$. For $|x|_p \leq R(f)$ we have

$$|a_n x^n|_p \leq |a_n (R(f))^n|_p \to 0,$$

and the uniform convergence on D follows. $\qquad \qquad \square$

Proposition 2.2.7. ([112, 3.2.]) *Every series* $f(x) = \sum_{n=0}^{\infty} a_n x^n$, $a_n \in \mathbb{Z}_p$, *converges in* $\{x \in \mathbb{Q}_p : |x|_p < 1\}$.

Proof. Let $|x|_p < 1$. Since for any $n \geq 0$ we have $|a_n|_p \leq 1$, it follows that $|a_n x^n|_p \leq |x|_p^n \to 0$ as $n \to \infty$. Hence the series converges. $\qquad \square$

2.2.3. Analytic functions

Definition 2.1. ([241, 2.2.]) A function $f(x)$ is called the *analytical* in a ball $B_\gamma \subset \mathbb{Q}_p$ if it can be represented by the power series (2.2.3) convergent in B_γ.

For $m = 0, 1, 2, \ldots$ by termwise differentiation and integration of the series (2.2.3), we obtain, respectively, the following series

$$f^{(m)}(x) = \sum_{n=m}^{\infty} n(n-1) \cdots (n-m+1) a_n x^{n-m}, \qquad (2.2.5)$$

$$f^{(-m)}(x) = \sum_{n=0}^{\infty} \frac{1}{(n+1)(n+2) \cdots (n+m)} a_n x^{n+m}. \qquad (2.2.6)$$

These series are called *derivative* and *primitive* of order m, respectively, $f^{(0)}(x) = f(x)$. If $R(f)$ is a radius of the domain of convergence of power series (2.2.3) defined in Proposition 2.2.6, then $R(f^{(-m)}) \leq R(f) \leq R(f^{(m)})$, $m = 1, 2, \ldots$.

Definition 2.2. ([214, 25.]) A function $f(x)$ is called the *analytical* in a ball $B_\gamma^n \subset \mathbb{Q}_p^n$ if it can be represented in the form of convergent series

$$f(x) = \sum_{k_1=0}^{\infty} \cdots \sum_{k_n=0}^{\infty} a_{k_1 \ldots k_n} x_1^{k_n} \cdots x_n^{k_n}, \qquad a_{k_1 \ldots k_n} \in \mathbb{Q}_p, \qquad (2.2.7)$$

$x = (x_1, \ldots, x_n) \in B_\gamma^n$.

Let U be an open subset of $\mathbb{Q}_p^n$. A function $f : U \to \mathbb{Q}_p$ is called the *locally analytical* in U if for each $a \in U$ there is a ball $B_\gamma^n(a) \subset U$ such that $f\big|_{B_\gamma^n(a)}$ is analytic.

2.3. Additive and multiplicative characters of the field $\mathbb{Q}_p$

2.3.1. Additive characters of the field $\mathbb{Q}_p$

An *additive character* of the field $\mathbb{Q}_p$ is defined as a continuous complex-valued function $\chi_p(x)$ defined on $\mathbb{Q}_p$ such that

$$\chi(x+y) = \chi(x)\chi(y), \qquad |\chi(x)| = 1, \qquad x, y \in \mathbb{Q}_p. \tag{2.3.1}$$

Similarly to this definition we define an *additive character* of the *abelian* group B_γ. It is clear that every additive character of the field $\mathbb{Q}_p$ is a character of any group B_γ.

If $\chi(x)$ is an arbitrary additive character, formula (2.3.1) implies the following relations

$$\begin{aligned}
\chi(0) &= 1, \\
\chi(-x) &= \overline{\chi(x)} = \chi^{-1}(x), \\
\chi(nx) &= \left(\chi(x)\right)^n, \quad n \in \mathbb{Z}.
\end{aligned} \tag{2.3.2}$$

It is clear that for every fixed $\xi \in \mathbb{Q}_p$ the function

$$\chi(x) = \chi_p(x\xi) = e^{2\pi i \{x\xi\}_p} \tag{2.3.3}$$

is an additive character of the field $\mathbb{Q}_p$ and the group B_γ, where $\{x\}_p$ is the fractional part of $x \in \mathbb{Q}_p$ defined by formula (1.6.4). Indeed, it follows from the obvious relation for fractional parts

$$\{x+y\}_p = \{x\}_p + \{y\}_p - n, \qquad x, y \in \mathbb{Q}_p, \quad n \in \mathbb{N}_0.$$

Now we will prove that formula (2.3.3) gives a general representation of additive characters of the field $\mathbb{Q}_p$ and the group B_γ.

At first we consider a case of the group B_γ.

Lemma 2.3.1. ([230, I.8.]) *If $\chi(x) \not\equiv 1$ is an additive character of $\mathbb{Q}_p$ then there exists $k \in \mathbb{Z}$ such that*

$$\chi(x) \equiv 1, \qquad x \in B_k. \tag{2.3.4}$$

Proof. Since χ is continuous there is $k \in \mathbb{Z}$ such that $|\chi(x) - 1| < \sqrt{2}$ for all $x \in B_k$. For all positive integers $n \in \mathbb{N}$ we have $nx \in B_k$ and $\chi(nx) = \left(\chi(x)\right)^n$. So $\left|\left(\chi(x)\right)^n - 1\right| < \sqrt{2}$ for all $n \in \mathbb{N}$. This is not possible unless $\chi(x) = 1$. $\square$

We assume that the ball B_k in (2.3.4) is maximal, i.e., if $\chi(x) \not\equiv 1$ in B_γ then $k < \gamma$.

The largest $k \in \mathbb{Z}$ for which the equality (2.3.4) holds is called the *rank of the additive character* $\chi(x)$.

Lemma 2.3.2. *Let k be the rank of an additive character $\chi(x)$. Then for any integer r, $k < r \leq \gamma$, we have*

$$\chi(p^{-r}) = \chi_p(p^{-r}\xi), \qquad k < r \leq \gamma \tag{2.3.5}$$

for some $\xi \in \mathbb{Q}_p$ such that $p^{-\gamma} < |\xi|_p \leq p^{-k}$.

Proof. 1. First, we prove that for $k < r \leq \gamma$

$$\chi(p^{-r}) = e^{2\pi i \frac{m}{p^{r-k}}}, \qquad \exists\, m = 1, 2, \dots, p^{\gamma-k} - 1, \tag{2.3.6}$$

where m does not depend on r.

Let us consider the case $r = \gamma$. Since $p^{-k} \in B_k$, $p^{-\gamma} \in B_\gamma \setminus B_k$, (2.3.2) and (2.3.4) imply

$$1 = \chi(p^{-k}) = \chi(p^{-\gamma+\gamma-k}) = \big(\chi(p^{-\gamma})\big)^{p^{\gamma-k}}.$$

Thus $\chi(p^{-\gamma}) = e^{2\pi i \frac{m}{p^{\gamma-k}}}$, where $m = 1, 2, \dots, p^{\gamma-k} - 1$.

For $k < r < \gamma$ we have

$$\chi(p^{-r}) = \chi(p^{-r+\gamma-\gamma}) = \big(\chi(p^{-\gamma})\big)^{p^{\gamma-r}} = \big(e^{2\pi i \frac{m}{p^{\gamma-k}}}\big)^{p^{\gamma-r}} = e^{2\pi i \frac{m}{p^{r-k}}}.$$

2. Now we represent the number m in (2.3.6) as

$$m = m_0 + m_1 p + \cdots + m_{\gamma-k-1} p^{\gamma-k-1}, \qquad m_j = 0, 1, 2, \dots, p - 1,$$

$j = 0, 1, 2, \dots, \gamma - k - 1$, and set $\xi = p^k(m + m')$, where $m' = p^{\gamma-k}(m_0' + m_1' p + \cdots)$. Here $|\xi|_p > p^{-k}|m|_p > p^{-k}p^{-\gamma+k} = p^{-\gamma}$ and $|\xi|_p \leq p^{-k}$. Taking into account that $p^{-r}\xi = p^{-r}p^k m + p^{-r}p^k m'$, where $p^{-r}p^k m' \in B_0$ for $k < r \leq \gamma$, we rewrite the representation (2.3.6) in the form (2.3.5) for some $\xi \in \mathbb{Q}_p$ such that $p^{-\gamma} < |\xi|_p \leq p^{-k}$. □

Lemma 2.3.3. *Let k be the rank of an additive character $\chi(x)$. If $x \in B_\gamma \setminus B_k$ then the following representation holds:*

$$\chi(x) = \chi_p(\xi x), \qquad \exists\, \xi \in \mathbb{Q}_p, \qquad |\xi|_p \geq p^{-\gamma+1}. \tag{2.3.7}$$

Thus, any additive character of the group B_γ has the form (2.3.3), where either $\xi = 0$ or $|\xi|_p \geq p^{-\gamma+1}$, $\xi \in \mathbb{Q}_p$.

Proof. If $x \in B_\gamma \setminus B_k$ then x can be represented in the form

$$x = x_0 p^{-r} + x_1 p^{-r+1} + \cdots + x_{r-k-1} p^{-k-1} + x', \qquad x' \in B_k, \qquad x_0 \neq 0$$

for some r, $k < r \leq \gamma$. Using (2.3.5) and (2.3.4) we get the representation (2.3.7):

$$
\begin{aligned}
\chi(x) &= \left(\chi(p^{-r})\right)^{x_0}\left(\chi(p^{-r+1})\right)^{x_1}\cdots\left(\chi(p^{-k-1})\right)^{x_{r-k-1}}\chi(x') \\
&= \left(\chi_p(p^{-r}\xi)\right)^{x_0}\left(\chi_p(p^{-r+1}\xi)\right)^{x_1}\cdots\left(\chi_p(p^{-k-1}\xi)\right)^{x_{r-k-1}}\chi_p(x'\xi) \\
&= \chi_p(x_0 p^{-r}\xi + x_1 p^{-r+1}\xi + \cdots + x_{r-k-1}p^{-k-1}\xi + x'\xi) = \chi_p(x\xi).
\end{aligned}
$$

The case $\xi = 0$ is impossible, since $\chi(x) = \chi_p(0) = 1$ in B_γ contradicts the definition of the number k. $\qquad\square$

Now we consider an additive character of the field $\mathbb{Q}_p$.

Lemma 2.3.4. *Any additive character of the field $\mathbb{Q}_p$ has the form (2.3.3) for some $\xi \in \mathbb{Q}_p$.*

Proof. 1. Let $\chi(\xi) \not\equiv 1$ be an additive character of the field $\mathbb{Q}_p$. From Lemmas 2.3.1 and 2.3.3, this character can be represented in the ball B_γ, $\gamma \geq 1$, in the form

$$
\chi(x) = \chi_p(\xi x), \qquad \xi \in \mathbb{Q}_p, \quad |\xi|_p \geq p^{-\gamma+1}, \tag{2.3.8}
$$

where

$$
\begin{aligned}
\xi &= \xi_m p^m + \cdots + \xi_{\gamma-2}p^{\gamma-2} + \xi_{\gamma-1}p^{\gamma-1} \\
&\quad + \xi_\gamma p^\gamma + \xi_{\gamma+1}p^{\gamma+1} + \cdots, \quad m \leq \gamma - 1. \tag{2.3.9}
\end{aligned}
$$

Since $|x|_p \leq p^\gamma$, we have $\xi x = \xi^{(0)}x + \eta^{(0)}$, where $\xi^{(0)} = \xi_m p^m + \cdots + \xi_{\gamma-2}p^{\gamma-2} + \xi_{\gamma-1}p^{\gamma-1}$, $m \leq \gamma - 1$, and $\eta^{(0)} \in \mathbb{Z}_p$. Consequently, in the ball B_γ, $\gamma \geq 1$,

$$
\chi(x) = \chi_p(\xi^{(0)}x), \quad \xi^{(0)} = \xi_m p^m + \cdots + \xi_{\gamma-2}p^{\gamma-2} + \xi_{\gamma-1}p^{\gamma-1}. \tag{2.3.10}
$$

2. Let us consider the character $\chi(x)$ in the ball $B_{\gamma+1}$. As $B_{\gamma+1} = B_\gamma \cup S_{\gamma+1}$ and $B_\gamma \cap S_{\gamma+1} = \emptyset$, any point $x \in S_{\gamma+1}$ can be represented in the form

$$
x = p^{-\gamma-1}x_0 + x', \quad \exists x_0 = 1, 2, \ldots, p-1, \quad x' \in B_\gamma.
$$

Then using (2.3.8), (2.3.9), we have in $B_{\gamma+1}$:

$$
\begin{aligned}
\chi(x) &= \left(\chi(x_0 p^{-\gamma-1})\chi_p(\xi x')\right) = \left(\chi(x_0 p^{-\gamma})\right)^{1/p}\chi_p(\xi x') \\
&= \left(\chi_p(\xi x_0 p^{-\gamma})\right)^{1/p}\chi_p(\xi x') = \chi_p(\xi x_0 p^{-\gamma-1})\chi_p(\xi x') \\
&= \chi_p\left(\xi(x_0 p^{-\gamma-1} + x')\right),
\end{aligned}
$$

where $\xi = \xi_m p^m + \cdots + \xi_{\gamma-2}p^{\gamma-2} + \xi_{\gamma-1}p^{\gamma-1} + \xi_\gamma p^\gamma + \cdots \xi \in \mathbb{Q}_p$, $m \leq \gamma - 1$. Since $|x|_p \leq p^{\gamma+1}$, we have $\xi x = \xi^{(1)}x + \eta^{(1)}$, where $\xi^{(1)} = \xi_m p^m +$

$\cdots + \xi_{\gamma-2} p^{\gamma-2} + \xi_{\gamma-1} p^{\gamma-1} + \xi_\gamma p^\gamma = \xi^{(0)} + \xi_\gamma p^\gamma$, $m \leq \gamma - 1$, and $\eta^{(0)} \in \mathbb{Z}_p$. Consequently, in the ball $B_{\gamma+1}$, $\gamma \geq 1$ the character $\chi(x)$ can be represented in the form

$$\chi(x) = \chi_p(\xi^{(1)} x), \quad \xi^{(1)} = \xi^{(0)} + \xi_\gamma p^\gamma,$$

for some $\xi_\gamma = 0, 1, \ldots, p - 1$.

3. Continuing this process, we obtain in the disc $B_{\gamma+2}$ the representation

$$\chi(x) = \chi_p(\xi^{(2)} x), \quad \xi^{(2)} = \xi^{(0)} + \xi_\gamma p^\gamma + \xi_{\gamma+1} p^{\gamma+1},$$

for some $\xi_{\gamma+1} = 0, 1, \ldots, p - 1$, and so on. As a result, for any additive character of the field $\mathbb{Q}_p$ we obtain the representation (2.3.3):

$$\chi(x) = \chi_p(\xi x), \quad \xi = \xi^{(0)} + \xi_\gamma p^\gamma + \xi_{\gamma+1} p^{\gamma+1} + \cdots \in \mathbb{Q}_p.$$

The lemma is thus proved. $\qquad\qquad\qquad\qquad\qquad\qquad\qquad\square$

If χ_1 and χ_2 are additive characters of the field $\mathbb{Q}_p$, we define their product in the pointwise sense: $(\chi_1 \chi_2)(x) = \chi_1(x) \chi_2(x)$.

The following evident statement holds.

Proposition 2.3.5. *The set of all additive characters of the field* $\mathbb{Q}_p$ *is an abelian group with the unit* 1, *where* $\chi^{-1} = \overline{\chi}$.

Thus, Lemma 2.3.4 can be reformulated as follows.

Theorem 2.3.6. *The group of additive characters of the field* $\mathbb{Q}_p$ *is isomorphic to its additive group* $\mathbb{Q}_p$, *where the isomorphism is given by the mapping* $\xi \to \chi_p(\xi x)$.

Proof. According to Lemma 2.3.4, the mapping $\xi \to \chi_p(\xi x)$ is a homomorphism of the additive group of the field $\mathbb{Q}_p$ onto the group of additive characters. This mapping is one-to-one, since the equality $\chi_p(\xi_1 x) = \chi_p(\xi_2 x)$ for all $x \in \mathbb{Q}_p$ implies $\xi_1 = \xi_2$. $\qquad\qquad\qquad\qquad\qquad\square$

2.3.2. Multiplicative characters of the field $\mathbb{Q}_p$

A *multiplicative character* of the multiplicative group $\mathbb{Q}_p^\times$ is defined as a continuous complex-valued function $\pi(x)$ such that

$$\pi(xy) = \pi(x) \pi(y), \quad x, y \in \mathbb{Q}_p^\times. \tag{2.3.11}$$

A multiplicative character of the field $\mathbb{Q}_p$ is a multiplicative character of its multiplicative group $\mathbb{Q}_p^*$.

Lemma 2.3.7. *Any* multiplicative character π *of the field* $\mathbb{Q}_p$ *can be represented as*

$$\pi(x) = \pi_\alpha(x) = |x|_p^{\alpha-1} \pi_1(|x|_p x), \quad \alpha \in \mathbb{C}, \qquad x \in \mathbb{Q}_p^*, \qquad (2.3.12)$$

where $\pi_1(x')$, $x' \in S_0$, *is a multiplicative character of the group* S_0 *(a so-called* normed multiplicative character) *such that*

$$\pi_1(1) = 1, \qquad |\pi_1(x)| = 1, \quad x \in S_0. \qquad (2.3.13)$$

Conversely, any multiplicative character of the group S_0 *is extended to a multiplicative character of the field* $\mathbb{Q}_p$ *by formula* (2.3.12).

Proof. Let π be a multiplicative character of the field $\mathbb{Q}_p$ defined by (2.3.11). Since any element $x \in \mathbb{Q}_p^*$ can be represented in the form (Section 1.6)

$$x = |x|_p^{-1} x', \qquad x' = |x|_p x \in S_0,$$

denoting $|x|_p = p^{-N}$ and $\pi(p) = p^{1-\alpha}$, we obtain

$$\pi(x) = \pi(|x|_p^{-1} x') = \pi(|x|_p^{-1})\pi(x') = \pi(p^N)\pi_1(x')$$
$$= \big(\pi(p)\big)^N \pi_1(x') = p^{(1-\alpha)N}\pi_1(x') = |x|_p^{\alpha-1}\pi_1(x').$$

For $\alpha = 1$ it follows from (2.3.11) and (2.3.12) that $\pi_1(xy) = \pi_1(x)\pi_1(y)$, $x, y \in S_0$. The last relation implies that

$$\pi_1(1) = 1, \quad \pi_1(x^{-1}) = \big(\pi_1(x)\big)^{-1} = \overline{\pi_1(x)}, \quad |\pi_1(x)| = 1.$$

$\square$

Next we define $\pi_1(x) \overset{def}{=} \pi_1(|x|_p x)$ for $x \in \mathbb{Q}_p^*$. Thus, $\pi_1(p) = \pi_1(1) = 1$. We set $\pi_0(x) = |x|_p^{-1}$.

Lemma 2.3.8. ([230, I.7.]) *Let* $A_0 = S_0 = \{x \in \mathbb{Q}_p : |x|_p = 1\}$, $A_k = B_{-k}(1) = \{x \in \mathbb{Q}_p : |x - 1|_p \le p^{-k}\}$, $k \in \mathbb{N}$. *If* $\pi_1(x) \not\equiv 1$ *is a normed multiplicative character* (2.3.12), (2.3.13), *of the group* S_0 *then there exists* $k \in \mathbb{N}_0$ *such that*

$$\pi_1(x) \equiv 1, \quad x \in A_k. \qquad (2.3.14)$$

Proof. Since π_1 is continuous and $\{A_k : k = 0, 1, \dots\}$ is a fundamental system of neighborhoods of 1 in S_0, there is $k \in \mathbb{N}$ such that if $x \in A_k$ then $|\pi_1(x) - 1| < \sqrt{2}$. Since A_k is a multiplicative group, $x^j \in A_k$ for any $j \in \mathbb{N}$ and $\pi_1(x^j) = \big(\pi_1(x)\big)^j$. So $\big|\big(\pi_1(x)\big)^j - 1\big| < \sqrt{2}$ for all $j \in \mathbb{N}$. This is not possible unless $\pi_1(x) = 1$. $\square$

The smallest $k \in \mathbb{N}_0$ for which the equality (2.3.14) holds is called the *rank of the normed multiplicative character* $\pi_1(x)$. There is only one *zero rank character*, namely, $\pi_1(x) \equiv 1$ for all $x \in S_0$.

Lemma 2.3.9. ([241, III, (2.3)]) *Let the rank k of the character $\pi_1(x)$ be positive. Then for $x = x_0 + x_1 p + \cdots \in S_0$*

$$\pi_1(x) = \pi_0(x_0 + x_1 p + \cdots + x_{k-1} p^{k-1}). \qquad (2.3.15)$$

Thus $\pi_1(x)$ depends only on $x_0, x_1, \ldots, x_{k-1}$.

Proof. Let us represent a number $x \in S_0$ in the form

$$x = x_0 + x_1 p + \cdots = (x_0 + x_1 p + \cdots + x_{k-1} p^{k-1})(1 + tp^k)$$

where $x_0 \neq 0$, $t \in B_0$ is some number. So $1 + tp^k \in B_{-k}(1)$. Then in view of (2.3.14), $\pi_1(1 + tp^k) = 1$ and consequently (2.3.15) holds. $\qquad \square$

Lemma 2.3.10. ([241, III, (2.4)]) *If the character $\pi_1(x)$ has a rank $k \geq 2$ then*

$$\sum_{x_{k-1}=0}^{p-1} \pi_1(x_0 + x_1 p + \cdots + x_{k-1} p^{k-1}) = 0. \qquad (2.3.16)$$

Proof. There exists a number $t = 0, 1, \ldots, p - 1$ such that

$$\rho = \pi_1(1 + tp^{k-1}) \neq 1.$$

(Otherwise, the rank of the character $\pi_{(x)}$ would be less than k.) Therefore,

$$\sum_{x_{k-1}=0}^{p-1} \pi_1(x_0 + x_1 p + \cdots + x_{k-1} p^{k-1})$$

$$= \frac{1}{\rho} \sum_{x_{k-1}=0}^{p-1} \pi_1(1 + tp^{k-1})\pi_1(x_0 + x_1 p + \cdots + x_{k-1} p^{k-1})$$

$$= \frac{1}{\rho} \sum_{x_{k-1}=0}^{p-1} \pi_1\big((x_0 + x_1 p + \cdots + x_{k-1} p^{k-1})(1 + tp^{k-1})\big).$$

From Lemma 2.3.9, we have

$$\sum_{x_{k-1}=0}^{p-1} \pi_1(x_0 + x_1 p + \cdots + x_{k-1} p^{k-1})$$

$$= \frac{1}{\rho} \sum_{x_{k-1}=0}^{p-1} \pi_1\big(x_0 + x_1 p + \cdots + x_{k-2} p^{k-2} + (x_{k-1} + x_0 t) p^{k-1}\big)$$

$$= \frac{1}{\rho} \sum_{x'_{k-1}=0}^{p-1} \pi_1\big(x_0 + x_1 p + \cdots + x_{k-2} p^{k-2} + x'_{k-1} p^{k-1}\big).$$

This implies the equality (2.3.16).   □

3

p-adic integration theory

3.1. Introduction

In this chapter the theory of integration of complex-valued functions of a *p*-adic argument is presented, which is based on the book [241, IV].

3.2. The Haar measure and integrals

A *locally compact group* is a topological group which is locally compact as a topological space.

Let G be a *locally compact group*. Denote by R and L the right and the left action of the group G on itself:

$$R_t(x) = xt^{-1}, \quad L_t(x) = tx, \quad t, x \in G.$$

A left *Haar measure* on a topological group G is a Radon measure which is *invariant* under the left action. A right *Haar measure* on a topological group G is a Radon measure which is *invariant* under the right action. By A. Weil's theorem [247] on any locally compact group G, there exists a non-zero left (right) Haar measure and this Haar measure is unique up to a positive factor. Since $\mathbb{Q}_p$ is a locally compact group, the Weil theorem holds for $\mathbb{Q}_p$. Nevertheless, another "natural" way to introduce a measure on $\mathbb{Q}_p$ is possible (see Appendix D which follows the book by A. V. Kosyak [161]).

Thus, since $\mathbb{Q}_p$ is a locally compact commutative group with respect to addition, in $\mathbb{Q}_p$ there exists the additive Haar measure, which is a positive measure dx invariant under shifts, $d(x + a) = dx, a \in \mathbb{Q}_p$. If the measure dx is normalized by the equality

$$\int_{\mathbb{Z}_p} dx = 1, \tag{3.2.1}$$

then dx is unique.

There is the following remarkable connection between the additive and multiplicative structures of $\mathbb{Q}_p$.

Proposition 3.2.1. ([241, IV, (2.1)]) *For $a \in \mathbb{Q}_p^*$ we have*

$$d(xa) = |a|_p \, dx. \tag{3.2.2}$$

Proof. For any $a \in \mathbb{Q}_p^*$ the measure $d(xa)$ is invariant with respect to shifts, and therefore it differs from measure dx by a factor $C(a) > 0$, i.e., $d(xa) = C(a)dx$. This implies that $C(a)$ is a continuous function satisfying the equation $C(ab) = C(a)C(b)$. According to definition (2.3.11), $C(a)$ is a multiplicative character, and from Lemma 2.3.7, it has the form (2.3.12):

$$C(a) = |a|_p^{\alpha-1} \pi_1(a'), \quad a' = |a|_p \, a \in S_0, \quad \alpha \in \mathbb{C}.$$

Since $C(a) > 0$, $\pi_1(a') = 1$. Thus $C(a) = |a|_p^{\alpha-1}$. Now we shall find a number α. Since the ball $B_0 = \mathbb{Z}_p$ is a union of disjoint balls $B_{-1}(k) = k + pB_0$, $k = 0, 1, 2, \ldots, p - 1$, whose measures are equal, then $\text{mes}(B_0) = p \cdot \text{mes}(pB_0)$, and, consequently, $d(px) = \frac{1}{p}dx$. Thus, $C(p) = \frac{1}{p} = |p|_p^{\alpha-1}$, i.e., $\alpha = 2$ and $C(a) = |a|_p$. $\quad\square$

There is also a measure $d^\times x$ of the multiplicative group $\mathbb{Q}_p^*$, which is invariant under multiplications, $d^\times(xa) = d^\times x, a \in \mathbb{Q}_p^*$. There is the following relation between the measures dx and $d^\times x$:

$$d^\times x = \frac{dx}{|x|_p}.$$

Indeed, in view of (3.2.2), we have $|xa|_p^{-1} d(xa) = |x|_p^{-1} \, dx$, i.e., the measure $|x|_p^{-1} \, dx$ is invariant with respect to multiplication. The normalized measure $d^\times x$ has the form

$$d^\times x = \frac{p}{p-1} \frac{dx}{|x|_p}, \quad \text{i.e.,} \quad \int_{\mathbb{Z}_p^\times} d^\times x = 1.$$

The invariant under shifts measure dx on the field $\mathbb{Q}_p$ is extended to an invariant under shifts measure $d^n x = dx_1 \cdots dx_n$ on $\mathbb{Q}_p^n$ in the standard way. We have

$$d^n(x + a) = d^n x, \quad a \in \mathbb{Q}_p^n, \quad d^n(Ax) = |\det A|_p \, d^n x,$$

where $A : \mathbb{Q}_p^n \to \mathbb{Q}_p^n$ is a linear isomorphism such that $\det A \neq 0$. The measure $d^{\times n} x$ invariant under multiplication by scalars $\alpha \in \mathbb{Q}_p^*$ is defined as

$$d^{\times n} x = \frac{d^n x}{|x|_p^n},$$

where the norm $|x|_p = \max_{1 \le j \le n} |x_j|_p$, $x \in \mathbb{Q}_p^n$ is given by (1.10.1).

Let A be a measurable subset in $\mathbb{Q}_p^n$. Denote by $\mathcal{L}^\rho(A)$ a set of all measurable functions $f : A \to \mathbb{C}$, $A \subset \mathbb{Q}_p^n$, such that

$$\int_A |f(x)|^\rho \, d^n x < \infty \quad (\rho \geq 1).$$

For $f \in \mathcal{L}^\rho(A)$ the norm is defined as

$$\|f\|_\rho = \left(\int_A |f(x)|^\rho \, d^n x \right)^{1/\rho}.$$

Proposition 3.2.2. *Let $\mathcal{O}$ be an open subset in $\mathbb{Q}_p^n$ and let $C_0(\mathcal{O})$ be the space of continuous functions ϕ with compact supports in $\mathcal{O}$. Then $C_0(\mathcal{O})$ is dense in $\mathcal{L}^\rho(\mathcal{O})$, $1 \leq \rho < \infty$.*

The proof of this proposition repeats the classical statement for the real case almost word for word (for example, see [237, §1.2.]).

Let $\mathcal{O} \subset \mathbb{Q}_p^n$ be an open subset. Denote by $\mathcal{L}_{loc}^\rho(\mathcal{O})$ the set of functions $f : \mathcal{O} \to \mathbb{C}$ such that $f \in \mathcal{L}^\rho(K)$ for any compact $K \subset \mathcal{O}$. It is clear that $\mathcal{L}^\rho(\mathcal{O}) \subset \mathcal{L}_{loc}^\rho(\mathcal{O})$.

A function $f \in \mathcal{L}^1(\mathbb{Q}_p^n)$ is called *integrable* on $\mathbb{Q}_p^n$ *in an improper sense* if there exists

$$\lim_{N \to \infty} \int_{B_N^n} f(x) \, d^n x.$$

We call this limit an improper integral and denote it as $\int_{\mathbb{Q}_p^n} f(x) \, d^n x$. In particular,

$$\int_{\mathbb{Q}_p} f(x) \, dx = \lim_{N \to \infty} \sum_{\gamma=-\infty}^{N} \int_{S_\gamma} f(x) \, dx = \sum_{\gamma=-\infty}^{\infty} \int_{S_\gamma} f(x) \, dx. \quad (3.2.3)$$

Similarly, if $f \in \mathcal{L}^1(\mathbb{Q}_p \setminus \{a\})$, $a \in \mathbb{Q}_p$, then one can define an improper integral

$$\int_{\mathbb{Q}_p} f(x) \, dx = \lim_{\substack{N \to \infty, \\ M \to -\infty}} \sum_{\gamma=M}^{N} \int_{S_\gamma(a)} f(x) \, dx. \quad (3.2.4)$$

For each integer $k > 0$ define

$$[f]_k(x) = \begin{cases} f(x), & p^{-k} \leq |x|_p \leq p^k \\ 0, & \text{otherwise} \end{cases}, \quad x \in \mathbb{Q}_p^n.$$

If each $[f]_k$ is integrable, we define the principal value integral of f as

$$P.V. \int_{\mathbb{Q}_p^n} f(x) \, d^n x \overset{def}{=} \lim_{k \to \infty} \int_{\mathbb{Q}_p^n} [f]_k(x) \, d^n x, \quad (3.2.5)$$

if this limit exists.

If f is an integrable function on $\mathbb{Q}_p$, then:

$$\int_{B_N} f(x)\,dx = \sum_{\gamma=-\infty}^{N} \int_{S_\gamma} f(x)\,dx,$$

$$\int_{S_\gamma} f(x)\,dx = \int_{B_\gamma} f(x)\,dx - \int_{B_{\gamma-1}} f(x)\,dx,$$

(3.2.6)

Let us formulate the Fubini theorem for the p-adic case.

Theorem 3.2.3. *If a function $f : \mathbb{Q}_p^{n+m} \to \mathbb{C}$ is such that the repeated integral*

$$\int_{\mathbb{Q}_p^n} \left(\int_{\mathbb{Q}_p^n} f(x, y)\,d^n y \right) d^n x$$

exists then $f \in \mathcal{L}^1(\mathbb{Q}_p^{n+m})$ and the the following equalities hold:

$$\int_{\mathbb{Q}_p^n} \left(\int_{\mathbb{Q}_p^n} f(x, y)\,d^n y \right) d^n x = \int_{\mathbb{Q}_p^{n+m}} f(x, y)\,d^n y\,d^n x$$

$$= \int_{\mathbb{Q}_p^n} \left(\int_{\mathbb{Q}_p^n} f(x, y)\,d^n x \right) d^n y.$$

Now we present the **Lebesgue dominated convergence theorem** for $\mathbb{Q}_p^n$.

Theorem 3.2.4. *If a sequence of functions $f_k \in \mathcal{L}^1(\mathbb{Q}_p^n)$, $k \to \infty$, converges almost everywhere (a.e.) in $\mathbb{Q}_p^n$ (with respect to the measure $d^n x$) to a function f, i.e.,*

$$f_k(x) \to f(x), \quad k \to \infty, \quad x \in \mathbb{Q}_p^n, \quad a.e.,$$

and there exists a function $\psi \in \mathcal{L}^1(\mathbb{Q}_p^n)$ such that

$$|f_k(x)| \le \psi(x), \quad k \in \mathbb{N}, \quad x \in \mathbb{Q}_p^n, \quad a.e.$$

then the following equality holds:

$$\lim_{k\to\infty} \int_{\mathbb{Q}_p^n} f_k(x)\,d^n x = \int_{\mathbb{Q}_p^n} f(x)\,d^n x.$$

3.3. Some simple integrals

In the next chapters, we will use the following simple integrals. It follows from (3.2.1), (3.2.2) and (3.2.6) that

$$\int_{B_\gamma} dx = p^\gamma,$$

$$\int_{S_\gamma} dx = p^\gamma \left(1 - \tfrac{1}{p}\right), \quad \gamma \in \mathbb{Z}. \tag{3.3.1}$$

By using (3.2.6) and (3.3.1), we find

$$\int_{B_0} |x|_p^{\alpha-1} dx = \frac{1 - p^{-1}}{1 - p^{-\alpha}}, \quad \operatorname{Re}\alpha > 0, \tag{3.3.2}$$

Let χ_p be an additive character (2.3.3). Then

$$\int_{B_\gamma} \chi_p(x\xi)\, d\xi = \begin{cases} p^\gamma, & |x|_p \le p^{-\gamma}, \\ 0, & |x|_p \ge p^{-\gamma+1}. \end{cases} \tag{3.3.3}$$

Indeed, if $|x|_p \le p^{-\gamma}$ then $|\xi x|_p \le 1$. Consequently, formulas (3.3.1) and (2.3.3) imply the first relation in (3.3.3). If $|x|_p \ge p^{-\gamma+1}$, then for some $\xi' \in S_\gamma$ we have $|\xi x|_p \ge p$. Therefore $\chi_p(x\xi') \ne 1$. Next, making the change of the variables $\xi = \eta - \xi'$, and taking into account Theorem 1.8.1 (1), we obtain

$$\int_{B_\gamma} \chi_p(x\xi)\, d\xi = \int_{B_\gamma(\xi')} \chi_p(x(\eta - \xi'))\, d\eta = \chi_p(-x\xi') \int_{B_\gamma} \chi_p(x\eta)\, d\eta.$$

This implies the second relation in (3.3.3).

It follows from formulas (3.2.6) and (3.3.3) that

$$\int_{S_\gamma} \chi_p(x\xi)\, d\xi = \begin{cases} p^\gamma (1 - p^{-1}), & |x|_p \le p^{-\gamma}, \\ -p^{\gamma-1}, & |x|_p = p^{-\gamma+1}, \\ 0, & |x|_p \ge p^{-\gamma+2}. \end{cases} \tag{3.3.4}$$

If $\pi(x) \not\equiv 1$ then

$$\int_{S_\gamma} \pi(x)\, dx = 0, \quad \gamma \in \mathbb{Z}. \tag{3.3.5}$$

Indeed, since $\pi(x) \not\equiv 1$, there exists $a \in \mathbb{Q}_p$ such that $|a|_p = 1$, $\pi(a) \ne 1$ (see Lemma 2.3.8). Making the change of variables $x = ax'$ in the integral (3.3.5), we obtain

$$\int_{S_\gamma} \pi(x)\, dx = \int_{S_\gamma} \pi(ax')\, d(ax') = \pi(a) \int_{S_\gamma} \pi(x')\, dx'.$$

This implies the relation (3.3.5).

3.4. Change of variables

The relation (3.2.2) can be considered as a special case of the general formula for the change of variables in integrals. The relation (3.2.2) implies that

$$\int_K f(x)\,dx = |a|_p \int_{\frac{K-b}{a}} f(ay+b)\,dy, \qquad (3.4.1)$$

where $K \subset \mathbb{Q}_p$ is a compact, $a, b \in \mathbb{Q}_p$, $a \neq 0$.

Now we present a general formula of change of variables in integrals.

Theorem 3.4.1. ([241, IV.2.]) *If $x(y)$ is an analytic diffeomorphism of a clopen set $K_1 \subset \mathbb{Q}_p$ onto a clopen set $K \subset \mathbb{Q}_p$, and $x'(y) \neq 0$, $y \in K_1$, then for any $f \in \mathcal{L}^1(K)$ we have*

$$\int_K f(x)\,dx = \int_{K_1} f(x(y))|x'(y)|_p\,dy, \qquad (3.4.2)$$

where $x'(y)$ is a derivative in the sense of (2.2.5).

Proof. According to Proposition 3.2.2, the space $C(K)$ of continuous functions on K is dense in $\mathcal{L}^1(K)$. Moreover, we will prove below by Proposition 4.3.2 that the space of locally constant functions on K is dense in $C(K)$. Thus, it is sufficient to prove the formula (3.4.2) for locally constant functions on K. Therefore, it is sufficient to prove (3.4.2) for $f(x) \equiv 1$, $x \in K$, i.e.,

$$\int_K dx = \int_{K_1} |x'(y)|_p\,dy.$$

Let us cover the compact K_1 by a finite number of disjoint balls $B_\rho(y_k)$ of a sufficient small radius p^ρ such that

$$|x'(y)|_p = |x'(y_k)|_p = p^{r_k}, \qquad y \in B_\rho(y_k),$$

and the ball $B_\rho(y_k)$ is mapped onto the ball $B_{\rho+r_k}(x_k)$, $x_k = x(y_k)$. This is possible in view of the theorem of inverse function [241, II.5.]. Next, using the formula (3.3.1), we obtain

$$\int_K dx = \sum_k \int_{B_{\rho+r_k}(x_k)} dx = \sum_k p^{\rho+r_k} = \sum_k \int_{B_\rho(y_k)} p^{r_k}\,dy$$

$$= \sum_k \int_{B_\rho(y_k)} |x'(y_k)|_p\,dy = \sum_k \int_{B_\rho(y_k)} |x'(y)|_p\,dy = \int_{K_1} |x'(y)|_p\,dy.$$

The theorem is thus proved. $\qquad\qquad\qquad\qquad\qquad\qquad\qquad\qquad\square$

If f is an integrable function on $\mathbb{Q}_p$, then Theorem 3.4.1 implies a formula for the change of variables:

$$\int_{\mathbb{Q}_p} f(x)\,dx = \int_{\mathbb{Q}_p} f\left(\frac{1}{\xi}\right)\frac{1}{|\xi|_p^2}\,d\xi, \qquad (3.4.3)$$

where the above integral is defined as an improper integral (3.2.4).

Theorem 3.4.2. ([241, IV.2.]) *Let* $x, y \in \mathbb{Q}_p^n$. *If* $x(y)$ *is an analytic diffeomorphism of a clopen set* $K_1 \subset \mathbb{Q}_p^n$ *onto* $K \subset \mathbb{Q}_p^n$, *and*

$$\det\left(\frac{\partial x(y)}{\partial y}\right) = \det\left(\frac{\partial x_k}{\partial y_j}\right) \neq 0, \quad y \in K_1,$$

then for any $f \in \mathcal{L}^1(K)$ *we have*

$$\int_K f(x)\,d^n x = \int_{K_1} f(x(y))\left|\det\left(\frac{\partial x(y)}{\partial y}\right)\right|_p d^n y. \qquad (3.4.4)$$

4

p-adic distributions

4.1. Introduction

The theory of *p*-adic distributions (generalized functions) plays an important role in solving mathematical problems of *p*-adic analysis and applications. Fundamental results in the *p*-adic theory of distributions can be found in [62], [96], [115], [230], [241]. In this chapter we collect some basic facts on the theory of *p*-adic distributions, especially following [241, VI]. In contrast to the book [241, VI], we consider the multidimensional case.

4.2. Locally constant functions

Definition 4.1. A complex-valued function ψ defined on an open set $\mathcal{O} \subset \mathbb{Q}_p^n$ is called *locally constant* on $\mathcal{O}$ if for any $x \in \mathcal{O}$ there exists an integer $l(x) \in \mathbb{Z}$ such that

$$\psi(x + x') = \psi(x), \quad x' \in B_{l(x)}^n, \quad x \in \mathcal{O}.$$

Typical examples of *locally constant* functions are additive characters (2.3.3) and also the characteristic function of the ball $B_l^n(a)$

$$\Delta_l(x - a) = \Omega(p^{-l}|x - a|_p), \quad a \in \mathbb{Q}_p^n, \quad x \in \mathbb{Q}_p^n, \quad (4.2.1)$$

where

$$\Omega(t) = \begin{cases} 1, & 0 \le t \le 1, \\ 0, & t > 1 \end{cases} \quad (4.2.2)$$

is a characteristic function of the interval [0, 1].

Denote by $\mathcal{E}(\mathcal{O})$ the linear space of locally constant $\mathbb{C}$-value functions on $\mathcal{O} \subset \mathbb{Q}_p^n$. It is clear that any function from $\mathcal{E}(\mathcal{O})$ is continuous on $\mathcal{O}$.

54

Convergence in $\mathcal{E}(\mathbb{Q}_p^n)$ is defined in the following way: $\psi_k \to 0, k \to \infty$ in $\mathcal{E}(\mathbb{Q}_p^n)$ if for any compact $K \subset \mathbb{Q}_p^n$

$$\psi_k \overset{x \in K}{\Longrightarrow} 0, \qquad k \to \infty,$$

where $\Rightarrow$ denotes uniform convergence.

Lemma 4.2.1. ([241, VI.1, Lemma 1]) *Let $\psi \in \mathcal{E}(\mathbb{Q}_p^n)$ and let K be a compact in $\mathbb{Q}_p^n$. then there exists $l \in \mathbb{Z}$ such that*

$$\psi(x + x') = \psi(x), \quad x' \in B_l^n, \quad x \in K.$$

Proof. Suppose that K is contained in the ball B_N^n. It is sufficient to prove the lemma for the compact B_N^n. By the Heine–Borel lemma (see Corollary 1.8.6 (3), from the infinite covering $\{B_{l(x)}^n(x) : x \in B_N^n\}$ of the compact B_N^n it is possible to choose a finite disjoint subcovering $\{B_{l(x^k)}^n(x^k) : x^k \in B_N^n, k = 1, 2, \ldots, M\}$. Let us denote $l = \min_k l(x^k)$. Then for any point $x \in B_{l(x^k)}^n(x^k)$ and for all $x' \in B_l^n$ we have

$$|x - x^k + x'|_p \leq \max(|x - x^k|_p, |x'|_p) \leq \max(p^{l(x^k)}, p^l) = p^{l(x^k)}.$$

Therefore,

$$\psi(x + x') = \psi(x^k + x - x^k + x') = \psi(x^k) = \psi(x).$$

$\square$

Lemma 4.2.2. ([241, VI.1, Lemma 2]) *If $\psi \in \mathcal{E}(\mathbb{Q}_p^n)$ then in every ball $B_N^n \subset \mathbb{Q}_p^n$ ψ is represented in the form*

$$\varphi(x) = \sum_{\nu=1}^{p^{n(N-l)}} \psi(a^\nu) \Delta_l(x - a^\nu), \quad x \in B_N^n, \tag{4.2.3}$$

where $\Delta_l(x - a^\nu) = \Omega(p^{-l}|x - a^\nu|_p)$ is the characteristic function of the ball $B_l^n(a^\nu)$ given by (4.2.1), $l \in \mathbb{Z}$ and the points $a^\nu \in B_N^n$ are such that the ball B_N^n is represented by the union of the disjoint balls $B_l^n(a^\nu)$, $\nu = 1, 2, \ldots, p^{n(N-l)}$.

Proof. By Lemma 4.2.1, there exists $l \in \mathbb{Z}$ such that $\psi(x + x') = \psi(x)$, $x \in B_N^n$, for all $x' \in B_l^n$, $l \leq N$. If $l = N$, setting $a^1 = 0$, one can see that the statement holds. Let $l < N$. From (1.10.2),

$$B_N^n = B_{N,1} \times \cdots \times B_{N,n},$$

where, according to Proposition 1.8.9, every one-dimensional ball $B_{N,k} = \{x_k \in \mathbb{Q}_p : |x_k|_p \leq p^N\}$ is represented by the union of p^{N-l} disjoint balls

$B_{l;k}(a_k^\mu)$ with the centers

$$a_k^\mu = 0, \quad a_k^\mu = a_{-r;k}p^{-r} + a_{-r+1;k}p^{-r+1} + \cdots + a_{-l-1;k}p^{-l-1},$$

where $0 \le a_{j;k} \le p - 1$; $j = -r, -r + 1, \ldots, -l - 1$; $a_{-r;k} \ne 0$; $r = l + 1$, $l + 2, \ldots, N - 1, N$; $k = 1, 2, \ldots, n$.

Now the ball B_N^n can be represented by the union of $p^{n(N-l)}$ disjoint balls $B_l(a^\nu)$ with the centers $a = 0$, $a^\nu = (a_1^\nu, \ldots a_n^\nu) \in B_N^n$. Here any ball $B_l(a^\nu)$ is a direct product of the corresponding one-dimensional balls $B_{l;k}(a_k^\mu)$, $k = 1, 2, \ldots, n$, $\mu = 1, 2, \ldots, p^{N-l}$; $\nu = 1, 2, \ldots, p^{n(N-l)}$. Thus,

$$\sum_{\nu=1}^{p^{n(N-l)}} \Delta_l(x - a^\nu) = \begin{cases} 1, & x \in B_N^n, \\ 0, & x \notin B_N^n. \end{cases}$$

Hence

$$\varphi(x) = \sum_{\nu=1}^{p^{n(N-l)}} \varphi(x)\Delta_l(x - a^\nu) = \sum_{\nu=1}^{p^{n(N-l)}} \psi(a^\nu)\Delta_l(x - a^\nu), \quad x \in B_N^n.$$

The lemma is proved.   □

4.3. The Bruhat–Schwartz test functions

Denote by $\mathcal{D}(\mathbb{Q}_p^n)$ the *Bruhat–Schwartz space of test functions* in $\mathbb{Q}_p^n$, which consists of all compactly supported functions $\varphi \in \mathcal{E}(\mathbb{Q}_p^n)$. If $\varphi \in \mathcal{D}(\mathbb{Q}_p^n)$, according to Lemma 4.2.1, there exists $l \in \mathbb{Z}$ such that

$$\varphi(x + x') = \varphi(x), \quad x' \in B_l^n, \quad x \in \mathbb{Q}_p^n. \tag{4.3.1}$$

The *largest* of such numbers $l = l(\varphi)$ is called the *parameter of constancy* of the function φ. Instead of the *parameter of constancy* one can consider the *vector of constancy* $\mathbf{l} = (l_1, \ldots, l_n)$, where l_j is the *parameter of constancy* of the function φ with respect to the variable x_j, $j = 1, \ldots, n$. In this case $l = \min_{1 \le j \le n}(l_j)$.

Let us denote by $\mathcal{D}_N^l(\mathbb{Q}_p^n)$ the finite-dimensional space of test functions from $\mathcal{D}(\mathbb{Q}_p^n)$ having supports in the ball B_N^n and with parameters of constancy $\ge l$. The following embedding holds: $\mathcal{D}_N^l(\mathbb{Q}_p^n) \subset \mathcal{D}_{N'}^{l'}(\mathbb{Q}_p^n)$, $N \le N'$, $l \ge l'$.

Convergence in $\mathcal{D}(\mathbb{Q}_p^n)$ is defined in the following way: $\varphi_k \to 0$, $k \to \infty$ in $\mathcal{D}(\mathbb{Q}_p^n)$ if and only if

(i) $\varphi_k \in \mathcal{D}_N^l(\mathbb{Q}_p^n)$ where N and l do not depend on k;

(ii) $\varphi_k \overset{x \in \mathbb{Q}_p^n}{\Longrightarrow} 0$, $k \to \infty$, where $\Rightarrow$ denotes uniform convergence.

Here

$$\mathcal{D}(\mathbb{Q}_p^n) = \lim_{N \to \infty} \text{ind} \, \mathcal{D}_N(\mathbb{Q}_p^n), \quad \mathcal{D}_N(\mathbb{Q}_p^n) = \lim_{l \to -\infty} \text{ind} \, \mathcal{D}_N^l(\mathbb{Q}_p^n).$$

The space $\mathcal{D}(\mathbb{Q}_p^n)$ is a complete locally convex topological vector space.

Let $\mathcal{O}$ be an open subset in $\mathbb{Q}_p^n$. By $\mathcal{D}(\mathcal{O})$ we denote the space of all test functions from $\mathcal{D}(\mathbb{Q}_p^n)$ with supports in $\mathcal{O}$. Thus

$$\mathcal{D}(\mathcal{O}) \subset \mathcal{D}(\mathbb{Q}_p^n). \tag{4.3.2}$$

Lemma 4.3.1. ([241, VI.5, (5.2)]) *Any function $\varphi \in \mathcal{D}_N^l(\mathbb{Q}_p^n)$ can be represented in the following way:*

$$\varphi(x) = \sum_{\nu=1}^{p^{nN-l_1-\cdots-l_n}} \varphi(a^\nu) \Delta_{l_1}(x_1 - a_1^\nu) \cdots \Delta_{l_n}(x_n - a_n^\nu), \quad x \in \mathbb{Q}_p^n, \tag{4.3.3}$$

where $\Delta_{l_j}(x_j - a_j^\nu)$ is the characteristic function of the ball $B_{l_j}(a_j^\nu)$, and the points $a^\nu = (a_1^\nu, \dots a_n^\nu) \in B_N^n$ do not depend on φ and are such that the balls $B_l^n(a^\nu) = B_l(a_1^\nu) \times \cdots \times B_l(a_n^\nu)$, $\nu = 1, \dots, p^{nN-l_1-\cdots-l_n}$, are disjoint and cover the ball B_N^n.

For $l_1 = l_2 = \cdots = l_n$ and using (1.10.2), the formula (4.3.3) takes the form

$$\varphi(x) = \sum_{\nu=1}^{p^{n(N-l)}} \varphi(a^\nu) \Delta_l(x - a^\nu), \quad x \in \mathbb{Q}_p^n, \tag{4.3.4}$$

where $\Delta_l(x - a^\nu)$ is the characteristic function of the ball $B_l^n(a^\nu)$.

Lemma 4.3.1 and representation (4.3.4) are a direct consequence of Lemma 4.2.2.

Proposition 4.3.2. ([241, VI.2.)]) *Let K be a compact in $\mathbb{Q}_p^n$. The space of test functions $\mathcal{D}(\mathbb{Q}_p^n)$ is dense in the space $C(K)$ of continuous functions on K.*

Proof. Let $f \in C(K)$ and $\varepsilon > 0$ be an arbitrary number. There exists $\gamma \in \mathbb{Z}$ such that $|f(x) - f(a)| < \varepsilon$ if $x \in B_\gamma(a) \cap K$ and $a \in K$. From Proposition 1.8.7, the compact K can be covered by a finite number of disjoint balls $B_\gamma(a^\nu)$. The characteristic functions of these balls $\Delta_\gamma(x - a^\nu) \in \mathcal{D}(\mathbb{Q}_p^n)$ and satisfy the equality

$$\sum_\nu \Delta_\gamma(x - a^\nu) = 1, \quad x \in K. \tag{4.3.5}$$

It is clear that

$$f_\gamma(x) = \sum_\nu f(a^\nu) \Delta_l(x - a^\nu) \in \mathcal{D}(\mathbb{Q}_p^n).$$

Next, taking (4.3.5) into account, we obtain

$$\| f(x) - f_\gamma(x) \|_{C(K)} \overset{def}{=} \max_{x \in K} | f(x) - f_\gamma(x) |$$

$$\leq \max_{x \in K} \left| \sum_v \left(f(x) - f(a^v) \right) \Delta_l(x - a^v) \right| < \varepsilon \sum_v \Delta_l(x - a^v) = \varepsilon.$$

$\square$

Proposition 4.3.3. ([241, VI.2.]) *The space of test functions $\mathcal{D}(\mathcal{O})$ is dense in $\mathcal{L}^\rho(\mathcal{O})$, $1 \leq \rho < \infty$, $\mathcal{O} \subset \mathbb{Q}_p^n$.*

Proof. The proof follows from the facts that, according to Proposition 4.3.2, $\mathcal{D}(\mathbb{Q}_p^n)$ is dense in $C(K)$ and, according to Proposition 3.2.2, $C(K)$ is dense in $\mathcal{L}^\rho(\mathcal{O})$. $\square$

Let $\mathcal{O}$ be an open subset in $\mathbb{Q}_p^n$. In $\mathcal{D}(\mathcal{O})$ the **theorem of "decomposition of the unity"** holds ([241, VI.2.]). Indeed, according to Proposition 1.8.3, the set $\mathcal{O}$ in $\mathbb{Q}_p^n$ is a union at most of a countable set of disjoint balls:

$$\mathcal{O} = \bigcup_k B_{\gamma_k}(a^k), \qquad B_{l_k}(a^k) \cap B_{l_j}(a^j) = \emptyset, \quad k \neq j.$$

Hence, the characteristic functions $\Delta_{l_k}(x - a^k)$ of these balls, $k = 1, 2, \ldots,$ form a decomposition of the unity in $\mathcal{O}$:

$$\sum_{k=1}^{\infty} \Delta_{l_k}(x - a^k) = 1, \qquad x \in \mathcal{O}. \tag{4.3.6}$$

4.4. The Bruhat–Schwartz distributions (generalized functions)

4.4.1. The space $\mathcal{D}'(\mathbb{Q}_p^n)$

Denote by $\mathcal{D}'(\mathbb{Q}_p^n)$ the set of all linear functionals on $\mathcal{D}(\mathbb{Q}_p^n)$, which is called the space of *Bruhat–Schwartz distributions* in $\mathbb{Q}_p^n$. Convergence in $\mathcal{D}'(\mathbb{Q}_p^n)$ is defined as the weak convergence: $f_k \to 0, k \to \infty$ in $\mathcal{D}'(\mathbb{Q}_p^n)$ if

$$\langle f_k, \varphi \rangle \to 0, \quad k \to \infty, \quad \forall \varphi \in \mathcal{D}(\mathbb{Q}_p^n).$$

Lemma 4.4.1. ([241, VI.3, Lemma]) *If A is a linear operator from $\mathcal{D}(\mathbb{Q}_p^n)$ into a linear topological space M then A is continuous from $\mathcal{D}(\mathbb{Q}_p^n)$ into M.*

Proof. Let $\varphi_k \to 0$, $k \to \infty$ in $\mathcal{D}(\mathbb{Q}_p^n)$. Then $\varphi_k \in \mathcal{D}_N^l(\mathbb{Q}_p^n)$ for all k. It follows from Lemma 4.3.1 (formula (4.3.4)) that

$$\varphi_k(x) = \sum_{\nu=1}^{p^{n(N-l)}} \varphi_k(a^\nu)\Delta_l(x - a^\nu), \quad x \in \mathbb{Q}_p^n,$$

where $\Delta_l(x - a^\nu)$ is the characteristic function of the ball $B_l^n(a^\nu)$, and $\varphi_k(a^\nu) \to 0$ as $k \to \infty$. Since A is a linear operator, the above representation implies

$$A\varphi_k(x) = \sum_{\nu=1}^{p^{n(N-l)}} \varphi_k(a^\nu)A\Delta_l(x - a^\nu) \to 0, \quad k \to \infty, \quad \text{in} \quad M,$$

i.e., A is continuous. $\qquad\square$

Corollary 4.4.2. $\mathcal{D}'(\mathbb{Q}_p^n)$ *is a space of linear continuous functionals on the space of test functions* $\mathcal{D}(\mathbb{Q}_p^n)$, *i.e.,* $\mathcal{D}'(\mathbb{Q}_p^n)$ *is a strongly conjugate space to the space* $\mathcal{D}(\mathbb{Q}_p^n)$.

Thus, it is possible to use general theorems of functional analysis for $\mathcal{D}'(\mathbb{Q}_p^n)$.

Proposition 4.4.3. *The space* $\mathcal{D}'(\mathbb{Q}_p^n)$ *is complete.*

Proof. Suppose that $\{f_k, \ k \to \infty\}$ is the Cauchy sequence of functionals $f_k \in \mathcal{D}'(\mathbb{Q}_p^n)$, i.e., $f_k - f_j \to 0$, $k, j \to \infty$. It means that $\langle f_k, \varphi \rangle - \langle f_j, \varphi \rangle \to 0$, $k, j \to \infty$, for all $\varphi \in \mathcal{D}(\mathbb{Q}_p^n)$. Thus $\{\langle f_k, \varphi \rangle, \ k \to \infty\}$ is the numerical Cauchy sequence for all $\varphi \in \mathcal{D}(\mathbb{Q}_p^n)$. Consequently, there exists a number $C(\varphi)$ such that

$$\lim_{k \to \infty} \langle f_k, \varphi \rangle = C(\varphi), \qquad \forall \varphi \in \mathcal{D}(\mathbb{Q}_p^n). \tag{4.4.1}$$

It is clear that the constructed functional is linear, i.e., belongs to $\mathcal{D}'(\mathbb{Q}_p^n)$. We put $C(\varphi) = \langle f, \varphi \rangle$, $f \in \mathcal{D}'(\mathbb{Q}_p^n)$, $\varphi \in \mathcal{D}(\mathbb{Q}_p^n)$. The equality (4.4.1) shows that $f_k \to f$, $k \to \infty$ in $\mathcal{D}'(\mathbb{Q}_p^n)$. $\qquad\square$

Let $\mathcal{O}$ be an open subset in $\mathbb{Q}_p^n$. By $\mathcal{D}'(\mathcal{O})$ we denote the set of all linear functionals on $\mathcal{D}(\mathcal{O})$. According to the well-known theorem of functional analysis, inclusion (4.3.2) implies that

$$\mathcal{D}'(\mathbb{Q}_p^n) \subset \mathcal{D}'(\mathcal{O}). \tag{4.4.2}$$

The convergence in $\mathcal{D}'(\mathcal{O})$ is defined similarly to that in $\mathcal{D}'(\mathbb{Q}_p^n)$.

Every function $f \in \mathcal{L}^1_{loc}(\mathbb{Q}^n_p)$ defines a distribution $f \in \mathcal{D}'(\mathbb{Q}^n_p)$ by the formula

$$\langle f, \varphi \rangle \overset{def}{=} \int_{\mathbb{Q}^n_p} f(x)\varphi(x)\, d^n x, \qquad \varphi \in \mathcal{D}(\mathbb{Q}^n_p). \tag{4.4.3}$$

Such distributions are called *regular distributions*.

4.4.2. Support of a distribution

Let $\mathcal{O}$ be an open subset in $\mathbb{Q}^n_p$. A distribution $f \in \mathcal{D}'(\mathcal{O})$ vanishes on $\mathcal{O}_1 \subset \mathcal{O}$ if $\langle f, \varphi \rangle = 0$ for all $\varphi \in \mathcal{D}(\mathcal{O}_1)$. It is written as $f(x) = 0, x \in \mathcal{O}_1$. Correspondingly two distributions $f, g \in \mathcal{D}'(\mathcal{O})$ are equal in $\mathcal{O}_1 \subset \mathcal{O}$, if $f(x) - g(x) = 0$, $x \in \mathcal{O}_1$.

Since in $\mathcal{D}(\mathcal{O})$ the theorem of "decomposition of the unity" (4.3.6) holds, the notion of the support of a distribution can be introduced by the standard definition as for the real case. Let $\mathcal{O}_f \subset \mathcal{O}$ be the maximal open set on which the distribution $f \in \mathcal{D}'(\mathcal{O})$. The *support* of f is the complement of $\mathcal{O}_f$ in $\mathcal{O}$. We denote it by $\mathrm{supp}\, f$. Thus the support of a distribution $f \in \mathcal{D}'(\mathcal{O})$ is the closed set $\mathrm{supp}\, f = \mathcal{O} \setminus \mathcal{O}_f$.

4.4.3. The Dirac delta-function

The Dirac δ-function is defined by

$$\langle \delta, \varphi \rangle = \varphi(0), \qquad \forall \varphi \in \mathcal{D}(\mathbb{Q}^n_p). \tag{4.4.4}$$

It is clear that $\delta \in \mathcal{D}'(\mathbb{Q}^n_p)$ and $\delta(x) = 0$ for all $x \neq 0$, i.e., $\mathrm{supp}\, \delta = \{0\}$.

The structure of a distribution with a point support is given by the following statement.

Proposition 4.4.4. *If the support of a distribution $f \in \mathcal{D}'(\mathbb{Q}^n_p)$ is the point $\{0\}$ then this distribution has a unique representation $f = C\delta$, where C is a constant.*

Proof. Let $\varphi \in \mathcal{D}(\mathbb{Q}^n_p)$ be a test function with a parameter of constancy l. Hence $\varphi(x) = \varphi(0)$ for all $|x|_p \leq p^l$. Let $\eta \in \mathcal{D}(\mathbb{Q}^n_p)$ be another function such that $\eta(x) = 1, |x|_p \leq 1$. Then we have

$$\langle f, \varphi \rangle = \langle f, \eta\varphi \rangle = \varphi(0)\langle f, \eta \rangle = C\langle \delta, \varphi \rangle,$$

where $C = \langle f, \eta \rangle$. If there is another representation $f = C_1\delta$ then $(C - C_1)\delta = 0$. Applying both sides of this equality to a test function φ such that $\varphi(0) \neq 0$, we conclude that $C = C_1$. $\square$

Let us introduce in $\mathcal{D}(\mathbb{Q}_p^n)$ a *canonical* 1-*sequence*

$$\Delta_k(x) \overset{def}{=} \Omega(p^{-k}|x|_p), \quad k \in \mathbb{Z}, \tag{4.4.5}$$

and a *canonical* δ-*sequence*

$$\delta_k(x) \overset{def}{=} p^{nk}\Omega(p^k|x|_p), \quad k \in \mathbb{Z}, \tag{4.4.6}$$

$x \in \mathbb{Q}_p^n$, where the function Ω is defined by (4.2.2). Here $\Delta_k(x)$ is the characteristic function of the ball B_k^n (4.2.1).

It is clear that

$$\Delta_k \to 1, \quad k \to \infty, \quad \text{in} \quad \mathcal{E}(\mathbb{Q}_p^n)$$

and

$$\delta_k \to \delta, \quad k \to \infty, \quad \text{in} \quad \mathcal{D}'(\mathbb{Q}_p^n). \tag{4.4.7}$$

Let us prove (4.4.7). Let $\varphi \in \mathcal{D}(\mathbb{Q}_p^n)$ be a test function with a parameter of constancy l. Then taking into account (3.3.1) and (1.10.2), for all $k \geq -l$ we obtain

$$\langle \delta_k, \varphi \rangle = \int_{\mathbb{Q}_p^n} p^{nk}\Omega(p^k|x|_p)\varphi(x)\varphi(x)\,d^n x = p^{nk} \int_{B_{-k}^n} \varphi(x)\varphi(x)\,d^n x$$

$$= \varphi(0)p^{nk} \int_{B_{-k}^n} \varphi(x)\,d^n x = \varphi(0) = \langle \delta, \varphi \rangle, \quad \forall\, \varphi \in \mathcal{D}(\mathbb{Q}_p^n).$$

4.4.4. Theorem of "piecewise sewing"

If $f \in \mathcal{D}'(\mathbb{Q}_p^n)$ and $\operatorname{supp} f \subset B_N^n$ then $f = \Delta_N(x)f$, where $\Delta_N(x)$ is a characteristic function of the ball B_N^n (4.2.1). Moreover, if $f \in \mathcal{D}'(\mathbb{Q}_p^n)$ and $\operatorname{supp} f$ is a clopen then

$$f = \chi_{\operatorname{supp} f}(x)f, \tag{4.4.8}$$

where $\chi_{\operatorname{supp} f}$ is a characteristic function of the set $\operatorname{supp} f$.

Let $\mathcal{O}_1 \subset \mathcal{O} \subset \mathbb{Q}_p^n$. Then any distribution $f \in \mathcal{D}'(\mathcal{O})$ admits a restriction $f|_{\mathcal{O}_1} \in \mathcal{D}'(\mathcal{O}_1)$ which is defined by the formula

$$\langle f|_{\mathcal{O}_1}, \varphi \rangle = \langle f, \varphi \rangle, \quad \forall\, \varphi \in \mathcal{D}(\mathcal{O}_1).$$

Conversely, in $\mathcal{D}'(\mathcal{O})$ the following *theorem of "piecewise sewing"* holds.

Theorem 4.4.5. ([241, VI.4.]) *Let $\mathcal{O}$ be an open subset in $\mathbb{Q}_p^n$ and let $\mathcal{O}$ be a union at most of a countable set of disjoint balls:*

$$\mathcal{O} = \bigcup_k B_{\gamma_k}(a^k), \qquad B_{l_k}(a^k) \cap B_{l_j}(a^j) = \emptyset, \quad k \neq j$$

(which is possible due to Proposition 1.8.3). Let $f_k \in \mathcal{D}'(B_{\gamma_k}(a^k))$. Then there exists a unique distribution $f \in \mathcal{D}'(\mathcal{O})$ such that its restriction $f_{B_{\gamma_k}(a^k)} = f_k$, $k = 1, 2, \ldots$.

Proof. This statement follows from the theorem of "decomposition of the unity" (4.3.6). The desired distribution is defined by the formula

$$f(x) = \sum_{k=1}^{\infty} \Delta_{l_k}(x - a^k) f_k(x), \qquad x \in \mathcal{O}.$$

$\square$

4.4.5. Linear operators in $\mathcal{D}'(\mathbb{Q}_p^n)$

A linear operator in $\mathcal{D}'(\mathbb{Q}_p^n)$ is defined as the corresponding conjugate linear operator in the space of test functions $\mathcal{D}(\mathbb{Q}_p^n)$. Let $A : \mathcal{D}(\mathbb{Q}_p^n) \to \mathcal{D}(\mathbb{Q}_p^n)$ be a linear operator. Then its conjugate operator $A^* : \mathcal{D}'(\mathbb{Q}_p^n) \to \mathcal{D}'(\mathbb{Q}_p^n)$ is introduced by the equality

$$\langle A^* f, \varphi \rangle = \langle f, A\varphi \rangle, \qquad f \in \mathcal{D}'(\mathbb{Q}_p^n), \quad \forall \varphi \in \mathcal{D}(\mathbb{Q}_p^n). \qquad (4.4.9)$$

It is clear that $A^* f \in \mathcal{D}'(\mathbb{Q}_p^n)$ and A^* is linear. Then, according to Lemma 4.4.1, A^* is a continuous operator from $\mathcal{D}'(\mathbb{Q}_p^n)$ to $\mathcal{D}'(\mathbb{Q}_p^n)$.

A specific expression for the operator A^* can be obtained in the following way: taking into account (4.4.3), we apply the formula (4.4.9) to the case $f \in \mathcal{L}_{loc}^1(\mathbb{Q}_p^n)$, and transform the integral

$$\langle f, A\varphi \rangle = \int_{\mathbb{Q}_p^n} f(x) A\varphi(x) \, d^n x$$

to the form $\langle A^* f, \varphi \rangle$.

According to (4.4.9), if $a \in \mathcal{E}(\mathbb{Q}_p^n)$, $f \in \mathcal{D}'(\mathbb{Q}_p^n)$, then the product af exists and it is defined as

$$\langle af, \varphi \rangle = \langle f, a\varphi \rangle, \qquad \forall \varphi \in \mathcal{D}(\mathbb{Q}_p^n).$$

Linear change of variables for distributions. Let A be a matrix, $\det A \neq 0$, $b \in \mathbb{Q}_p^n$, $f \in \mathcal{D}'(\mathbb{Q}_p^n)$. Similarly to the real case (see [236, §1.9.]), it is easy to prove that for $f \in \mathcal{L}_{loc}^1(\mathbb{Q}_p^n)$, by using (4.4.3) and the change of variables formula (3.4.4), we obtain

$$\langle f(Ax + b), \varphi \rangle = |\det A|_p^{-1} \langle f(y), \varphi(A^{-1}(y - b)) \rangle, \qquad (4.4.10)$$

for all $\varphi \in \mathcal{D}(\mathbb{Q}_p^n)$. Then according to (4.4.9), a distribution $f(y)$, $y = Ax + b$, is defined by the same relation (4.4.10).

In particular, if $f \in \mathcal{D}'(\mathbb{Q}_p)$, $a \in \mathbb{Q}_p^\times$, $b \in \mathbb{Q}_p$ then

$$\langle f(ax + b), \varphi \rangle = |a|_p^{-1} \langle f(y), \varphi(a^{-1}(y - b)) \rangle, \qquad \forall \varphi \in \mathcal{D}(\mathbb{Q}_p^n). \quad (4.4.11)$$

We have $\langle \delta(x - b), \varphi \rangle = \varphi(b)$, $b \in \mathbb{Q}_p^n$.

4.5. The direct product of distributions

Let $f \in \mathcal{D}'(\mathbb{Q}_p^n)$ and $g \in \mathcal{D}'(\mathbb{Q}_p^m)$. Their *direct product* is defined by the formula

$$\langle f(x) \times g(y), \varphi \rangle \overset{def}{=} \langle f(x), \langle g(x), \varphi(x, y) \rangle \rangle, \qquad \forall \varphi \in \mathcal{D}(\mathbb{Q}_p^{n+m}). \quad (4.5.1)$$

It is easy to see that definition (4.5.1) is correct. In view of the representation (4.3.3), a test function $\varphi \in \mathcal{D}(\mathbb{Q}_p^{n+m})$ is represented by a finite sum of the form

$$\varphi(x, y) = \sum_k \phi_k(x) \psi_k(y), \qquad \phi_k \in \mathcal{D}(\mathbb{Q}_p^n), \qquad \psi_k \in \mathcal{D}(\mathbb{Q}_p^m). \quad (4.5.2)$$

Then the operator $\varphi \to \langle g(x), \varphi(x, y) \rangle$ is linear from $\mathcal{D}(\mathbb{Q}_p^{n+m})$ into $\mathcal{D}(\mathbb{Q}_p^n)$, and thus the functional on the right-hand side of (4.5.1) is linear on $\mathcal{D}(\mathbb{Q}_p^{n+m})$. Hence $f(x) \times g(y) \in \mathcal{D}'(\mathbb{Q}_p^{n+m})$.

Proposition 4.5.1. *The direct product (4.5.1) is commutative:*

$$f(x) \times g(y) = g(y) \times f(x). \qquad (4.5.3)$$

Proof. From (4.5.1) and (4.5.2), we have

$$\langle f(x) \times g(y), \varphi \rangle = \sum_k \langle f(x), \phi_k \rangle \langle g(y), \psi_k \rangle$$

$$= \langle g(y) \times f(x), \varphi \rangle, \qquad \varphi \in \mathcal{D}(\mathbb{Q}_p^{n+m}). \quad (4.5.4)$$

Thus the direct product is commutative. □

It follows from (4.5.4) that the direct product $f(x) \times g(y)$ is continuous with respect to the joint factors f and g: if $f_k \to 0$, $k \to \infty$ in $\mathcal{D}'(\mathbb{Q}_p^n)$ and $g_k \to 0$, $k \to \infty$ in $\mathcal{D}'(\mathcal{D}_p^m)$ then $f_k \times g_k \to 0$, $k \to \infty$ in $\mathcal{D}'(\mathbb{Q}_p^{n+m})$.

Note that for $g = 1$ the equality (4.5.3) is equivalent to the equality

$$\left\langle f(x), \int_{\mathbb{Q}_p^m} \varphi(x, y)\, d^m y \right\rangle = \int_{\mathbb{Q}_p^m} \langle f(x), \varphi(x, y) \rangle\, d^m y, \qquad (4.5.5)$$

$\varphi \in \mathcal{D}(\mathbb{Q}_p^{n+m},)$.

For example, $\delta(x) = \delta(x_1) \times \delta(x_2) \times \cdots \times \delta(x_n)$.

The direct product $f(x) \times g(y)$ of distributions $f \in \mathcal{D}'(\mathcal{O})$, $\mathcal{O} \subset \mathbb{Q}_p^n$ and $g \in \mathcal{D}'(\mathcal{O}_1)$, $\mathcal{O}_1 \subset \mathbb{Q}_p^m$ is defined in a similar way.

For example, $\delta(x) = \delta(x_1) \times \cdots \times \delta(x_x)$, $x = (x_1 \ldots, x_n) \in \mathbb{R}^n$.

4.6. The Schwartz "kernel" theorem

Every distribution $f \in \mathcal{D}'(\mathbb{Q}_p^{n+m})$ defines the linear operator $A : \mathcal{D}(\mathbb{Q}_p^n) \to \mathcal{D}'(\mathbb{Q}_p^m)$ by the formula

$$\langle A(\phi), \psi \rangle = \langle f(x, y), \phi(x)\psi(y) \rangle, \qquad \phi \in \mathcal{D}(\mathbb{Q}_p^n), \quad \psi \in \mathcal{D}(\mathbb{Q}_p^m). \quad (4.6.1)$$

According to Lemma 4.4.1 and Corollary 4.4.2, every linear operator and functional defined on $\mathcal{D}$ is continuous.

The inverse statement is also true.

Theorem 4.6.1. ([241, VII.7.]) *Let $B(\phi, \psi)$ be a bilinear functional, $\phi \in \mathcal{D}(\mathbb{Q}_p^n)$, $\psi \in \mathcal{D}(\mathbb{Q}_p^m)$. Then there exists a unique distribution $f \in \mathcal{D}'(\mathbb{Q}_p^{n+m})$ such that*

$$\langle f, \phi(x)\psi(y) \rangle = B(\phi, \psi), \qquad \phi \in \mathcal{D}(\mathbb{Q}_p^n), \quad \psi \in \mathcal{D}(\mathbb{Q}_p^m). \quad (4.6.2)$$

Proof. Since every test function $\varphi(x, y)$ from $\mathcal{D}(\mathbb{Q}_p^{n+m})$ is represented by a finite sum of the form (4.5.2), the bilinear functional $B(\phi, \psi)$ defines some linear functional

$$f : \varphi \to \sum_k B(\phi_k, \psi_k)$$

on $\mathcal{D}(\mathbb{Q}_p^{n+m})$ and, consequently, $f \in \mathcal{D}'(\mathbb{Q}_p^{n+m})$. The distribution f satisfies the equality (4.6.2). It is clear that f is unique. $\qquad \square$

Corollary 4.6.2. ([241, VII.7.]) (**Schwartz "kernels" theorem**) *Let $\phi \to A(\phi)$ be a linear operator from $\mathcal{D}(\mathbb{Q}_p^n)$ into $\mathcal{D}'(\mathbb{Q}_p^m)$. Then there exists a unique distribution $f \in \mathcal{D}'(\mathbb{Q}_p^{n+m})$ such that the equality (4.6.1) holds.*

The distribution $f(x, y)$ is called the *kernel* of the operator A.

Corollary 4.6.3. ([241, VII.7.]) *Every bilinear form* $B(\phi, \psi)$, $\phi \in \mathcal{D}(\mathbb{Q}_p^n)$, $\psi \in \mathcal{D}(\mathbb{Q}_p^m)$, *is continuous.*

4.7. The convolution of distributions

Let $f, g \in \mathcal{D}'(\mathbb{Q}_p^n)$. The convolution $f * g$ of these distributions is defined as the functional

$$\langle f * g, \varphi \rangle = \lim_{k \to \infty} \langle f(x) \times g(y), \Delta_k(x)\varphi(x + y) \rangle, \tag{4.7.1}$$

if the limit exists for all $\varphi \in \mathcal{D}(\mathbb{Q}_p^n)$, where $\Delta_k(x)$ is a canonical 1-sequence (4.4.5). The right-hand side of the equality (4.7.1) defines a linear functional on $\mathcal{D}(\mathbb{Q}_p^n)$, and thus $f * g \in \mathcal{D}'(\mathbb{Q}_p^n)$.

The convolution $g * f$ is defined in a similar way:

$$\langle g * f, \varphi \rangle = \lim_{k \to \infty} \langle g(y) \times f(x), \Delta_k(y)\varphi(x + y) \rangle, \quad \varphi \in \mathcal{D}(\mathbb{Q}_p^n). \tag{4.7.2}$$

Theorem 4.7.1. ([241, VII.1.]) *Let* $f, g \in \mathcal{D}'(\mathbb{Q}_p^n)$. *If the convolution* $f * g$ *exists then the convolution* $g * f$ *exists, and they are equal:*

$$f * g = g * f.$$

Proof. By the condition of the theorem the convolution $f * g$ exists. Let $\varphi \in \mathcal{D}(\mathbb{Q}_p^n)$ so that $\varphi \in \mathcal{D}_N^l(\mathbb{Q}_p^n)$ for some $N, l \in \mathbb{Z}$. By formula (4.3.4) the function φ is represented in the form of the finite sum

$$\varphi(x) = \sum_{\nu=1}^{p^{n(N-l)}} \varphi(a^\nu)\Delta_l(x - a^\nu),$$

for some $a^\nu \in B_N^n$, where $\Delta_l(x - a^\nu)$ is the characteristic function of the ball $B_l^n(a^\nu)$. Therefore it is sufficient to prove this theorem for the test functions of the form

$$\varphi_\nu(x) = \Delta_l(x - a^\nu).$$

It is clear that for sufficiently large k we have

$$\Delta_k(x) = \Omega(p^{-k}|x|_p) = \Omega(p^{-k}|x - a^\nu|_p) = \Delta_k(x - a^\nu), \quad x \in \mathbb{Q}_p^n. \tag{4.7.3}$$

Thus (4.7.1) and (4.7.3) imply

$$\begin{aligned}
\langle f * g, \varphi_\nu \rangle &= \lim_{k \to \infty} \langle f(x) \times g(y), \Delta_k(x - a^\nu)\varphi_\nu(x + y) \rangle \\
&= \lim_{k \to \infty} \langle f(x) \times g(y), \Delta_k(x - a^\nu)\Delta_l(x + y - a^\nu) \rangle.
\end{aligned}$$

Taking into account the following identity [241, VII.1], [163]:

$$\Omega\big(|p^l x - a|_p\big)\Omega\big(|p^k x - b|_p\big) = \Omega\big(|p^l x - a|_p\big)\Omega\big(|p^{k-l} a - b|_p\big), \quad (4.7.4)$$

$x, a, b \in \mathbb{Q}_p^n$ for $k \geq l$, $k, l \in \mathbb{Z}$, we obtain that

$$\Delta_k(x - a^\nu)\Delta_l(x + y - a^\nu) = \Delta_k(y)\Delta_l(x + y - a^\nu).$$

Consequently, we have

$$\begin{aligned}
\langle f * g, \varphi_\nu \rangle &= \lim_{k \to \infty} \langle f(x) \times g(y), \Delta_k(x - a^\nu)\varphi_\nu(x + y) \rangle \\
&= \lim_{k \to \infty} \langle f(x) \times g(y), \Delta_k(y)\varphi_\nu(x + y) \rangle = \langle g * f, \varphi_\nu \rangle.
\end{aligned}$$

Here we used the commutativity of the direct product (4.5.3). □

Lemma 4.7.2. ([241, VII.1.].) *If $f, g \in \mathcal{D}'(\mathbb{Q}_p^n)$ and* supp $g \subset B_N^n$ *then the convolution $f * g$ exists and*

$$\langle f * g, \varphi \rangle = \langle f(x) \times g(y), \Delta_N(y)\varphi(x + y) \rangle, \quad \forall \varphi \in \mathcal{D}(\mathbb{Q}_p^n). \quad (4.7.5)$$

Proof. Formulas (4.7.1) and (4.4.8) imply that for all $\varphi \in \mathcal{D}(\mathbb{Q}_p^n)$

$$\begin{aligned}
\langle f * g, \varphi \rangle &= \lim_{k \to \infty} \langle f(x) \times g(y), \Delta_k(y)\varphi(x + y) \rangle \\
&= \lim_{k \to \infty} \langle f(x) \times \Delta_N(y)g(y), \Delta_k(y)\varphi(x + y) \rangle \\
&= \lim_{k \to \infty} \langle f(x) \times g(y), \Delta_N(y)\Delta_k(y)\varphi(x + y) \rangle \\
&= \langle f(x) \times g(y), \Delta_N(y)\varphi(x + y) \rangle.
\end{aligned}$$

Here we used the relation

$$\Delta_k(y)\Delta_N(y)\varphi(x + y) \to \Delta_N(y)\varphi(x + y), \quad k \to \infty, \quad \text{in} \quad \mathcal{D}(\mathbb{Q}_p^{2n}).$$

  □

If the conditions of Lemma 4.7.2 hold, the convolution $f * g$ is continuous with respect to the joint factors f and g: if $f_k \to f, k \to \infty$ in $\mathcal{D}'(\mathbb{Q}_p^n)$; $g_k \to g$, $k \to \infty$ in $\mathcal{D}'(\mathbb{Q}_p^n)$, supp $g \subset B_N$ then $f_k * g_k \to 0, k \to \infty$ in $\mathcal{D}'(\mathbb{Q}_p^n)$.

This result follows from the representation (4.7.5) and the continuity of the direct product (see Section 4.5).

Example 4.7.1.

$$f * \delta = \delta * f = f, \quad f \in \mathcal{D}'(\mathbb{Q}_p^n).$$

This follows from representation (4.7.5) and formula (4.5.1):

$$\begin{aligned}
\langle f * \delta, \varphi \rangle &= \langle f(x) \times \delta(y), \Delta_N(y)\varphi(x + y) \rangle \\
&= \big(f(x), \langle \delta(y), \Delta_N(y)\varphi(x + y) \rangle \big) = \langle f, \varphi \rangle.
\end{aligned}$$

In view of the continuity of the convolution, it follows from Example 4.7.1 and formula (4.4.7) that if $f \in \mathcal{D}'(\mathbb{Q}_p^n)$ then

$$f * \delta_k \to f, \quad k \to \infty, \quad \text{in} \quad \mathcal{D}'(\mathbb{Q}_p^n), \tag{4.7.6}$$

where δ_k is a canonical δ-sequence (4.4.6).

Proposition 4.7.3. ([241, VII.1.]) *If $f \in \mathcal{D}'(\mathbb{Q}_p^n)$, $\psi \in \mathcal{D}(\mathbb{Q}_p^n)$ then*

$$(f * \psi)(x) = \langle f(y), \psi(x - y) \rangle \in \mathcal{E}(\mathbb{Q}_p^n), \quad x \in \mathbb{Q}_p^n, \tag{4.7.7}$$

where the parameter of constancy of the function $f * \psi$ does not exceed the parameter of constancy of the function ψ.

Proof. Let $\psi \in \mathcal{D}_N^l(\mathbb{Q}_p^n)$. Using (4.7.5) and (4.5.5), we obtain the representation (4.7.7):

$$\langle f * \psi, \varphi \rangle = \langle f(x), \langle \psi(y) \Delta_N(y) \varphi(x + y) \rangle \rangle$$

$$= \left\langle f(x), \int_{\mathbb{Q}_p^n} \psi(y) \varphi(x + y) \, d^n y \right\rangle = \left\langle f(x), \int_{\mathbb{Q}_p^n} \psi(\xi - x) \varphi(\xi) \, d^n \xi \right\rangle$$

$$= \int_{\mathbb{Q}_p^n} \langle f(x), \psi(\xi - x) \rangle \varphi(\xi) \, d^n \xi, \quad \forall \varphi \in \mathcal{D}'(\mathbb{Q}_p^n).$$

Here we take into account that $\psi(\xi - x)\varphi(\xi) \in \mathcal{D}(\mathbb{Q}_p^{2n})$. Since φ is locally constant then $\langle f(y), \varphi(x - y) \rangle \in \mathcal{E}(\mathbb{Q}_p^n)$ is also locally constant. $\square$

Proposition 4.7.4. *Every distribution $f \in \mathcal{D}'$ is a weak limit of test functions:*

$$\mathcal{D}(\mathbb{Q}_p^n) \ni \Delta_k(f * \delta_k) \to f, \quad k \to \infty, \quad \text{in} \quad \mathcal{D}'(\mathbb{Q}_p^n). \tag{4.7.8}$$

Thus, $\mathcal{D}(\mathbb{Q}_p^n)$ is dense in $\mathcal{D}'(\mathbb{Q}_p^n)$.

In view of formula (4.7.5) we have

$$\langle f(x) \times g(y), \Delta_k(x)\varphi(x + y) \rangle$$

$$= \langle \Delta_k(x) f(x) \times g(y), \varphi(x + y) \rangle = \langle (\Delta_k f) * g, \varphi \rangle,$$

and, consequently, definition (4.7.1) of the convolution $f * g$ is equivalent to the following one:

$$(\Delta_k f) * g \to f * g, \quad k \to \infty, \quad \text{in} \quad \mathcal{D}'(\mathbb{Q}_p^n). \tag{4.7.9}$$

It follows from definition (4.7.1) of the convolution that $\operatorname{supp}(f * g)$ is contained in the closure of the set

$$\{\xi : \xi \in \mathbb{Q}_p^n, \quad \xi = x + y, \, x \in \operatorname{supp} f, \, y \in \operatorname{supp} g\}.$$

In particular, if supp $f \subset B_N$ and supp $g \subset B_N$ then supp$(f * g) \subset B_N$. Thus, the set of distributions with supports in B_N forms the commutative and associative convolution algebra, where the δ function plays the role of the unit element.

4.8. The Fourier transform of test functions

The Fourier transform of a test function $\varphi \in \mathcal{D}(\mathbb{Q}_p^n)$ is defined by the formula

$$\widehat{\varphi}(\xi) = F[\varphi](\xi) = \int_{\mathbb{Q}_p^n} \chi_p(\xi \cdot x)\varphi(x) \, d^n x, \quad \xi \in \mathbb{Q}_p^n, \tag{4.8.1}$$

where $\chi_p(\xi \cdot x) = \chi_p(\xi_1 x_1) \cdots \chi_p(\xi_n x_n) = e^{2\pi i \sum_{j=1}^{n} \{\xi_j x_j\}_p}$, $\xi \cdot x$ is the scalar product of vectors, and the function $\chi_p(\xi_j x_j) = e^{2\pi i \{\xi_j x_j\}_p}$ for every fixed $\xi_j \in \mathbb{Q}_p$ is an additive character of the field $\mathbb{Q}_p$ (2.3.3), $\{\xi_j x_j\}_p$ is the fractional part of a number $\xi_j x_j$ defined by (1.6.4), $j = 1, \ldots, n$.

Lemma 4.8.1. ([96, Ch. II,§2.4.], [230, III,(3.2)], [241, VII.2.]) *If $\varphi \in \mathcal{D}_N^l(\mathbb{Q}_p^n)$ then $\widehat{\varphi}(\xi) = F[\varphi](\xi) \in \mathcal{D}_{-l}^{-N}(\mathbb{Q}_p^n)$. Thus the map $\varphi \to F[\varphi]$ is linear and continuous from $\mathcal{D}(\mathbb{Q}_p^n)$ in $\mathcal{D}(\mathbb{Q}_p^n)$.*

Proof. Let $\varphi \in \mathcal{D}_N^l(\mathbb{Q}_p^n)$. Making the change of variable $x = x' + a$ in (4.8.5), where $a = (a_1, \ldots, a_n) \in \mathbb{Q}_p^n$, $|a_k|_p = p^l, k = 1, 2, \ldots, n$, we obtain

$$\widehat{\varphi}(\xi) = \int_{\mathbb{Q}_p^n} \chi_p(\xi \cdot (x' + a))\varphi(x' + a) \, d^n x'$$

$$= \chi_p(\xi \cdot a) \int_{\mathbb{Q}_p^n} \chi_p(\xi \cdot x')\varphi(x') \, d^n x' = \chi_p(\xi \cdot a)\widehat{\varphi}(\xi). \tag{4.8.2}$$

If $|\xi|_p > p^{-l}$ then there is a $k \in \{1, 2, \ldots, n\}$ such that $|a_k \xi_k|_p = |a_k|_p |\xi_k|_p > 1$. Thus from formula (2.3.3) we have $\chi_p(\xi \cdot a) = \chi_p(\xi_1 a_1) \cdots \chi_p(\xi_n a_n) \neq 1$, and, consequently, (4.8.2) implies that $\widehat{\varphi}(\xi) = 0$ for $|\xi|_p > p^{-l}$, i.e., supp $\widehat{\varphi} \subset B_{-l}^n$.

If $\xi' \in B_{-N}^n$ then for $|x|_p \leq p^N$ we have $|\xi' \cdot x|_p \leq 1$ and $\chi_p(\xi' \cdot x) = 1$. Consequently,

$$\widehat{\varphi}(\xi + \xi') = \int_{B_N^n} \chi_p((\xi + \xi') \cdot x)\varphi(x) \, d^n x$$

$$= \int_{B_N^n} \chi_p(\xi \cdot x)\chi_p(\xi' \cdot x)\varphi(x) \, d^n x = \int_{B_N^n} \chi_p(\xi \cdot x)\varphi(x) \, d^n x = \widehat{\varphi}(\xi).$$

Thus for $\widehat{\varphi}$ the parameter of constancy $\geq -N$. $\qquad \square$

Theorem 4.8.2. ([96, Ch. II, §2.4.], [241, VII.2.]) *The Fourier transform is an isomorphism of the linear spaces* $\mathcal{D}(\mathbb{Q}_p^n) \to \mathcal{D}(\mathbb{Q}_p^n)$ *and the following inversion formula holds:*

$$\varphi(x) = \int_{\mathbb{Q}_p^n} \chi_p(-x \cdot \xi) F[\varphi](\xi) \, d^n\xi = F^{-1}[\widehat{\varphi}]$$

$$= F[F[\varphi](-\xi)](x) = F[F[\varphi]](-x), \quad \varphi \in \mathcal{D}(\mathbb{Q}_p^n). \quad (4.8.3)$$

Proof. From Lemma 4.8.1, the Fourier transform (4.8.1) maps $\mathcal{D}(\mathbb{Q}_p^n)$ into $\mathcal{D}(\mathbb{Q}_p^n)$. Let $\varphi \in \mathcal{D}_N^l(\mathbb{Q}_p^n)$. Taking into account that $\widehat{\varphi}(\xi) \in \mathcal{D}_{-l}^{-N}(\mathbb{Q}_p^n)$, we calculate

$$\int_{\mathbb{Q}_p^n} \chi_p(-x \cdot \xi) \widehat{\varphi}(\xi) \, d^n\xi$$

$$= \int_{B_{-l}^n} \chi_p(-x \cdot \xi) \left(\int_{B_N^n} \chi_p(\xi \cdot x') \varphi(x') \, d^n x' \right) d^n\xi$$

$$= \int_{\mathbb{Q}_p^n} \varphi(x') \left(\int_{B_{-l}^n} \chi_p(\xi \cdot (x' - x)) \, d^n\xi \right) d^n x'$$

$$= \int_{|x'-x|_p \leq p^l} \varphi(x') \left(\int_{B_{-l}^n} \chi_p(\xi \cdot (x' - x)) \, d^n\xi \right) d^n x'$$

$$+ \int_{|x'-x|_p \geq p^{l+1}} \varphi(x') \left(\int_{B_{-l}^n} \chi_p(\xi \cdot (x' - x)) \, d^n\xi \right) d^n x'. \quad (4.8.4)$$

Let us consider the first integral in (4.8.4). If $|x' - x|_p \leq p^l$, $|\xi|_p \leq p^{-l}$ then $\chi_p(\xi \cdot (x' - x)) = 1$. Now taking into account that $\varphi \in \mathcal{D}_N^l(\mathbb{Q}_p^n)$ and formula (3.3.1), we have

$$\int_{|x'-x|_p \leq p^l} \varphi(x + (x' - x)) \left(\int_{B_{-l}^n} \chi_p(\xi \cdot (x' - x)) \, d^n\xi \right) d^n x'$$

$$= \varphi(x) \int_{B_l^n(x)} \int_{B_{-l}^n} d^n\xi \, d^n x' = \varphi(x) p^{nl} p^{-nl} = \varphi(x).$$

Now we consider the second integral in (4.8.4). Since $|x' - x|_p \geq p^{l+1}$ and $|\xi|_p \leq p^{-l}$, from (3.3.3), $\chi_p(\xi \cdot a) = \chi_p(\xi_1 a_1) \cdots \chi_p(\xi_n a_n) = 0$. Consequently, the second integral in (4.8.4) is equal to 0.

Thus the inversion formula (4.8.3) holds and the map $\varphi \to \widehat{\varphi}$ is one-to-one. □

Lemma 4.8.1 and Theorem 4.8.2 imply the following statement.

Lemma 4.8.3.

$$\varphi \in \mathcal{D}'_N(\mathbb{Q}_p^n) \quad iff \quad F[\varphi] \in \mathcal{D}_{-l}^{-N}(\mathbb{Q}_p^n). \tag{4.8.5}$$

Proposition 4.8.4. ([241, VII.2.]) *The* Parseval–Steklov *equality*

$$\int_{\mathbb{Q}_p^n} \varphi(x)\,\overline{\psi(x)}\,d^n x = \int_{\mathbb{Q}_p^n} F[\varphi](\xi)\,\overline{F[\psi](\xi)}\,d^n\xi, \quad \varphi,\psi \in \mathcal{D}(\mathbb{Q}_p^n), \tag{4.8.6}$$

and the equivalent equality

$$\int_{\mathbb{Q}_p^n} \varphi(x)\,F[\psi](x)\,d^n x = \int_{\mathbb{Q}_p^n} F[\varphi](\xi)\,\psi(\xi)\,d^n\xi, \quad \varphi,\psi \in \mathcal{D}(\mathbb{Q}_p^n), \tag{4.8.7}$$

hold.

Proof. First let us prove (4.8.7):

$$\begin{aligned}
\int_{\mathbb{Q}_p^n} \varphi(x)\,F[\psi](x)\,d^n x &= \int_{\mathbb{Q}_p^n} \varphi(x)\left(\int_{\mathbb{Q}_p^n} \psi(\xi)\chi_p(\xi \cdot x)\,d^n\xi\right)d^n x \\
&= \int_{\mathbb{Q}_p^n} \psi(\xi)\left(\int_{\mathbb{Q}_p^n} \varphi(x)\chi_p(\xi \cdot x)\,d^n x\right)d^n\xi \\
&= \int_{\mathbb{Q}_p^n} F[\varphi](\xi)\,\psi(\xi)\,d^n\xi.
\end{aligned}$$

If we denote $\eta(\xi) = \overline{F[\psi](\xi)}$ then by (4.8.3) $\overline{\psi(x)} = F[\eta](x)$. Next, we can see that (4.8.7) is equivalent to (4.8.6). □

Proposition 4.8.5. *Let A be a matrix such that $\det A \neq 0$, and $b \in \mathbb{Q}_p^n$. Then for a test function $\varphi \in \mathcal{D}(\mathbb{Q}_p^n)$ the following relation holds:*

$$\begin{aligned}
&F[\varphi(Ax + b)](\xi) \\
&= |\det A|_p^{-1}\chi_p\big(-A^{-1}b \cdot \xi\big)F[\varphi(x)]\big((A^T)^{-1}\xi\big), \quad x,\xi \in \mathbb{Q}_p^n, \tag{4.8.8}
\end{aligned}$$

where A^{-1} and A^T are the inverse and transpose of matrix A, respectively. In particular, if $\varphi \in \mathcal{D}(\mathbb{Q}_p)$, $a \in \mathbb{Q}_p^\times$, $b \in \mathbb{Q}_p$ then

$$F[\varphi(ax + b)](\xi) = |a|_p^{-1}\chi_p\left(-\frac{b}{a}\xi\right)F[\varphi(x)]\left(\frac{\xi}{a}\right), \quad x,\xi \in \mathbb{Q}_p. \tag{4.8.9}$$

Proof. By (4.8.1) and (4.4.10) we have

$$F[\varphi(Ax + b)](\xi)$$

$$= \int_{\mathbb{Q}_p^n} \chi_p(\xi \cdot x)\varphi(Ax + b)\, d^n x$$

$$= |\det A|_p^{-1} \int_{\mathbb{Q}_p^n} \chi_p\big(\xi \cdot (A^{-1}(y - b))\big)\varphi(y)\, d^n y$$

$$= |\det A|_p^{-1} \chi_p\big(-A^{-1}b \cdot \xi\big) \int_{\mathbb{Q}_p^n} \chi_p\big(\xi \cdot (A^{-1}y)\big)\varphi(y)\, d^n y$$

$$= |\det A|_p^{-1} \chi_p\big(-A^{-1}b \cdot \xi\big) \int_{\mathbb{Q}_p^n} \chi_p\big(((A^T)^{-1}\xi) \cdot y\big)\varphi(y)\, d^n y.$$

$\square$

It is easy to calculate that

$$F[\Delta_k](x) = \delta_k(x), \quad k \in \mathbb{Z}, \quad x \in \mathbb{Q}_p^n, \tag{4.8.10}$$

where Δ_k and δ_k are the canonical 1- and δ-sequences (4.4.5) and (4.4.6), respectively. In particular,

$$F[\Omega](x) = \Omega(x). \tag{4.8.11}$$

4.9. The Fourier transform of distributions

According to (4.4.9), the Fourier transform $F[f]$ of a distribution $f \in \mathcal{D}'(\mathbb{Q}_p^n)$ is defined by the relation

$$\langle F[f], \varphi \rangle = \langle f, F[\varphi] \rangle, \tag{4.9.1}$$

for all $\varphi \in \mathcal{D}(\mathbb{Q}_p^n)$.

From Lemma 4.8.1, the map $\varphi \to F[\varphi]$ is linear and continuous from $\mathcal{D}(\mathbb{Q}_p^n)$ in $\mathcal{D}(\mathbb{Q}_p^n)$. Hence the functional in the right-hand side of (4.9.1) is linear on $\mathcal{D}(\mathbb{Q}_p^n)$, and from Lemma 4.4.1 it is continuous. Thus $F[f] \in \mathcal{D}'(\mathbb{Q}_p^n)$ and the map $f \to F[f]$ is continuous from $\mathcal{D}'(\mathbb{Q}_p^n)$ in $\mathcal{D}'(\mathbb{Q}_p^n)$.

Proposition 4.9.1. *The inversion formula holds:*

$$f = F\big[F[f](-\xi)\big], \qquad f \in \mathcal{D}'(\mathbb{Q}_p^n). \tag{4.9.2}$$

Thus the Fourier transform $f \to F[f]$ is the linear isomorphism $\mathcal{D}'(\mathbb{Q}_p^n)$ onto $\mathcal{D}'(\mathbb{Q}_p^n)$.

Proof. By using definition (4.9.1) and the inversion formula (4.8.3) for test functions, we obtain

$$\langle F[F[f](-\xi)], \varphi \rangle = \langle F[f](-\xi), F[\varphi] \rangle$$
$$= \langle F[f], F[\varphi](-\xi) \rangle = \langle f, F[F[\varphi](-\xi)] \rangle = \langle f, \varphi \rangle.$$

$\square$

Example 4.9.1.

$$F[\delta] = 1, \qquad F[1] = \delta.$$

Indeed, according to (4.9.1) we have the first equality

$$\langle F[\delta], \varphi \rangle = \langle \delta, F[\varphi] \rangle = F[\varphi](0) = \int_{\mathbb{Q}_n} \varphi(x) \, d^n x = \langle 1, \varphi \rangle.$$

The second one follows from Proposition 4.9.1.

Definition (4.9.1) and formulas (4.4.10), (4.4.11), (4.8.8) and (4.8.9) imply the following statement for distributions.

Proposition 4.9.2. *Let A be a matrix such that* $\det A \neq 0$, *and* $b \in \mathbb{Q}_p^n$. *Then for a distribution* $f \in \mathcal{D}'(\mathbb{Q}_p^n)$ *the following relation holds:*

$$F[f(Ax + b)](\xi)$$
$$= |\det A|_p^{-1} \chi_p\big(-A^{-1}b \cdot \xi\big) F[f(x)]\big((A^T)^{-1}\xi\big), \quad x, \xi \in \mathbb{Q}_p^n. \quad (4.9.3)$$

In particular, if $f \in \mathcal{D}'(\mathbb{Q}_p)$, $a \in \mathbb{Q}_p^\times$, $b \in \mathbb{Q}_p$, *then*

$$F[f(ax + b)](\xi) = |a|_p^{-1} \chi_p\Big(-\frac{b}{a}\xi\Big) F[f(x)]\Big(\frac{\xi}{a}\Big), \quad x, \xi \in \mathbb{Q}_p. \quad (4.9.4)$$

Now we derive a formula for the Fourier transform for a distribution with a compact support.

Theorem 4.9.3. (see [241, VII.3.]) *Let* $f \in \mathcal{D}'(\mathbb{Q}_p^n)$. *We have* $\operatorname{supp} f \subset B_N^n$ *if and only if* $F[f] \in \mathcal{E}(\mathbb{Q}_p^n)$, *where for* $F[f]$ *a parameter of constancy is* $\geq -N$. *In addition*

$$F[f](\xi) = \langle f(x), \Delta_N(x)\chi_p(\xi \cdot x) \rangle. \quad (4.9.5)$$

Proof. Let $f \in \mathcal{D}'(\mathbb{Q}_p^n)$ and $\operatorname{supp} f \subset B_N^n$. Definition (4.9.1) and formulas (4.4.8) and (4.5.3) imply that

$$
\begin{aligned}
\langle F[f], \varphi \rangle &= \langle \Delta_N(x) f(x), F[\varphi](x) \rangle \\
&= \left\langle f(x), \Delta_N(x) \int_{\mathbb{Q}_p^n} \chi_p(x \cdot \xi) \varphi(\xi) \, d^n \xi \right\rangle \\
&= \int_{\mathbb{Q}_p^n} \left\langle f(x), \Delta_N(x) \chi_p(x \cdot \xi) \right\rangle \varphi(\xi) \, d^n \xi,
\end{aligned}
$$

for all $\varphi \in \mathcal{D}(\mathbb{Q}_p^n)$. Thus the last relation implies (4.9.5).

Let $\xi' \in B_{-N}^n$. Hence

$$
F[f](\xi + \xi') = \left\langle f(x), \Delta_N(x) \chi_p((\xi + \xi') \cdot x) \right\rangle.
$$

Since for $x \in B_N^n$ we have $|\xi' \cdot x|_p \le 1$, then $\chi_p(\xi' \cdot x) = 1$ and

$$
F[f](\xi + \xi') = \left\langle f(x), \Delta_N(x) \chi_p(\xi \cdot x) \right\rangle = F[f](\xi).
$$

Thus, for $F[f]$ the parameter of constancy $\ge -N$.

Conversely, let $F[f] \in \mathcal{E}(\mathbb{Q}_p^n)$ and its parameter of constancy is $\ge -N$. Applying the inverse Fourier transform to the equality

$$
F[f](\xi + \xi') = F[f](\xi), \qquad \xi \in \mathbb{Q}_p^n, \quad \forall \xi' \in B_{-N}^n,
$$

and using (4.9.3), we obtain

$$
\chi_p(\xi' \cdot x) f(x) = f(x), \qquad x \in \mathbb{Q}_p^n, \quad \forall \xi' \in B_{-N}^n. \tag{4.9.6}
$$

However, for any $|x|_p > p^N$ there exists some ξ' such that $|\xi|_p = p^{-N}$, and $\chi_p(\xi' \cdot x) \ne 1$. Therefore, (4.9.6) implies that $f(x) = 0$ for $|x|_p > p^N$, i.e., $\operatorname{supp} f \subset B_N^n$. $\qquad\square$

Theorem 4.9.4. ([241, VII, (5.3)]) *If $f, g \in \mathcal{D}'(\mathbb{Q}_p^n)$, $\operatorname{supp} g \subset B_N^n$ then*

$$
F[f * g] = F[f] F[g]. \tag{4.9.7}
$$

Proof. Using definitions (4.7.1) and (4.9.1), and also formulas (4.7.5) and (4.9.5), we obtain

$$
\begin{aligned}
\langle F[f * g], \varphi \rangle &= \langle f * g, F[\varphi] \rangle = \langle f(x) \times g(y), \Delta_N(y) F[\varphi](x + y) \rangle \\
&= \left\langle f(x), \left\langle g(y), \Delta_N(y) \int_{\mathbb{Q}_p^n} \varphi(\xi) \chi_p(\xi \cdot (x + y)) \, d^n\xi \right\rangle \right\rangle \\
&= \left\langle f(x), \int_{\mathbb{Q}_p^n} \left\langle g(y), \Delta_N(y) \chi_p(\xi \cdot y) \right\rangle \varphi(\xi) \chi_p(\xi \cdot x) \, d^n\xi \right\rangle \\
&= \left\langle f(x), \int_{\mathbb{Q}_p^n} F[g](\xi) \varphi(\xi) \chi_p(\xi \cdot x) \, d^n\xi \right\rangle \\
&= \langle f(x), F[F[g]\varphi] \rangle = \langle F[f], F[g]\varphi \rangle = \langle F[f] F[g], \varphi \rangle,
\end{aligned}
$$

for all $\varphi \in \mathcal{D}(\mathbb{Q}_p^n)$. Thus the formula (4.9.7) holds. $\qquad\square$

Theorem 4.9.5. ([241, VII, (5.4)]) *If for distributions $f, g \in \mathcal{D}'(\mathbb{Q}_p^n)$ the convolution $f * g$ exists, then*

$$
F[f * g] = F[f]F[g]. \tag{4.9.8}
$$

Proof. From Theorem 4.9.4 and formula (4.8.10),

$$
F[f * (\Delta_k g)] = F[f] F[\Delta_k g] = F[f] (F[g] * \delta_k), \quad k \in \mathbb{Z},
$$

where $\Delta_k(x) = \Omega(p^{-k}|x|_p)$ is a canonical 1-sequence (4.4.5). Next, using relations (4.7.9) and (4.7.6), we conclude that (4.9.8) holds. $\qquad\square$

5

Some results from p-adic $\mathcal{L}^1$- and $\mathcal{L}^2$-theories

5.1. Introduction

In this chapter we present some results related with $\mathcal{L}^1$- and $\mathcal{L}^2$-spaces, which we need below.

5.2. $\mathcal{L}^1$-theory

If $f \in \mathcal{L}^1(\mathbb{Q}_p^n)$ then formula (4.9.1) is equivalent to formula (4.8.1):

$$\widehat{f}(\xi) = F[f](\xi) = \int_{\mathbb{Q}_p^n} \chi_p(\xi \cdot x) f(x) \, d^n x, \quad \xi \in \mathbb{Q}_p^n. \tag{5.2.1}$$

There is an analog of the real Riemann-Lebesgue theorem.

Theorem 5.2.1. ([230, III, (1.6)], [241, VII.3.]) *If* $f \in \mathcal{L}^1(\mathbb{Q}_p^n)$ *then* $F[f](\xi)$ *is continuous on* $\mathbb{Q}_p^n$ *and*

$$F[f](\xi) = \int_{\mathbb{Q}_p^n} f(x) \chi_p(\xi \cdot x) \, d^n x \to 0, \quad \xi \to \infty.$$

Proof. Since $|\chi_p(\xi \cdot x) f(x)| = |f(x)|$, $x \in \mathbb{Q}_p^n$, by applying the Lebesgue dominated convergence theorem 3.2.4, we conclude that function $F[f](\xi)$ (which is given by (5.2.1)) is continuous.

Let us prove that $F[f](\xi) \to 0$, as $\xi \to \infty$. From Proposition 4.3.3, the space $\mathcal{D}(\mathbb{Q}_p^n)$ is dense in $\mathcal{L}^1(\mathbb{Q}_p^n)$. Consequently, for any ε there exists $\varphi \in \mathcal{D}(\mathbb{Q}_p^n)$ such that $\int_{\mathbb{Q}_p^n} |f(x) - \varphi(x)| \, d^n x < \varepsilon$. Moreover, according to Lemma 4.8.3, there are $l, N \in \mathbb{Z}$ such that $\varphi \in \mathcal{D}_N^l(\mathbb{Q}_p^n)$, i.e., $F[f](\xi) = 0$

75

for $|\xi|_p > p^N$. Therefore, if $|\xi|_p > p^N$ we have

$$
|F[f](\xi)| = \left| \int_{\mathbb{Q}_p^n} (f(x) - \varphi(x)) \chi_p(\xi \cdot x) d^n x + \int_{\mathbb{Q}_p^n} \varphi(x) \chi_p(\xi \cdot x) d^n x \right|
$$

$$
\leq \int_{\mathbb{Q}_p^n} |f(x) - \varphi(x)| d^n x + |F[\varphi](\xi)| < \varepsilon.
$$

□

Using the standard scheme of a proof for the real setting (see [236, Ch. I, §4.1.]), one can obtain the p-adic version of the Young inequality.

Theorem 5.2.2. *Let $f \in \mathcal{L}^\rho(\mathbb{Q}_p^n)$, $1 \leq \rho \leq \infty$, $g \in \mathcal{L}^\mu(\mathbb{Q}_p^n)$, where $\frac{1}{\rho} + \frac{1}{\mu} \geq 1$. Then $f * g \in \mathcal{L}^r(\mathbb{Q}_p^n)$, where $\frac{1}{r} = \frac{1}{\rho} + \frac{1}{\mu} - 1$, and the Young inequality holds:*

$$
\|f * g\|_r \leq \|f\|_\rho \|g\|_\mu. \tag{5.2.2}
$$

In particular [230, III, (1.7)], if $f \in \mathcal{L}^\rho(\mathbb{Q}_p^n)$, $1 \leq \rho \leq \infty$, and $g \in \mathcal{L}^1(\mathbb{Q}_p^n)$ then $f * g \in \mathcal{L}^\rho(\mathbb{Q}_p^n)$ and

$$
\|f * g\|_\rho \leq \|f\|_\rho \|g\|_1. \tag{5.2.3}
$$

Proof. Let us choose $\alpha, \beta \geq 0$ and $s, t \geq 1$ such that

$$
\frac{1}{r} + \frac{1}{s} + \frac{1}{t} = 1, \quad \alpha r = \rho = (1 - \alpha)s, \quad \beta r = \mu = (1 - \beta)t.
$$

Then $\rho + \frac{\rho r}{s} = r = \mu + \frac{\mu r}{t}$. Next, using the Hölder inequality and the Fubini theorem 3.2.3, we obtain

$$
(\|f * g\|_r)^r = \int_{\mathbb{Q}_p^n} \left| \int_{\mathbb{Q}_p^n} f(y) g(x - y) d^n y \right|^r d^n x
$$

$$
\leq \int_{\mathbb{Q}_p^n} \left(\int_{\mathbb{Q}_p^n} |f(y)|^\alpha |g(x - y)|^\beta |f(y)|^{1-\alpha} |g(x - y)|^{1-\beta} d^n y \right)^r d^n x
$$

$$
\leq \int_{\mathbb{Q}_p^n} \left(\int_{\mathbb{Q}_p^n} |f(y)|^{\alpha r} |g(x - y)|^{\beta r} d^n y \right) \left(\int_{\mathbb{Q}_p^n} |f(y)|^{(1-\alpha)s} d^n y \right)^{r/s}
$$

$$
\times \left(\int_{\mathbb{Q}_p^n} |g(x - y)|^{(1-\beta)t} d^n y \right)^{r/t} d^n x \leq (\|f\|_\rho)^r (\|g\|_\mu)^r.
$$

□

5.3. $\mathcal{L}^2$-theory

The set $\mathcal{L}^2(\mathbb{Q}_p^n)$ is the Hilbert space with the scalar product

$$(f, g) = \int_{\mathbb{Q}_p^n} f(x)\overline{g}(x)\, d^n x, \qquad f, g \in \mathcal{L}^2(\mathbb{Q}_p^n), \qquad (5.3.1)$$

so that $\|f\|_2 = \sqrt{(f, f)}$.

In $\mathcal{L}^2(\mathbb{Q}_p^n)$ the Cauchy–Bunjakovsky inequality holds:

$$|(f, g)| \leq \|f\|_2 \|g\|_2, \qquad f, g \in \mathcal{L}^2(\mathbb{Q}_p^n). \qquad (5.3.2)$$

Using the standard scheme of the proof for the real setting, one can prove the following statement.

Theorem 5.3.1. ([241, VII.4.]) *The Fourier transform $f \to F[f]$ maps $\mathcal{L}^2(\mathbb{Q}_p^n)$ onto $\mathcal{L}^2(\mathbb{Q}_p^n)$ one-to-one and mutually continuous, where*

$$F[f](\xi) = \widehat{f}(\xi) = \lim_{\gamma \to \infty} \int_{B_\gamma^n} f(x)\chi_p(\xi \cdot x)\, d^n x \quad in \quad \mathcal{L}^2(\mathbb{Q}_p^n); \qquad (5.3.3)$$

$$F^{-1}[\widehat{f}](x) = f(x) = \lim_{\gamma \to \infty} \int_{B_\gamma^n} \widehat{f}(\xi)\chi_p(-x \cdot \xi)\, d^n \xi \quad in \quad \mathcal{L}^2(\mathbb{Q}_p^n). \qquad (5.3.4)$$

Moreover, the Parseval–Steklov equality holds:

$$(f, g) = (F[f], F[g]), \qquad \|f\|_2 = \|F[f]\|_2, \qquad f, g \in \mathcal{L}^2(\mathbb{Q}_p^n), \qquad (5.3.5)$$

cf. (4.8.6).

Proof. Let $f \in \mathcal{L}^2(\mathbb{Q}_p^n)$. Then $f_\gamma = \Delta_\gamma \in \mathcal{L}^1(\mathbb{Q}_p^n)$ for all $\gamma \in \mathbb{Z}$, where $\Delta_\gamma(x) = \Omega(p^{-\gamma}|x|_p)$ is a canonical 1-sequence (4.4.5). From (5.2.1), there exists the Fourier transform $\widehat{f}_\gamma(\xi) = F[f_\gamma](\xi)$. It is clear that

$$\lim_{\gamma \to \infty} \|f - f_\gamma\| = \lim_{\gamma \to \infty} \int_{|x|_p \geq p^\gamma} |f(x)|^2\, d^n x = 0, \qquad (5.3.6)$$

and hence

$$\lim_{\gamma \to \infty} \|f_\gamma\| = \|f\|. \qquad (5.3.7)$$

The Parseval–Steklov equality (4.8.6) for test functions $\varphi, \psi \in \mathcal{D}(\mathbb{Q}_p^n)$ can be rewritten in the form of the scalar product (5.3.1) as

$$(\varphi, \psi) = (F[\varphi], F[\psi]), \qquad \|\varphi\|_2 = \|F[\varphi]\|_2, \qquad \varphi, \psi \in \mathcal{D}(\mathbb{Q}_p^n). \qquad (5.3.8)$$

According to Proposition 4.3.3, the space of test functions $\mathcal{D}(\mathcal{O})$ is dense in $\mathcal{L}^2(\mathcal{O})$. Thus, taking into account (5.3.8) and (5.3.7), we conclude that

$$\|\widehat{f_\gamma}\|_2 = \|f_\gamma\|_2 \to \|f\|, \quad \gamma \to \infty, \tag{5.3.9}$$

$$\|\widehat{f_\gamma} - \widehat{f_{\gamma'}}\|_2 = \|f_\gamma - f_{\gamma'}\|_2 \to 0, \quad \gamma, \gamma' \to \infty. \tag{5.3.10}$$

From relation (5.3.10), $\{\widehat{f_\gamma}, \gamma \to \infty\}$ is the Cauchy sequence in $\mathcal{L}^2(\mathbb{Q}_p^n)$. Then, according to the Riesz–Fisher theorem, there exists a function $g \in \mathcal{L}^2(\mathbb{Q}_p^n)$ such that $\widehat{f_\gamma} \to g$ as $\gamma \to \infty$ in $\mathcal{L}^2(\mathbb{Q}_p^n)$. On the other hand $\widehat{f_\gamma} \to \widehat{f}$ as $\gamma \to \infty$ in $\mathcal{D}'(\mathbb{Q}_p^n)$. Hence, $\widehat{f} = g \in \mathcal{L}^2(\mathbb{Q}_p^n)$ and $\widehat{f_\gamma} \to \widehat{f}$ as $\gamma \to \infty$ in $\mathcal{L}^2(\mathbb{Q}_p^n)$. Thus the representation (5.3.3) holds. Since $\|\widehat{f_\gamma}\|_2 \to \|\widehat{f}\|$, $\gamma \to \infty$, taking into account (5.3.9), we conclude that the equalities in (5.3.5) hold.

From (5.3.3), the inversion formula $f = F[\widehat{f}(-\xi)]$, which is given by (4.9.2), takes the form (5.3.4).

Thus, $f \to \widehat{f}$ is one-to-one and a continuous map from $\mathcal{L}^2(\mathbb{Q}_p^n)$ onto $\mathcal{L}^2(\mathbb{Q}_p^n)$. □

Corollary 5.3.2. ([241, VII.4.]) *The Fourier transform $f \to \widehat{f}$ is a unitary operator in $\mathcal{L}^2(\mathbb{Q}_p^n)$.*

Lemma 5.3.3. ([241, VII.4.]) *If $f \in \mathcal{L}^2(\mathbb{Q}_p^n)$ then*

$$\lim_{\gamma \to \infty} p^{-n\gamma/2} \int_{B_\gamma^n} |f(x)|\, d^n x = 0.$$

Proof. Let $f \in \mathcal{L}^2(\mathbb{Q}_p^n)$ and $\varepsilon > 0$. Then there exists $N \in \mathbb{Z}$ such that

$$\int_{\mathbb{Q}_p^n \setminus B_N^n} |f(x)|^2\, d^n x < \frac{\varepsilon^2}{4}.$$

Assuming that $\gamma > N$ and applying the Cauchy–Bunjakovsky inequality, we obtain

$$p^{-n\gamma/2} \int_{B_\gamma^n} |f(x)|\, d^n x = p^{-n\gamma/2} \int_{B_\gamma^n \setminus B_N^n} |f(x)|\, d^n x + p^{-n\gamma/2} \int_{B_N^n} |f(x)|\, d^n x$$

$$\leq p^{-n\gamma/2} \left(\int_{B_\gamma^n \setminus B_N^n} |f(x)|^2\, d^n x \right)^{1/2} \left(\int_{B_\gamma^n \setminus B_N^n} d^n x \right)^{1/2}$$

$$+ p^{-n\gamma/2} \left(\int_{B_N^n} |f(x)|^2\, d^n x \right)^{1/2} \left(\int_{B_N^n} d^n x \right)^{1/2}$$

$$\leq \left(\int_{\mathbb{Q}_p^n \setminus B_N^n} |f(x)|^2\, d^n x \right)^{1/2} + p^{-n\gamma/2} p^{nN/2} \left(\int_{\mathbb{Q}_p^n} |f(x)|^2\, d^n x \right)^{1/2}$$

$$< \frac{\varepsilon}{2} + p^{-n(\gamma-N)/2} \|f\|_2.$$

For all γ such that $p^{-n(\gamma-N)/2}||f||_2 < \frac{\varepsilon}{2}$ we have

$$p^{-n\gamma/2} \int_{B_\gamma^n} |f(x)| \, d^n x < \varepsilon.$$

$\square$

Remark 5.1. In Chapter 8 we construct an infinite set of different wavelet bases in $\mathcal{L}^2(\mathbb{Q}_p^n)$.

6

The theory of associated and quasi associated homogeneous p-adic distributions

6.1. Introduction

In this chapter we construct and study *associated homogeneous distributions* (*AHD*) and *quasi associated homogeneous distributions* (*QAHD*) in the p-adic case. These results are based on the papers [16], [17], [223]. The results of this chapter will be used below in Chapters 9, 10 and 14.

In Appendix A, Sections A.2–A.7, we recall the theory of QAHDs in the better known case where the underlying field consists of the real numbers ("real QAHD"). By analogy with the theory of real QAHDs, in Sections 6.2–6.5 we develop the theory of *p-adic associated and quasi associated homogeneous distributions*. In Section 6.2, we recall the facts on p-adic *homogeneous distributions* from classical books [96, Ch.II, §2], [241, VIII]. In Section 6.3, Definition 6.2 of a p-adic *QAHD* is introduced. We prove Theorems 6.3.3 (Section 6.3) and 6.4.1 (Section 6.4) which give a description of all quasi associated homogeneous distributions and their Fourier transform respectively. In Section 6.5, a new type of p-adic Γ-functions is introduced. These Γ-functions are generated by QAHDs.

6.2. p-adic homogeneous distributions

6.2.1. Definition and characterization

Let π_α be a multiplicative character of the field $\mathbb{Q}_p$. We recall some facts from the theory of π_α-homogeneous Bruhat–Schwartz p-adic distributions, where π_α is a multiplicative character of the field $\mathbb{Q}_p$ [96, Ch.II, §2.3.], [241, VIII.1.].

Definition 6.1. (a) ([96, Ch.II, §2.3.], [241, VIII.1.]) A distribution $f \in \mathcal{D}'(\mathbb{Q}_p)$ is called *homogeneous* of degree π_α if for all $\varphi \in \mathcal{D}(\mathbb{Q}_p)$ and $t \in \mathbb{Q}_p^\times$ we have the relation

$$\left\langle f, \varphi\left(\frac{x}{t}\right)\right\rangle = \pi_\alpha(t)|t|_p\langle f, \varphi\rangle,$$

i.e., $f(tx) = \pi_\alpha(t)f(x)$, $t \in \mathbb{Q}_p^\times$.

(b) ([21]) We say that a distribution $f \in \mathcal{D}'(\mathbb{Q}_p^n)$ is *homogeneous* of degree π_α if for all $t \in \mathbb{Q}_p^\times$ we have

$$f(tx) = f(tx_1, \ldots, tx_n) = \pi_\alpha(t)f(x), \quad x = (x_1, \ldots, x_n) \in \mathbb{Q}_p^n. \quad (6.2.1)$$

A *homogeneous* distribution of degree $\pi_\alpha(x) = |x|_p^{\alpha-1}$ ($\alpha \neq 0$) is called homogeneous of degree $\alpha - 1$.

For every multiplicative character $\pi_\alpha(x) \neq \pi_0 = |x|_p^{-1}$, $x \neq 0$ a *homogeneous* distribution $\pi_\alpha \in \mathcal{D}'(\mathbb{Q}_p)$ of degree $\pi_\alpha(x)$ is defined by the formula [241, VIII, (1.6)]

$$\langle \pi_\alpha, \varphi \rangle = \int_{B_0} |x|_p^{\alpha-1}\pi_1(x)\big(\varphi(x) - \varphi(0)\big)\, dx$$

$$+ \int_{\mathbb{Q}_p \backslash B_0} |x|_p^{\alpha-1}\pi_1(x)\varphi(x)\, dx + \varphi(0)I_0(\alpha), \quad (6.2.2)$$

for all $\varphi \in \mathcal{D}(\mathbb{Q}_p)$. Here, from (3.3.2) and (3.2.3), for Re $\alpha > 0$ we have

$$I_0(\alpha) = \int_{B_0} |x|_p^{\alpha-1}\pi_1(x)\, dx = \begin{cases} 0, & \pi_1(x) \not\equiv 1, \\ \frac{1-p^{-1}}{1-p^{-\alpha}}, & \pi_1(x) \equiv 1. \end{cases} \quad (6.2.3)$$

For $\alpha \neq \mu_j = \frac{2\pi i}{\ln p}j$, $j \in \mathbb{Z}$, we define $I_0(\alpha)$ by means of an analytic continuation.

In the one-dimensional case the following characterization theorem holds.

Theorem 6.2.1. ([96, Ch.II, §2.3.], [241, VIII.1.]) Every homogeneous distribution $f \in \mathcal{D}'(\mathbb{Q}_p)$ of degree $\pi_\alpha(x) = |x|_p^{\alpha-1}\pi_1(x)$ has the form

(a) $C\pi_\alpha$, if either $\pi_1(x) \not\equiv 1$ or $\alpha \neq 0$;

(b) $C\delta$, if $\pi_1(x) \equiv 1$ and $\alpha = 0$, where C is a constant.

The multidimensional ($n \geq 2$) homogeneous distribution $|x|_p^{\alpha-n} \in \mathcal{D}'(\mathbb{Q}_p^n)$ of degree $\alpha - n$ is constructed as follows. If Re $\alpha > 0$ then the function $|x|_p^{\alpha-n}$ generates a regular functional

$$\langle |x|_p^{\alpha-n}, \varphi \rangle = \int_{\mathbb{Q}_p^n} |x|_p^{\alpha-n}\varphi(x)\, d^n x, \quad \forall \varphi \in \mathcal{D}(\mathbb{Q}_p^n). \quad (6.2.4)$$

If Re $\alpha \leq 0$ this distribution is defined by means of analytic continuation [229, (*)], [230, III, (4.3)], [241, VIII, (4.2)]:

$$\langle |x|_p^{\alpha-n}, \varphi \rangle = \int_{B_0^n} |x|_p^{\alpha-n} (\varphi(x) - \varphi(0)) \, d^n x$$

$$+ \int_{\mathbb{Q}_p^n \backslash B_0^n} |x|_p^{\alpha-n} \varphi(x) \, d^n x + \varphi(0) \frac{1 - p^{-n}}{1 - p^{-\alpha}}, \qquad (6.2.5)$$

for all $\varphi \in \mathcal{D}(\mathbb{Q}_p^n)$, $\alpha \neq \mu_j = \frac{2\pi i}{\ln p} j$, $j \in \mathbb{Z}$, where $|x|_p$, $x \in \mathbb{Q}_p^n$, is given by (1.10.1). The distribution $|x|_p^{\alpha-n}$ is meromorphic on the whole complex plane α, and it has at the points μ_j, $j \in \mathbb{Z}$ simple poles with residua $\frac{1-p^{-n}}{\log p} \delta(x)$.

6.2.2. The Fourier transform

Let us recall the theorem describing the Fourier transform of a homogeneous distribution.

Theorem 6.2.2. ([96, Ch.II, §2.5], [241, VIII.2.]) Let $f \in \mathcal{D}'(\mathbb{Q}_p)$ be a homogeneous distribution of degree $\pi_\alpha(x) = |x|_p^{\alpha-1} \pi_1(x)$. Then its Fourier transform $F[f]$ is a homogeneous distribution of degree $\pi_\alpha^{-1}(x)|x|_p^{-1} = |x|_p^{-\alpha} \pi_1^{-1}(x)$.

Consequently, the Fourier transform of the homogeneous distribution $\pi_\alpha(x)$, $\pi_\alpha \neq \pi_0 = |x|_p^{-1}$, is proportional to the homogeneous distribution $|\xi|_p^{-\alpha} \pi_1^{-1}(\xi)$, and the coefficient of proportionality $\Gamma_p(\pi_\alpha)$ is called the Γ-function. Here the Γ-functions are given by the following integrals (see [241, VIII, (2.2), (2.17)]):

$$\Gamma_p(\alpha) \overset{def}{=} \Gamma_p(|x|_p^{\alpha-1}) = \int_{\mathbb{Q}_p} |x|_p^{\alpha-1} \chi_p(x) \, dx = \frac{1 - p^{\alpha-1}}{1 - p^{-\alpha}}, \qquad (6.2.6)$$

$$\Gamma_p(\pi_\alpha) \overset{def}{=} F[\pi_\alpha](1) = \int_{\mathbb{Q}_p} |x|_p^{\alpha-1} \pi_1(x) \chi_p(x) \, dx. \qquad (6.2.7)$$

The integrals on the right-hand sides of (6.2.6) and (6.2.7) are defined by means of analytic continuation with respect to the parameter α, or as improper integrals (see [229, §1])

$$\int_{\mathbb{Q}_p} f(x) \, dx = \lim_{k \to \infty} \int_{p^{-k} \leq |x|_p \leq p^k} f(x) \, dx.$$

Thus,

$$F[\pi_\alpha(x)](\xi) = \Gamma_p(\pi_\alpha)|\xi|_p^{-\alpha} \pi_1^{-1}(\xi), \qquad \text{if } \pi_\alpha \neq \pi_0 = |x|_p^{-1},$$

$$F[\delta(x)](\xi) = 1, \qquad \text{if } \pi_\alpha = \pi_0 = |x|_p^{-1}, \qquad (6.2.8)$$

where $\Gamma_p(\pi_\alpha)$ is the Γ-function (6.2.7).

If $\pi_\alpha^1(x)$, $\pi_\beta^2(x)$ are homogeneous distributions then the following relation holds [241, VIII, (3.6)]:

$$\left(\pi_\alpha^1 * \pi_\beta^2\right)(x) = \mathcal{B}_p(\pi_\alpha^1, \pi_\beta^2)|x|_p^{\alpha+\beta-1}\pi_1^1(x)\pi_1^2(x), \quad x \in \mathbb{Q}_p, \qquad (6.2.9)$$

where

$$\mathcal{B}_p(\pi_\alpha^1, \pi_\beta^2) = \frac{\Gamma_p(\pi_\alpha^1)\Gamma_p(\pi_\beta^2)}{\Gamma_p(\pi_\alpha^1\pi_\beta^2|x|_p)} \qquad (6.2.10)$$

is the $\mathcal{B}$-function [241, VIII, (3.5)].

Similarly, the Fourier transform of the multidimensional homogeneous distribution (6.2.5) is proportional to the homogeneous distribution $|x|_p^{-\alpha}$, and the coefficient of proportionality $\Gamma_p^{(n)}(\alpha)$ is called the multidimensional Γ-function.

In the n-dimensional case the Fourier transform of the distribution (6.2.5) $|x|_p^{\alpha-n}$ is given by the formula [228], [229, Theorem 2.], and [230, III, Theorem (4.5)], [241, VIII, (4.3)]

$$F[|x|_p^{\alpha-n}] = \Gamma_p^{(n)}(\alpha)|\xi|_p^{-\alpha}, \quad \alpha \neq 0, n, \qquad (6.2.11)$$

where $\Gamma_p^{(n)}(\alpha)$ is the n-dimensional Γ-*function* given by the following formula (see [228], [229, Theorem 1.], [230, III, Theorem (4.2)], [241, VIII, (4.4)]):

$$\Gamma_p^{(n)}(\alpha) \overset{def}{=} \lim_{k\to\infty} \int_{p^{-k}\leq|x|_p\leq p^k} |x|_p^{\alpha-n}\chi_p(u\cdot x)\,d^nx$$

$$= \int_{\mathbb{Q}_p^n} |x|_p^{\alpha-n}\chi_p(x_1)\,d^nx = \frac{1-p^{\alpha-n}}{1-p^{-\alpha}} \qquad (6.2.12)$$

in a sense of the principal value integral (3.2.5), where $|u|_p = 1$, and the distribution $|x|_p^{\alpha-n}$ is given by (6.2.5). Here $\Gamma_p^{(1)}(\alpha) = \Gamma_p(\alpha)$.

6.3. *p*-adic quasi associated homogeneous distributions

6.3.1. Definition and characterization

In order to introduce the *p*-adic analog of QAHDs, we adapt Definitions A.10 and A.12 of the real QAHDs from Appendix A.

Definition 6.2. ([16], [17], [20], [21]) (a) A distribution $f_m \in \mathcal{D}'(\mathbb{Q}_p)$ is said to be *quasi associated homogeneous* of degree π_α and order m, $m \in \mathbb{N}_0 = \{0\} \cup \mathbb{N}$, if

$$\left\langle f_m, \varphi\left(\frac{x}{t}\right)\right\rangle = \pi_\alpha(t)|t|_p\langle f_m, \varphi\rangle + \sum_{j=1}^m \pi_\alpha(t)|t|_p \log_p^j |t|_p \langle f_{m-j}, \varphi\rangle$$

for all $\varphi \in \mathcal{D}(\mathbb{Q}_p)$ and $t \in \mathbb{Q}_p^\times$, where $f_{m-j} \in \mathcal{D}'(\mathbb{Q}_p)$ is a quasi associated homogeneous distribution of degree π_α and order $m - j$, $j = 1, 2, \ldots, m$, i.e.,

$$f_m(tx) = \pi_\alpha(t)f_m(x) + \sum_{j=1}^{m} \pi_\alpha(t)\log_p^j |t|_p f_{m-j}(x), \quad t \in \mathbb{Q}_p^\times.$$

If $m = 0$ we say that the above sum is empty.

(b) We say that a distribution $f \in \mathcal{D}'(\mathbb{Q}_p^n)$ is *quasi associated homogeneous* of degree π_α and order m, $m \in \mathbb{N}_0$, if for all $t \in \mathbb{Q}_p^\times$ we have

$$f_m(tx) = f_m(tx_1, \ldots, tx_n)$$

$$= \pi_\alpha(t)f_m(x) + \sum_{j=1}^{m} \pi_\alpha(t)\log_p^j |t|_p f_{m-j}(x), \qquad (6.3.1)$$

where $f_{m-j} \in \mathcal{D}'(\mathbb{Q}_p^n)$ is a quasi associated homogeneous distribution of degree π_α and order $m - j$, $j = 1, 2, \ldots, m$.

A *quasi associated homogeneous* distribution of degree $\pi_\alpha(t) = |t|_p^{\alpha-1}$ and order m is called *quasi associated homogeneous* of degree $\alpha - 1$ and order m.

(c) Quasi associated homogeneous distributions of order $m = 1$ are called *associated homogeneous* distributions (see [95] and [16], [17]).

Remark 6.1. The sum of a QAHD of degree $\pi_\alpha(x)$ and order m and a QAHD of degree $\pi_\alpha(x)$ and order $k < m$ is a QAHD of degree $\pi_\alpha(x)$ and order m.

For every multiplicative character $\pi_\alpha(x) \neq \pi_0(x) = |x|_p^{-1}$, $x \neq 0$, and for every $m \in \mathbb{N}$, we define the distribution $\pi_\alpha(x)\log_p^m |x|_p \in \mathcal{D}'(\mathbb{Q}_p)$ by an analytical continuation with respect to α:

$$\langle \pi_\alpha(x)\log_p^m |x|_p, \varphi(x) \rangle$$

$$= \int_{B_0} |x|_p^{\alpha-1}\pi_1(x)\log_p^m |x|_p \big(\varphi(x) - \varphi(0)\big)\, dx$$

$$+ \int_{\mathbb{Q}_p \setminus B_0} |x|_p^{\alpha-1}\pi_1(x)\log_p^m |x|_p \varphi(x)\, dx + \varphi(0)I_0(\alpha; m), \qquad (6.3.2)$$

for all $\varphi \in \mathcal{D}(\mathbb{Q})$. Here for Re $\alpha > 0$

$$I_0(\alpha; m) = \int_{B_0} |x|_p^{\alpha-1}\pi_1(x)\log_p^m |x|_p\, dx = \sum_{\gamma=-\infty}^{0} \gamma^m p^{\gamma(\alpha-1)} \int_{S_\gamma} \pi_1(x)\, dx$$

$$= \begin{cases} 0, & \pi_1(x) \not\equiv 1, \\ \log_p^m e(1 - p^{-1})\frac{d^m}{d\alpha^m}\left(\frac{1}{1-p^{-\alpha}}\right), & \pi_1(x) \equiv 1, \end{cases} \qquad (6.3.3)$$

and for all $\alpha \neq \alpha_j = \frac{2\pi i}{\ln p} j$, $j \in \mathbb{Z}$, we define the integral $I_0(\alpha; m)$ by means of an analytic continuation with respect to α. Relation (6.3.2) can also be obtained by differentiation of (6.2.2) with respect to α.

For the multiplicative character $\pi_0(x) = |x|_p^{-1}$, $x \neq 0$, we define the distribution $P\left(\frac{1}{|x|_p}\right) \in \mathcal{D}'(\mathbb{Q}_p)$ (the principal value of the function $|x|_p^{-1}$) by the formula [241, VIII, (1.10)]:

$$\left\langle P\left(\frac{1}{|x|_p}\right), \varphi \right\rangle = \int_{B_0} |x|_p^{-1}\left(\varphi(x) - \varphi(0)\right) dx$$
$$+ \int_{\mathbb{Q}_p \setminus B_0} |x|_p^{-1}\varphi(x)\, dx, \quad \varphi(x) \in \mathcal{D}. \quad (6.3.4)$$

By analogy with formula (6.3.4), for $m \in \mathbb{N}$ we introduce the distribution

$$\left\langle P\left(\frac{\log_p^m |x|_p}{|x|_p}\right), \varphi \right\rangle = \int_{B_0} \frac{\log_p^m |x|_p}{|x|_p}\left(\varphi(x) - \varphi(0)\right) dx$$
$$+ \int_{\mathbb{Q}_p \setminus B_0} \frac{\log_p^m |x|_p}{|x|_p}\varphi(x)\, dx, \quad \varphi(x) \in \mathcal{D}. \quad (6.3.5)$$

Lemma 6.3.1. *The distribution* $\pi_\alpha(x)\log_p^m |x|_p$, *where either* $\pi_1(x) \not\equiv 1$ *or* $\alpha \neq 0$, *and the distribution* $P\left(|x|_p^{-1}\log_p^{m-1}|x|_p\right)$ *are QAHDs of degree* $\pi_\alpha(x) = |x|_p^{\alpha-1}\pi_1(x)$ *and order* $m \in \mathbb{N}$.

Proof. 1. For all $\varphi \in \mathcal{D}(\mathbb{Q})$, $t \in \mathbb{Q}_p^\times$ we have the following.

(a) If $|t|_p < 1$ then

$$\left\langle P\left(\frac{1}{|x|_p}\right), \varphi\left(\frac{x}{t}\right) \right\rangle$$
$$= \int_{B_0} |x|_p^{-1}\left(\varphi\left(\frac{x}{t}\right) - \varphi(0)\right) dx + \int_{\mathbb{Q}_p \setminus B_0} |x|_p^{-1}\varphi\left(\frac{x}{t}\right) dx$$
$$= \int_{|\xi|_p \leq |t|_p^{-1}} |\xi|_p^{-1}\left(\varphi(\xi) - \varphi(0)\right) d\xi + \int_{|\xi|_p > |t|_p^{-1}} |\xi|_p^{-1}\varphi(\xi)\, d\xi$$
$$= \left\langle P\left(\frac{1}{|x|_p}\right), \varphi(x) \right\rangle - \varphi(0)\int_{p \leq |\xi|_p \leq |t|_p^{-1}} |\xi|_p^{-1}\, d\xi, \quad (6.3.6)$$

where (see (3.3.1), (3.2.6), and (B.2) from Appendix B),

$$\int_{p \leq |\xi|_p \leq |t|_p^{-1}} |\xi|_p^{-1}\, d\xi = \sum_{\gamma=1}^{\gamma_0}\int_{S_\gamma} |\xi|_p^{-1}\, d\xi = \left(1 - p^{-1}\right)\sum_{\gamma=1}^{\gamma_0} 1 = \left(1 - p^{-1}\right)\gamma_0,$$
$$(6.3.7)$$

and $\sum_{\gamma=1}^{\gamma_0} 1 = \mathbf{S}_0(\gamma_0) = \gamma^0$, $|t|_p = p^{-\gamma_0}$, $\gamma_0 = -\log_p |t|_p \geq 1$.

(b) If $|t|_p = 1$ then

$$\left\langle P\left(\frac{1}{|x|_p}\right), \varphi(x/t) \right\rangle = \left\langle P\left(\frac{1}{|x|_p}\right), \varphi(x) \right\rangle.$$

(c) If $|t|_p > 1$ then

$$\left\langle P\left(\frac{1}{|x|_p}\right), \varphi\left(\frac{x}{t}\right)\right\rangle$$

$$= \int_{|\xi|_p \le |t|_p^{-1}} |\xi|_p^{-1}(\varphi(\xi) - \varphi(0)) \, d\xi + \int_{|\xi|_p > |t|_p^{-1}} |\xi|_p^{-1}\varphi(\xi) \, d\xi$$

$$= \left\langle P\left(\frac{1}{|x|_p}\right), \varphi(x)\right\rangle + \varphi(0) \int_{|t|_p^{-1} < |\xi|_p \le 1} |\xi|_p^{-1} \, d\xi, \qquad (6.3.8)$$

where (see (3.3.1), (3.2.6), and (B.2) from Appendix B)

$$\int_{|t|_p^{-1} < |\xi|_p \le 1} |\xi|_p^{-1} \, d\xi = \sum_{\gamma = \gamma_0 + 1}^{0} \int_{S_\gamma} |\xi|_p^{-1} \, d\xi$$

$$= (1 - p^{-1}) \sum_{\gamma = \gamma_0 + 1}^{0} 1 = -(1 - p^{-1})\gamma_0, \quad (6.3.9)$$

and $\sum_{\gamma=\gamma_0+1}^{0} 1 = -\gamma^0 = -S_0(\gamma_0)$, $|t|_p = p^{-\gamma_0}$, $\gamma_0 = -\log_p |t|_p \le -1$.
Thus from (6.3.6)–(6.3.9) we obtain

$$\left\langle P\left(\frac{1}{|x|_p}\right), \varphi\left(\frac{x}{t}\right)\right\rangle = |t|_p|t|_p^{-1}\left\langle P\left(\frac{1}{|x|_p}\right), \varphi(x)\right\rangle$$

$$+ |t|_p|t|_p^{-1}\log_p |t|_p(1 - p^{-1})\langle \delta(x), \varphi(x)\rangle, \quad (6.3.10)$$

i.e., according to Definition 6.2 (a), $P\left(\frac{1}{|x|_p}\right)$ is a QAHD of degree $|x|_p^{-1}$ and order 1.

2. For $m \in \mathbb{N}$, all $\varphi(x) \in \mathcal{D}(\mathbb{Q})$, $t \in \mathbb{Q}_p^\times$, from (6.3.5) we have the following.
(a) If $|t|_p < 1$ then

$$\left\langle P\left(\frac{\log_p^m |x|_p}{|x|_p}\right), \varphi\left(\frac{x}{t}\right)\right\rangle$$

$$= \int_{B_0} \frac{\log_p^m |x|_p}{|x|_p}\left(\varphi\left(\frac{x}{t}\right) - \varphi(0)\right) dx + \int_{\mathbb{Q}_p \setminus B_0} \frac{\log_p^m |x|_p}{|x|_p}\varphi\left(\frac{x}{t}\right) dx$$

$$= \int_{|\xi|_p \le |t|_p^{-1}} \frac{\log_p^m |t\xi|_p}{|\xi|_p}(\varphi(\xi) - \varphi(0)) \, d\xi$$

$$+ \int_{|\xi|_p > |t|_p^{-1}} \frac{\log_p^m |t\xi|_p}{|\xi|_p}\varphi(\xi) \, d\xi$$

$$= \sum_{j=0}^{m} C_m^j \log_p^j |t|_p \left\{\left\langle P\left(\frac{\log_p^{m-j} |\xi|_p}{|\xi|_p}\right), \varphi(\xi)\right\rangle\right.$$

$$\left. - \varphi(0) \int_{p \le |\xi|_p \le |t|_p^{-1}} \frac{\log_p^{m-j} |\xi|_p}{|\xi|_p} \, d\xi\right\}. \qquad (6.3.11)$$

Here $|t|_p = p^{-\gamma_0}$, $\gamma_0 = -\log_p |t|_p \geq 1$, and (see (3.3.1), (3.2.6), and (B.2) from Appendix B)

$$\int_{p \leq |\xi|_p \leq |t|_p^{-1}} \frac{\log_p^{m-j} |\xi|_p}{|\xi|_p} \, d\xi = \sum_{\gamma=1}^{\gamma_0} \int_{S_\gamma} \frac{\log_p^{m-j} |\xi|_p}{|\xi|_p} \, d\xi$$

$$= (1 - p^{-1}) \sum_{\gamma=1}^{\gamma_0} \gamma^{m-j}, \qquad (6.3.12)$$

where $\sum_{\gamma=1}^{\gamma_0} \gamma^{m-j} = \mathbf{S}_{m-j}(\gamma_0)$, $j = 0, 1, \ldots, m$.

(b) If $|t|_p = 1$ then

$$\left\langle P\left(\frac{\log_p^m |x|_p}{|x|_p}\right), \varphi(x/t) \right\rangle = \left\langle P\left(\frac{\log_p^m |x|_p}{|x|_p}\right), \varphi(x) \right\rangle.$$

(c) If $|t|_p > 1$ then

$$\left\langle P\left(\frac{\log_p^m |x|_p}{|x|_p}\right), \varphi\left(\frac{x}{t}\right) \right\rangle$$

$$= \int_{|\xi|_p \leq |t|_p^{-1}} \frac{\log_p^m |t\xi|_p}{|\xi|_p} \big(\varphi(\xi) - \varphi(0)\big) \, d\xi$$

$$+ \int_{|\xi|_p > |t|_p^{-1}} \frac{\log_p^m |t\xi|_p}{|\xi|_p} \varphi(\xi) \, d\xi$$

$$= \sum_{j=0}^{m} C_m^j \log_p^j |t|_p \left\{ \left\langle P\left(\frac{\log_p^{m-j} |\xi|_p}{|\xi|_p}\right), \varphi(\xi) \right\rangle \right.$$

$$\left. + \varphi(0) \int_{|t|_p^{-1} < |\xi|_p \leq 1} \frac{\log_p^{m-j} |\xi|_p}{|\xi|_p} \, d\xi \right\}. \qquad (6.3.13)$$

Here $|t|_p = p^{-\gamma_0}$, $\gamma_0 = -\log_p |t|_p \leq -1$, and (see (3.3.1), (3.2.6), and (B.2) from Appendix B)

$$\int_{|t|_p^{-1} < |\xi|_p \leq 1} \frac{\log_p^{m-j} |\xi|_p}{|\xi|_p} \, d\xi = \sum_{\gamma=\gamma_0+1}^{0} \int_{S_\gamma} \frac{\log_p^{m-j} |\xi|_p}{|\xi|_p} \, d\xi$$

$$= (1 - p^{-1}) \sum_{\gamma=\gamma_0+1}^{0} \gamma^{m-j}, \qquad j = 0, 1, \ldots, m.$$

$$(6.3.14)$$

Using equality (B.2) from Appendix B, and taking into account that according to Lemma B.2, $\sum_{\gamma=\gamma_0+1}^{0} \gamma^{m-j} = -\mathbf{S}_{m-j}(\gamma_0)$, we obtain from

(6.3.11)–(6.3.14)

$$\left\langle P\left(\frac{\log_p^m |x|_p}{|x|_p}\right), \varphi\left(\frac{x}{t}\right)\right\rangle$$

$$= |t|_p |t|_p^{-1} \left\langle P\left(\frac{\log_p^m |x|_p}{|x|_p}\right), \varphi(x)\right\rangle$$

$$+ \sum_{j=1}^{m} |t|_p |t|_p^{-1} \log_p^j |t|_p \left\langle C_m^j P\left(\frac{\log_p^{m-j} |x|_p}{|x|_p}\right), \varphi(x)\right\rangle$$

$$+ \log_p^{m+1} |t|_p \frac{1}{m+1}\left(1 - p^{-1}\right)\langle\delta(x), \varphi(x)\rangle + \langle\delta(x), \varphi(x)\rangle \sum_{j=1}^{m} a_j \log_p^j |t|_p,$$

for all $t \in \mathbb{Q}_p^\times$, where

$$a_m = -\left(1 - p^{-1}\right)\mathbf{B}_1 \frac{C_{m+1}^1}{m+1},$$

$$a_{m+1-j} = \left(1 - p^{-1}\right)\mathbf{B}_j \frac{C_{m+1}^j}{m+1}, \quad j = 2, 3, \ldots, m. \tag{6.3.15}$$

We have thus proved that for all $t \in \mathbb{Q}_p^\times$

$$P\left(\frac{\log_p^m |tx|_p}{|tx|_p}\right) = |t|_p^{-1} P\left(\frac{\log_p^m |x|_p}{|x|_p}\right) + \sum_{j=1}^{m+1} |t|_p^{-1} \log_p^j |t|_p f_{m+1-j}(x), \tag{6.3.16}$$

where

$$f_{m+1-j}(x) = C_m^j \left(\frac{\log_p^{m-j} |x|_p}{|x|_p}\right) + a_j \delta(x), \quad j = 1, 2, \ldots, m,$$

$$f_0(x) = \frac{1}{m+1}\left(1 - p^{-1}\right)\delta(x). \tag{6.3.17}$$

Taking into account Remark 6.1, we can prove by induction that a distribution $P\left(|x|_p^{-1} \log_p^m |x|_p\right)$ is a QAHD of degree $|x|_p^{-1}$ and order $m + 1$, $m \in \mathbb{N}$.

3. For Re $\alpha > 0$, $m \in \mathbb{N}$ and for all $\varphi \in \mathcal{D}(\mathbb{Q})$, $t \in \mathbb{Q}_p^\times$ from (6.3.2) we have

$$\langle \pi_\alpha(x) \log_p^m |x|_p, \varphi(x)\rangle = \int_{\mathbb{Q}_p} |x|_p^{\alpha-1} \pi_1(x) \log_p^m |x|_p \varphi(x)\, dx$$

and

$$\left\langle \pi_\alpha(x) \log_p^m |x|_p, \varphi\left(\frac{x}{t}\right) \right\rangle$$

$$= \int_{\mathbb{Q}_p} |t\xi|_p^{\alpha-1} \pi_1(t\xi) \log_p^m |t\xi|_p \varphi(\xi) \, d(t\xi)$$

$$= \pi_\alpha(t)|t|_p \sum_{j=0}^m C_m^j \log_p^j |t|_p \int_{\mathbb{Q}_p} |\xi|_p^{\alpha-1} \pi_1(\xi) \log_p^{m-j} |\xi|_p \varphi(\xi) \, d(\xi)$$

$$= \pi_\alpha(t)|t|_p \langle \pi_\alpha(x) \log_p^m |x|, \varphi(x) \rangle$$

$$+ \sum_{j=1}^m \pi_\alpha(t)|t|_p \log_p^j |t|_p \langle f_{m-j}(x), \varphi(x) \rangle,$$

where $f_{m-j}(x) = C_m^j \pi_\alpha(x) \log_p^{m-j} |\xi|_p$, $j = 1, 2, \ldots, m$. Thus, for Re $\alpha > 0$ we can prove by induction that $\pi_\alpha(x) \log_p^m |x|_p$ is a QAHD of degree $\pi_\alpha(x)$ and order m in the sense of Definition 6.2 (a), $m \in \mathbb{N}$. For other $\alpha \neq \alpha_j = \frac{2\pi i}{\ln p} j$, $j \in \mathbb{Z}$, this statement follows by the principle of analytical continuation. $\square$

Lemma 6.3.2. *Let $f_s(x)$ be a QAHD of degree $\pi_{\alpha_s}(x) = |x|_p^{\alpha_s-1} \pi_1(x)$ and order m_s, $s = 1, 2, \ldots, r$, where all α_s or m_s are different. If $\sum_{s=1}^r f_s(x) = 0$ then $f_s(x) = 0$, $s = 1, 2, \ldots, r$. Thus QAHDs of different degrees and orders are linearly independent.*

This lemma is proved in the same way as the corresponding lemma on linear independent homogeneous distributions from [241, VIII.1.].

Let us prove the following theorem which describes all QAHDs.

Theorem 6.3.3. *Every QAHD $f \in \mathcal{D}'(\mathbb{Q}_p)$ of degree $\pi_\alpha(x) = |x|_p^{\alpha-1} \pi_1(x)$ and order $m \in \mathbb{N}$ (with accuracy up to QAHD of order $\leq m - 1$) has the form*
(a) $C\pi_\alpha(x) \log_p^m |x|_p$, *if $\pi_\alpha(x) \neq \pi_0(x) = |x|_p^{-1}$;*
(b) $CP\left(|x|_p^{-1} \log_p^{m-1} |x|_p\right)$, *if $\pi_\alpha(x) = \pi_0(x) = |x|_p^{-1}$, where C is a constant.*

Proof. (a) Let $f \not\equiv 0$ be a QAHD of degree $\pi_\alpha(x) \neq \pi_0(x) = |x|_p^{-1}$ and order $m \in \mathbb{N}$. It is clear that supp $f \not\equiv \{0\}$. Therefore there exists a function $\psi \in \mathcal{D}(\mathbb{Q})$, $x \in \mathbb{Q}_p$, $\psi(0) = 0$ such that

$$A_0 = \langle f(x), \psi(x) \rangle, \qquad B_0 = \langle \pi_\alpha(\xi) \log_p^m |\xi|_p, \psi(\xi) \rangle \neq 0,$$

$$A_j = \langle f_{m-j}(x), \psi(x) \rangle, \qquad B_j = \langle \pi_\alpha(\xi) \log_p^{m-j} |\xi|_p, \psi(\xi) \rangle \neq 0,$$

$j = 1, \ldots, m$, for which by virtue of Definition 6.2 (a) we have

$$\left\langle f(x), \psi\left(\frac{x}{t}\right)\right\rangle = \pi_\alpha(t)|t|_p \langle f(x), \psi(x)\rangle$$

$$+ \sum_{j=1}^{m} \pi_\alpha(t)|t|_p \log_p^j |t|_p \langle f_{m-j}(x), \psi(x)\rangle, \quad t \in \mathbb{Q}_p^\times,$$

where $f_{m-j}(x)$ is a QAHD of degree $\pi_{\alpha_s}(x)$ and order $m - j$, $j = 1, 2, \ldots, m$.
Hence we have

$$\int_{\mathbb{Q}_p} \left\langle f(x), \psi\left(\frac{x}{t}\right)\right\rangle \frac{\varphi(t)}{|t|_p}\, dt$$

$$= A_0 \langle \pi_\alpha(t), \varphi(t)\rangle + \sum_{j=1}^{m} A_j \langle \pi_\alpha(t) \log_p^j |t|_p, \varphi(t)\rangle, \qquad (6.3.18)$$

for all $\varphi \in \mathcal{D}(\mathbb{Q})$, $\varphi(0) = 0$.

Since $\psi(0) = 0$ and $\varphi(0) = 0$, we have $\psi\left(\frac{x}{t}\right)\frac{\varphi(t)}{|t|_p} \in \mathcal{D}(\mathbb{Q}^2)$, and (6.3.18)
implies

$$\left\langle f(x), \int_{\mathbb{Q}_p} \psi\left(\frac{x}{t}\right)\frac{\varphi(t)}{|t|_p}\, dt\right\rangle$$

$$= A_0 \langle \pi_\alpha(t), \varphi(t)\rangle + \sum_{j=1}^{m} A_j \langle \pi_\alpha(t) \log_p^j |t|_p, \varphi(t)\rangle. \qquad (6.3.19)$$

Making the change of variables (for $x \neq 0$) $t = \frac{x}{\xi}$, $dt = \frac{|x|_p}{|\xi|_p^2}\, d\xi$ in the inner
integral on the left-hand side of (6.3.19), using relation (3.4.3), and taking into
account that $\frac{\psi(\xi)}{|\xi|_p}\varphi(x/\xi) \in \mathcal{D}(\mathbb{Q}^2)$, we have

$$\int_{\mathbb{Q}_p} \frac{\psi(\xi)}{|\xi|_p}\left\langle f(x), \varphi\left(\frac{x}{\xi}\right)\right\rangle d\xi$$

$$= A_0 \langle \pi_\alpha(t), \varphi(t)\rangle + \sum_{j=1}^{m} A_j \langle \pi_\alpha(t) \log_p^j |t|_p, \varphi(t)\rangle. \qquad (6.3.20)$$

By using Definition 6.2 (a), from (6.3.20) we have

$$\langle \pi_\alpha(\xi), \psi(\xi)\rangle \langle f(x), \varphi(x)\rangle + \sum_{j=1}^{m} \langle \pi_\alpha(\xi) \log_p^j |\xi|_p, \psi(\xi)\rangle \langle f_{m-j}(x), \varphi(x)\rangle$$

$$= A_0 \langle \pi_\alpha(t), \varphi(t)\rangle + \sum_{j=1}^{m} A_j \langle \pi_\alpha(t) \log_p^j |t|_p, \varphi(t)\rangle,$$

for all $\varphi \in \mathcal{D}(\mathbb{Q})$, $\varphi(0) = 0$. The latter relation can be rewritten as

$$\Big\langle B_m f(x) - A_m \pi_\alpha(x) \log_p^m |x|_p$$
$$+ \sum_{j=1}^m \Big(B_{m-j} f_{m-j}(x) - A_{m-j} \pi_\alpha(x) \log_p^{m-j} |x|_p \Big), \varphi(x) \Big\rangle = 0. \quad (6.3.21)$$

It follows from equality (6.3.21) that either

$$B_m f(x) - A_m \pi_\alpha(x) \log_p^m |x|_p$$
$$+ \sum_{j=1}^m \Big(B_{m-j} f_{m-j}(x) - A_{m-j} \pi_\alpha(x) \log_p^{m-j} |x|_p \Big) = 0 \quad (6.3.22)$$

or

$$B_m f(x) - A_m \pi_\alpha(x) \log_p^m |x|_p$$
$$+ \sum_{j=1}^m \Big(B_{m-j} f_{m-j}(x) - A_{m-j} \pi_\alpha(x) \log_p^{m-j} |x|_p \Big) = C\delta(x), \quad (6.3.23)$$

where C is a constant.

Since $\pi_\alpha(x) \neq \pi_0(x) = |x|_p^{-1}$, according to Lemmas 6.3.1, 6.3.2 and Theorem 6.2.1, relation (6.3.23) holds only for $C = 0$. Thus we have relation (6.3.22).

Suppose that any QAHD $f_{m-j}(x)$ is a linear combination of the following QAHDs $\pi_\alpha(x) \log_p^k |x|_p$, $k = 0, 1, \ldots, m-j-1$, $j = 1, 2, \ldots, m$. Then by induction we have from (6.3.22) that

$$f(x) = \frac{A_m}{B_m} \pi_\alpha(x) \log_p^m |x|_p$$

up to a QAHD of order $\leq m-1$.

(b) Let $\pi_\alpha(x) = \pi_0(x) = |x|_p^{-1}$. It is clear that supp $f \neq \{0\}$. Then repeating literally the previous arguments, we obtain the following relation:

$$B_m f(x) - A_m P\Big(\frac{\log_p^{m-1} |\xi|_p}{|\xi|_p}\Big)$$
$$+ \sum_{j=1}^{m-1} \Big(B_{m-j} f_{m-j}(x) - A_{m-j} P\Big(\frac{\log_p^{m-1-j} |\xi|_p}{|\xi|_p}\Big) \Big) = C\delta(x), \quad (6.3.24)$$

where

$$A_1 = \langle f(x), \psi(x) \rangle, \qquad B_j = \langle P\Big(\frac{\log_p^{m-j} |\xi|_p}{|\xi|_p}\Big), \psi(\xi) \rangle \neq 0,$$
$$A_{j+1} = \langle f_{m-j}(x), \psi(x) \rangle, \qquad B_m = \langle P\Big(\frac{1}{|\xi|_p}\Big), \psi(\xi) \rangle \neq 0,$$

$j = 1, 2, \ldots, m - 1$, C is a constant. Similarly to the previous case, by induction we have from (6.3.24) that

$$f(x) = \frac{A_m}{B_m} P\left(\frac{\log_p^{m-1} |\xi|_p}{|\xi|_p}\right).$$

up to a QAHD of order $\leq m - 1$. The theorem is thus proved. $\qquad\square$

Lemma 6.3.4. *A QAHD of degree $\pi_\alpha(x)$ and order m is periodic in the variable α with period $T = \alpha_1 = \frac{2\pi i}{\ln p}$, i.e.,*

$$\pi_{\alpha+T}(x) \log_p^m |x|_p = \pi_\alpha(x) \log_p^m |x|_p, \qquad \pi_\alpha \neq |x|_p^{-1},$$

$$P\left(\frac{\log_p^m |x|_p}{|x|_p^{1+T}}\right) = P\left(\frac{\log_p^m |x|_p}{|x|_p^1}\right), \qquad m \in \mathbb{N}_0.$$

Proof. The integrals in (6.3.2) and (6.3.5) can be rewritten as

$$\int_{B_0} |x|_p^{\alpha-1} \pi_1(x) \log_p^m |x|_p \big(\varphi(x) - \varphi(0)\big)\, dx$$

$$= \sum_{\gamma=-\infty}^{0} p^{\gamma(\alpha-1)} \gamma^m \int_{S_\gamma} \pi_1(x) \big(\varphi(x) - \varphi(0)\big)\, dx,$$

$$\int_{\mathbb{Q}_p \backslash B_0} |x|_p^{\alpha-1} \pi_1(x) \log_p^m |x|_p \varphi(x)\, dx = \sum_{\gamma=1}^{\infty} p^{\gamma(\alpha-1)} \gamma^m \int_{S_\gamma} \pi_1(x) \varphi(x)\, dx.$$

Since $p^T = 1$, one can see that all terms on the right-hand side of the latter relation are T-periodic in the variable α, $T = \alpha_1 = \frac{2\pi i}{\ln p}$. The integral $I_0(\alpha; m)$ given by (6.3.3) is also T-periodic. Thus from Theorem 6.3.3, the lemma is proved. $\qquad\square$

6.3.2. Multidimensional case

Similarly to the one-dimensional case (6.3.4), one can construct the distribution $P\left(\frac{1}{|x|_p^n}\right)$ called the principal value of the function $\frac{1}{|x|_p^n}$:

$$\left\langle P\left(\frac{1}{|x|_p^n}\right), \varphi \right\rangle = \int_{B_0^n} \frac{\varphi(x) - \varphi(0)}{|x|_p^n}\, d^n x + \int_{\mathbb{Q}_p^n \backslash B_0^n} \frac{\varphi(x)}{|x|_p^n}\, d^n x, \qquad (6.3.25)$$

for all $\varphi \in \mathcal{D}(\mathbb{Q}_p^n)$. It is easy to show that according to Definition 6.2 (b), this distribution is *quasi associated homogeneous* of degree $-n$ and order 1.

6.4. The Fourier transform of p-adic quasi associated homogeneous distributions

Now we prove the theorem describing the Fourier transform of p-adic QAHDs.

Theorem 6.4.1. *Let* $f \in \mathcal{D}'(\mathbb{Q}_p)$ *be a QAHD of degree* $\pi_\alpha = |x|_p^{\alpha-1}\pi_1(x)$ *and order* m. *Then its Fourier transform* $F[f]$ *is a QAHD of degree* $\pi_\alpha^{-1}|x|_p^{-1} = |x|_p^{-\alpha}\pi_1^{-1}(x)$ *and order* m, $m \in \mathbb{N}$.

Proof. If $m = 1$ then using (4.9.4) and Definition 6.2 (a), for all $t \in \mathbb{Q}_p^\times$ we have

$$
\begin{aligned}
F[f(x)](t\xi) &= |t|_p^{-1}F\left[f\left(\frac{x}{t}\right)\right](\xi) \\
&= |t|_p^{-1}\pi_\alpha(t^{-1})F[f(x)](\xi) + |t|_p^{-1}\pi_\alpha(t^{-1})\log_p|t^{-1}|_p F[f_0(x)](\xi) \\
&= |t|_p^{-\alpha}\pi_1^{-1}(t)F[f(x)](\xi) - |t|_p^{-\alpha}\pi_1^{-1}(t)\log_p|t|_p F[f_0(x)](\xi),
\end{aligned}
$$

where f_0 is a homogeneous distribution of degree $\pi_\alpha(x)$. From Theorem 6.2.2, $F[f_0(x)](\xi) = C_0|\xi|_p^{-\alpha}\pi_1^{-1}(\xi)$ is a homogeneous distribution of degree $|\xi|_p^{-\alpha}\pi_1^{-1}(\xi)$, where C_0 is a constant.

Thus for $t \in \mathbb{Q}_p^\times$

$$
\begin{aligned}
&F[f(x)](t\xi) \\
&= |t|_p^{-\alpha}\pi_1^{-1}(t)F[f(x)](\xi) - |t|_p^{-\alpha}\pi_1^{-1}(t)\log_p|t|_p C_0|\xi|_p^{-\alpha}\pi_1^{-1}(\xi),
\end{aligned}
$$

i.e., according to Definition 6.2 (a), the distribution $F[f(x)](\xi)$ is a QAHD of degree $|\xi|_p^{-\alpha}\pi_1^{-1}(\xi)$ and order 1.

Let f be a QAHD of degree $\pi_\alpha(x)$ and order m, $m = 2, 3, \ldots$. By using (4.9.4) and Definition 6.2 (a), for all $t \in \mathbb{Q}_p^\times$ we have

$$
\begin{aligned}
F[f(x)](t\xi) &= |t|_p^{-1}F\left[f\left(\frac{x}{t}\right)\right](\xi) \\
&= |t|_p^{-1}\pi_\alpha(t^{-1})F[f(x)](\xi) + \sum_{j=1}^{m}|t|_p^{-1}\pi_\alpha(t^{-1})\log_p^j|t^{-1}|_p F[f_{m-j}(x)](\xi) \\
&= |t|_p^{-\alpha}\pi_1^{-1}(t)F[f(x)](\xi) + \sum_{j=1}^{m}|t|_p^{-\alpha}\pi_1^{-1}(t)\log_p^j|t|_p(-1)^j F[f_{m-j}(x)](\xi),
\end{aligned}
$$

where $f_{m-j}(x)$ is a QAHD of degree $\pi_\alpha(x)$ and order $m - j$, $j = 1, 2, \ldots, m$.

Suppose that the theorem holds for a QAHD of degree $\pi_\alpha(x)$ and order $k = 1, 2 \ldots, m - 1$. Then, by induction the last relation implies that $F[f](\xi)$ is a QAHD of degree $|\xi|_p^{-\alpha}\pi_1^{-1}(\xi)$ and order m.

The theorem is thus proved.

Taking into account Theorem 6.3.3 and Remark 6.1, one can prove this theorem directly, by calculating the Fourier transform of the distributions $\pi_\alpha(x) \log_p^m |x|_p$, $P\left(\frac{\log_p^m |x|_p}{|x|_p}\right)$. $\qquad\qquad\qquad\qquad\qquad\qquad\qquad\square$

Example 6.4.1. According to [241, IX, (2.8)],

$$F\left[P\left(\frac{1}{|x|_p}\right)\right](\xi) = \frac{1-p}{p} \log_p |\xi|_p - \frac{1}{p}.$$

Here $P\left(\frac{1}{|x|_p}\right)$ is a QAHD of degree $\pi_0(x) = |x|_p^{-1}$ and order 1, and $\log |\xi|_p$ is a QAHD of degree 0 and order 1.

6.5. New type of p-adic Γ-functions

According to Remark 6.1, and Theorems 6.3.3 and 6.4.1, for the QAHD $\pi_\alpha(x) \log_p^m |x|_p$, $\pi_\alpha(x) \neq \pi_0(x) = |x|_p^{-1}$, $m \in \mathbb{N}$, we have

$$F\left[\pi_\alpha(x) \log_p^m |x|_p\right](\xi) = \sum_{k=0}^m A_{m-k} |\xi|_p^{-\alpha} \pi_1^{-1}(\xi) \log_p^{m-k} |\xi|_p, \qquad (6.5.1)$$

and for the QAHD $P\left(|x|_p^{-1} \log_p^{m-1} |x|_p\right)$, $\pi_\alpha(x) = \pi_0(x) = |x|_p^{-1}$, $m \in \mathbb{N}$, we have

$$F\left[P\left(|x|_p^{-1} \log_p^{m-1} |x|_p\right)\right](\xi) = \sum_{k=0}^m B_{m-k} \log_p^{m-k} |\xi|_p, \qquad (6.5.2)$$

where A_k, B_k are constants, $k = 1, \ldots, m$.

By analogy with (A.7.5) and (A.7.6), the coefficients A_k and B_k from relations (6.5.1) and (6.5.2) we will call Γ-functions of the types $(k; m, \pi_\alpha)$ and $(k; m, \pi_0)$, respectively, where $\pi_\alpha(x) \neq \pi_0(x) = |x|_p^{-1}$, $m \in \mathbb{N}$, $k = 0, 1, \ldots, m$. We denote A_k by $\Gamma_k(\pi_\alpha; m)$, and B_k by $\Gamma_k(\pi_0; m)$.

By setting $\xi = 1$ in (6.5.1) and using (2.3.13), we obtain

$$F\left[\pi_\alpha(x) \log_p^m |x|_p\right](1) = A_0 = \Gamma_0(\pi_\alpha; m).$$

Using (6.2.7), we obtain

$$\Gamma_0(\pi_\alpha; m) \stackrel{def}{=} \int_{\mathbb{Q}_p} |x|_p^{\alpha-1} \pi_1(x) \log_p^m |x|_p \chi_p(x)\, dx$$

$$= \frac{d^m \Gamma_p(\pi_\alpha)}{d\alpha^m} \log_p^m e. \qquad (6.5.3)$$

Here the integrals, just as above, are defined by means of an analytic continuation with respect to the parameter α, or as improper integrals.

Next, by successively substituting $p, p^2, \ldots, p^m$ into (6.5.1), we obtain a linear system of equations for $A_1, \ldots, A_m$. Solving this system, we can calculate the Γ-functions $\Gamma_k(\pi_\alpha; m)$, $k = 1, \ldots, m-1$.

By setting $\xi = 1$ in (6.5.2), we obtain

$$F\big[P\big(|x|_p^{-1} \log_p^{m-1} |x|_p\big)\big](1) = B_0 = \Gamma_0(\pi_0; m-1).$$

Here the integral is defined as an improper integral. Thus, by definitions (6.3.4) and (6.3.5),

$$
\begin{aligned}
\Gamma_0(\pi_0; m-1) &= \int_{\mathbb{Q}_p} P\Big(\frac{\log_p^{m-1} |x|_p}{|x|_p}\Big) \chi_p(x)\,dx \\
&\stackrel{def}{=} \lim_{k \to \infty} \int_{p^{-k} \le |x|_p \le p^k} P\Big(\frac{\log_p^{m-1} |x|_p}{|x|_p}\Big) \chi_p(x)\,dx \\
&= \lim_{k \to \infty} \Big(\int_{p^{-k} \le |x|_p \le 1} \frac{\log_p^{m-1} |x|_p}{|x|_p} \big(\chi_p(x) - 1\big)\,dx \\
&\quad + \int_{p \le |x|_p \le p^k} \frac{\log_p^{m-1} |x|_p}{|x|_p} \chi_p(x)\,dx \Big).
\end{aligned}
$$

(6.5.4)

Since $\chi_p(x) = 1$ for $|x|_p \le 1$, we see that the first integral in (6.5.4) is equal to zero. Let us calculate the second integral in (6.5.4), $\int_{p \le |x|_p \le p^k}$. According to (3.3.1), (3.2.6) and (3.3.4), we have

$$\int_{p \le |x|_p \le p^k} \frac{\log_p^{m-1} |x|_p}{|x|_p} \chi_p(x)\,dx = \int_{S_1} \frac{\log_p^{m-1} |x|_p}{|x|_p} \chi_p(x)\,dx = -\frac{1}{p}.$$

Thus

$$\Gamma_0(\pi_0; m-1) = -\frac{1}{p}. \tag{6.5.5}$$

Let us calculate $\Gamma_1(\pi_0; 0)$. According to (6.5.2) and (6.5.5), we have

$$F\big[P\big(|x|_p^{-1}\big)\big](\xi) = B_1 \log_p |\xi|_p - \frac{1}{p}.$$

By substituting p into the latter relation, we have

$$F\big[P\big(|x|_p^{-1}\big)\big](p) = -B_1 - \frac{1}{p}.$$

Just as above,

$$\Gamma_1(\pi_0; 0) = B_1 = -\int_{\mathbb{Q}_p} P\left(\frac{1}{|x|_p}\right)\chi_p(px)\,dx - \frac{1}{p}$$

$$= -\lim_{k\to\infty}\left(\int_{p^{-k}\le|x|_p\le 1}\frac{1}{|x|_p}(\chi_p(px)-1)\,dx\right.$$

$$+ \int_{p\le|x|_p\le p^k}\frac{1}{|x|_p}\chi_p(px)\,dx - \frac{1}{p}. \tag{6.5.6}$$

Since according to (3.3.1), (3.2.6) and (3.3.4),

$$\int_{S_\gamma}\chi_p(p\xi)\,d\xi = \begin{cases} p^\gamma\left(1-p^{-1}\right), & \gamma \le 1, \\ -p, & \gamma = 2, \\ 0, & \gamma \ge 3, \end{cases}$$

and $\chi_p(x) = 1$ for $|x|_p \le 1$, relation (6.5.6) implies that

$$\Gamma_1(\pi_0; 0) = -p\left(1-p^{-1}\right)p^{-1} + pp^{-2} - \frac{1}{p} = \frac{1}{p} - 1. \tag{6.5.7}$$

Relations (6.5.5) and (6.5.7) are in accordance with Example 6.4.1, since

$$\Gamma_0(\pi_0; 0) = F\left[P\left(\frac{1}{|x|_p}\right)\right](1) = -\frac{1}{p}.$$

$$\Gamma_1(\pi_0; 0) = F\left[P\left(\frac{1}{|x|_p}\right)\right](1) = \frac{1}{p} - 1.$$

By successively substituting $p, p^2, \ldots, p^{m-1}$ into (6.5.2), we obtain a linear system of equations for $B_1, \ldots, B_m$. Solving this system, we can calculate the Γ-functions $\Gamma_k(\pi_0; m)$, $k = 1, \ldots, m-1$.

7

p-adic Lizorkin spaces of test functions and distributions

7.1. Introduction

It is well known that fractional operators play a key role in applications of *p*-adic analysis. However, the Bruhat–Schwartz space of distributions $\mathcal{D}'(\mathbb{Q}_p^n)$ is not invariant under fractional operators (see Chapter 9, Section 9.1). To deal with fractional operators we need "natural" definition domains for them. Similar problems also arise for the usual fractional operators on functions from $\mathbb{R}^n$ into $\mathbb{C}$ ([206], [210], [211]): in general, the space of Schwartzian test function $\mathcal{S}(\mathbb{R}^n)$ *is not invariant* under fractional operators. A solution of this problem (in the real case) was suggested by P. I. Lizorkin in the excellent papers [180]–[182], where a new type of spaces *invariant* under fractional operators was introduced.

In this chapter using classical Lizorkin's ideas, we introduce and study a *p*-adic analog of the well-known Lizorkin spaces. These results are based on [21] and [22].

At the beginning, in Section 7.2, we recall some facts about the well-known Lizorkin spaces of test functions and distributions on $\mathbb{R}^n$. These spaces were introduced by P. I. Lizorkin in the papers [180]–[182] (see also [208], [209]). One of the their important properties is the following: they are *invariant* under fractional operators.

The aim of Section 7.3 is to introduce and study the *p*-adic Lizorkin spaces of test functions and distributions. In Section 7.3.1 we introduce the spaces of test functions $\Phi_\times(\mathbb{Q}_p^n)$ and distributions $\Phi'_\times(\mathbb{Q}_p^n)$ of the first kind, and in Section 7.3.2 we introduce the spaces of test functions $\Phi(\mathbb{Q}_p^n)$ and distributions $\Phi'(\mathbb{Q}_p^n)$ of the second kind. The Lizorkin spaces $\Phi_\times(\mathbb{Q}_p^n)$ and $\Phi(\mathbb{Q}_p^n)$ admit the characterizations (7.3.1) and (7.3.3), respectively.

In Section 7.4, by Theorems 7.4.3, 7.4.4, we prove that the Lizorkin spaces of test functions $\Phi_\times(\mathbb{Q}_p^n)$ and $\Phi(\mathbb{Q}_p^n)$ are dense in $\mathcal{L}^\rho(\mathbb{Q}_p^n)$, $1 < \rho < \infty$. In fact, for $n = 1$ and $\rho = 2$ this statement was already proved in [241, IX.4.]. We point out that for $\rho = 2$ the statements of Theorems 7.4.3 and 7.4.4 are almost obvious, but for $\rho \neq 2$, as for the real case [210], these statements are nontrivial. Our proofs of these theorems follow closely the proofs developed for the real case in [209] and [210].

In fact, the one-dimensional Lizorkin space of test functions $\Phi(\mathbb{Q}_p)$ was introduced in [241, IX.2]. More exactly, in order to define the one-dimensional fractional Vladimirov operator D^{-1} in [241, IX.2] the subspace of test functions ϕ in $\mathcal{D}(\mathbb{Q}_p)$ which are such that $\int_{\mathbb{Q}_p} \varphi(x)\,dx = 0$ was constructed (cf. (7.3.3)). A similar idea was used in [96, Ch.II, §3.8, (3)]. More recently, in [25], a Lizorkin type space was introduced for the case of ultrametric spaces.

The Lizorkin spaces are useful for some applications in p-adic analysis. It turns out that the p-adic wavelets constructed in Chapter 8 belong to the Lizorkin spaces of test functions $\Phi(\mathbb{Q}_p^n)$. Moreover, there are characterizations of the p-adic Lizorkin spaces $\Phi(\mathbb{Q}_p^n)$ and $\Phi'(\mathbb{Q}_p^n)$ in the terms of the wavelet functions (see Lemma 4.3.1 and Proposition 8.14.3). Next, in Chapters 9 and 10, the Lizorkin spaces play a key role in the theory of pseudo-differential operators. The point is that the Lizorkin spaces are invariant under these p-adic operators, i.e., these spaces are "natural" definition domains for them. These two facts are used in Chapter 10 to solve pseudo-differential equations in Lizorkin spaces. The Lizorkin spaces are also used in Chapters 12 and 14. Note that in [159] the Lizorkin space is used to solve the Cauchy problem for a p-adic wave equation.

7.2. The real case of Lizorkin spaces

For $\gamma = (\gamma_1, \dots, \gamma_n) \in \mathbb{N}_0^n$ and $x = (x_1, \dots, x_n) \in \mathbb{R}^n$ we set $|\gamma| = \sum_{k=1}^n \gamma_k$ and $x^\gamma = x_1^{\gamma_1} \cdots x_n^{\gamma_n}$. We shall denote partial derivatives of the order $|\gamma|$ by $\partial_x^\gamma = \frac{\partial^{|\gamma|}}{\partial x_1^{\gamma_1} \cdots \partial x_n^{\gamma_n}}$.

Now we recall the definition of one type of Lizorkin spaces for the usual (real) functions (for details, see [182], [210], [211]). Let us consider the following subspace of test functions:

$$\Psi(\mathbb{R}^n) = \{\psi(\xi) : \psi \in \mathcal{S}(\mathbb{R}^n) : (\partial_\xi^\gamma \psi)(0) = 0, \ |\gamma| = 1, 2, \dots \}. \quad (7.2.1)$$

The space of functions

$$\Phi(\mathbb{R}^n) = \{\phi : \phi = F[\psi], \ \psi \in \Psi(\mathbb{R}^n)\} \subset \mathcal{S}(\mathbb{R}^n), \quad (7.2.2)$$

where F is the Fourier transform, is called the *Lizorkin space*.

This space admits a simple characterization: $\phi \in \Phi(\mathbb{R}^n)$ if and only if $\phi \in \mathcal{S}(\mathbb{R}^n)$ and

$$\int_{\mathbb{R}^n} x^\gamma \phi(x)\, d^n x = 0, \quad |\gamma| = 0, 1, 2, \ldots. \tag{7.2.3}$$

Thus $\Phi(\mathbb{R}^n)$ is the subspace of Schwartzian test functions, for which all the moments are equal to zero.

The Lizorkin space has many "useful" properties. One of them is the following: the Lizorkin space $\Phi(\mathbb{R}^n)$ is invariant under the Riesz fractional operator D^α, $\alpha \in \mathbb{C}$, given by the formula

$$\left(D^\alpha \phi\right)(x) \overset{def}{=} (-\Delta)^{\alpha/2} \phi(x) = \left(\kappa_{-\alpha} * \phi\right)(x), \quad \phi \in \Phi(\mathbb{R}^n), \tag{7.2.4}$$

where the *Riesz kernel* is defined as

$$\kappa_\alpha(x) = \frac{\Gamma(\frac{n-\alpha}{2})}{2^\alpha \pi^{\frac{n}{2}} \Gamma(\frac{\alpha}{2})} |x|^{\alpha-n},$$

where $|x| = \sqrt{x_1^2 + \cdots + x_n^2}$, and $|x|^\alpha \in \mathcal{S}'(\mathbb{R}^n)$ is a distribution, and Δ is the Laplacian on $\mathbb{R}^n$ (see [210, Lemma 2.9.], [211, (25.19)]). This property shows that the Lizorkin type spaces are "natural" definition domains for fractional operators.

Moreover, these spaces are "natural" definition domains for some types of pseudo-differential operators and are used in applications [210] and [211].

7.3. *p*-adic Lizorkin spaces

7.3.1. Lizorkin space of the first kind

Consider the subspace of the space of test functions $\mathcal{D}(\mathbb{Q}_p^n)$ given by

$$\Psi_\times(\mathbb{Q}_p^n) = \big\{\psi(\xi) \in \mathcal{D}(\mathbb{Q}_p^n) :$$
$$\psi(\xi_1, \ldots, \xi_{j-1}, 0, \xi_{j+1}, \ldots, \xi_n) = 0,\ j = 1, 2, \ldots, n\big\}.$$

Obviously, $\Psi_\times(\mathbb{Q}_p^n) \neq \emptyset$.

Definition 7.1. The space

$$\Phi_\times = \Phi_\times(\mathbb{Q}_p^n) = \big\{\phi : \phi = F[\psi],\ \psi \in \Psi_\times(\mathbb{Q}_p^n)\big\}$$

is called the *p*-adic *Lizorkin space of test functions of the first kind*.

Since the Fourier transform is a linear isomorphism $\mathcal{D}(\mathbb{Q}_p^n)$ into $\mathcal{D}(\mathbb{Q}_p^n)$, we have $\Phi_\times(\mathbb{Q}_p^n) \neq \emptyset$, $\Phi_\times(\mathbb{Q}_p^n) \subset \mathcal{D}(\mathbb{Q}_p^n)$.

By analogy with the real case ([210, 2.2.], [211, §25.1.]), $\Phi_\times(\mathbb{Q}_p^n)$ can be equipped with the topology of the space $\mathcal{D}(\mathbb{Q}_p^n)$ which makes $\Phi_\times(\mathbb{Q}_p^n)$ a complete space.

The Lizorkin space $\Phi_\times$ admits the following characterization.

Lemma 7.3.1. $\phi \in \Phi_\times(\mathbb{Q}_p^n)$ *if and only if* $\phi \in \mathcal{D}(\mathbb{Q}_p^n)$ *and*

$$\int_{\mathbb{Q}_p} \phi(x_1, \ldots, x_{j-1}, x_j, x_{j+1}, \ldots, x_n) \, dx_j = 0, \qquad (7.3.1)$$

where $(x_1, \ldots, x_{j-1}, x_{j+1}, \ldots, x_n) \in \mathbb{Q}_p^{n-1}$, $j = 1, 2, \ldots, n$.

The space $\Phi'_\times(\mathbb{Q}_p^n)$ is called the *p-adic Lizorkin space of distributions of the first kind.*

Let $\Psi_\times^\perp(\mathbb{Q}_p^n) = \{ f \in \mathcal{D}'(\mathbb{Q}_p^n) : \langle f, \psi \rangle = 0, \forall \psi \in \Psi_\times(\mathbb{Q}_p^n) \}$, i.e., let $\Psi_\times^\perp(\mathbb{Q}_p^n)$ be the set of functionals from $\mathcal{D}'(\mathbb{Q}_p^n)$ concentrated on the set $\cup_{j=1}^n \{ x \in \mathbb{Q}_p^n : x_j = 0 \}$. Let $\Phi_\times^\perp(\mathbb{Q}_p^n) = \{ f \in \mathcal{D}'(\mathbb{Q}_p^n) : \langle f, \phi \rangle = 0, \forall \phi \in \Phi_\times(\mathbb{Q}_p^n) \}$. Thus $\Phi_\times^\perp(\mathbb{Q}_p^n)$ and $\Psi_\times^\perp(\mathbb{Q}_p^n)$ are subspaces of functionals in $\mathcal{D}'(\mathbb{Q}_p^n)$ orthogonal to $\Phi_\times(\mathbb{Q}_p^n)$ and $\Psi_\times(\mathbb{Q}_p^n)$, respectively. It is clear that the set $\Psi_\times^\perp(\mathbb{Q}_p^n)$ consists of linear combinations of functionals of the form $f(\xi_1, \ldots, \widehat{\xi}_j, \ldots, \xi_n)$, $j = 1, 2, \ldots, n$, where the hat $\widehat{}$ over ξ_j denotes deletion of the corresponding variable from the vector $\xi = (\xi_1, \ldots, \xi_n)$. The set $\Phi_\times^\perp(\mathbb{Q}_p^n)$ consists of linear combinations of functionals of the form $g(x_1, \ldots, \widehat{x}_j, \ldots, x_n) \times \delta(x_j)$, $j = 1, 2, \ldots, n$.

Theorem 7.3.2. *The spaces of linear and continuous functionals* $\Phi'_\times(\mathbb{Q}_p^n)$ *and* $\Psi'_\times(\mathbb{Q}_p^n)$ *can be identified with the quotient spaces*

$$\Phi'_\times(\mathbb{Q}_p^n) = \mathcal{D}'(\mathbb{Q}_p^n) / \Phi_\times^\perp(\mathbb{Q}_p^n), \qquad \Psi'_\times(\mathbb{Q}_p^n) = \mathcal{D}'(\mathbb{Q}_p^n) / \Psi_\times^\perp(\mathbb{Q}_p^n)$$

modulo the subspaces $\Phi_\times^\perp(\mathbb{Q}_p^n)$ *and* $\Psi_\times^\perp(\mathbb{Q}_p^n)$, *respectively.*

Proof. The statement follows from the well-known assertion (for example, see [213, Ch.IV, 4.1, Corollary 1.]): if E is a topological vector space with a closed subspace M and E' is a topological dual of E, then the space M' can be identified with the quotient space $M' = E'/M^\perp$, where $M^\perp = \{ f \in E' : \langle f, \varphi \rangle = 0, \forall \varphi \in M \}$ is a subspace of all functionals in E' orthogonal to M. □

It is natural to define the Fourier transform of distributions $f \in \Phi'_\times(\mathbb{Q}_p^n)$ and $g \in \Psi'_\times(\mathbb{Q}_p^n)$ by the relations

$$\langle F[f], \psi \rangle = \langle f, F[\psi] \rangle, \quad \forall \psi \in \Psi_\times(\mathbb{Q}_p^n),$$

$$\langle F[g], \phi \rangle = \langle g, F[\phi] \rangle, \quad \forall \phi \in \Phi_\times(\mathbb{Q}_p^n). \qquad (7.3.2)$$

By definition, $F[\Phi_\times(\mathbb{Q}_p^n)] = \Psi_\times(\mathbb{Q}_p^n)$, $F[\Psi_\times(\mathbb{Q}_p^n)] = \Phi_\times(\mathbb{Q}_p^n)$, i.e., (7.3.2) gives well-defined objects. Moreover, we have

$$F[\Phi'_\times(\mathbb{Q}_p^n)] = \Psi'_\times(\mathbb{Q}_p^n), \qquad F[\Psi'_\times(\mathbb{Q}_p^n)] = \Phi'_\times(\mathbb{Q}_p^n).$$

7.3.2. Lizorkin space of the second kind

Now we consider the space

$$\Psi(\mathbb{Q}_p^n) = \left\{ \psi(\xi) \in \mathcal{D}(\mathbb{Q}_p^n) : \psi(0) = 0 \right\}.$$

It is clear that $\Psi(\mathbb{Q}_p^n) \neq \emptyset$.

Definition 7.2. The space

$$\Phi(\mathbb{Q}_p^n) = \left\{ \phi : \phi = F[\psi], \ \psi \in \Psi(\mathbb{Q}_p^n) \right\}$$

is called the *p*-adic *Lizorkin space of test functions of the second kind*.

We have $\Phi(\mathbb{Q}_p^n) \neq \emptyset$, $\Phi(\mathbb{Q}_p^n) \subset \mathcal{D}(\mathbb{Q}_p^n)$. Similarly to $\Phi_\times(\mathbb{Q}_p^n)$, the space $\Phi(\mathbb{Q}_p^n)$ can be equipped with the topology of the space $\mathcal{D}(\mathbb{Q}_p^n)$ which makes $\Phi(\mathbb{Q}_p^n)$ a complete space.

It is easy to see that the *p*-adic Lizorkin space $\Phi(\mathbb{Q}_p^n)$ is an analog of the Lizorkin space $\Phi(\mathbb{R}^n)$ defined by (7.2.2).

Since the Fourier transform is an automorphism of the linear space $\mathcal{D}(\mathbb{Q}_p^n)$, from (4.8.5) the space Φ admits the following characterization.

Lemma 7.3.3. (a) $\phi \in \Phi(\mathbb{Q}_p^n)$ *iff* $\phi \in \mathcal{D}(\mathbb{Q}_p^n)$ *and*

$$\int_{\mathbb{Q}_p^n} \phi(x)\, d^n x = 0. \tag{7.3.3}$$

(b) $\phi \in \mathcal{D}_N^l(\mathbb{Q}_p^n) \cap \Phi(\mathbb{Q}_p^n)$, *i.e.,* $\int_{B_N^n} \phi(x)\, d^n x = 0$, *iff* $\psi = F^{-1}[\phi] \in \mathcal{D}_{-l}^{-N}(\mathbb{Q}_p^n) \cap \Psi(\mathbb{Q}_p^n)$, *i.e.,* $\psi(\xi) = 0$ *for* $\xi \notin B_{-l}^n \setminus B_{-N}^n$.

In fact, for $n = 1$, this lemma was proved in [241, IX.2.]. Unlike the real case (7.2.1), (7.2.2), any function $\psi(\xi) \in \Phi$ is equal to zero not only at $\xi = 0$ but in a ball $B^n \ni 0$ as well.

It follows from (7.3.3) that the space $\Phi(\mathbb{Q}_p^n)$ does not contain real-valued functions which are everywhere different from zero.

Let $\Phi' = \Phi'(\mathbb{Q}_p^n)$ denote the topological dual of the space $\Phi(\mathbb{Q}_p^n)$. We call it the *p*-adic *Lizorkin space of distributions of the second kind*.

By $\Psi^\perp(\mathbb{Q}_p^n)$ and $\Phi^\perp(\mathbb{Q}_p^n)$ we denote the subspaces of functionals in $\mathcal{D}'(\mathbb{Q}_p^n)$ orthogonal to $\Psi(\mathbb{Q}_p^n)$ and $\Phi(\mathbb{Q}_p^n)$, respectively. Thus

$$\Psi^\perp(\mathbb{Q}_p^n) = \{ f \in \mathcal{D}'(\mathbb{Q}_p^n) : f = C\delta, \ C \in \mathbb{C} \}$$

and

$$\Phi^\perp(\mathbb{Q}_p^n) = \{f \in \mathcal{D}'(\mathbb{Q}_p^n) : f = C, \ C \in \mathbb{C}\}.$$

Theorem 7.3.4.

$$\Phi'(\mathbb{Q}_p^n) = \mathcal{D}'(\mathbb{Q}_p^n)/\Phi^\perp(\mathbb{Q}_p^n), \qquad \Psi'(\mathbb{Q}_p^n) = \mathcal{D}'(\mathbb{Q}_p^n)/\Psi^\perp(\mathbb{Q}_p^n).$$

This assertion is proved in the same way as Theorem 7.3.2.

The space $\Phi'(\mathbb{Q}_p^n)$ can be obtained from $\mathcal{D}'(\mathbb{Q}_p^n)$ by "sifting out" constants. Thus two distributions in $\mathcal{D}'(\mathbb{Q}_p^n)$ differing by a constant are indistinguishable as elements of $\Phi'(\mathbb{Q}_p^n)$.

We define the Fourier transform of distributions $f \in \Phi'(\mathbb{Q}_p^n)$ and $g \in \Psi'(\mathbb{Q}_p^n)$ by an analog of formula (7.3.2):

$$\begin{aligned}
\langle F[f], \psi \rangle = \langle f, F[\psi] \rangle, && \forall \psi \in \Psi(\mathbb{Q}_p^n), \\
\langle F[g], \phi \rangle = \langle g, F[\phi] \rangle, && \forall \phi \in \Phi(\mathbb{Q}_p^n).
\end{aligned} \tag{7.3.4}$$

It is clear that

$$F[\Phi'(\mathbb{Q}_p^n)] = \Psi'(\mathbb{Q}_p^n), \qquad F[\Psi'(\mathbb{Q}_p^n)] = \Phi'(\mathbb{Q}_p^n).$$

7.4. Density of the Lizorkin spaces of test functions in $\mathcal{L}^\rho(\mathbb{Q}_p^n)$

The proofs of the assertions for the p-adic case follow closely the proofs developed for the real case in [209] and [210, 2.2, 2.4.].

Lemma 7.4.1. *Let* $g(\cdot) \in \mathcal{L}^1(\mathbb{Q}_p^n)$ *and* $f(\cdot) \in \mathcal{L}^\rho(\mathbb{Q}_p^n)$, $1 < \rho < \infty$. *Then for* $x \in \mathbb{Q}_p^n$

$$\begin{aligned}
h_t(x) &= \int_{\mathbb{Q}_p^n} g(y) f(x - ty) \, d^n y \\
&= \frac{1}{|t|_p^n} \int_{\mathbb{Q}_p^n} g\left(\frac{\xi}{t}\right) f(x - \xi) \, d^n \xi \xrightarrow{\mathcal{L}^\rho} 0, \quad |t|_p \to \infty. \quad t \in \mathbb{Q}_p^\times, \quad (7.4.1)
\end{aligned}$$

Proof. For the case $\rho = 2$, taking into account the Parseval–Steklov equality (4.8.6), formula (4.9.3), and using the Riemann–Lebesgue Theorem 5.2.1, we have

$$\begin{aligned}
\|h_t\|_2 &= \|F[h_t]\|_2 \\
&= \left(\int_{\mathbb{Q}_p^n} \left| F[g](ty) \, F[f](y) \right|^2 d^n y \right)^{\frac{1}{2}} \to 0, \quad |t|_p \to \infty. \quad (7.4.2)
\end{aligned}$$

Here the passage to the limit under the integral sign is justified by the Lebesgue-dominated convergence Theorem 3.2.4.

Let now $\rho \neq 2$. In view of the Young inequality (5.2.3), we have $h_t \in \mathcal{L}^\rho(\mathbb{Q}_p^n)$ and

$$\|h_t\|_\rho \leq \|g\|_1 \|f\|_\rho, \tag{7.4.3}$$

where the last estimate is uniform.

Clearly, it is sufficient to prove (7.4.1) for $f \in \mathcal{D}(\mathbb{Q}_p^n)$. Let $r > 1$ be such that ρ is located between 2 and r. Using the Hölder inequality and taking into account that $f \in \mathcal{D}(\mathbb{Q}_p^n)$, we obtain

$$\|h_t\|_\rho \leq \|h_t\|_r^{1-\lambda} \|h_t\|_2^\lambda, \tag{7.4.4}$$

where $\frac{1}{\rho} = \frac{1-\lambda}{r} + \frac{\lambda}{2}$ (i.e., $\lambda = \frac{2(\rho-r)}{\rho(2-r)}$). Since the lemma holds for $\rho = 2$, i.e., $\|h_t\|_2 \to 0$, $|t|_p \to \infty$, by (7.4.2) and (7.4.3), we have

$$\|h_t\|_\rho \leq \left(\|g\|_1 \|f\|_r\right)^{1-\lambda} \|h_t\|_2^\lambda \to 0, \quad |t|_p \to \infty, \quad t \in \mathbb{Q}_p^\times.$$

The lemma is thus proved. $\qquad\qquad\qquad\qquad\qquad\qquad\qquad\qquad\square$

Lemma 7.4.2. *Let* $g(\cdot) \in \mathcal{L}^1(\mathbb{Q}_p^{n-m})$, $m \leq n - 1$ *and* $f(\cdot) \in \mathcal{L}^\rho(\mathbb{Q}_p^n)$, $1 < \rho < \infty$. *Then*

$$h_t(x) = \int_{\mathbb{Q}_p^{n-m}} g(y) f(x', x'' - ty) \, d^{n-m}y \xrightarrow{\mathcal{L}^\rho} 0, \quad |t|_p \to \infty, \quad t \in \mathbb{Q}_p^\times, \tag{7.4.5}$$

where $x' = (x_1, \ldots, x_m) \in \mathbb{Q}_p^m$, $x'' = (x_{m+1}, \ldots, x_n) \in \mathbb{Q}_p^{n-m}$, $1 \leq m \leq n - 1$.

Proof. If $\rho = 2$, just as above, using the Parseval–Steklov equality from [241, VII, (4.4)], and formula (4.9.3), we have

$$\|h_t\|_2 = \|F[h_t]\|_2 = \left(\int_{\mathbb{Q}_p^n} \left|F[g](ty'') F[f](y)\right|^2 d^n y\right)^{\frac{1}{2}} \to 0, \tag{7.4.6}$$

as $|t|_p \to \infty$.

Let now $\rho \neq 2$. In view of the Young inequality, we have the uniform estimate

$$\|h_t\|_{\mathcal{L}^\rho(\mathbb{Q}_p^n)} \leq \|g\|_{\mathcal{L}^1(\mathbb{Q}_p^{n-m})} \|f\|_{\mathcal{L}^\rho(\mathbb{Q}_p^n)}. \tag{7.4.7}$$

Let $r > 1$ be such that ρ is located between 2 and r. Setting $f \in \mathcal{D}(\mathbb{Q}_p^n)$, using inequality (7.4.4), and taking into account that $\|h_t\|_2 \to 0$, $|t|_p \to \infty$, we obtain

$$\|h_t\|_{\mathcal{L}^p(\mathbb{Q}_p^n)} \le \left(\|g\|_{\mathcal{L}^1(\mathbb{Q}_p^{n-m})} \|f\|_{\mathcal{L}^r(\mathbb{Q}_p^n)}\right)^{1-\lambda} \|h_t\|_{\mathcal{L}^2(\mathbb{Q}_p^n)}^{\lambda} \to 0, \quad |t|_p \to \infty.$$

The lemma is thus proved. □

Theorem 7.4.3. *The space* $\Phi(\mathbb{Q}_p^n)$ *is dense in* $\mathcal{L}^p(\mathbb{Q}_p^n)$, $1 < \rho < \infty$.

Proof. Since $\mathcal{D}(\mathbb{Q}_p^n)$ is dense in $\mathcal{L}^p(\mathbb{Q}_p^n)$, $1 < \rho < \infty$ (see [241, VI.2.]), it is sufficient to approximate the function $\varphi \in \mathcal{D}(\mathbb{Q}_p^n)$ by functions $\phi_t \in \Phi(\mathbb{Q}_p^n)$ in the norm of $\mathcal{L}^p(\mathbb{Q}_p^n)$.

Consider the family of functions

$$\psi_t(\xi) = (1 - \Delta_t(\xi))F^{-1}[\varphi](\xi) \in \Psi(\mathbb{Q}_p^n),$$

where $\Delta_t(\xi) = \Omega(|t\xi|_p)$ is the characteristic function of the ball $B^n_{\log_p |t|_p^{-1}}$, $x \in \mathbb{Q}_p^n$, $t \in \mathbb{Q}_p^{\times}$, the function Ω being defined by (4.2.2). From (4.9.8), we have

$$\phi_t(x) = F[\psi_t](x) = F[\left(1 - \Delta_t(\xi)\right)](x) * \varphi(x)$$
$$= \delta(x) * \varphi(x) - F[\Delta_t(\xi)](x) * \varphi(x) \in \Phi(\mathbb{Q}_p^n).$$

According to (4.8.10), $F[\Delta_t(\xi)](x) = \frac{1}{|t|_p^n}\Omega\left(\frac{|x|_p}{|t|_p}\right)$, i.e., the latter relation can be rewritten as follows:

$$\phi_t(x) = \varphi(x) - \int_{\mathbb{Q}_p^n} \Omega(|y|_p)\varphi(x - ty)d^n y.$$

Applying Lemma 7.4.1 to the latter relation, we see that $\|\phi_t - \varphi\|_\rho \to 0$ as $|t|_p \to \infty$. □

Theorem 7.4.4. *The space* $\Phi_{\times}(\mathbb{Q}_p^n)$ *is dense in* $\mathcal{L}^p(\mathbb{Q}_p^n)$, $1 < \rho < \infty$.

Proof. The proof of this lemma is based on calculations similar to those carried out above. In this case we set $\varphi \in \mathcal{D}(\mathbb{Q}_p^n)$ and

$$\psi_t(\xi) = (1 - \Delta_t(\xi_1)\cdots - \Delta_t(\xi_n))F^{-1}[\varphi](\xi) \in \Psi_{\times}(\mathbb{Q}_p^n),$$

where $\Delta_t(\xi_j) = \Omega(|t\xi_j|_p)$ is the characteristic function of the ball $B_{\log_p |t|_p^{-1}}$, $x_j \in \mathbb{Q}_p$, $t \in \mathbb{Q}_p^{\times}$, $j = 1, \ldots, n$. By (4.9.8), we obtain

$$\phi_t(x) = \varphi(x) - \left(\delta(x_2, \ldots, x_n) \times F[\Delta_t(\xi_1)](x_1)\right) * \varphi(x)$$
$$\cdots - \left(\delta(x_1, \ldots, x_{n-1}) \times F[\Delta_t(\xi_n)](x_n)\right) * \varphi(x) \in \Phi_{\times}(\mathbb{Q}_p^n).$$

Since $F[\Delta_t(\xi_j)](x_j) = \frac{1}{|t|_p}\Omega\left(\frac{|x_j|_p}{|t|_p}\right)$, $x_j \in \mathbb{Q}_p$, $j = 1, \ldots, n$, the latter relation can be rewritten as

$$\phi_t(x) = \varphi(x) - \int_{\mathbb{Q}_p} \Omega(|y_1|_p)\varphi(x_1 - ty_1, x_2, \ldots, x_n)dy_1$$

$$\cdots - \int_{\mathbb{Q}_p} \Omega(|y_n|_p)\varphi(x_1, \ldots, x_{n-1}, x_n - ty_n)dy_n.$$

According to Lemma 7.4.2,

$$h_{j,t}(x) = \int_{\mathbb{Q}_p} \Omega(|y_j|_p)\varphi(x_1, \ldots, x_{j-1}, x_j - ty_j, x_{j+1}, \ldots, x_n)dy_j \xrightarrow{\mathcal{L}^\rho} 0$$

as $|t|_p \to \infty$, $j = 1, \ldots, n$. Thus $||\phi_t - \varphi||_\rho \le ||h_{1,t}||_\rho + \cdots + ||h_{n,t}||_\rho \to 0$ as $|t|_p \to \infty$. $\square$

For $n = 1$ and $\rho = 2$ the statements of Theorems 7.4.3 and 7.4.4 coincide with the lemma from [241, IX.4.]. In this case the statement is practically obvious, and is proved by using the Parseval–Steklov equality.

8

The theory of p-adic wavelets

8.1. Introduction

Nowadays it is difficult to find an engineering area where wavelets are not applied. The first wavelet basis was introduced by Haar [101] in 1910. In this paper Haar constructed an orthogonal basis for $\mathcal{L}^2(\mathbb{R})$ consisting of the dyadic shifts and scales of one piecewise constant function:

$$\psi_{jn}^H(t) = 2^{-j/2}\psi^H\left(2^{-j}x - n\right), \quad t \in \mathbb{R}, \quad j \in \mathbb{Z}, \quad n \in \mathbb{Z}, \qquad (8.1.1)$$

where

$$\psi^H(t) = \begin{cases} 1, & 0 \le t < \frac{1}{2}, \\ -1, & \frac{1}{2} \le t < 1, \\ 0, & t \notin [0, 1), \end{cases} = \chi_{[0,\frac{1}{2}]}(t) - \chi_{[\frac{1}{2},1]}(t), \quad t \in \mathbb{R}, \qquad (8.1.2)$$

is called the *Haar wavelet function* (whose shifts and scales form the Haar basis (8.1.1)). Here $\chi_A(x)$ is a characteristic function of the set $A \subset \mathbb{R}$.

A lot of mathematicians actively studied the Haar basis (8.1.1), different kinds of generalizations were introduced, but for almost a whole century nobody could find another wavelet function (a function whose shifts and scales form an orthogonal basis). Only in the early 1990s a method for more general construction of wavelet functions appeared. This method is based on the notion of *multiresolution analysis* (MRA below) introduced by Y. Meyer [186] and S. Mallat [184], [185]. Smooth compactly supported wavelet functions were found in this way, which have been very important for various engineering applications. Now the wavelet theory is intensively developed.

Compared with the theory of wavelets for the reals, p-adic wavelets theory is still at an early stage of investigation. In 2002, S. V. Kozyrev [163] found a compactly supported p-adic wavelet basis for $\mathcal{L}^2(\mathbb{Q}_p)$ which is an analog of the real Haar basis:

$$\theta_{k;ja}(x) = p^{-j/2}\chi_p\big(p^{-1}k(p^jx - a)\big)\Omega\big(|p^jx - a|_p\big), \quad x \in \mathbb{Q}_p, \quad (8.1.3)$$

$k \in J_p = \{1, 2, \ldots, p - 1\}, j \in \mathbb{Z}, a \in I_p = \mathbb{Q}_p/\mathbb{Z}_p$. Here shifts and scales of the wavelet function

$$\theta_k(x) = \chi_p\big(p^{-1}kx\big)\Omega\big(|x|_p\big), \quad x \in \mathbb{Q}_p, \quad (8.1.4)$$

form the Kozyrev basis (8.1.3). A multidimensional p-adic basis obtained by direct multiplying out the wavelets (8.1.3) was considered in [21]. The Haar wavelet basis (8.1.3) was extended to the ultrametric spaces by S. V. Kozyrev [164], [165], A. Yu. Khrennikov, and S. V. Kozyrev [131], [132]. The authors of the cited papers have not yet provided a general theory describing common properties of p-adic wavelet bases and giving methods for their systematic finding.

J. J. Benedetto and R. L. Benedetto [50], and R. L. Benedetto [52] suggested a method for finding wavelet bases on the locally compact abelian groups with compact open subgroups. This method is applicable to the p-adic setting. These authors did not however develop the MRA approach; their method is rather based on a *theory of wavelet sets*. In fact their method only allows the construction of wavelet functions whose Fourier transforms are the characteristic functions of some sets (see [50, Proposition 5.1.]). Moreover, they doubted that the development of the MRA approach would be possible: "Kozyrev [163] produced one specific set of wavelet generators, analogous to Haar wavelets on $\mathbb{R}^d$, using a discrete set of translation operators which do not form a group. However, ... those operators allow the Haar wavelets, but they preclude the possibility of general theory of wavelet sets or multiresolution analysis in $\mathcal{L}^2(\mathbb{Q}_p)$" [52], and "there are substantial obstacles to generalizing Kozyrev's method to produce other wavelets, even for $\mathbb{Q}_p$," [50].

In this chapter we develop systematically the p-adic wavelet theory. In spite of the above opinions and arguments [50], [52], we develop the p-adic *MRA theory* in $\mathcal{L}^2(\mathbb{Q}_p)$ and find new p-adic wavelet bases. These results are based on [21], [141], [142], [143], [144], [146], [147], [148] and [225].

To construct a p-adic analog of a classical MRA we need a corresponding p-adic *refinement equation*. In [142] the following conjecture was proposed: the equality

$$\phi(x) = \sum_{r=0}^{p-1} \phi\Big(\frac{1}{p}x - \frac{r}{p}\Big), \quad x \in \mathbb{Q}_p, \quad (8.1.5)$$

may be considered as a *p-adic "natural" refinement equation*. A solution ϕ to this equation (a *refinable function*) is the characteristic function $\Omega(|x|_p)$ of the unit ball.

Equation (8.1.5) reflects a *natural* "self-similarity" of the space $\mathbb{Q}_p$: according to (1.8.5), the unit ball $B_0(0) = \{x : |x|_p \leq 1\}$ is represented as the union of p mutually *disjoint* balls $B_{-1}(r) = \{x : |x - r|_p \leq p^{-1}\}$ of radius p^{-1}:

$$B_0(0) = B_{-1}(0) \cup \left(\cup_{r=1}^{p-1} B_{-1}(r) \right).$$

For $p = 2$ equation (8.1.5) is the 2-*adic refinement equation*

$$\phi(x) = \phi\left(\frac{1}{2}x\right) + \phi\left(\frac{1}{2}x - \frac{1}{2}\right), \quad x \in \mathbb{Q}_2, \tag{8.1.6}$$

which is an analog of the *refinement equation*

$$\phi(t) = \phi(2t) + \phi(2t - 1), \quad t \in \mathbb{R}, \tag{8.1.7}$$

generating the Haar MRA and the Haar wavelet basis (8.1.1) and (8.1.2) in the real case.

A solution of equation (8.1.6) (a 2-adic refinable function) $\phi(x) = \Omega(|x|_2)$ is the characteristic function of the unit ball, whereas a solution of equation (8.1.7) (a real refinable function)

$$\phi^H(t) = \begin{cases} 1, & t \in [0, 1], \\ 0, & t \notin [0, 1], \end{cases} = \chi_{[0,1]}(x), \quad t \in \mathbb{R}, \tag{8.1.8}$$

is the characteristic function of the unit interval $[0, 1]$, an analog of the refinable function.

First, in Section 8.2 (which follows Kozyrev's paper [163]), we consider a connection between the *p*-adic Haar type wavelet basis (8.1.3) in $\mathcal{L}^2(\mathbb{Q}_p)$ and the real Haar wavelet basis (8.1.1) and (8.1.2) in $\mathcal{L}^2(\mathbb{R}_+)$. This connection is established by the Monna map (1.9.3) from $\mathbb{Q}_p$ to $\mathbb{R}_+$.

Next, in Section 8.3, we introduce Definition 8.1 of MRA in $\mathcal{L}^2(\mathbb{Q}_p)$. In Section 8.4, using equation (8.1.5) as the *refinement equation* we construct a concrete *p*-adic MRA which is an analog of Haar MRA in $\mathcal{L}^2(\mathbb{R})$. However, the *p*-adic Haar MRA is not an identical copy of its real analog. In Section 8.5, we prove that, in contrast to the Haar MRA in $\mathcal{L}^2(\mathbb{R})$, in the *p*-adic setting (for $p = 2$) there exist *infinity many different Haar orthogonal bases* for $\mathcal{L}^2(\mathbb{Q}_2)$ generated by the same MRA. By Theorem 8.5.1 explicit formulas for generating wavelet functions are given. In Section 8.6, we realize our scheme for arbitrary *p*: namely, for arbitrary *p* we prove Theorem 8.6.1, which gives a description of all Haar bases generated by wavelet-functions. It turns out that the Kozyrev wavelet basis (8.1.3) (see (8.4.8) and (8.4.6)) is one of our bases, which are given by Theorems 8.5.1 and 8.6.1. This basis can be found also in the framework of the approach [50, Proposition 5.1.]. According to Remarks 8.2 and 8.3, except for some particular cases, our other bases (8.5.4), (8.5.5) and (8.6.2), (8.6.3) *cannot be constructed* by Benedettos' method [50]. Thus, by Theorems 8.5.1 and 8.6.1 we construct wavelet functions which generate an *infinite family of new Haar wavelet bases.*

Note that an MRA theory was also developed for the Cantor dyadic group [170], [171] and for the half-line with the dyadic addition [91]. These setting may seem very similar to the p-adics at first sight, but the existence of infinitely many different orthogonal wavelet bases generated by the same MRA does not hold there. This effect is provided by the non-Archimedean metric in $\mathbb{Q}_p$.

In Section 8.7, we study the p-adic *refinement equations* (8.7.1) and their solutions. One of them coincides with the "natural" refinement equation (8.1.5). A wide class of p-adic *refinable functions* generating a MRA is described. All of these functions are 1-periodic and such that their translations are mutually orthogonal (orthogonal refinable functions). It was proved in [4]–[6], [90] that orthogonal test refinable functions different from those described in this section do not exist. Moreover, it was proved in [4]–[6], [90] that all these functions generate the same p-adic Haar MRA. It was proved in [4] that there are no orthogonal MRA-based wavelet bases except for those described in Sections 8.5 and 8.6.

In Section 8.9, using a standard approach of Y. Meyer and S. Mallat (see, e.g., [192, §2.1]), we construct a multidimensional MRA by means of a tensor product of one-dimensional MRAs. In Section 8.8, using results of Section 8.9, the Haar multidimensional wavelet bases are constructed. These bases are described by Theorems 8.9.2 and 8.9.4.

In Section 8.10, the *non-Haar* p-adic compactly supported one-dimensional wavelet basis (8.10.3) is constructed. In contrast to (8.1.3), the number of generating wavelet functions for (8.10.3) is not minimal; for example, if $p = 2$, then we have 2^{m-1} wavelet functions (instead of one wavelet function as it is for the basis (8.1.3) and for real wavelet bases obtained by the classical MRA scheme). According to Remark 8.4, Kozyrev's wavelet basis (8.1.3) is a particular case of the basis (8.10.3) for $m = 1$. According to the same remark, our non-Haar wavelet basis (8.10.3) can be obtained by using the algorithm developed by Benedettos [50]. However, using our approach, we obtain the *explicit formulas* (8.10.3) for this basis. Moreover, our technique allows us to produce new wavelet bases (see below). In Section 8.11, we give explicit formulas for generating wavelet functions (8.11.8), (8.11.2), (8.11.3), which generate an *infinite family of different non-Haar wavelet bases* (which are distinct from the non-Haar wavelet basis (8.10.3)). According to Theorems 8.10.1, 8.10.2 and 8.11.1, these bases cannot be obtained in the framework of the standard scheme of the MRA developed in Sections 8.3–8.9. According to Remark 8.5, except for some particular cases, our bases (8.11.8), (8.11.2), (8.11.3) *cannot be constructed* by Benedettos' method [50]. Thus, by Theorem, 8.11.1, we construct wavelet functions which generate *an infinite family of new wavelet bases.* In Section 8.12, n-dimensional non-Haar wavelet bases (8.12.1) and (8.12.7) are introduced as n-direct products of the corresponding one-dimensional non-Haar wavelet bases. In particular, we construct n-dimensional wavelet basis generated by the one-dimensional Kozyrev wavelet basis (8.1.3) as the n-direct product of the one-dimensional wavelets.

The *non-Haar* basis cannot be constructed by classical MRA techniques, i.e., by using Definition 8.1 and the *refinement equation* of the form (8.3.5)

$$\phi = \sum_{a \in I_p} \alpha_a \phi(p^{-1} \cdot -a), \quad \alpha_a \in \mathbb{C}.$$

Due to this fact, the multidimensional wavelet basis (8.12.1) is constructed as the n-direct product of one-dimensional wavelets (8.10.3).

All the above-mentioned wavelets belong to the Lizorkin space of test functions $\Phi(\mathbb{Q}_p^n)$ which was introduced in Chapter 7, Section 7.3.2.

It is well known that the theory of p-adic pseudo-differential operators (in particular, fractional operators) and equations is closely connected with the theory of p-adic wavelets. It is typical that p-adic compactly supported wavelets are eigenfunctions of p-adic pseudo-differential operators [21]–[23], [131], [132], [142], [163]–[165]. Thus the *p-adic wavelet analysis is related to the spectral analysis of pseudo-differential operators.* Therefore, the *wavelet theory plays a key role in applications of p-adic analysis and gives a powerful new technique for solving p-adic problems.* Below, in Chapter 9, Section 9.4, we prove that the above-mentioned p-adic wavelets (under some conditions) are eigenfunctions of p-adic pseudo-differential operators. In Chapter 10, we use the wavelet bases to construct solutions of pseudo-differential equations.

In Section 8.13, we prove the multidimensional p-adic version of the Shannon–Kotelnikov theorem (in the paper [141] one-dimensional case was proved). It is well known that the series reconstructing a signal in the standard version of the Shannon–Kotelnikov theorem (8.13.1) converges *rather slowly.* Unlike the standard case (8.13.1), in the p-adic case given by Theorem 8.13.2, a series reconstructing a signal has only *one term* at any point $x \in \mathbb{Q}_p$. It turned out that the Shannon–Kotelnikov basis (8.13.4) is generated by dilatations and shifts of the Haar refinable function $\Phi(x) = \Omega(|x|_p)$, $x \in \mathbb{Q}_p^n$. Thus the p-adic Shannon–Kotelnikov theorem gives the representation (8.13.3) of any element from the space $V_j = F[\mathcal{L}^2(B_j^n)]$ (see (8.8.3)). According to Theorem 8.13.2 and

Corollary 8.13.3, in the p-adic case, the Shannon–Kotelnikov MRA coincides with the Haar MRA.

The problem of reconstructing band limited signals and images from regular and irregular samples has attracted many mathematicians and engineers. Many theoretical results and efficient algorithms have been derived in recent years (see [51], [253] and the references cited therein). So it is tempting to assume that the p-adic Shannon–Kotelnikov theorem can also be used in such applications.

In Section 8.14 characterizations of the p-adic Lizorkin spaces are given in terms of the wavelet functions. Namely, we prove an analog of Lemma 4.3.1 for test functions from the Lizorkin space $\Phi(\mathbb{Q}_p^n)$ (see Lemma 8.14.1). By Proposition 8.14.3 we also prove that any distribution $f \in \Phi'(\mathbb{Q}_p^n)$ can be realized as an *infinite* linear combination of wavelets of the form (8.14.4). In [25], assertions of the type of Lemma 8.14.1 and Proposition 8.14.3 were proved for ultrametric Lizorkin spaces.

8.2. p-adic Haar type wavelet basis via the real Haar wavelet basis

Now we consider the connection between the p-adic Haar type wavelet basis (Kozyrev's basis) (8.1.3) and the real Haar wavelet basis (8.1.1), (8.1.2) in $\mathcal{L}^2(\mathbb{R}_+)$. For simplicity we consider the case $p = 2$.

In Section 1.9.4, the Monna surjective map $\rho : \mathbb{Q}_p \to \mathbb{R}_+$, which is given by (1.9.3), is introduced. Its properties are described by Lemmas 1.9.9–1.9.12. Lemma 1.9.12 shows that if $\rho : \mathbb{Q}_p \to \mathbb{R}_+$ then the corresponding map

$$\rho^* : \mathcal{L}^2(\mathbb{R}_+) \to \mathcal{L}^2(\mathbb{Q}_p), \qquad \rho^* f(x) \overset{def}{=} f(\rho(x)), \quad x \in \mathbb{Q}_p, \quad (8.2.1)$$

is a unitary operator.

The following statement is a direct corollary of Lemma 1.9.11.

Lemma 8.2.1. ([163]) *Let $p = 2$. The Monna map ρ induces the transformation (8.2.1) of the real Haar wavelet function (8.1.2) to the Kozyrev's wavelet function (8.1.4) (p-adic Haar type wavelet function):*

$$\rho^* \psi^H(x) \overset{def}{=} \psi^H(\rho(x)) = \theta_1(x), \quad x \in \mathbb{Q}_p, \quad (8.2.2)$$

where this formula is understood as an equality in $\mathcal{L}^2$: it may fail on a set of measure zero.

Moreover, the following theorem holds.

Theorem 8.2.2. ([163]) *Let* $p = 2$. *The Monna map* ρ *induces the transformation* (8.2.1) *of the real Haar wavelet basis* (8.1.1) *to the Kozyrev's wavelet basis* (8.1.3) *(p-adic Haar type wavelet basis):*

$$\rho^* \psi_{jn}^H(x) \overset{def}{=} \psi_{j\rho(a)}^H(\rho(x)) = \theta_{1; ja}(x), \quad x \in \mathbb{Q}_p, \qquad (8.2.3)$$

where $j \in \mathbb{Z}, n \in \mathbb{N}, a \in I_p = \mathbb{Q}_p/\mathbb{Z}_p$.

Proof. Using the properties of the Monna map (1.9.5), (1.9.8), and (8.2.2), we obtain for the real Haar wavelet function (8.1.1):

$$\rho^* \psi_{jn}^H(x) \overset{def}{=} \psi_{j\rho(a)}^H(\rho(x)) = 2^{-j/2} \psi^H \left(2^{-j} \rho(x) - \rho(a) \right)$$
$$= 2^{-j/2} \psi^H \left(\rho(2^j x - a) \right) = 2^{-j/2} \theta_{1; ja}(2^j x - a) = \theta_{1; ja}(x).$$

$\square$

For arbitrary p the above connection between p-adic and real Haar wavelet bases was considered in Kozyrev's book [162, 2.4.].

8.3. *p*-adic multiresolution analysis (one-dimensional case)

8.3.1. Definition of the *p*-adic multiresolution analysis

Let us consider the set

$$I_p = \{x \in \mathbb{Q}_p : \{x\}_p = x\}$$
$$= \{a = p^{-\gamma}(a_0 + a_1 p + \cdots + a_{\gamma-1} p^{\gamma-1}) :$$
$$\gamma \in \mathbb{N}; a_j = 0, 1, \ldots, p-1; j = 0, 1, \ldots, \gamma - 1\}. \quad (8.3.1)$$

This set can be identified with a set of elements of the factor group $\mathbb{Q}_p/\mathbb{Z}_p$.

It is well known that $\mathbb{Q}_p = B_0(0) \cup \bigcup_{\gamma=1}^{\infty} S_\gamma$, where $S_\gamma = \{x \in \mathbb{Q}_p : |x|_p = p^\gamma\}$. Because of (1.6.2), $x \in S_\gamma$, $\gamma \geq 1$, if and only if $x = x_{-\gamma} p^{-\gamma} + x_{-\gamma+1} p^{-\gamma+1} + \cdots + x_{-1} p^{-1} + \xi$, where $x_{-\gamma} \neq 0$, $\xi \in B_0(0)$. Since $x_{-\gamma} p^{-\gamma} + x_{-\gamma+1} p^{-\gamma+1} + \cdots + x_{-1} p^{-1} \in I_p$, we have a "natural" decomposition of $\mathbb{Q}_p$ into the union of mutually disjoint balls:

$$\mathbb{Q}_p = \bigcup_{a \in I_p} B_0(a).$$

So I_p is a *"natural" set of shifts* for $\mathbb{Q}_p$, which will be used below.

Now we introduce the p-adic adaptation of the standard definition of *multiresolution analysis* (MRA) (see, e.g., [192, §1.3]).

Definition 8.1. A collection of closed spaces $V_j \subset \mathcal{L}^2(\mathbb{Q}_p)$, $j \in \mathbb{Z}$, is called a *multiresolution analysis (MRA)* in $\mathcal{L}^2(\mathbb{Q}_p)$ if the following axioms hold:

(a) $V_j \subset V_{j+1}$ for all $j \in \mathbb{Z}$;

(b) $\bigcup_{j \in \mathbb{Z}} V_j$ is dense in $\mathcal{L}^2(\mathbb{Q}_p)$;

(c) $\bigcap_{j \in \mathbb{Z}} V_j = \{0\}$;

(d) $f(\cdot) \in V_j \iff f(p^{-1}\cdot) \in V_{j+1}$ for all $j \in \mathbb{Z}$;

(e) there exists a function $\phi \in V_0$ such that the system $\{\phi(\cdot - a), a \in I_p\}$ is an orthonormal basis for V_0.

The function ϕ from axiom (e) is called *refinable* or *scaling*. One also says that an *MRA is generated by its scaling function*.

It follows immediately from axioms (d) and (e) that the functions $p^{j/2}\phi(p^{-j} \cdot -a)$, $a \in I_p$, form an orthonormal basis for V_j, $j \in \mathbb{Z}$.

According to the standard scheme (see, e.g., [76] or [192, §1.3]) for the construction of MRA-based wavelets, for each j, we define a space W_j *(wavelet space)* as the orthogonal complement of V_j in V_{j+1}, i.e.,

$$V_{j+1} = V_j \oplus W_j, \qquad j \in \mathbb{Z}, \tag{8.3.2}$$

where $W_j \perp V_j$, $j \in \mathbb{Z}$. It is not difficult to see that

$$f \in W_j \iff f(p^{-1}\cdot) \in W_{j+1}, \quad \text{for all } j \in \mathbb{Z} \tag{8.3.3}$$

and $W_j \perp W_k$, $j \neq k$. Taking into account axioms (b) and (c), we obtain

$$\bigoplus_{j \in \mathbb{Z}} W_j = \mathcal{L}^2(\mathbb{Q}_p) \quad \text{(orthogonal direct sum).} \tag{8.3.4}$$

If we now find a finite number of functions $\psi_\nu \in W_0$, $\nu \in A$ such that the system $\{\psi_\nu(x - a), a \in I_p, \nu \in A\}$ is an orthonormal basis for W_0, then, due to (8.3.3) and (8.3.4), the system

$$\{p^{j/2}\psi_\nu(p^{-j} \cdot -a), a \in I_p, j \in \mathbb{Z}, \nu \in A\},$$

is an orthonormal basis for $\mathcal{L}^2(\mathbb{Q}_p)$. Such functions ψ_ν, $\nu \in A$, are called *wavelet functions* and the corresponding basis is called a *wavelet basis*.

8.3.2. *p*-adic refinement equation

Let ϕ be a refinable function for an MRA. As was mentioned above, the system $\{p^{1/2}\phi(p^{-1} \cdot -a), a \in I_p\}$ is a basis for V_1. It follows from axiom (a) that

$$\phi = \sum_{a \in I_p} \alpha_a \phi(p^{-1} \cdot -a), \qquad \alpha_a \in \mathbb{C}. \tag{8.3.5}$$

We see that the function ϕ is a solution of a special kind of functional equation. Such equations are called *refinement equations*, and their solutions are called *refinable functions*[1]. The investigation of refinement equations and their solutions constitutes the most difficult part of the wavelet theory in real analysis.

A natural way of constructing an MRA (see, e.g., [192, §1.2]) is the following. We start with an appropriate function ϕ whose I_p-shifts form an orthonormal system and set

$$V_j = \overline{\text{span}\{\phi(p^{-j} \cdot -a) : a \in I_p\}}, \quad j \in \mathbb{Z}. \tag{8.3.6}$$

It is clear that axioms (d) and (e) of Definition 8.1 are fulfilled.

Of course, not every such function ϕ meets axiom (a). In the *real setting*, the relation $V_0 \subset V_1$ holds if and only if the refinable function satisfies a refinement equation. A different situation arises in the p-adic case. Generally speaking, a refinement equation (8.3.5) *does not imply the including property* $V_0 \subset V_1$. Indeed, we need all the functions $\phi(\cdot - b)$, $b \in I_p$, to belong to the space V_1, i.e., the identities

$$\phi(x - b) = \sum_{a \in I_p} \alpha_{a,b} \phi(p^{-1}x - a)$$

should be fulfilled for all $b \in I_p$. Since $p^{-1}b + a$ is not in I_p in general, we *cannot state* that

$$\phi(x - b) = \sum_{a \in I_p} \alpha_a \phi(p^{-1}x - p^{-1}b - a) \in V_1$$

for all $b \in I_p$. Nevertheless, it may happen that some refinable equations *imply the including property*.

The *refinement equation* (8.3.5) reflects *some "self-similarity"*. The structure of the space $\mathbb{Q}_p$ has a *natural* "self-similarity" property which is given by formulas (1.8.4) and (1.8.5). By (1.8.5), the characteristic function $\Omega(|x|_p)$ of the unit ball $B_0(0)$ is represented as a sum of p characteristic functions of the mutually disjoint balls $B_{-1}(r)$, $r = 0, 1, \ldots, p - 1$, i.e.,

$$\Omega(|x|_p) = \sum_{r=0}^{p-1} \Omega(p|x - r|_p) = \sum_{r=0}^{p-1} \Omega\left(\left|\frac{1}{p}x - \frac{r}{p}\right|_p\right), \quad x \in \mathbb{Q}_p. \tag{8.3.7}$$

---

[1] Usually the terms "scaling function" and "refinable function" are synonyms in the literature, and they are used in both senses: as a solution to a refinement equation and as a function generating MRA. We separate the meanings of these terms.

Thus, in p-adics, we have a *natural refinement equation* (8.1.5) whose solution is $\phi(x) = \Omega(|x|_p)$. This equation is an analog of the *refinement equation* (8.1.7) generating the Haar MRA in real analysis.

The *refinement equation* (8.1.5) is a particular case of (8.3.5).

8.4. Construction of the p-adic Haar multiresolution analysis

8.4.1. Construction of the p-adic MRA generated by the refinable function $\phi(x) = \Omega(|x|_p)$

To construct the p-adic multiresolution analysis, we use the *refinement equation* (8.1.5), and define a collection of closed spaces $V_j \subset \mathcal{L}^2(\mathbb{Q}_p)$, $j \in \mathbb{Z}$, by (8.3.6), where $\phi(x) = \Omega(|x|_p)$ (the refinable function) is the solution of (8.1.5).

Theorem 8.4.1. *In $\mathcal{L}^2(\mathbb{Q}_p)$ there exists an MRA generated by the refinable function $\phi(x) = \Omega(|x|_p)$.*

Proof. 1. It is clear that axioms (d) and (e) of Definition 8.1 are fulfilled and the system $\{p^{j/2}\phi(p^{-j} \cdot -a), a \in I_p\}$ is an orthonormal basis for V_j, $j \in \mathbb{Z}$. Since the numbers $\frac{a}{p}$, $\frac{a}{p} + \frac{r}{p}$ are in I_p for all $a \in I_p$ and $r = 0, 1, \ldots, p-1$, it follows from the refinement equation (8.1.6) that $V_0 \subset V_1$. By the definition (8.3.6) of the spaces V_j, this yields axiom (a). From the *refinement equation* (8.1.6), we obtain that $V_j \subset V_{j+1}$, i.e., axiom (a) from Definition 8.1 also holds.

Note that the characteristic function of the unit ball $\Omega(|x|_p)$ has a wonderful feature: $\Omega(|\cdot + \xi|_p) = \Omega(|\cdot|_p)$ for all $\xi \in \mathbb{Z}_p$ because the p-adic norm is non-Archimedean. In particular, $\Omega(|\cdot \pm 1|_p) = \Omega(|\cdot|_p)$, i.e.,

$$\phi(x \pm 1) = \phi(x), \quad \forall x \in \mathbb{Q}_p. \tag{8.4.1}$$

Thus ϕ is a 1-*periodic* function.

2. Let us prove that axiom (b) of Definition 8.1 holds, i.e., $\overline{\cup_{j \in \mathbb{Z}} V_j} = \mathcal{L}^2(\mathbb{Q}_p)$.

According to (4.3.4), any function $\varphi \in \mathcal{D}(\mathbb{Q}_p)$ belongs to one of the spaces $\mathcal{D}_N^l(\mathbb{Q}_p)$, and consequently, is represented in the form

$$\varphi(x) = \sum_{\nu=1}^{p^{N-l}} \varphi(c^\nu)\Omega(p^{-l}|x - c^\nu|_p), \quad x \in \mathbb{Q}_p, \tag{8.4.2}$$

where $c^\nu \in B_N(0)$, $\nu = 1, 2, \ldots p^{N-l}$; $l = l(\varphi)$, $N = N(\varphi)$; $l \in \mathbb{Z}$. Taking into account that $\Omega(p^{-l}|x - c^\nu|_p) = \Omega(|p^l x - p^l c^\nu|_p) = \phi(p^l x - p^l c^\nu)$, we can rewrite (8.4.2) as

$$\varphi(x) = \sum_{\nu=1}^{p^{N-l}} \alpha_\nu \phi(p^l x - p^l c^\nu), x \in \mathbb{Q}_p, \quad c^\nu \in B_N(0), \quad \alpha_\nu \in \mathbb{C}.$$

Since any number $p^l c^\nu$ can be represented in the form $p^l c^\nu = a^\nu + b^\nu, a^\nu \in I_p$, $b^\nu \in \mathbb{Z}_p$, using (8.4.1), we have

$$\varphi(x) = \sum_{\nu=1}^{p^{N-l}} \alpha_\nu \phi(p^l x - a^\nu), \quad x \in \mathbb{Q}_p, \quad a^\nu \in I_p, \quad \alpha_\nu \in \mathbb{C},$$

i.e., $\varphi(x) \in V_{-l}$. Thus any test function φ belongs to one of the space V_j, where $j = j(\varphi), \ j \in \mathbb{Z}$.

From Proposition 4.3.3, the space $\mathcal{D}(\mathbb{Q}_p)$ is dense in $\mathcal{L}^2(\mathbb{Q}_p)$, therefore approximating any function from $\mathcal{L}^2(\mathbb{Q}_p)$ by test functions $\varphi \in \mathcal{D}(\mathbb{Q}_p)$ we prove our assertion.

3. Let us prove that axiom (c) of Definition 8.1 holds, i.e., $\cap_{j \in \mathbb{Z}} V_j = \{0\}$.

Assume that $\cap_{j \in \mathbb{Z}} V_j \neq \{0\}$. Then there exists a function $f \in \mathcal{D}(\mathbb{Q}_p)$ such that $\|f\| \neq 0$ and $f \in V_j$ for all $j \in \mathbb{Z}$. Hence, from (8.3.6), we have

$$f(x) = \sum_{a \in I_p} c_{ja} \phi\left(p^{-j}x - a\right), \quad \forall j \in \mathbb{Z}.$$

Let $x = p^{-N}(x_0 + x_1 p + x_2 p^2 + \cdots)$. If $j \leq -N$, then $p^{-j}x \in \mathbb{Z}_p$, which implies that $|p^{-j}x - a|_p > 1$ for all $a \in I_p, a \neq 0$. Thus, $\phi(p^{-j}x - a) = 0$ for all $a \in I_p$, $a \neq 0$, and $\phi(p^{-j}x) = 1$, whenever $j \leq -N$. So we have $f(x) = c_{j0}$ for all $j \leq -N$. Similarly, for another $x' = p^{-N'}(x_0' + x_1' p + x_2' p^2 + \cdots)$, we have $f(x') = c_{j'0}$ for all $j \leq -N'$. This yields that $f(x) = f(x')$. Consequently, $f(x) \equiv C$, where C is a constant. However, if $C \neq 0$, then $f \notin \mathcal{L}^2(\mathbb{Q}_p)$. Thus, $C = 0$ which was to be proved. $\qquad \square$

The p-adic refinement equation (8.1.5) for $p = 2$ is an analog of the real Haar refinement equation (8.1.7). Thus, the MRA constructed by Theorem 8.4.1 is a p-adic analog of the real Haar MRA and we will call it the *p-adic Haar MRA*. But in contrast to the real setting, the refinable function $\phi(x) = \Omega(|x|_p)$ generating our *Haar MRA* is *periodic* with the period 1 (see (8.4.1)), which *never holds* for real refinable functions. It will be shown below that due to this specific property of ϕ, there exist infinitely many different orthonormal wavelet bases in the same Haar MRA (see Section 8.4.2 below, and Sections 8.5 and 8.6).

8.4.2. The Haar wavelet basis

According to the above scheme, we introduce the space W_0 as the orthogonal complement of V_0 in V_1.

Set

$$\psi_k^{(0)}(x) = \sum_{r=0}^{p-1} e^{2\pi i \frac{kr}{p}} \phi\left(\frac{1}{p}x - \frac{r}{p}\right), \qquad x \in \mathbb{Q}_p, \tag{8.4.3}$$

where $k = 1, 2, \ldots, p - 1$.

Proposition 8.4.2. *The shift system* $\{\psi_k^{(0)}(\cdot - a), k = 1, 2, \ldots, p - 1; a \in I_p\}$ *is an orthonormal basis of the space* W_0.

Proof. Let us prove that $W_0 \perp V_0$. It follows from (8.1.5) and (8.4.3) that

$$\left(\psi_k^{(0)}(\cdot - a), \phi(\cdot - b)\right) = \int_{\mathbb{Q}_p} \psi^{(0)}(x - a)\phi(x - b)\, dx$$

$$= \int_{\mathbb{Q}_p} \left(\sum_{r=0}^{p-1} e^{2\pi i \frac{kr}{p}} \phi\left(\frac{1}{p}x - \frac{r}{p} - \frac{a}{p}\right)\right)\left(\sum_{s=0}^{p-1} \phi\left(\frac{1}{p}x - \frac{s}{p} - \frac{b}{p}\right)\right) dx$$

for all $a, b \in I_p$. Let $a \neq b$. Since it is impossible that $a + r = b + s$, $r, s = 1, 2, \ldots, p - 1$, taking into account that the functions $p^{1/2}\phi(p^{-1} \cdot -c), c \in I_p$ are mutually orthogonal, we conclude that $\left(\psi_k^{(0)}(x - a), \phi(x - b)\right) = 0$. If $a = b$, again due to the orthonormality of the system $\{p^{1/2}\phi(p^{-1} \cdot -c), c \in I_p\}$, taking into account that $\frac{a}{p}, \frac{a}{p} + \frac{r}{p} \in I_p$ for $r = 1, 2, \ldots, p - 1$, and the first formula in (3.3.1), we have

$$\left(\psi_k^{(0)}(\cdot - a), \phi(\cdot - a)\right) = \int_{\mathbb{Q}_p} \left(\sum_{r=0}^{p-1} e^{2\pi i \frac{kr}{p}} \phi^2\left(\frac{1}{p}x - \frac{r}{p} - \frac{a}{p}\right)\right) dx$$

$$= \sum_{r=0}^{p-1} e^{2\pi i \frac{kr}{p}} \int_{\mathbb{Q}_p} \phi^2\left(\frac{1}{p}x - \frac{r}{p} - \frac{a}{p}\right) dx$$

$$= \frac{1}{p} \sum_{r=0}^{p-1} e^{2\pi i \frac{kr}{p}} = \frac{1 - e^{2\pi i kr}}{p\left(1 - e^{2\pi i \frac{k}{p}}\right)} = 0.$$

Thus, $\psi^{(0)}(\cdot + a) \perp \phi(\cdot + b)$ for all $a, b \in I_p$.

Similarly, computing for $k, k' = 1, 2, \ldots, p - 1$ the integrals

$$\left(\psi_k^{(0)}(\cdot - a), \psi_{k'}^{(0)}(\cdot - b)\right)$$

$$= \int_{\mathbb{Q}_p} \psi_k^{(0)}(x - a)\psi_{k'}^{(0)}(x - b)\, dx$$

$$= \int_{\mathbb{Q}_p} \left(\sum_{r=0}^{p-1} e^{2\pi i \frac{kr}{p}} \phi\left(\frac{1}{p}x - \frac{r}{p} - \frac{a}{p}\right)\right)\left(\sum_{s=0}^{p-1} e^{2\pi i \frac{k'r}{p}} \phi\left(\frac{1}{p}x - \frac{s}{p} - \frac{b}{p}\right)\right) dx,$$

we establish that the system $\{\psi_k^{(0)}(\cdot - a), k = 1, 2, \ldots, p - 1; a \in I_p\}$ is orthonormal.

According to (8.1.5) and (8.4.3) we have the system of equations with respect of functions $\phi\left(\frac{x}{p} - \frac{r}{p}\right)$, $r = 0, 1, \ldots, p - 1$,

$$
\begin{aligned}
\phi\left(\frac{x}{p}\right) + \phi\left(\frac{x}{p} - \frac{1}{p}\right) + \cdots + \phi\left(\frac{x}{p} - \frac{p-1}{p}\right) &= \phi(x), \\
h_0\phi\left(\frac{x}{p}\right) + h_1\phi\left(\frac{x}{p} - \frac{1}{p}\right) + \cdots + h_{p-1}\phi\left(\frac{x}{p} - \frac{p-1}{p}\right) &= \psi_1^{(0)}(x), \\
h_0^2\phi\left(\frac{x}{p}\right) + h_1^2\phi\left(\frac{x}{p} - \frac{1}{p}\right) + \cdots + h_{p-1}^2\phi\left(\frac{x}{p} - \frac{p-1}{p}\right) &= \psi_2^{(0)}(x),
\end{aligned}
$$

$$\cdots\cdots\cdots\cdots\cdots\cdots\cdots\cdots\cdots\cdots\cdots\cdots\cdots\cdots\cdots\cdots\cdots\cdots$$

$$h_0^{p-1}\phi\left(\frac{x}{p}\right) + h_1^{p-1}\phi\left(\frac{x}{p} - \frac{1}{p}\right) + \cdots + h_{p-1}^{p-1}\phi\left(\frac{x}{p} - \frac{p-1}{p}\right) = \psi_{p-1}^{(0)}(x),$$

where $h_r = e^{2\pi i \frac{r}{p}}$, $r = 0, 1, \ldots, p - 1$. It is easy to see that a determinant of the above system is the Vandermonde's determinant $\Delta \neq 0$, and consequently this system has a unique solution

$$\phi\left(\frac{x}{p} - \frac{r}{p}\right) = A_{0,r}\phi(x) + A_{1,r}\psi_1^{(0)}(x) + \cdots + A_{p-1,r}\psi_{p-1}^{(0)}(x),$$

$r = 0, 1, \ldots, p - 1$.

If $a \in I_p$, then $a = \frac{b}{p} + \frac{r}{p}$, $b \in I_p$, $r = 0, 1, \ldots, p - 1$. Hence, the last relations imply that

$$
\begin{aligned}
\phi\left(\frac{x}{p} - a\right) &= \phi\left(\frac{x}{p} - \frac{b}{p} - \frac{r}{p}\right) \\
&= A_{0,r}\phi(x - b) + A_{1,r}\psi_1^{(0)}(x - b) + \cdots + A_{p-1,r}\psi_{p-1}^{(0)}(x - b), \quad \text{(8.4.4)}
\end{aligned}
$$

$r = 0, 1, \ldots, p - 1$. Since $\{p^{1/2}\phi(p^{-1} \cdot -a) : a \in I_p\}$ is a basis for V_1, we obtain that the system of functions $\{\phi(\cdot - b), \psi_k^{(0)}(\cdot - b), k = 1, 2, \ldots, p - 1; b \in I_p\}$ is also a basis for V_1, i.e., the functions $\psi_k^{(0)}(\cdot - b), k = 1, 2, \ldots, p - 1, b \in I_2$, form a basis for the space $W_0 = V_1 \ominus V_0$. $\qquad\square$

Since according to Theorem 8.4.1 the collection $\{V_j : j \in \mathbb{Z}\}$ is the p-adic Haar MRA in $\mathcal{L}^2(\mathbb{Q}_p)$, the functions $\psi_k^{(0)}$, $k = 1, 2, \ldots, p - 1$, defined by (8.4.3) are wavelet functions. The wavelet basis generated by the wavelet functions $\psi_k^{(0)}$, $k = 1, 2, \ldots, p - 1$ (for $p = 2$) is an analog of the real Haar basis (8.1.1) which is generated by the wavelet functions (8.1.2). But in the *p-adic setting* the Haar basis generated by the Haar MRA is *not unique*.

From (2.3.3) and (1.8.5), the functions (8.4.3) can be rewritten in the form

$$\psi_k^{(0)}(x) = \chi_p(p^{-1}kx)\Omega(|x|_p), \quad k = 1, 2, \ldots, p - 1, \qquad x \in \mathbb{Q}_p. \quad \text{(8.4.5)}$$

Thus, the constructed wavelet functions (8.4.5) coincide with Kozyrev's wavelet functions (8.1.4) (i.e., $\psi_k^{(0)} = \theta_k$), and the corresponding Haar wavelet basis

$$\psi_{k;ja}^{(0)}(x) = p^{-j/2}\psi_k^{(0)}(p^j x - a)$$

$$= p^{-j/2}\chi_p\left(\frac{k}{p}(p^j x - a)\right)\Omega\big(|p^j x - a|_p\big), \quad j \in \mathbb{Z}, \quad a \in I_p,$$

$$(8.4.6)$$

coincides with the Kozyrev wavelet basis (8.1.3) (i.e., $\psi_{k;ja}^{(0)} = \theta_{k;ja}$).

It is easy to see that for a locally constant function $\psi_{k;ja}^{(0)}(x)$ we have $\int_{\mathbb{Q}_p} \psi_{k;ja}^{(0)}(x)\,dx = 0$. Therefore, according to (7.3.3), wavelet function $\psi_{k;ja}^{(0)}(x)$ belongs to the Lizorkin space $\Phi(\mathbb{Q}_p)$, $k = 1, 2, \ldots, p - 1$, $j \in \mathbb{Z}$, $a \in I_p$.

Remark 8.1. Because of the periodicity (8.4.1) of the refinable function ϕ, we can use the shifts $\psi_k^{(0)}(\cdot + a)$, $a \in I_p$, instead of $\psi_k^{(0)}(\cdot - a)$, $a \in I_p$, $k = 1, 2, \ldots, p - 1$.

For the case $p = 2$ the refinement equation (8.1.5) has the form (8.1.6), and a 2-adic MRA in $\mathcal{L}^2(\mathbb{Q}_2)$ is defined by a collection of the spaces

$$V_j = \overline{\text{span}\{\phi(2^{-j}x - a) : a \in I_2\}}, \quad j \in \mathbb{Z}.$$

where $\phi(x) = \Omega(|x|_2)$ is a solution of (8.1.6). In this case the wavelet function (which defines an orthonormal basis of the space W_0) is

$$\psi^{(0)}(x) = \phi\left(\frac{x}{2}\right) - \phi\left(\frac{x}{2} - \frac{1}{2}\right) = \chi_2(2^{-1}x)\Omega(|x|_2), \quad x \in \mathbb{Q}_2. \quad (8.4.7)$$

The corresponding Haar wavelet basis is

$$\psi_{ja}^{(0)}(x) = 2^{-j/2}\psi^{(0)}(2^j x - a)$$

$$= 2^{-j/2}\chi_2\big(2^{-1}(2^j x - a)\big)\Omega\big(|2^j x - a|_2\big), \quad j \in \mathbb{Z}, \quad a \in I_2. \quad (8.4.8)$$

Now for simplicity we consider the case $p = 2$ and show that there is another function $\psi^{(1)}$ whose shifts form an orthonormal basis for W_0 (different from the basis generated by $\psi^{(0)}$). Set

$$\psi^{(1)}(x) = \frac{1}{\sqrt{2}}\left(\phi\left(\frac{x}{2}\right) + \phi\left(\frac{x}{2} - \frac{1}{2^2}\right)\right.$$

$$\left. - \phi\left(\frac{x}{2} - \frac{1}{2}\right) - \phi\left(\frac{x}{2} - \frac{1}{2^2} - \frac{1}{2}\right)\right) \quad (8.4.9)$$

and prove that the functions $\psi^{(1)}(\cdot - a)$, $a \in I_2$, are mutually orthonormal. If $a \in I_2$, $a \neq 0, \frac{1}{2}$, then each of the numbers $0, \frac{1}{2^2}, \frac{1}{2}, \frac{1}{2^2} + \frac{1}{2}$ differs modulo 1 from each of the numbers $\frac{a}{2}, \frac{1}{2^2} + \frac{a}{2}, \frac{1}{2} + \frac{a}{2}, \frac{1}{2^2} + \frac{1}{2} + \frac{a}{2}$. Due to the orthonormality of the system $\{2^{1/2}\phi(2^{-1}x - a), a \in I_2\}$ and (8.4.1), it follows that $\psi^{(1)}$ is orthogonal to $\psi^{(1)}(\cdot - a)$ whenever $a \in I_2, a \neq 0, \frac{1}{2}$. Again due to the orthonormality of the system $\{2^{1/2}\phi(2^{-1}x - a), a \in I_2\}$ and (8.4.1), we have

$$
\begin{aligned}
\left(\psi^{(1)}, \psi^{(1)}(\cdot - 2^{-1})\right) &= \int_{Q_2} \psi^{(1)}(x)\psi^{(1)}(x - 2^{-1})\,dx \\
&= 2^{-1}\int_{Q_2}\left(-\phi^2\left(\frac{x}{2}\right) + \phi^2\left(\frac{x}{2} - \frac{1}{2^2}\right) - \phi^2\left(\frac{x}{2} - \frac{1}{2}\right)\right. \\
&\quad \left. + \phi^2\left(\frac{x}{2} - \frac{1}{2^2} - \frac{1}{2}\right)\right)dx = 0, \\
\left(\psi^{(1)}, \psi^{(1)}\right) &= \int_{Q_2} \psi^{(1)}(x)\psi^{(1)}(x)\,dx \\
&= 2^{-1}\int_{Q_2}\left(\phi^2\left(\frac{x}{2}\right) + \phi^2\left(\frac{x}{2} - \frac{1}{2}\right) + \phi^2\left(\frac{x}{2} - \frac{1}{2^2} - \frac{1}{2}\right)\right. \\
&\quad \left. + \phi^2\left(\frac{x}{2} - \frac{1}{2^2}\right)\right)dx = 1.
\end{aligned}
$$

Thus we have proved that the system $\{\psi^{(1)}(\cdot + a), a \in I_2\}$ is orthonormal. It is not difficult to see that

$$
\psi^{(1)}(x) = \frac{1}{\sqrt{2}}\left(\psi^{(0)}(x) + \psi^{(0)}\left(x - \frac{1}{2}\right)\right),
$$
$$
\psi^{(1)}\left(x - \frac{1}{2}\right) = \frac{1}{\sqrt{2}}\left(-\psi^{(0)}(x) + \psi^{(0)}\left(x - \frac{1}{2}\right)\right). \tag{8.4.10}
$$

This yields that

$$
\psi^{(0)}(x) = \frac{1}{\sqrt{2}}\left(\psi^{(1)}(x) - \psi^{(1)}\left(x - \frac{1}{2}\right)\right).
$$

Since the system $\{\psi^{(0)}(\cdot - a), a \in I_2\}$ is a basis for W_0, it follows that the system $\{\psi^{(1)}(\cdot - a), a \in I_2\}$, is another orthonormal basis for W_0.

So we have showed that a wavelet basis generated by the Haar MRA is *not unique*.

8.5. Description of one-dimensional 2-adic Haar wavelet bases

8.5.1. Wavelet functions

Now we are going to show that there exist infinitely many different Haar type wavelet functions $\psi^{(s)}$, $s \in \mathbb{N}$, in W_0 generating different bases for $\mathcal{L}^2(\mathbb{Q}_2)$. In what follows, we shall write the 2-adic number $a = 2^{-s}(a_0 + a_1 2 + \cdots + a_{s-1} 2^{s-1}) \in I_2, a_j = 0, 1, j = 0, 1, \ldots, s - 1$, in the form $a = \frac{m}{2^s}$, where $m = a_0 + a_1 2 + \cdots + a_{s-1} 2^{s-1}$.

Since the refinable function ϕ of the Haar MRA is 1-*periodic* (see (8.4.1)), evidently, the wavelet function $\psi^{(0)}$ has the following property:

$$\psi^{(0)}(x \pm 1) = -\psi^{(0)}(x). \tag{8.5.1}$$

Before we prove a general result, let us consider a simple special case. Set

$$\psi^{(1)}(x) = \alpha_0 \psi^{(0)}(x) + \alpha_1 \psi^{(0)}\left(x - \frac{1}{2}\right), \quad \alpha_0, \alpha_1 \in \mathbb{C}, \tag{8.5.2}$$

and find all the complex numbers α_0, α_1 for which $\{\psi^{(1)}(x - a), a \in I_2\}$ is an orthonormal basis for W_0.

Taking into account the orthonormality of the system $\{\psi^{(0)}(\cdot - a), a \in I_2\}$ and (8.5.1), we can easily see that $\psi^{(1)}$ is orthogonal to $\psi^{(1)}(\cdot - a)$ whenever $a \in I_2, a \neq 0, \frac{1}{2}$. Thus the system $\{\psi^{(1)}(x - a), a \in I_2\}$ is orthonormal if and only if the system consisting of the functions (8.5.2) and

$$\psi^{(1)}\left(x - \frac{1}{2}\right) = -\alpha_1 \psi^{(0)}(x) + \alpha_0 \psi^{(0)}\left(x - \frac{1}{2}\right) \tag{8.5.3}$$

is orthonormal, which is equivalent to the unitary property of the matrix

$$D = \begin{pmatrix} \alpha_0 & \alpha_1 \\ -\alpha_1 & \alpha_0 \end{pmatrix}$$

It is clear that D is a unitary matrix whenever $|\alpha_0|^2 + |\alpha_1|^2 = 1$ and $\alpha_0 \overline{\alpha_1} = \overline{\alpha_0} \alpha_1$. In this case the system $\{\psi^{(1)}(\cdot - a), a \in I_2\}$ is a basis for W_0 because $\{\psi^{(0)}(\cdot - a), a \in I_2\}$ is a basis for W_0 and we have

$$\psi^{(0)}(x) = \overline{\alpha_0} \psi^{(1)}(x) - \overline{\alpha_1} \psi^{(1)}\left(x - \frac{1}{2}\right).$$

So, $\psi^{(1)}$ is a Haar wavelet function if and only if $|\alpha_0|^2 + |\alpha_1|^2 = 1$ and $\alpha_0 \overline{\alpha_1} = \overline{\alpha_0} \alpha_1$. In particular, we obtain (8.4.10) for $\alpha_0 = \alpha_1 = \frac{1}{\sqrt{2}}$.

Theorem 8.5.1. *Let $\psi^{(0)}$ be a wavelet function given by (8.4.7). For every $s = 0, 1, 2, \ldots$ the function*

$$\psi^{(s)}(x) = \sum_{k=0}^{2^s - 1} \alpha_k \psi^{(0)}\left(x - \frac{k}{2^s}\right) \tag{8.5.4}$$

is a compactly supported wavelet function for the Haar MRA if and only if

$$\alpha_k = 2^{-s}\sum_{r=0}^{2^s-1}\gamma_r e^{-i\pi\frac{2r-1}{2^s}k}, \quad k = 0, \ldots, 2^s - 1, \tag{8.5.5}$$

where $\gamma_r \in \mathbb{C}$ is an arbitrary constant such that $|\gamma_r| = 1, r = 0, \ldots, 2^s - 1$.

Proof. 1. Let $\psi^{(s)} \in W_0$ be a wavelet function such that we have supp $\psi^{(s)} \subset B_s(0)$, $s \geq 0$. In this case $\psi^{(s)}$ is defined by (8.5.4). Indeed, since according to Section 8.4 $\{\psi^{(0)}(\cdot - a),\ a \in I_2\}$ is a basis for W_0, then

$$\psi^{(s)}(x) = \sum_{a\in I_2} g_a \psi^{(0)}(x-a)$$

$$= \sum_{\substack{a\in I_2 \\ |a|_2 \leq 2^s}} g_a^\mu \psi^{(0)}(x-a) + \sum_{\substack{a\in I_2 \\ |a|_2 > 2^s}} g_a^\mu \psi^{(0)}(x-a).$$

It is easy to see that the second sum on the right-hand side vanishes whenever $|x|_2 \leq 2^s$, and the first sums on the right-hand side and the left-hand side vanish whenever $|x|_2 > 2^s$. So the second sum equals zero for all $x \in \mathbb{Q}_2$, and after renaming the coefficients we obtain (8.5.4).

Let $\psi^{(s)}$ be defined by (8.5.4). Since $\{\psi^{(0)}(\cdot - a),\ a \in I_2\}$ is an orthonormal system, taking into account (8.5.1), we see that the function $\psi^{(s)}$ is orthogonal to the function $\psi^{(s)}(\cdot - a)$ whenever $a \in I_2, a \neq \frac{k}{2^s}, k = 0, 1, \ldots 2^s - 1$. Thus the system $\{\psi^{(s)}(x - a),\ a \in I_2\}$ is orthonormal if and only if the system consisting of the functions

$$\psi^{(s)}\left(x - \frac{r}{2^s}\right) = -\alpha_{2^s-r}\psi^{(0)}(x)$$

$$- \alpha_{2^s-r+1}\psi^{(0)}\left(x - \frac{1}{2^s}\right) - \cdots - \alpha_{2^s-1}\psi^{(0)}\left(x - \frac{r-1}{2^s}\right)$$

$$+ \alpha_0\psi^{(0)}\left(x - \frac{r}{2^s}\right) + \cdots + \alpha_{2^s-r-1}\psi^{(0)}\left(x - \frac{2^s-1}{2^s}\right), \tag{8.5.6}$$

$r = 0, \ldots, 2^s - 1$, is orthonormal. Set

$$\Xi^{(0)} = \left(\psi^{(0)}, \psi^{(0)}\left(\cdot - \frac{1}{2^s}\right), \ldots, \psi^{(0)}\left(\cdot - \frac{2^s-1}{2^s}\right)\right)^T,$$

$$\Xi^{(s)} = \left(\psi^{(s)}, \psi^{(s)}\left(\cdot - \frac{1}{2^s}\right), \ldots, \psi^{(s)}\left(\cdot - \frac{2^s-1}{2^s}\right)\right)^T.$$

By (8.5.6), we have $\Xi^{(s)} = D\,\Xi^{(0)}$, where

$$D = \begin{pmatrix} \alpha_0 & \alpha_1 & \alpha_2 & \ldots & \alpha_{2^s-2} & \alpha_{2^s-1} \\ -\alpha_{2^s-1} & \alpha_0 & \alpha_1 & \ldots & \alpha_{2^s-3} & \alpha_{2^s-2} \\ -\alpha_{2^s-2} & -\alpha_{2^s-1} & \alpha_0 & \ldots & \alpha_{2^s-4} & \alpha_{2^s-3} \\ \multicolumn{6}{c}{\dotfill} \\ -\alpha_2 & -\alpha_3 & -\alpha_4 & \ldots & \alpha_0 & \alpha_1 \\ -\alpha_1 & -\alpha_2 & -\alpha_3 & \ldots & -\alpha_{2^s-1} & \alpha_0 \end{pmatrix} \qquad (8.5.7)$$

is a $2^s \times 2^s$ matrix. Due to the orthonormality of $\{\psi^{(0)}(\cdot - a), a \in I_2\}$, the coordinates of $\Xi^{(s)}$ form an orthonormal system if and only if matrix D is unitary.

2. Let $u = (\alpha_0, \alpha_1, \ldots, \alpha_{2^s-1})^T$ be a vector and

$$A = \begin{pmatrix} 0 & 0 & \ldots & 0 & 0 & -1 \\ 1 & 0 & \ldots & 0 & 0 & 0 \\ 0 & 1 & \ldots & 0 & 0 & 0 \\ \multicolumn{6}{c}{\dotfill} \\ 0 & 0 & \ldots & 1 & 0 & 0 \\ 0 & 0 & \ldots & 0 & 1 & 0 \end{pmatrix}.$$

be a $2^s \times 2^s$ matrix. It is not difficult to see that

$$A^r u = \big(-\alpha_{2^s-r}, -\alpha_{2^s-r+1}, \ldots, -\alpha_{2^s-1}, \alpha_0, \alpha_1, \ldots, \alpha_{2^s-r-1} \big)^T,$$

where $r = 1, 2, \ldots, 2^s - 1$. Thus

$$D = \big(u, Au, \ldots, A^{2^s-1}u \big)^T.$$

Hence, to describe all unitary matrixes D, we should find all vectors $u = (\alpha_0, \alpha_1, \ldots, \alpha_{2^s-1})^T$ such that the system of vectors $\{A^r u, r = 0, \ldots, 2^s - 1\}$ is orthonormal. We already have one such vector $u_0 = (1, 0, \ldots, 0, 0)^T$ because the matrix $D_0 = \big(u_0, Au_0, \ldots, A^{2^s-1}u_0 \big)^T$ is the identity matrix. Let us prove that the system $\{A^r u, r = 0, \ldots, 2^s - 1\}$ is orthonormal if and only if $u = Bu_0$, where B is a unitary matrix such that $AB = BA$. Indeed, let $u = Bu_0$, where B is a unitary matrix, and $AB = BA$. Then $A^r u = BA^r u_0$, $r = 0, 1, \ldots, 2^s - 1$. Since the system $\{A^r u_0, r = 0, 1, \ldots, 2^s - 1\}$ is orthonormal and matrix B is unitary, the vectors $A^r u$, $r = 0, 1, \ldots, 2^s - 1$, are also orthonormal. Conversely, if the system $A^r u$, $r = 0, 1, \ldots, 2^s - 1$, is orthonormal, taking into account that $\{A^r u_0, r = 0, 1, \ldots, 2^s - 1\}$ is also an orthonormal system, we conclude that there exists a unitary matrix B such that $A^r u = B(A^r u_0)$, $r = 0, 1, \ldots, 2^s - 1$. Since $A^{2^s} u = -u$ and $A^{2^s} u_0 = -u_0$, we obtain the additional relation $A^{2^s} u = B(A^{2^s} u_0)$. It follows from the above relations that $(AB - BA)(A^r u_0) = 0$,

$r = 0, 1, \ldots, 2^s - 1$. Since the vectors $A^r u_0, r = 0, 1, \ldots, 2^s - 1$, form a basis in the 2^s-dimensional space, we conclude that $AB = BA$.

Thus all unitary matrixes D are given by

$$D = \left(Bu_0, BAu_0, \ldots, BA^{2^s-1} u_0 \right)^T,$$

where B is a unitary matrix such that $AB = BA$. It remains for us to describe all such matrixes B.

3. It is not difficult to see that the eigenvalues and the corresponding normalized eigenvectors of A are respectively

$$\lambda_r = e^{i\pi \frac{2r-1}{2^s}} \tag{8.5.8}$$

and $v_r = \left((v_r)_0, \ldots, (v_r)_{2^s-1} \right)^T$, where

$$(v_r)_l = 2^{-s/2} e^{-i\pi \frac{2r-1}{2^s} l}, \quad l = 0, 1, 2, \ldots, 2^s - 1, \tag{8.5.9}$$

$r = 0, 1, \ldots, 2^s - 1$. Hence the matrix A can be represented as $A = C \tilde{A} C^{-1}$, where

$$\tilde{A} = \begin{pmatrix} \lambda_0 & 0 & \cdots & 0 \\ 0 & \lambda_1 & \cdots & 0 \\ \vdots & \vdots & \ddots & \vdots \\ 0 & 0 & \cdots & \lambda_{2^s-1} \end{pmatrix}$$

is a diagonal matrix, and $C = \left(v_0, \ldots, v_{2^s-1} \right)$ is a unitary matrix. It follows that the matrix $B = C \tilde{B} C^{-1}$ is unitary if and only if $\tilde{B}$ is unitary. On the other hand, $AB = BA$ if and only if $\tilde{A}\tilde{B} = \tilde{B}\tilde{A}$. Moreover, since according to (8.5.8), $\lambda_k \neq \lambda_l$ whenever $k \neq l$, all unitary matrices $\tilde{B}$ such that $\tilde{A}\tilde{B} = \tilde{B}\tilde{A}$ are given by

$$\tilde{B} = \begin{pmatrix} \gamma_0 & 0 & \cdots & 0 \\ 0 & \gamma_1 & \cdots & 0 \\ \vdots & \vdots & \ddots & \vdots \\ 0 & 0 & \cdots & \gamma_{2^s-1} \end{pmatrix},$$

where $\gamma_k \in \mathbb{C}$, $|\gamma_k| = 1$. Hence all unitary matrices B such that $AB = BA$ are given by $B = C\tilde{B}C^{-1}$.

4. Using (8.5.9), we can calculate

$$\alpha_k = (Bu_0)_k = (C\tilde{B}C^{-1}u_0)_k = \sum_{r=0}^{2^s-1} \gamma_r (v_r)_k (\overline{v}_r)_0$$

$$= 2^{-s} \sum_{r=0}^{2^s-1} \gamma_r e^{-i\pi \frac{2r-1}{2^s} k}, \quad k = 0, 1, \ldots, 2^s - 1,$$

where $\gamma_k \in \mathbb{C}$, $|\gamma_k| = 1$.

5. It remains for us to prove that the $\{\psi^{(s)}(\cdot - a), a \in I_2\}$ form a basis for W_0 whenever $\psi^{(s)}$ is defined by (8.5.4) and (8.5.5). Since $\{\psi^{(0)}(\cdot - a), a \in I_2\}$ is a basis for W_0, it suffices to check that any function $\psi^{(0)}(\cdot - c), c \in I_2$, can be decomposed with respect to the functions $\psi^{(s)}(\cdot - a), a \in I_2$. Any $c \in I_2$, $c \neq 0$, can be represented in the form $c = \frac{r}{2^s} + b$, where $r = 0, 1, \ldots, 2^s - 1$, $|b|_2 \geq 2^{s+1}$. Taking into account that $\Xi^{(0)} = D^{-1}\Xi^{(s)}$, i.e.,

$$\psi^{(0)}\left(x - \frac{r}{2^s}\right) = \sum_{k=0}^{2^s-1} \beta_k^{(r)} \psi^{(s)}\left(x - \frac{k}{2^s}\right), \quad r = 0, 1, \ldots, 2^s - 1,$$

we have

$$\psi^{(0)}\left(x - c\right) = \psi^{(0)}\left(x - \frac{r}{2^s} - b\right) = \sum_{k=0}^{2^s-1} \beta_k^{(r)} \psi^{(s)}\left(x - \frac{k}{2^s} - b\right),$$

and $\frac{k}{2^s} + b \in I_2, k = 0, 1, \ldots, 2^s - 1$. $\qquad\square$

All dilatations and shifts of any wavelet function (8.5.4), (8.5.5) form a 2-*adic orthonormal Haar wavelet basis* in $\mathcal{L}^2(\mathbb{Q}_2)$.

Remark 8.2. From formulas (8.5.4), (8.5.5) and (4.9.4), (4.8.10), we have

$$F[\psi^{(s)}(x - a)](\xi) = \chi_2(a\xi)\Omega\left(\left|\xi + \frac{1}{2}\right|_2\right) \sum_{k=0}^{2^s-1} \alpha_k \chi_2\left(\frac{k}{2^s}\xi\right)$$

$$= \chi_2(a\xi)\Omega\left(\left|\xi + \frac{1}{2}\right|_2\right) 2^{-s} \sum_{r=0}^{2^s-1} \gamma_r \sum_{k=0}^{2^s-1} e^{2i\pi\frac{\xi+\frac{1}{2}-r}{2^s}k}, \quad a \in I_2,$$

where $\gamma_r \in \mathbb{C}, |\gamma_r| = 1, r = 0, 1, \ldots, 2^s$. It is clear that the right-hand side of the last relation is not equal to zero only if $\xi = -\frac{1}{2} + \eta, \eta = \eta_0 + \eta_1 2 + \cdots \in \mathbb{Z}_p, \xi \in B_0(-\frac{1}{2})$. Hence, for $\xi \in B_0(-\frac{1}{2})$ we have

$$F[\psi^{(s)}(x - a)](\xi) = \chi_2\left(a\left(-\frac{1}{2} + \eta\right)\right) 2^{-s} \sum_{r=0}^{2^s-1} \gamma_r \sum_{k=0}^{2^s-1} e^{2i\pi\frac{\tilde\eta-r}{2^s}k},$$

where $\tilde\eta = \eta_0 + \eta_1 2 + \cdots \eta_{s-1} 2^{s-1}$. According to Proposition 1.8.9, the ball $B_0(-\frac{1}{2})$ is represented by the sum of 2^s disjoint balls $B_{-s}(-\frac{1}{2} + \tilde\eta)$. If $\xi \in B_{-s}(-\frac{1}{2} + \tilde\eta)$ then $\xi = -\frac{1}{2} + \tilde\eta + 2^s \tilde\eta_0, \tilde\eta_0 \in \mathbb{Z}_p$. Thus on any ball

$B_{-s}(-\frac{1}{2} + \widetilde{\eta})$ we have

$$F[\psi^{(s)}(x - a)](\xi)$$

$$= \chi_2\left(a\left(-\frac{1}{2} + \eta\right)\right)\left(\gamma_{\widetilde{\eta}} + 2^{-s}\sum_{\substack{r=0 \\ r \neq \widetilde{\eta}}}^{2^s-1} \gamma_r \sum_{k=0}^{2^s-1} e^{2i\pi\frac{\widetilde{\eta}-r}{2^s}k}\right)$$

$$= \chi_2\left(a\left(-\frac{1}{2} + \eta\right)\right)\left(\gamma_{\widetilde{\eta}} + 2^{-s}\sum_{\substack{r=0 \\ r \neq \widetilde{\eta}}}^{2^s-1} \gamma_r \sum_{k=0}^{2^s-1} \frac{1 - e^{2i\pi(\widetilde{\eta}-r)}}{1 - e^{2i\pi\frac{\widetilde{\eta}-r}{2^s}}}\right)$$

$$= \chi_2\left(a\left(-\frac{1}{2} + \eta\right)\right)\gamma_{\widetilde{\eta}}, \quad \widetilde{\eta} = \eta_0 + \eta_1 2 + \cdots \eta_{s-1}2^{s-1}, \quad a \in I_2.$$

$$(8.5.10)$$

According to [50, Proposition 5.1.], *only* the wavelet functions whose Fourier transforms are the characteristic functions of some sets can be obtained by Benedettos' method. It is easy to see if in (8.5.10)

$$\gamma_{\widetilde{\eta}} = 1, \quad \forall \widetilde{\eta} = \eta_0 + \eta_1 2 + \cdots \eta_{v-1}2^{s-1}, \tag{8.5.11}$$

where $\eta_j = 0, 1, j = 0, 1, \ldots, s - 1$, then for $a = 0$ the right-hand side of (8.5.10) *is the characteristic function* of the ball $B_0(-\frac{1}{2})$. Consequently, in this case the wavelet basis generated by wavelet functions $\psi^{(s)}$ can be obtained by Benedettos' method [50]. If (8.5.11) does not hold, the right-hand side of (8.5.10) *is not a characteristic function of any set* for all shifts $a \in I_2$. Consequently, in this case the wavelet basis generated by wavelet functions $\psi^{(s)}$ *cannot be constructed* by the method of [50].

It is easy to see that $\int_{\mathbb{Q}_2} \psi^{(s)}(2^j x - a)\, dx = 0$, $j \in \mathbb{Z}, a \in I_2$, and in view of (7.3.3), wavelet functions $\psi^{(s)}(2^j \cdot -a)$ belong to the Lizorkin space $\Phi(\mathbb{Q}_2)$.

8.5.2. Real wavelet functions

Using formulas (8.5.5), one can extract all *real* wavelet functions (8.5.4).

Let $s = 1$. According to (8.5.2) and (8.5.3),

$$\psi^{(1)}(x) = \cos\theta\, \psi^{(0)}(x) + \sin\theta\, \psi^{(0)}\left(x - \frac{1}{2}\right) \tag{8.5.12}$$

is the *real* wavelet function.

Let $s = 2$. Set $\gamma_r = e^{i\theta_r}$, $r = 0, 1, \ldots, 2^s - 1$. It follows from (8.5.5) that the wavelet function $\psi^{(1)}$ is real if and only if

$$\sin\theta_1 + \sin\theta_2 + \sin\theta_3 + \sin\theta_4 = 0,$$
$$\cos\theta_1 - \cos\theta_2 + \cos\theta_3 - \cos\theta_4 = 0,$$
$$\sin\theta_1 - \sin\theta_2 - \sin\theta_3 + \sin\theta_4 =$$
$$\cos\theta_1 + \cos\theta_2 - \cos\theta_3 - \cos\theta_4,$$
$$\sin\theta_1 - \sin\theta_2 - \sin\theta_3 + \sin\theta_4 =$$
$$-(\cos\theta_1 + \cos\theta_2 - \cos\theta_3 - \cos\theta_4).$$

These relations are equivalent to the system

$$\sin\theta_1 = -\sin\theta_4, \qquad \cos\theta_1 = \cos\theta_4,$$
$$\sin\theta_2 = -\sin\theta_3, \qquad \cos\theta_2 = \cos\theta_3.$$

Thus the real wavelet functions (8.5.4) with $s = 2$ are given by

$$\psi^{(1)}(x) = \frac{1}{2}(\cos\theta_1 + \cos\theta_2)\psi^{(0)}(x)$$
$$+ \frac{1}{2\sqrt{2}}(\cos\theta_1 - \cos\theta_2 + \sin\theta_1 + \sin\theta_2)\psi^{(0)}\left(x - \frac{1}{2^2}\right)$$
$$+ \frac{1}{2}(\sin\theta_1 - \sin\theta_2)\psi^{(0)}\left(x - \frac{1}{2}\right)$$
$$+ \frac{1}{2\sqrt{2}}(\cos\theta_1 - \cos\theta_2 - \sin\theta_1 - \sin\theta_2)\psi^{(0)}\left(x - \frac{1}{2^2} - \frac{1}{2}\right).$$
$$(8.5.13)$$

In particular, for the special cases $\theta_1 = \theta_2 = \theta$, $\theta_1 = -\theta_2 = \theta$, $\theta_1 = \theta_2 + \frac{\pi}{2} = \theta$, we obtain respectively the following one-parameter families of real wavelet functions

$$\psi^{(1)}(x) = \cos\theta\,\psi^{(0)}(x) + \sin\theta\,\psi^{(0)}\left(x - \tfrac{1}{2}\right),$$
$$\psi^{(1)}(x) = \cos\theta\,\psi^{(0)}(x) + \tfrac{1}{\sqrt{2}}\sin\theta\,\psi^{(0)}\left(x - \tfrac{1}{2^2}\right)$$
$$- \tfrac{1}{\sqrt{2}}\sin\theta\,\psi^{(0)}\left(x - \tfrac{1}{2^2} - \tfrac{1}{2}\right),$$
$$\psi^{(1)}(x) = \tfrac{1}{2}(\cos\theta - \sin\theta)\psi^{(0)}(x)$$
$$+ \tfrac{1}{2\sqrt{2}}(\cos\theta + \sin\theta)\psi^{(0)}\left(x - \tfrac{1}{2^2}\right)$$
$$- \tfrac{1}{2}(\cos\theta - \sin\theta)\psi^{(0)}\left(x - \tfrac{1}{2}\right).$$
$$(8.5.14)$$

For $s = 3, 4, \ldots$ families of real wavelet functions can be obtained in the same way.

8.6. Description of one-dimensional p-adic Haar wavelet bases

It was proved in Section 8.4 that shifts of wavelet functions (8.4.5) $\psi_\nu^{(0)}(x) = \chi_p(p^{-1}\nu x)\Omega(|x|_p)$, $\nu = 1, \ldots, p - 1$, generated an orthonormal basis in the space W_0. The corresponding wavelet basis (8.4.6)

$$\{p^{j/2}\psi_\nu^{(0)}(p^{-j}x - a), \quad a \in I_p, \ j \in \mathbb{Z}, \ \nu = 1, \ldots, p - 1\}$$

coincides with the Kozyrev basis (8.1.3).

Here we construct an infinite family of wavelet functions ψ_ν, $\nu = 1, \ldots, p - 1$, in W_0 generating different wavelet bases for $\mathcal{L}^2(\mathbb{Q}_p)$, where p is an arbitrary prime number.

Note that the functions ψ_ν^0, $\nu = 1, \ldots, p - 1$, have the following obvious property:

$$\psi_\nu^{(0)}(x - 1) = \chi_p\left(-\frac{\nu}{p}\right)\psi_\nu^{(0)}(x), \quad \nu = 1, \ldots, p - 1. \tag{8.6.1}$$

Theorem 8.6.1. *All sets of compactly supported wavelet functions are given by*

$$\psi_\mu(x) = \sum_{\nu=1}^{p-1} \sum_{k=0}^{p^s-1} \alpha_{\nu;k}^\mu \psi_\nu^{(0)}\left(x - \frac{k}{p^s}\right), \quad \mu = 1, 2, \ldots, p - 1, \tag{8.6.2}$$

where wavelet functions $\psi_\nu^{(0)}$ are given by (8.4.5), $s = 0, 1, 2, \ldots$, and

$$\alpha_{\nu;k}^\mu = \begin{cases} -p^{-s}\sum_{m=0}^{p^s-1} e^{-2\pi i \frac{-\frac{\nu}{p}+m}{p^s}k}\sigma_{\mu m}z_{\mu\mu}, & \mu = \nu, \\[4mm] p^{-2s}\sum_{m=0}^{p^s-1}\sum_{n=0}^{p^s-1} e^{-2\pi i \frac{-\frac{\nu}{p}+m}{p^s}k}\dfrac{1 - e^{2\pi i \frac{\mu-\nu}{p}}}{e^{2\pi i \frac{\frac{\mu-\nu}{p}+m-n}{p^s}} - 1}\sigma_{\nu m}z_{\nu\mu}, & \mu \neq \nu, \end{cases}$$

$$\tag{8.6.3}$$

$|\sigma_{\mu m}| = 1$, $z_{\mu\nu}$ *are entries of an arbitrary unitary $(p - 1) \times (p - 1)$ matrix Z.*

Proof. 1. Let $\psi_\mu \in W_0$, $\nu = 1, \ldots, p - 1$, be a set of wavelet functions such that $\operatorname{supp} \psi_\mu \subset B_s(0)$, $s \geq 0$. First of all we prove that

$$\psi_\mu(x) = \sum_{\nu=1}^{p-1} \sum_{k=0}^{p^s-1} \alpha_{\nu;k}^\mu \psi_\nu^{(0)}\left(x - \frac{k}{p^s}\right), \quad \mu = 1, \ldots, p - 1, \tag{8.6.4}$$

Since $\{\psi_\nu^{(0)}(\cdot - a), \nu = 1, \ldots, p-1, a \in I_p\}$ is a basis for W_0 and $\psi_\mu \in W_0$, we have

$$\psi_\mu(x) = \sum_{\nu=1}^{p-1} \sum_{a \in I_p} g_a^\mu \psi_\nu^{(0)}(x-a)$$

$$= \sum_{\nu=1}^{p-1} \sum_{\substack{a \in I_p \\ |a|_p \leq p^s}} g_a^\mu \psi_\nu^{(0)}(x-a) + \sum_{\nu=1}^{p-1} \sum_{\substack{a \in I_p \\ |a|_p > p^s}} g_a^\mu \psi_\nu^{(0)}(x-a).$$

The second sum on the right-hand side vanishes whenever $|x|_p \leq p^s$, the first sums on the right-hand side and the left-hand side vanish whenever $|x|_p > p^s$. So the second sum equals zero for all $x \in \mathbb{Q}_p$, and after renaming the coefficients we obtain (8.6.4).

2. Since $\{\psi_\nu^{(0)}(\cdot - a), \nu = 1, \ldots, p-1, a \in I_p\}$ is an orthonormal system, taking (8.6.1) into account, we see that ψ_μ is orthogonal to $\psi_\nu(\cdot - a)$ whenever $a \in I_p$, $|a|_p > p^s$, $\nu, \mu = 1, \ldots, p-1$. Thus the system $\{\psi_\mu(x-a), \mu = 1, \ldots, p-1, a \in I_p\}$ is orthonormal if and only if the system consisting of the functions $\psi_\mu\left(\cdot - \frac{r}{p^s}\right)$, $r = 0, \ldots, p^s-1$, $\mu = 1, \ldots, p-1$ forms an orthonormal system.

Using (8.6.1), we have

$$\psi_\mu\left(x - \frac{r}{p^s}\right)$$

$$= \sum_{\nu=1}^{p-1} \left(\chi_p\left(-\frac{\nu}{p}\right) \alpha_{\nu;p^s-r}^\mu \psi_\nu^{(0)}(x) + \chi_p\left(-\frac{\nu}{p}\right) \alpha_{\nu;p^s-r+1}^\mu \psi_\nu^{(0)}\left(x - \frac{1}{p^s}\right) \right.$$

$$+ \cdots + \chi_p\left(-\frac{\nu}{p}\right) \alpha_{\nu;p^s-1}^\mu \psi_\nu^{(0)}\left(x - \frac{r-1}{p^s}\right)$$

$$\left. + \alpha_{\nu;0}^\mu \psi_\nu^{(0)}\left(x - \frac{r}{p^s}\right) + \cdots + \alpha_{\nu;p^s-r-1}^\mu \psi_\nu^{(0)}\left(x - \frac{p^s-1}{p^s}\right) \right). \quad (8.6.5)$$

Let us introduce the vectors

$$\Xi^{(0)} = \left(\psi_1^{(0)}, \psi_1^{(0)}\left(\cdot - \frac{1}{p^s}\right), \ldots, \psi_1^{(0)}\left(\cdot - \frac{p^s-1}{p^s}\right), \right.$$

$$\left. \ldots, \psi_{p-1}^{(0)}, \psi_{p-1}^{(0)}\left(\cdot - \frac{1}{p^s}\right), \ldots, \psi_{p-1}^{(0)}\left(\cdot - \frac{p^s-1}{p^s}\right) \right)^T,$$

$$\Xi = \left(\psi_1, \psi_1\left(\cdot - \frac{1}{p^s}\right), \ldots, \psi_1\left(\cdot - \frac{p^s-1}{p^s}\right), \right.$$

$$\left. \ldots, \psi_{p-1}, \psi_{p-1}\left(\cdot - \frac{1}{p^s}\right), \ldots, \psi_{p-1}\left(\cdot - \frac{p^s-1}{p^s}\right) \right)^T$$

of length $p^s(p-1)$, and the $p^s(p-1) \times p^s(p-1)$ matrix

$$D = \begin{pmatrix} D_{11} & D_{12} & \cdots & D_{1\,p-1} \\ \cdots & \cdots & \cdots & \cdots \\ D_{p-11} & D_{p-12} & \cdots & D_{p-1\,p-1} \end{pmatrix}, \qquad (8.6.6)$$

where

$$D_{\mu\nu}$$

$$= \begin{pmatrix} \alpha^{\mu}_{\nu;0} & \alpha^{\mu}_{\nu;1} & \cdots & \alpha^{\mu}_{\nu;p^s-2} & \alpha^{\mu}_{\nu;p^s-1} \\ \chi_p(-\frac{\nu}{p})\alpha^{\mu}_{\nu;p^s-1} & \alpha^{\mu}_{\nu;0} & \cdots & \alpha^{\mu}_{\nu;p^s-3} & \alpha^{\mu}_{\nu;p^s-2} \\ \chi_p(-\frac{\nu}{p})\alpha^{\mu}_{\nu;p^s-2} & \chi_p(-\frac{\nu}{p})\alpha^{\mu}_{\nu;p^s-1} & \cdots & \alpha^{\mu}_{\nu;p^s-4} & \alpha^{\mu}_{\nu;p^s-3} \\ \cdots & \cdots & \cdots & \cdots & \cdots \\ \chi_p(-\frac{\nu}{p})\alpha^{\mu}_{\nu;2} & \chi_p(-\frac{\nu}{p})\alpha^{\mu}_{\nu;3} & \cdots & \alpha^{\mu}_{\nu;0} & \alpha^{\mu}_{\nu;1} \\ \chi_p(-\frac{\nu}{p})\alpha^{\mu}_{\nu;1} & \chi_p(-\frac{\nu}{p})\alpha^{\mu}_{\nu;2} & \cdots & \chi_p(-\frac{\nu}{p})\alpha^{\mu}_{\nu;p^s-1} & \alpha^{\mu}_{\nu;0} \end{pmatrix}.$$

By (8.6.5), we have $\Xi = D\Xi^{(0)}$. Because of the orthonormality of the system $\{\psi^{(0)}_\nu(\cdot - a), \nu = 1, \ldots, p-1, a \in I_p\}$, the coordinates of Ξ form an orthonormal system if and only if the matrix D is unitary.

Let

$$u^\mu = (\alpha^\mu_{1;0}, \ldots, \alpha^\mu_{1;p^s-1}, \ldots\ldots, \alpha^\mu_{p-1;0}, \ldots, \alpha^\mu_{p-1;p^s-1})^T \qquad (8.6.7)$$

be a vector of the length $p^s(p-1)$, $\mu = 1, \ldots, p-1$, and let

$$A = \begin{pmatrix} A_1 & O & \cdots & O \\ O & A_2 & \cdots & O \\ \vdots & \vdots & \ddots & \vdots \\ O & O & \cdots & A_{p-1} \end{pmatrix} \qquad (8.6.8)$$

be a $p^s(p-1) \times p^s(p-1)$ matrix, where

$$A_\nu = \begin{pmatrix} 0 & 0 & \cdots & 0 & 0 & \chi_p(-\frac{\nu}{p}) \\ 1 & 0 & \cdots & 0 & 0 & 0 \\ 0 & 1 & \cdots & 0 & 0 & 0 \\ \cdots & \cdots & \cdots & \cdots & \cdots & \cdots \\ 0 & 0 & \cdots & 1 & 0 & 0 \\ 0 & 0 & \cdots & 0 & 1 & 0 \end{pmatrix} \qquad (8.6.9)$$

is a $p^s \times p^s$ matrix, and O is the $p^s \times p^s$ zero matrix.

It is easy to see that

$$(A)^r u^\mu = \left(\chi_p\left(-\frac{1}{p}\right)\alpha_{1;p^s-r}^\mu, \chi_p\left(-\frac{1}{p}\right)\alpha_{1;p^s-r+1}^\mu, \ldots, \chi_p\left(-\frac{1}{p}\right)\alpha_{1;p^s-1}^\mu,$$

$$\alpha_{1;0}^\mu, \alpha_{1;1}^\mu, \ldots, \alpha_{1;p^s-r-1}^\mu, \ldots,$$

$$\chi_p\left(-\frac{p-1}{p}\right)\alpha_{p-1;p^s-r}^\mu, \chi_p\left(-\frac{p-1}{p}\right)\alpha_{p-1;p^s-r+1}^\mu, \ldots,$$

$$\chi_p\left(-\frac{p-1}{p}\right)\alpha_{p-1;p^s-1}^\mu, \alpha_{p-1;0}^\mu, \alpha_{p-1;1}^\mu, \ldots, \alpha_{p-1;p^s-r-1}^\mu\right)^T,$$

where $r = 0, \ldots, p^s - 1$. Thus, the matrix (8.6.6) can be represented in the form

$$D = \left(u^1, Au^1, \ldots, A^{p^s-1}u^1, \ldots\ldots u^{p-1}, Au^{p-1}, \ldots, A^{p^s-1}u^{p-1}\right)^T.$$

Hence, to describe all unitary matrixes D, we should find all families of vectors (8.6.7) such that the system of vectors $\{A^r u^\mu : r = 0, \ldots, p^s - 1, \mu = 1, \ldots, p - 1\}$ is orthonormal. We have already one such family of vectors

$$u_0^\mu = (\underbrace{0, \ldots, 0}_{p^s(\mu-1)}, 1, \underbrace{0, \ldots, 0}_{p^s(p-\mu)-1})^T, \qquad \mu = 1, \ldots, p - 1, \qquad (8.6.10)$$

and the corresponding matrix

$$D_0 = \left(u_0^1, Au_0^1, \ldots, A^{p^s-1}u_0^1, \ldots\ldots u_0^{p-1}, Au_0^{p-1}, \ldots, A^{p^s-1}u_0^{p-1}\right)^T,$$

which is the identity matrix.

3. Now we prove that the system of vectors $\{A^r u^\mu, r = 0, \ldots, p^s - 1, \mu = 1, \ldots, p - 1\}$ is orthonormal if and only if

$$u^\mu = Bu_0^\mu, \qquad (8.6.11)$$

where B is a $p^s(p-1) \times p^s(p-1)$ unitary matrix such that

$$[A, B](A^r u_0^\mu) = 0, \quad r = 0, \ldots, p^s - 2, \quad \mu = 1, \ldots, p - 1, \quad (8.6.12)$$

where $[A, B] = AB - BA$ is the commutator of matrixes A and B. Indeed, let $u^\mu = Bu_0^\mu$, where B is a unitary matrix satisfying (8.6.12). Then it is easy to see that $A^r u^\mu = BA^r u_0^\mu, r = 0, \ldots, p^s - 1, \mu = 1, \ldots, p - 1$. Since the system $\{A^r u_0^\mu, r = 0, \ldots, p^s - 1, \mu = 1, \ldots, p - 1\}$ is orthonormal and the matrix B is unitary, the vectors $A^r u^\mu, r = 0, \ldots, p^s - 1, \mu = 1, \ldots, p - 1$, are also orthonormal. Conversely, if the system $\{A^r u^\mu, r = 0, \ldots, p^s - 1, \mu = 1, \ldots, p - 1\}$ is orthonormal, taking into account that $\{A^r u_0^\mu, r = 0, \ldots, p^s - 1, \mu = 1, \ldots, p - 1\}$ is also an orthonormal system, we conclude that there exists a unitary $p^s(p-1) \times p^s(p-1)$ matrix B such that $A^r u^\mu = B(A^r u_0^\mu)$,

$r = 0, \ldots, p^s - 1$, $\mu = 1, \ldots, p - 1$. Substituting $A^{l-1}u^\mu = B(A^{l-1}u_0^\mu)$ into $A^l u^\mu = B(A^l u_0^\mu)$, $l = 1, \ldots, p^s - 1$ we prove (8.6.12) by induction on l.

4. It is easy to see that

$$A^{p^s}u^\mu = \Lambda u^\mu, \qquad A^{p^s}u_0^\mu = \Lambda u_0^\mu, \tag{8.6.13}$$

where

$$\Lambda = \begin{pmatrix} \Lambda_1 & O & \ldots & O \\ O & \Lambda_2 & \ldots & O \\ \vdots & \vdots & \ddots & \vdots \\ O & O & \ldots & \Lambda_{p-1} \end{pmatrix}$$

is a $p^s(p-1) \times p^s(p-1)$ matrix, and

$$\Lambda_\nu = \begin{pmatrix} \chi_p(-\frac{\nu}{p}) & 0 & \ldots & 0 \\ 0 & \chi_p(-\frac{\nu}{p}) & \ldots & 0 \\ \vdots & \vdots & \ddots & \vdots \\ 0 & 0 & \ldots & \chi_p(-\frac{\nu}{p}) \end{pmatrix}$$

is a diagonal $p^s \times p^s$ matrix, $\mu, \nu = 1, \ldots, p - 1$.

Let us write a $p^s(p-1) \times p^s(p-1)$ unitary matrix B in the form

$$B = \begin{pmatrix} B_{11} & B_{12} & \ldots & B_{1p-1} \\ \ldots & \ldots & \ldots & \ldots \\ B_{p-11} & B_{p-12} & \ldots & B_{p-1p-1} \end{pmatrix}, \tag{8.6.14}$$

where

$$B_{\mu\nu} = \begin{pmatrix} b_{00}^{\mu\nu} & b_{01}^{\mu\nu} & \ldots & b_{0p^s-1}^{\mu\nu} \\ \ldots & \ldots & \ldots & \ldots \\ b_{p^s-10}^{\mu\nu} & b_{p^s-11}^{\mu\nu} & \ldots & b_{p^s-1p^s-1}^{\mu\nu} \end{pmatrix}. \tag{8.6.15}$$

It is easy to calculate that

$$\Lambda B u_0^\mu = \left(\chi_p\left(-\frac{1}{p}\right) b_{00}^{1\mu}, \ldots, \chi_p\left(-\frac{1}{p}\right) b_{p^s-10}^{1\mu}, \ldots \right.$$
$$\left. \ldots, \chi_p\left(-\frac{p-1}{p}\right) b_{00}^{p-1\mu}, \ldots, \chi_p\left(-\frac{p-1}{p}\right) b_{p^s-10}^{p-1\mu} \right)^T,$$
$$B \Lambda u_0^\mu = \left(\chi_p\left(-\frac{\mu}{p}\right) b_{00}^{1\mu}, \ldots, \chi_p\left(-\frac{\mu}{p}\right) b_{p^s-10}^{1\mu}, \ldots \right.$$
$$\left. \ldots, \chi_p\left(-\frac{\mu}{p}\right) b_{00}^{p-1\mu}, \ldots, \chi_p\left(-\frac{\mu}{p}\right) b_{p^s-10}^{p-1\mu} \right)^T.$$

Consequently, we have

$$[\Lambda, B]u_0^\mu = d_\mu \qquad (8.6.16)$$

where

$$d_\mu = \left(d_{\mu;1}, \ldots, d_{\mu;\mu-1}, \underbrace{0, \ldots, 0}_{p^s}, d_{\mu;\mu+1}, \ldots, d_{\mu;p-1}\right)^T, \qquad (8.6.17)$$

$$d_{\mu;\nu} = \left(\chi_p\left(-\frac{\nu}{p}\right) - \chi_p\left(-\frac{\mu}{p}\right)\right)\left(b_{00}^{\nu\mu}, \ldots, b_{p^s-10}^{\nu\mu}\right)$$
$$\nu - 1, 2, \ldots, p-1, \nu \neq \mu. \qquad (8.6.18)$$

Relations (8.6.11), (8.6.13) and (8.6.16) imply

$$A^{p^s} B u_0^\mu = A^{p^s} u^\mu = \Lambda B u_0^\mu = B\Lambda u_0^\mu + d_\mu = B A^{p^s} u_0^\mu + d_\mu. \qquad (8.6.19)$$

Using (8.6.12), we have

$$A^{p^s} B u_0^\mu = A^{p^s-1}(AB)u_0^\mu = A^{p^s-2}(AB)Au_0^\mu = \cdots = (AB)(A^{p^s-1}u_0^\mu).$$

Thus relation (8.6.19) can be rewritten as

$$[A, B](A^{p^s-1}u_0^\mu) = d_\mu, \qquad (8.6.20)$$

where d_μ is given by (8.6.17), $\mu = 1, \ldots, p-1$.

5. Let

$$G = [A, B] = \begin{pmatrix} G_{11} & G_{12} & \cdots & G_{1p-1} \\ \cdots & \cdots & \cdots & \cdots \\ G_{p-11} & G_{p-12} & \cdots & G_{p-1p-1} \end{pmatrix},$$

where

$$G_{\mu\nu} = \begin{pmatrix} g_{00}^{\mu\nu} & g_{01}^{\mu\nu} & \cdots & g_{0p^s-1}^{\mu\nu} \\ \cdots & \cdots & \cdots & \cdots \\ g_{p^s-10}^{\mu\nu} & g_{p^s-11}^{\mu\nu} & \cdots & g_{p^s-1p^s-1}^{\mu\nu} \end{pmatrix} \qquad (8.6.21)$$

is a $p^s \times p^s$ matrix, $\nu, \mu = 1, \ldots, p-1$. According to (8.6.12) and (8.6.20) we have

$$G_{\mu\nu} = (\underbrace{\widehat{0}, \ldots, \widehat{0}}_{p^s-1}, d_{\mu;\nu}^T), \qquad (8.6.22)$$

where $\widehat{0}$ is a matrix column consisting of zeros, and d_μ is given by (8.6.17).

6. It is easy to see that the matrix (8.6.9) has eigenvalues

$$\lambda_{\nu r} = e^{2\pi i \frac{-\frac{\nu}{p}+r}{p^s}} \qquad (8.6.23)$$

and the corresponding normalized eigenvectors

$$v_{\nu r} = \big((v_{\nu r})_0, \dots, (v_{\nu r})_{p^s-1}\big)^T,$$

where

$$(v_{\nu r})_l = p^{-s/2} e^{-2\pi i \frac{-\frac{\nu}{p}+r}{p^s} l}, \quad l = 0, 1, 2, \dots, p^s - 1, \quad (8.6.24)$$

$r = 0, 1, \dots, p^s - 1$, $\nu = 1, 2, \dots, p - 1$. In view of (8.6.8), (8.6.23) and (8.6.24), the eigenvalues of A and the corresponding normalized eigenvectors are respectively

$$\{\lambda_{1\,0}, \dots, \lambda_{1\,p^s-1}, \dots\dots, \lambda_{p-1\,0}, \dots, \lambda_{p-1\,p^s-1}\}$$

and

$$V_{\nu r} = (\underbrace{0, \dots, 0}_{p^s(\nu-1)}, \underbrace{(v_{\nu r})_0, \dots, (v_{\nu r})_{p^s-1}}_{p^s}, \underbrace{0, \dots, 0}_{p^s(p-1-\nu)})^T, \quad (8.6.25)$$

where $\nu = 1, 2, \dots, p - 1$, $r = 0, 1, \dots, p^s - 1$.

Consequently, the matrix A can be represented as $A = C\widetilde{A}C^{-1}$, where

$$\widetilde{A} = \begin{pmatrix} \widetilde{A}_1 & 0 & \cdots & 0 \\ 0 & \widetilde{A}_2 & \cdots & 0 \\ \vdots & \vdots & \ddots & \vdots \\ 0 & 0 & \cdots & \widetilde{A}_{p-1} \end{pmatrix}, \quad \widetilde{A}_\nu = \begin{pmatrix} \lambda_{\nu 0} & 0 & \cdots & 0 \\ 0 & \lambda_{\nu 1} & \cdots & 0 \\ \vdots & \vdots & \ddots & \vdots \\ 0 & 0 & \cdots & \lambda_{\nu\,p^s-1} \end{pmatrix},$$

$\nu = 1, 2, \dots, p - 1$, and

$$C = \big(V_{1\,0}, \dots, V_{1\,p^s-1}, \dots\dots, V_{p-1\,0}, \dots, V_{p-1\,p^s-1}\big) \quad (8.6.26)$$

is a unitary matrix. In view of (8.6.25),

$$C = \begin{pmatrix} C_1 & O & \cdots & O \\ O & C_2 & \cdots & O \\ \vdots & \vdots & \ddots & \vdots \\ O & O & \cdots & C_{p-1} \end{pmatrix}, \quad C^{-1} = \begin{pmatrix} C_1^{-1} & O & \cdots & O \\ O & C_2^{-1} & \cdots & O \\ \vdots & \vdots & \ddots & \vdots \\ O & O & \cdots & C_{p-1}^{-1} \end{pmatrix},$$

where

$$C_\nu = \begin{pmatrix} (v_{\nu 0})_0 & (v_{\nu 1})_0 & \cdots & (v_{\nu\,p^s-1})_0 \\ (v_{\nu 0})_1 & (v_{\nu 1})_1 & \cdots & (v_{\nu\,p^s-1})_1 \\ \cdots & \cdots & \cdots & \cdots \\ (v_{\nu 0})_{p^s-1} & (v_{\nu 1})_{p^s-1} & \cdots & (v_{\nu\,p^s-1})_{p^s-1} \end{pmatrix}, \quad (8.6.27)$$

and

$$
C_\nu^{-1} = \begin{pmatrix} \overline{(v_{\nu 0})_0} & \overline{(v_{\nu 0})_1} & \cdots & \overline{(v_{\nu 0})_{p^s-1}} \\ \overline{(v_{\nu 1})_0} & \overline{(v_{\nu 1})_1} & \cdots & \overline{(v_{\nu 1})_{p^s-1}} \\ \cdots & \cdots & \cdots & \cdots \\ \overline{(v_{\nu\, p^s-1})_0} & \overline{(v_{\nu\, p^s-1})_1} & \cdots & \overline{(v_{\nu\, p^s-1})_{p^s-1}} \end{pmatrix}, \tag{8.6.28}
$$

where $\bar{z}$ is a complex conjugate to z, $\nu = 1, 2, \ldots, p - 1$.

Let us calculate $\widetilde{G} = C^{-1}GC$. On the one hand, according to (8.6.21), (8.6.22), (8.6.25), (8.6.26), and (8.6.17), (8.6.18), (8.6.27), the block $(C^{-1}G)_{\mu\nu} = C_\mu^{-1}G_{\mu\nu}$ of the matrix $C^{-1}G$ consists of the following entries

$$
\big((C^{-1}G)_{\mu\nu}\big)_{mn} = \sum_{r=0}^{p^s-1} \overline{(v_{\mu m})_r}\, g_{rn}^{\mu\nu}
$$

$$
= \delta_{n\, p^s-1}\left(\chi_p\left(-\frac{\nu}{p}\right) - \chi_p\left(-\frac{\mu}{p}\right)\right) p^{-s/2} \sum_{r=0}^{p^s-1} e^{2\pi i \frac{-\frac{\mu}{p}+m}{p^s} r}\, b_{r0}^{\nu\mu}.
$$

Taking (8.6.28) and (8.6.24) into account, we obtain that the matrix $\widetilde{G} = C^{-1}GC$ consists of the $p^s \times p^s$ blocks $\widetilde{G}_{\mu\nu}$, $\nu, \mu = 1, \ldots, p - 1$, defined by

$$
(\widetilde{G}_{\mu\nu})_{mn} = \left(\chi_p\left(-\frac{\nu}{p}\right) - \chi_p\left(-\frac{\mu}{p}\right)\right) p^{-s} e^{-2\pi i \frac{-\frac{\nu}{p}+n}{p^s}(p^s-1)} \sum_{r=0}^{p^s-1} e^{2\pi i \frac{-\frac{\mu}{p}+m}{p^s} r}\, b_{r0}^{\nu\mu}.
$$

$$
\tag{8.6.29}
$$

On the other hand, $\widetilde{G} = [\widetilde{A}, \widetilde{B}]$, where

$$
\widetilde{B} = C^{-1}BC = \begin{pmatrix} \widetilde{B}_{11} & \widetilde{B}_{12} & \cdots & \widetilde{B}_{1p-1} \\ \cdots & \cdots & \cdots & \cdots \\ \widetilde{B}_{p-11} & \widetilde{B}_{p-12} & \cdots & \widetilde{B}_{p-1p-1} \end{pmatrix}, \tag{8.6.30}
$$

where

$$
\widetilde{B}_{\mu\nu} = \begin{pmatrix} \widetilde{b}_{00}^{\mu\nu} & \widetilde{b}_{01}^{\mu\nu} & \cdots & \widetilde{b}_{0p^s-1}^{\mu\nu} \\ \cdots & \cdots & \cdots & \cdots \\ \widetilde{b}_{p^s-10}^{\mu\nu} & \widetilde{b}_{p^s-11}^{\mu\nu} & \cdots & \widetilde{b}_{p^s-1p^s-1}^{\mu\nu} \end{pmatrix}. \tag{8.6.31}
$$

It is clear that

$$\left((\widetilde{A}\widetilde{B})_{\mu\nu}\right)_{mn} = \sum_{k=0}^{p^s-1} \left(\widetilde{A}_\mu\right)_{mk}\left(\widetilde{B}_{\mu\nu}\right)_{kn} = \lambda_{\mu m}\widetilde{b}_{mn}^{\mu\nu},$$

$$\left((\widetilde{B}\widetilde{A})_{\mu\nu}\right)_{mn} = \sum_{k=0}^{p^s-1} \left(\widetilde{B}_{\mu\nu}\right)_{mk}\left(\widetilde{A}_\nu\right)_{kn} = \lambda_{\nu n}\widetilde{b}_{mn}^{\mu\nu}.$$

Thus, we have

$$\left(\widetilde{G}_{\mu\nu}\right)_{mn} = (\lambda_{\mu m} - \lambda_{\nu n})\widetilde{b}_{mn}^{\mu\nu}, \qquad (8.6.32)$$

where $\lambda_{\mu m}, \lambda_{\nu n}$ are given by (8.6.23). Set

$$q_m^{\mu\nu} = \begin{cases} \sum_{r=0}^{p^s-1} e^{2\pi i \frac{-\frac{\mu}{p}+m}{p^s}r} b_{r0}^{\nu\mu}, & \text{if } \mu \neq \nu, \\ -\widetilde{b}_{mm}^{\mu\mu}, & \text{if } \mu = \nu. \end{cases} \qquad (8.6.33)$$

Combining (8.6.29) and (8.6.32), we obtain

$$\widetilde{b}_{mn}^{\mu\nu} = p^{-s}\frac{\chi_p\left(-\frac{\nu}{p}\right) - \chi_p\left(-\frac{\mu}{p}\right)}{\lambda_{\mu m} - \lambda_{\nu n}} e^{-2\pi i \frac{-\frac{\nu}{p}+n}{p^s}(p^s-1)} q_m^{\mu\nu}, \qquad (8.6.34)$$

whenever $\mu \neq \nu$ and

$$\widetilde{b}_{mn}^{\mu\mu} = -q_m^{\mu\mu}\delta_{mn} \qquad (8.6.35)$$

for all $\mu, \nu = 1, 2, \ldots, p-1; m = 0, 1, \ldots, p^s - 1$. Substituting (8.6.23) into (8.6.34), it is easy to rewrite (8.6.34) as

$$\widetilde{b}_{mn}^{\mu\nu} = p^{-s}\frac{1 - e^{2\pi i \frac{\nu-\mu}{p}}}{e^{2\pi i \frac{\nu-\mu}{p}+m-n}{p^s}} - 1} q_m^{\mu\nu}, \qquad (8.6.36)$$

where $\mu \neq \nu$.

7. Now we prove that

$$\sum_{n=0}^{p^s-1} \widetilde{b}_{mn}^{\mu\nu}\overline{\widetilde{b}_{m'n}^{\mu'\nu}} = \left(C_\mu^{-1}C_{\mu'}\right)_{mm'} q_m^{\mu\nu}\overline{q_{m'}^{\mu'\nu}} \qquad (8.6.37)$$

for all $\mu, \mu', \nu = 1, 2, \ldots, p-1; \; m, m' = 0, 1, \ldots, p^s - 1$. In particular, since $C_\mu^{-1}C_\mu = E$, this formula yields

$$\sum_{n=0}^{p^s-1} \widetilde{b}_{mn}^{\mu\nu}\overline{\widetilde{b}_{m'n}^{\mu\nu}} = 0, \quad m \neq m'. \qquad (8.6.38)$$

(a) Let $\mu \neq \mu'$, $\mu' = \nu$. In this case, using (8.6.36), (8.6.33) and (8.6.35), we have

$$\sum_{n=0}^{p^s-1} \widetilde{b}_{mn}^{\mu\nu}\overline{\widetilde{b}_{m'n}^{\mu'\nu}} = \sum_{n=0}^{p^s-1} \widetilde{b}_{mn}^{\mu\mu'}\overline{\widetilde{b}_{m'n}^{\mu'\mu'}} = \widetilde{b}_{mm'}^{\mu\mu'}\overline{\widetilde{b}_{m'm'}^{\mu'\mu'}}$$

$$= -p^{-s}\frac{\chi_p\left(-\frac{\mu'}{p}\right) - \chi_p\left(-\frac{\mu}{p}\right)}{\lambda_{\mu m} - \lambda_{\mu' m'}} e^{-2\pi i\frac{-\frac{\mu'}{p}+m'}{p^s}(p^s-1)} q_m^{\mu\mu'}\overline{q_{m'}^{\mu'\mu'}}$$

$$= -p^{-s}\frac{e^{-2\pi i\frac{\mu'}{p}} - e^{-2\pi i\frac{\mu}{p}}}{e^{2\pi i\frac{-\frac{\mu}{p}+m}{p^s}} - e^{2\pi i\frac{-\frac{\mu'}{p}+m'}{p^s}}} e^{-2\pi i\frac{-\frac{\mu'}{p}+m'}{p^s}(p^s-1)} q_m^{\mu\mu'}\overline{q_{m'}^{\mu'\mu'}}$$

$$= p^{-s}\frac{1 - e^{2\pi i\frac{\mu'-\mu}{p}}}{1 - e^{2\pi i\frac{\frac{\mu'-\mu}{p}+m-m'}{p^s}}} q_m^{\mu\mu'}\overline{q_{m'}^{\mu'\mu'}}.$$

According to (8.6.27), (8.6.28) and (8.6.24),

$$(C_\mu^{-1}C_{\mu'})_{mm'} = \sum_{n=0}^{p^s-1}(C_\mu^{-1})_{mn}(C_{\mu'})_{nm'} = \sum_{n=0}^{p^s-1}\overline{(v_{\mu m})_n}(v_{\mu' m'})_n$$

$$= p^{-s}\sum_{n=0}^{p^s-1}e^{2\pi i\frac{\frac{\mu'-\mu}{p}+m-m'}{p^s}n} = p^{-s}\frac{1 - e^{2\pi i\frac{\mu'-\mu}{p}}}{1 - e^{2\pi i\frac{\frac{\mu'-\mu}{p}+m-m'}{p^s}}}.$$

Thus, we have

$$\sum_{n=0}^{p^s-1}\widetilde{b}_{mn}^{\mu\mu'}\overline{\widetilde{b}_{m'n}^{\mu'\mu'}} = (C_\mu^{-1}C_{\mu'})_{mm'} q_m^{\mu\mu'}\overline{q_{m'}^{\mu'\mu'}}.$$

(b) Let $\mu = \mu' = \nu$, and $m \neq m'$. In this case, using (8.6.36), (8.6.33) and (8.6.35), we have

$$\sum_{n=0}^{p^s-1}\widetilde{b}_{mn}^{\mu\nu}\overline{\widetilde{b}_{m'n}^{\mu'\nu}} = \sum_{n=0}^{p^s-1}\widetilde{b}_{mn}^{\mu\mu}\overline{\widetilde{b}_{m'n}^{\mu\mu}} = 0 = (C_\mu^{-1}C_\mu)_{mm'} q_m^{\mu\mu}\overline{q_{m'}^{\mu\mu}}.$$

(c) Let $\mu = \mu' = \nu$, and $m = m'$. In this case, using (8.6.36), (8.6.33) and (8.6.35), we have

$$\sum_{n=0}^{p^s-1}\widetilde{b}_{mn}^{\mu\nu}\overline{\widetilde{b}_{m'n}^{\mu'\nu}} = \sum_{n=0}^{p^s-1}\widetilde{b}_{mn}^{\mu\mu}\overline{\widetilde{b}_{mn}^{\mu\mu}} = |q_m^{\mu\mu}|^2 = (C_\mu^{-1}C_\mu)_{mm} q_m^{\mu\mu}\overline{q_m^{\mu\mu}}.$$

(d) Let $\mu \neq \nu$, $\mu' \neq \nu$. In this case, using (8.6.36) and (8.6.33), we have

$$\sum_{n=0}^{p^s-1} \widetilde{b}_{mn}^{\mu\nu} \overline{\widetilde{b}_{m'n}^{\mu'\nu}} = p^{-2s} \left(\chi_p\left(-\frac{\nu}{p}\right) - \chi_p\left(-\frac{\mu}{p}\right) \right) \left(\overline{\chi_p\left(-\frac{\nu}{p}\right)} - \overline{\chi_p\left(-\frac{\mu'}{p}\right)} \right)$$

$$\times q_m^{\mu\nu} \overline{q_{m'}^{\mu'\nu}} \sum_{n=0}^{p^s-1} \frac{1}{(\lambda_{\mu m} - \lambda_{\nu n})(\overline{\lambda_{\mu' m'} - \lambda_{\nu n}})}. \qquad (8.6.39)$$

This formula is applicable for $\mu = \mu'$ and for $\mu \neq \mu'$. To complete the proof of (8.6.37) it remains to check that

$$\sum_{n=0}^{p^s-1} \frac{1}{(\lambda_{\mu m} - \lambda_{\nu n})(\overline{\lambda_{\mu' m'} - \lambda_{\nu n}})}$$

$$= \frac{p^{2s} \overline{(C_\mu^{-1} C_{\mu'})_{mm'}}}{\left(\chi_p\left(-\frac{\nu}{p}\right) - \chi_p\left(-\frac{\mu}{p}\right) \right) \left(\overline{\chi_p\left(-\frac{\nu}{p}\right)} - \overline{\chi_p\left(-\frac{\mu'}{p}\right)} \right)}. \qquad (8.6.40)$$

8. To compute the sum $\sum_{n=0}^{p^s-1} (\lambda_{\mu m} - \lambda_{\nu n})^{-1} (\overline{\lambda_{\mu' m'} - \lambda_{\nu n}})^{-1}$, let us consider the functions

$$\psi_\mu^{(s)}(x) = \sum_{\nu=1}^{p-1} \alpha_{\nu;0}^\mu \psi_\nu^{(0)}(x), \qquad \mu = 1, \ldots, p-1,$$

where $\alpha_{\nu;0}^\mu = \omega_{\mu\nu}$, $\omega_{\mu\nu}$ are the entries of a unitary $(p-1) \times (p-1)$ matrix.

According to (8.6.6) and (8.6.7), we have $D_{\mu\nu} = \omega_{\mu\nu} E$, where E is the identity $p^s \times p^s$ matrix, which implies that D is a unitary matrix, and

$$u^\mu = (\omega_{\mu 1}, 0, \ldots, 0, \omega_{\mu 2}, 0, \ldots, 0, \ldots\ldots, \omega_{\mu p-1}, 0, \ldots, 0)^T.$$

Let $\mu \neq \nu$. It follows from (8.6.11), (8.6.12) and (8.6.33) that

$$b_{mn}^{\mu\nu} = \omega_{\nu\mu} \delta_{mn}, \qquad q_m^{\mu\nu} = \omega_{\mu\nu}, \qquad (8.6.41)$$

$\mu, \nu = 1, 2, \ldots, p-1$; $m, n = 0, 1, \ldots, p^s - 1$. Due to (8.6.41), (8.6.30) and (8.6.31), $B_{\mu\nu} = D_{\mu\nu} = \omega_{\mu\nu} E$, and

$$\widetilde{B}_{\mu\nu} = (C^{-1} B C)_{\mu\nu} = \sum_{\alpha=1}^{p-1} \sum_{\beta=1}^{p-1} C_{\mu\alpha}^{-1} B_{\alpha\beta} C_{\beta\nu}$$

$$= \sum_{\alpha=1}^{p-1} \sum_{\beta=1}^{p-1} C_\mu^{-1} \delta_{\mu\alpha} \omega_{\alpha\beta} C_\beta \delta_{\beta\nu} = \omega_{\mu\nu} C_\mu^{-1} C_\nu. \qquad (8.6.42)$$

Thus,

$$\widetilde{b}_{mn}^{\mu\nu} = \omega_{\mu\nu} (C_\mu^{-1} C_\nu)_{mn}. \qquad (8.6.43)$$

Substituting (8.6.41) and (8.6.43) into (8.6.39) with $\mu \neq \nu$, $\mu' \neq \nu$, we obtain (8.6.40).

9. Since C is a unitary matrix, the matrix B is unitary if and only if $\widetilde{B} = C^{-1}BC$ is unitary, i.e., $\sum_{\nu=1}^{p-1} \sum_{n=0}^{p^s-1} \widetilde{b}_{mn}^{\mu\nu} \overline{\widetilde{b}_{m'n}^{\mu'\nu}} = \delta_{\mu\mu'}\delta_{mm'}$. According to (8.6.37) and (8.6.38), the last relation holds if and only if

$$\sum_{\nu=1}^{p-1} q_m^{\mu\nu} \overline{q_{m'}^{\mu'\nu}} = \begin{cases} 1, & \text{if } \mu = \mu', \; m = m', \\ 0, & \text{if } \mu \neq \mu'. \end{cases} \qquad (8.6.44)$$

Set $Q_m^\mu = (q_m^{\mu 1}, \ldots, q_m^{\mu p-1})$. It follows from (8.6.44) that the vectors Q_0^μ, $\mu = 1, \ldots, p-1$, are the rows of a unitary $(p-1) \times (p-1)$ matrix $Z = (z_{\mu\nu})$. Any vector Q_m^μ is orthogonal to all vectors $Q_0^{\mu'}$ with $\mu' \neq \mu$. Hence, $Q_m^\mu = \sigma_{\mu m} Q_0^\mu$, where $|\sigma_{\mu m}| = 1$. So, we have

$$q_m^{\mu\nu} = \sigma_{\mu m} z_{\mu\nu}. \qquad (8.6.45)$$

10. From (8.6.11), (8.6.7) and (8.6.10), $\alpha_{\nu;k}^\mu = b_{k0}^{\nu\mu}$, where $b_{mm}^{\nu\mu}$ are elements of the block $B_{\nu\mu}$ of the matrix B (see (8.6.15), (8.6.14)). Since $B = C\widetilde{B}C^{-1}$, we have $B_{\nu\mu} = (C\widetilde{B}C^{-1})_{\nu\mu} = C_\nu \widetilde{B}_{\nu\mu} C_\mu^{-1}$. Thus,

$$\alpha_{\nu;k}^\mu = b_{k0}^{\nu\mu} = \sum_{m=0}^{p^s-1} \sum_{n=0}^{p^s-1} (C_\nu)_{km} (\widetilde{B}_{\nu\mu})_{mn} (C_\mu^{-1})_{n0}. \qquad (8.6.46)$$

Using (8.6.27), (8.6.28), (8.6.36), (8.6.35) and (8.6.45), we can rewrite (8.6.46) as

$$\alpha_{\nu;k}^\mu = b_{k0}^{\nu\mu} = \sum_{m=0}^{p^s-1} \sum_{n=0}^{p^s-1} (\nu_\nu{}_m)_n \widetilde{b}_{mn}^{\nu\mu} \overline{(\nu_\mu{}_n)_0}$$

$$= p^{-2s} \sum_{m=0}^{p^s-1} \sum_{n=0}^{p^s-1} e^{-2\pi i \frac{-\frac{\nu}{p}+m}{p^s} k} \frac{1 - e^{2\pi i \frac{\mu-\nu}{p}}}{e^{2\pi i \frac{\mu-\nu}{p}+m-n} - 1} \sigma_{\nu m} z_{\nu\mu}, \qquad (8.6.47)$$

whenever $\mu \neq \nu$ and

$$\alpha_{\mu;k}^\mu = b_{k0}^{\mu\mu} = -p^{-s} \sum_{m=0}^{p^s-1} e^{-2\pi i \frac{-\frac{\mu}{p}+m}{p^s} k} \sigma_{\mu m} z_{\mu\mu}, \qquad (8.6.48)$$

for all $\mu, \nu = 1, 2, \ldots, p-1$; $k = 0, 1, \ldots, p^s - 1$. Relations (8.6.47) and (8.6.48) imply (8.6.3).

11. Conversely, assume that the functions ψ_μ, $\mu = 1, \ldots, p-1$, are given by (8.6.2) and (8.6.3). Define $q_m^{\mu\nu}$ by (8.6.45). Evidently, (8.6.44) is satisfied, which together with (8.6.38) implies the unitarity of $\widetilde{B}$ whose entries are given by (8.6.35) and (8.6.34). Hence the matrix B defined by (8.6.30) is also unitary. Taking (8.6.32) into account, it is not difficult to check (8.6.12). It follows

that the functions $\psi_\mu\left(x - \frac{r}{p^s}\right)$, $r = 0, \ldots, p^s - 1$, $\mu = 1, \ldots, p - 1$, form an orthonormal system, and, from (8.6.5), the functions $\psi_\nu^{(0)}\left(x - \frac{r}{p^s}\right)$, $r = 0, 1, \ldots, p^s - 1$, $\nu = 1, \ldots, p - 1$, may be decomposed with respect to this system. Now it is easy to see that each $\psi_\nu^{(0)}(x - b)$, $b \in I_p$, may be decomposed with respect to the functions $\psi_\mu(x - a)$, $a \in I_p$, $\mu = 1, \ldots, p - 1$. So we have proved that ψ_μ, $\nu = 1, \ldots, p - 1$, is a set of wavelet functions. □

All dilatations and shifts of any set of wavelet functions (8.6.2), (8.6.3) form a *p-adic orthonormal Haar wavelet basis* in $\mathcal{L}^2(\mathbb{Q}_p)$.

Remark 8.3. Repeating the reasoning of Remark 8.2 almost word for word, we conclude that there are *infinitely many bases* generated by the wavelet functions (8.6.2), (8.6.3) such that they *cannot be constructed* by the Benedetto method [50].

Just as above, we obtain that all wavelets which are generated by wavelet functions (8.6.2), (8.6.3) belong to belongs to the Lizorkin space $\Phi(\mathbb{Q}_p)$.

8.7. *p*-adic refinable functions and multiresolution analysis

8.7.1. Construction of refinable functions

Now we are going to study *p*-adic refinement equations and their solutions. We restrict ourselves by the refinement equations (8.3.5) with a *finite* number of the terms on the right-hand side:

$$\phi(x) = \sum_{k=0}^{p^s-1} \beta_k \phi\left(\frac{1}{p}x - \frac{k}{p^s}\right). \tag{8.7.1}$$

If $\phi \in \mathcal{L}^2(\mathbb{Q}_p)$, taking the Fourier transform and using (4.9.4), we can rewrite (8.7.1) as

$$\widehat{\phi}(\xi) = m_0\left(\frac{\xi}{p^{s-1}}\right)\widehat{\phi}(p\xi), \tag{8.7.2}$$

where

$$m_0(\xi) = \frac{1}{p}\sum_{k=0}^{p^s-1} \beta_k \chi_p(k\xi) \tag{8.7.3}$$

is a trigonometric polynomial (which is called the mask). It is clear that $m_0(0) = 1$ whenever $\widehat{\phi}(0) \neq 0$.

Theorem 8.7.1. *If $\phi \in \mathcal{L}^2(\mathbb{Q}_p)$ is a solution of refinement equation* (8.7.1) *and* supp $\widehat{\phi} \subset B_0(0)$, *then axiom* (a) *from Definition* 8.1 *holds for the spaces* (8.3.6).

Proof. Since $\chi_p(-\xi) = 1$ for $\xi \in B_0(0)$, we have $\chi_p(-\xi)\widehat{\phi}(\xi) = \widehat{\phi}(\xi)$. Applying the Fourier transform, we obtain $\phi(x + 1) = \phi(x)$. So ϕ is a 1-periodic function. Since either $\frac{a}{p} + \frac{k}{p^s} \in I_p$ or $\frac{a}{p} + \frac{k}{p^s} - 1 \in I_p$ for any $a \in I_p$ and any $k = 0, \dots, p^s - 1$, because of the 1-*periodicity* of ϕ, it follows from (8.7.1) that $\phi(x - b) \in V_1$ for all $b \in I_p$. This implies $V_0 \subset V_1$, and, similarly, $V_j \subset V_{j+1}$ for any $j \in \mathbb{Z}$. $\square$

Theorem 8.7.2. *Let $\phi \in \mathcal{L}^2(\mathbb{Q}_p)$ be a solution of the refinement equation* (8.7.1) *and* supp $\widehat{\phi} \subset B_0(0)$. *Axiom* (b) *of Definition* 8.1 *holds for the spaces* (8.3.6) *(i.e.,* $\overline{\cup_{j \in \mathbb{Z}} V_j} = \mathcal{L}^2(\mathbb{Q}_p)$) *if and only if*

$$\bigcup_{j \in \mathbb{Z}} \text{supp } \widehat{\phi}(p^j \cdot) = \mathbb{Q}_p. \tag{8.7.4}$$

Proof. First of all we note that, from axioms (d) and (e), each space V_j is invariant with respect to the shifts $t = p^j a$, $a \in I_p$. We shall show that the space $\overline{\cup_{j \in \mathbb{Z}} V_j}$ is invariant with respect to any shift $t \in \mathbb{Q}_p$. Every $t \in \mathbb{Q}_p$ may be approximated by a vector $p^j a$, $a \in I_p$, with arbitrary large $j \in \mathbb{Z}$. If $f \in \cup_{j \in \mathbb{Z}} V_j$, by axiom (a) (which holds due to Theorem 8.7.1), then $f \in V_j$ for all $j \geq j_1$. It follows from the continuity of the function $\|f(\cdot + t)\|_2$ that $f(\cdot + t) \in \overline{\cup_{j \in \mathbb{Z}} V_j}$. Now let $t \in \mathbb{Q}_p$. If $g \in \overline{\cup_{j \in \mathbb{Z}} V_j}$, then approximating g by the functions $f \in \cup_{j \in \mathbb{Z}} V_j$, again using the continuity of the shift operator and the invariance of $\mathcal{L}^2$ norm with respect to the shifts, we derive $g(\cdot + t) \in \overline{\cup_{j \in \mathbb{Z}} V_j}$. For $X \subset \mathcal{L}^2(\mathbb{Q}_p)$, set $\widehat{X} = \{ \widehat{f} : f \in X \}$. By the Wiener theorem for $\mathcal{L}^2$ (see, e.g., [192, Appendix A.8.], [207]; all the arguments of the proof given there may be repeated word for word replacing $\mathbb{R}$ by $\mathbb{Q}_p$), a closed subspace X of the space $\mathcal{L}^2(\mathbb{Q}_p)$ is invariant with respect to the shifts if and only if $\widehat{X} = \mathcal{L}^2(\Omega)$ for some set $\Omega \subset \mathbb{Q}_p$. Let $X = \overline{\cup_{j \in \mathbb{Z}} V_j}$, then $\widehat{X} = L^2(\Omega)$. Thus $X = \mathcal{L}^2(\mathbb{Q}_p)$ if and only if $\Omega = \mathbb{Q}_p$. We shall set $\phi_j = \phi(p^{-j} \cdot)$, $\Omega_0 = \cup_{j \in \mathbb{Z}} \text{supp } \widehat{\phi}_j$ and prove that $\Omega = \Omega_0$. Indeed, since $\phi_j \in V_j$, $j \in \mathbb{Z}$, we have supp $\widehat{\phi}_j \subset \Omega$, and hence $\Omega_0 \subset \Omega$. Now assume that $\Omega \backslash \Omega_0$ contains a set of positive measure Ω_1. If $f \in V_j$, taking the Fourier transform from the expansion $f = \sum_{a \in I_p} h_a \phi(p^{-j} \cdot -a)$ we see that $\widehat{f} = 0$ almost everywhere on Ω_1. Hence the same is true for any $f \in \cup_{j \in \mathbb{Z}} V_j$. Passing to the limit we deduce that the Fourier transform of any $f \in X$ is equal to zero almost everywhere on Ω_1, i.e., $\mathcal{L}^2(\Omega) = \mathcal{L}^2(\Omega_0)$. It remains only to note that supp $\widehat{\phi}_j = \text{supp } \widehat{\phi}(p^j \cdot)$ $\square$

A real analog of Theorem 8.7.2 was proved by C. de Boor, R. DeVore and A. Ron in [56].

Theorem 8.7.3. *If* $\phi \in \mathcal{L}^2(\mathbb{Q}_p)$ *and the system* $\{\phi(x - a) : a \in I_p\}$ *is orthonormal, then axiom (c) of Definition 8.1 holds, i.e.,* $\cap_{j \in \mathbb{Z}} V_j = \{0\}$.

Proof. First, using the standard scheme (e.g., see [192, Lemma 1.2.8.]), we prove that for any $f \in \mathcal{L}^2(\mathbb{Q}_p)$

$$\lim_{j \to -\infty} \sum_{a \in I_p} \left| \left(f, p^{j/2} \phi(p^{-j} \cdot - a) \right) \right|^2 = 0, \tag{8.7.5}$$

where $(\cdot, \cdot)$ is the scalar product in $\mathcal{L}^2(\mathbb{Q}_p)$. Since according to Proposition 4.3.3 the space $\mathcal{D}(\mathbb{Q}_p)$ is dense in $\mathcal{L}^2(\mathbb{Q}_p)$, it suffices to prove (8.7.5) for any $\varphi \in \mathcal{D}(\mathbb{Q}_p)$. If $\varphi \in \mathcal{D}(\mathbb{Q}_p)$, then there exists N such that $\varphi(x) = 0$ for all $|\xi|_p > p^N$. Since $|\varphi(x)| \leq M$ for all $|\xi|_p \leq p^N$, we have

$$\sum_{a \in I_p} \left| \left(\varphi(\cdot), p^{j/2} \phi(p^{-j} \cdot - a) \right) \right|^2 \leq p^j \sum_{a \in I_p} \left(\int_{|x|_p \leq p^N} |\varphi(x)| |\phi(p^{-j}x - a)| \, dx \right)^2$$

$$\leq p^{j+N} M^2 \sum_{a \in I_p} \int_{|x|_p \leq p^N} |\phi(p^{-j}x - a)|^2 \, dx.$$

By the change of variables $\eta = p^{-j}x - a$, we obtain

$$\sum_{a \in I_p} \left| \left(\varphi, p^{j/2} \phi(p^{-j} \cdot - a) \right) \right|^2 \leq M^2 p^N \int_{A_{Nj}} |\phi(\eta)|^2 \, d\eta$$

$$= M^2 p^N \int_{\mathbb{Q}_p} \theta_{Nj}(\eta) |\phi(\eta)|^2 \, d\eta,$$

where θ_{Nj} is the characteristic function of the set

$$A_{Nj} = \cup_{a \in I_p} \{ \eta : |\eta + a|_p \leq 2^{N+j} \}.$$

Since $\lim_{j \to -\infty} \theta_{Nj}(\eta) = 0$ for any $\eta \neq -a$, using Lebesgue's dominated convergence Theorem 3.2.4, we obtain

$$\lim_{j \to -\infty} \int_{\mathbb{Q}_p} \theta_{Nj}(\eta) |\phi(\eta)|^2 \, d\eta = 0.$$

If we now assume that $f \in \cap_{j \in \mathbb{Z}} V_j$, then $f \in V_j$ for all $j \in \mathbb{Z}$ and, from (8.7.5),

$$\|f\| = \left(\sum_{a \in I_p} \left| \left(f, p^{j/2} \phi(p^{-j} \cdot - a) \right) \right|^2 \right)^{1/2} \to 0, \quad j \to -\infty,$$

i.e., $\|f\| = 0$, which was to be proved. $\qquad \square$

A real analog of Theorem 8.7.3 is well known.

Theorem 8.7.4. *Let ϕ be a refinable function such that* supp $\widehat{\phi} \subset B_0(0)$. *If* $|\widehat{\phi}(\xi)| = 1$ *for all* $\xi \in B_0(0)$ *then the system* $\{\phi(x-a) : a \in I_p\}$ *is orthonormal.*

Proof. Taking (4.8.10) into account, using the inclusion supp $\widehat{\phi} \subset B_0(0)$ and the Parseval–Steklov formula (4.8.6), we have for any $a \in I_p$

$$\left(\phi(\cdot), \phi(\cdot - a)\right) = \int_{\mathbb{Q}_p} \phi(x)\overline{\phi}(x-a)\,dx = \int_{B_0(0)} |\widehat{\phi}(\xi)|^2 \chi_p(a\xi)\,d\xi$$

$$= \int_{B_0(0)} \chi_p(a\xi)\,d\xi = \int_{\mathbb{Q}_p} \Omega(|\xi|_p)\chi_p(a\xi)\,d\xi = \Omega(|a|_p) = \delta_{a0}.$$

$\square$

So, to construct a MRA we can take a function ϕ for which the hypotheses of Theorems 8.7.4 and (8.7.4) are fulfilled. Next we are going to describe all such functions.

Proposition 8.7.5. *If $\phi \in \mathcal{L}^2(\mathbb{Q}_p)$ is a solution of refinement equation (8.7.1), $\widehat{\phi}(\xi)$ is continuous at the point 0 and $\widehat{\phi}(0) \neq 0$, then*

$$\widehat{\phi}(\xi) = \widehat{\phi}(0) \prod_{j=1}^{\infty} m_0\left(\frac{\xi}{p^{s-j}}\right). \tag{8.7.6}$$

Proof. Iterating (8.7.2) N times, $N \geq 1$, we have

$$\widehat{\phi}(\xi) = \prod_{j=1}^{N} m_0\left(\frac{\xi}{p^{s-j}}\right)\widehat{\phi}(p^N\xi). \tag{8.7.7}$$

Taking into account that $\widehat{\phi}(\xi)$ is continuous at the point 0 and the fact that $|p^N\xi|_p = p^{-N}|\xi|_p \to 0$ as $N \to +\infty$ for any $\xi \in \mathbb{Q}_p$, we obtain (8.7.6). $\square$

Proposition 8.7.6. *If $\widehat{\phi}$ is defined by (8.7.6), where m_0 is a trigonometric polynomial (8.7.3), $m_0(0) = 1$, then (8.7.2) holds. Furthermore, if $\xi \in \mathbb{Q}_p$ such that $|\xi|_p = p^{-n}$, then $\widehat{\phi}(\xi) = \widehat{\phi}(0)$ for $n \geq s-1$, and*

$$\widehat{\phi}(\xi) = \widehat{\phi}(0) \prod_{j=1}^{s-n-1} m_0\left(\frac{\xi}{p^{s-j}}\right) = \widehat{\phi}(0) \prod_{j=1}^{s-n-1} m_0\left(\frac{\widetilde{\xi}}{p^{s-j}}\right) = \widehat{\phi}(\widetilde{\xi}) \tag{8.7.8}$$

where $\widetilde{\xi} = \xi_n p^n + \xi_{n+1} p^{n+1} + \cdots + \xi_{s-2} p^{s-2}$, $\xi_n \neq 0$, for $n \leq s-2$.

Proof. Relation (8.7.6) implies $\widehat{\phi}(p\xi) = \widehat{\phi}(0) \prod_{j=1}^{\infty} m_0\left(\frac{\xi}{p^{s-j-1}}\right)$ and, consequently, (8.7.2). Let $n \leq s-2$, $|\xi|_p = p^{-n}$, i.e., $\xi = \xi_n p^n + \xi_{n+1} p^{n+1} + \cdots$, $\xi_n \neq 0$. Since $\chi_p\left(\frac{k\xi'}{p^{s-j}}\right) = 1$ whenever $\xi' \in B_{-s+1}(0)$, $j \in \mathbb{N}$, $k = 0, \ldots, p^s - 1$, and $\chi_p\left(\frac{k\xi}{p^{s-j}}\right) = 1$ whenever $j \geq s-n$, we have (8.7.8). It is clear that $\widehat{\phi}(\xi) = \widehat{\phi}(0)$ for $n \geq s-1$. $\square$

Corollary 8.7.7. *The function $\widehat{\phi}$ from Proposition 8.7.6 is locally constant. Moreover, if $M \geq -s + 2$, supp $\widehat{\phi} \subset B_M(0)$, then for all $k = 0, \ldots, p^{M+s-1} - 1$ and for all $x \in B_{s-1}\left(\frac{k}{p^M}\right)$ we have $\widehat{\phi}(x) = \widehat{\phi}\left(\frac{k}{p^M}\right)$.*

Proposition 8.7.8. *Let $\widehat{\phi}$ be defined by (8.7.6), where m_0 is a trigonometric polynomial (8.7.3). If $m_0(0) = 1$, $m_0\left(\frac{k}{p^s}\right) = 0$ for all $k = 1, \ldots, p^s - 1$ which are not divisible by p, then supp $\widehat{\phi} \subset B_0(0)$, $\widehat{\phi} \in \mathcal{L}^2(\mathbb{Q}_p)$. If, furthermore, $\left|m_0\left(\frac{k}{p^s}\right)\right| = 1$ for all $k = 1, \ldots, p^s - 1$ which are divisible by p, then $|\widehat{\phi}(x)| = |\widehat{\phi}(0)|$ for any $x \in B_0(0)$.*

Proof. By Proposition 8.7.6, $\widehat{\phi}$ satisfies (8.7.2) and (8.7.8). Let us check that $\widehat{\phi}(\xi) = 0$ for all ξ such that $|\xi|_p = p^M$, $M \geq 1$. It follows from (8.7.2) that it suffices to consider only $M = 1$. Let $|\xi|_p = p$, i.e., $\xi = \frac{1}{p}\xi_{-1} + \xi_0 + \xi_1 p + \cdots + \xi_{s-2}p^{s-2} + \xi'$, where $\xi_{-1} \neq 0$, $\xi' \in B_{-s+1}(0)$. In view of (8.7.8), we have $\widehat{\phi}(\xi) = \widehat{\phi}(\tilde{\xi}) = \widehat{\phi}\left(\frac{k}{p}\right)$, where $k = \xi_{-1} + \xi_0 p + \xi_1 p^2 + \cdots + \xi_{s-2}p^{s-1}$, $\xi_{-1} \neq 0$. Note that the first factor of the product in (8.7.8) is $m_0\left(\frac{k}{p^s}\right) = 0$. Thus $\widehat{\phi}(\xi) = 0$ for all ξ such that $|\xi|_p = p$. The rest of the statements follow from Propositions 8.7.5, 8.7.6. $\square$

From Theorems 8.7.1–8.7.4, the refinable functions with masks satisfying the hypotheses of Proposition 8.7.8 generate MRAs. Next, we will see that all properties of a mask m_0 described in Proposition 8.7.8 are necessary for the corresponding refinable function ϕ to be such that supp $\widehat{\phi} \subset B_0(0)$ and the system $\{\phi(x - a) : a \in I_p\}$ is orthonormal.

Theorem 8.7.9. *Let $\widehat{\phi}$ be defined by (8.7.6), where m_0 is a trigonometric polynomial (8.7.3). If supp $\widehat{\phi} \subset B_0(0)$ and the system $\{\phi(x - a) : a \in I_p\}$ is orthonormal, then $\left|m_0\left(\frac{k}{p^s}\right)\right| = 0$ whenever k is not divisible by p, and $\left|m_0\left(\frac{k}{p^s}\right)\right| = 1$ whenever k is divisible by p, $k = 1, \ldots, p^s - 1$.*

Proof. Let $a \in I_p$. Due to the orthonormality of $\{\phi(x - a) : a \in I_p\}$, using the Plancherel formula and Corollary 8.7.7, we have

$$\delta_{a0} = \big(\phi(\cdot), \phi(\cdot - a)\big) = \int_{\mathbb{Q}_p} \phi(x)\overline{\phi}(x - a)\, dx$$

$$= \int_{B_0(0)} |\widehat{\phi}(\xi)|^2 \chi_p(a\xi)\, d\xi = \sum_{k=0}^{p^{s-1}-1} \int_{|\xi - k|_p \leq p^{-s+1}} |\widehat{\phi}(\xi)|^2 \chi_p(a\xi)\, d\xi$$

$$= \sum_{k=0}^{p^{s-1}-1} |\hat{\phi}(k)|^2 \int_{|\xi-k|_p \le p^{-s+1}} \chi_p(a\xi)\,d\xi$$

$$= \sum_{k=0}^{p^{s-1}-1} |\hat{\phi}(k)|^2 \chi_p(ak) \int_{|\xi|_p \le p^{-s+1}} \chi_p(a\xi)\,d\xi$$

$$= \frac{1}{p^{s-1}}\Omega(|p^{s-1}a|_p) \sum_{k=0}^{p^{s-1}-1} |\hat{\phi}(k)|^2 \chi_p(ak).$$

Since $\Omega(|p^{s-1}a|_p) \ne 0$ for all $a \in B_{-s+1}(0)$, this yields

$$\frac{1}{p^{s-1}} \sum_{k=0}^{p^{s-1}-1} |\hat{\phi}(k)|^2 \chi_p(ak) = \delta_{a0}, \quad a = 0, \frac{1}{p^{s-1}}, \dots, \frac{p^{s-1}-1}{p^{s-1}}.$$

Consider these equalities as a linear system with respect to the unknowns $z_k = |\hat{\phi}(k)|^2$. It is easy to verify that this system has a unique solution $z_k = 1$, $k = 0, \dots, p^{s-1} - 1$, i.e., $|\hat{\phi}(k)| = 1$ for all $k = 0, \dots, p^{s-1} - 1$. In particular, it follows that $|\hat{\phi}(0)|^2 = 1$, which reduces (8.7.6) to

$$\hat{\phi}(\xi) = m_0\left(\frac{\xi}{p^{s-1}}\right) m_0\left(\frac{\xi}{p^{s-2}}\right) \cdots m_0(\xi). \qquad (8.7.9)$$

Let us check that $\left|m_0\left(\frac{k}{p^s}\right)\right| = 1$ for all k divisible by p, $k = 1, \dots, p^{s-1} - 1$. This is equivalent to $\left|m_0\left(\frac{k}{p^N}\right)\right| = 1$ whenever $N = 1, \dots, p^{s-1}$, $k = 1, \dots, p^N - 1$, k not divisible by p. We will prove this statement by induction on N. For the inductive base with $N = 1$, note that $1 = |\hat{\phi}(p^{s-2})| = |m_0(\frac{1}{p})|$. For the inductive step, assume that $\left|m_0\left(\frac{k}{p^n}\right)\right| = 1$ for all $n = 1, \dots, N$, $N \le s - 2$, $k = 1, \dots, p^n - 1$, k not divisible by p. Using (8.7.9) and 1-periodicity of m_0, we have

$$1 = \left|\hat{\phi}\left(lp^{s-N-2}\right)\right| = \left|m_0\left(\frac{l}{p^{N+1}}\right) m_0\left(\frac{l}{p^N}\right) \cdots m_0\left(\frac{l}{p^{N-s+2}}\right)\right|$$

$$= \left|m_0\left(\frac{l}{p^{N+1}}\right)\right|,$$

for all $l = 1, \dots, p^{N+1} - 1$, l not divisible by p. Now assume that $m_0\left(\frac{k}{p^s}\right) \ne 0$ for some $k = 1, \dots, p^s - 1$ not divisible by p. Since $\hat{\phi}(\frac{k}{p}) = 0$, from (8.7.9), there exists $n = 1, \dots, s$ such that $m_0\left(\frac{k}{p^{s-n}}\right) = 0$, which contradicts to the assumption $\left|m_0\left(\frac{k}{p^{s-n}}\right)\right| = 1$. $\qquad\square$

We have investigated the refinable functions whose Fourier transforms are supported in the unit disk $B_0(0)$. Such functions provide axiom (a) of Definition 8.1 because of a trivial argument given in Theorem 8.7.1. If supp $\widehat{\phi} \not\subset B_0(0)$, generally speaking, the relation $\phi(\cdot - a) \in V_1$ does not follow from the refinability of ϕ for all $a \in I_p$ because I_p is not a group. Nevertheless, we observed that some of such refinable functions also satisfy axiom (a).

Let $p = 2$, $s = 3$, ϕ be defined by (8.7.6), where m_0 is given by (8.7.3), $m_0(1/4) = m_0(3/8) = m_0(7/16) = m_0(15/16) = 0$. It is not difficult to see that supp $\widehat{\phi} \subset B_1(0)$, supp $\widehat{\phi} \not\subset B_0(0)$. Evidently, axiom (a) will be fulfilled whenever

$$\phi\left(x - \frac{k}{4}\right) = \sum_{r=0}^{7} \gamma_{kr}\phi\left(\frac{1}{2}x - \frac{r}{8}\right), \quad x \in \mathbb{Q}_2, \quad k = 1, 2, 3,$$

which is equivalent to

$$\widehat{\phi}(\xi)\chi_2\left(\frac{k\xi}{4}\right) = m_k\left(\frac{\xi}{4}\right)\widehat{\phi}(2\xi), \quad \xi \in \mathbb{Q}_2, \ k = 1, 2, 3,$$

where $m_k(\xi) = \frac{1}{2}\sum_{r=0}^{7}\gamma_{k,r}\chi_2(r\xi)$. Combining this with (8.7.2) we have

$$\widehat{\phi}(8\xi)(m_0(\xi)\chi_2(k\xi)) - m_k(\xi)) = 0, \quad \xi \in \mathbb{Q}_2, \ k = 1, 2, 3,$$

These equalities will be fulfilled for any $\xi \in \mathbb{Q}_2$ whenever they are fulfilled for $\xi = l/16$, $l = 0, \ldots, 15$. The desired polynomials m_k, $k = 1, 2, 3$, exist because we have

$$\widehat{\phi}\left(\frac{1}{2}\right) = \widehat{\phi}\left(\frac{3}{2}\right) = \widehat{\phi}\left(\frac{5}{2}\right) = \widehat{\phi}\left(\frac{9}{2}\right)$$
$$= \widehat{\phi}\left(\frac{11}{2}\right) = \widehat{\phi}\left(\frac{13}{2}\right) = \widehat{\phi}(1) = \widehat{\phi}(5) = 0.$$

So we have succeeded in proving axiom (a) of Definition 8.1, but, unfortunately, ϕ is not a scaling function generating MRA because axiom (e) is not valid. Moreover, it is possible to show that for any refinable function whose Fourier transform is in $B_1(0)$ but not in $B_0(0)$ the shift system $\{\phi(x - a) : a \in I_p\}$ is not orthogonal.

Evidently, any function $\phi \in \mathcal{D}(\mathbb{Q}_p)$ is p^M-periodic for some $M \in \mathbb{Z}$. Let us denote by $\mathcal{D}_{M;N}(\mathbb{Q}_p)$ the set of all p^M-periodic functions supported in $B_N(0)$. Now we summarize the above results.

Theorem 8.7.10. *Let $\widehat{\phi}$ be defined by (8.7.6):*

$$\widehat{\phi}(\xi) = \widehat{\phi}(0)\prod_{j=1}^{\infty} m_0\left(\frac{\xi}{p^{N-j}}\right), \quad \widehat{\phi}(0) = 1,$$

where m_0 is a trigonometric polynomial (8.7.3):

$$m_0(\xi) = \frac{1}{p} \sum_{k=0}^{p^{N+1}-1} \beta_k \chi_p(k\xi), \quad m_0(0) = 1.$$

If $m_0\left(\frac{k}{p^{N+1}}\right) = 0$ for all $k = 1, \ldots, p^{N+1} - 1$ not divisible by p, then $\phi \in \mathcal{D}_{0;N}$. If, furthermore, $\left|m_0\left(\frac{k}{p^{N+1}}\right)\right| = 1$ for all $k = 1, \ldots, p^{N+1} - 1$ divisible by p, then $\{\phi(x - a) : a \in I_p\}$ is an orthonormal system. Conversely, if $\mathrm{supp}\,\widehat{\phi} \subset B_0(0)$ and the system $\{\phi(x - a) : a \in I_p\}$ is orthonormal, then $\left|m_0\left(\frac{k}{p^{N+1}}\right)\right| = 0$ whenever k is not divisible by p, $\left|m_0\left(\frac{k}{p^{N+1}}\right)\right| = 1$ whenever k is divisible by p, $k = 1, 2, \ldots, p^{N+1} - 1$, and $|\widehat{\phi}(x)| = 1$ for any $x \in B_0(0)$.

8.7.2. Refinable functions and wavelet bases

If we already have a p-adic MRA generating by some scaling function, then we can find wavelet functions. Let the refinement equation for ϕ be (8.7.1). We will search wavelet functions $\psi^{(\nu)}$, $\nu = 1, \ldots, p - 1$, in the form

$$\psi^{(\nu)}(x) = \sum_{k=0}^{p^s-1} \gamma_{\nu k} \phi\left(\frac{1}{p}x - \frac{k}{p^s}\right), \tag{8.7.10}$$

where the coefficients $\gamma_{\nu k}$ are chosen such that

$$(\psi^{(\nu)}, \phi(\cdot - a)) = 0, \qquad (\psi^{(\nu)}, \psi^{(\mu)}(\cdot - a)) = \delta_{\nu\mu}\delta_{0a}, \tag{8.7.11}$$

$\nu, \mu = 1, \ldots, p - 1$, for any $a \in I_p$. It is clear that (8.7.11) is fulfilled for all $a \neq 0, \frac{1}{p^{s-1}}, \ldots, \frac{p^{s-1}-1}{p^{s-1}}$. Let

$$S = \begin{pmatrix} 0 & 0 & \ldots & 0 & 0 & 1 \\ 1 & 0 & \ldots & 0 & 0 & 0 \\ 0 & 1 & \ldots & 0 & 0 & 0 \\ \multicolumn{6}{c}{\ldots\ldots\ldots\ldots\ldots\ldots} \\ 0 & 0 & \ldots & 1 & 0 & 0 \\ 0 & 0 & \ldots & 0 & 1 & 0 \end{pmatrix}$$

be a $p^s \times p^s$ matrix, $B = \frac{1}{\sqrt{p}}(\beta_0, \ldots, \beta_{p^s-1})^T$, $G_\nu = \frac{1}{\sqrt{p}}(\gamma_{\nu 0}, \ldots, \gamma_{\nu p^s-1})^T$, where β_r and $\gamma_{\nu r}$ are coefficients in (8.7.1) and (8.7.10), respectively, $r = 0, 1 \ldots, p^s - 1$, $\nu = 1, \ldots, p - 1$,

To provide (8.7.11) for $a = 0, \frac{1}{p^{s-1}}, \ldots, \frac{p^{s-1}-1}{p^{s-1}}$ we should find vectors $G_1, \ldots, G_{p-1}$ so that the $p^s \times p^s$ matrix

$$U = \left(S^0 B, S^1 B, \ldots, S^{p^{s-1}-1} B, S^0 G_1, S^1 G_1, \ldots, S^{p^{s-1}-1} G_1, \ldots, \right.$$
$$\left. \ldots, S^0 G_{p-1}, S^1 G_{p-1}, \ldots, S^{p^{s-1}-1} G_{p-1} \right)$$

is unitary.

Example 8.7.1. Let $s = 1$. According to Proposition 8.7.8, set $m_0(0) = 1$, $m_0\left(\frac{k}{p}\right) = 0$ for all odd $k = 1, \ldots, p - 1$. Such a mask is

$$m_0(\xi) = \frac{1}{p} \sum_{k=0}^{p-1} \chi_p(k\xi)$$

(here $\beta_k = 1$ in (8.7.3) and $B = \frac{1}{\sqrt{p}}(1, \ldots, 1)^T$). The corresponding refinement equation (8.7.1) coincides with the "natural" refinement equation (8.1.5), and its solution $\Omega(|\cdot|_p)$ is a refinable function generating an MRA because of Theorems 8.7.1–8.7.4.

To find wavelets we observe that the unitary matrix

$$\left\{ \frac{1}{\sqrt{p}} e^{\frac{2\pi i k l}{p}} \right\}_{k,l=0,\ldots,p-1}$$

may be taken as U^T. Computing the wavelet functions corresponding to this matrix U, we derive the formulas (8.1.4) which were found by Kozyrev in [163]:
$\psi^{(v)}(x) = \chi_p\left(\frac{v}{p}x\right)\Omega(|x|_p), v = 1, \ldots, p - 1$.

Fundamental results in p-adic wavelet theory were obtained in [4]–[6]. In [5], a theorem which gives a *description of all test functions generating MRA* was proved. It turned out that the class of the orthogonal scaling functions is essentially smaller.

Theorem 8.7.11. ([5]) *If ϕ is an orthogonal scaling test function for some MRA such that $\widehat{\phi}(0) \neq 0$, then supp $\widehat{\phi} \subset B_0(0)$.*

So, the support of the Fourier transform of each *orthogonal scaling test function* is contained in $B_0(0)$, which is equivalent to its 1-periodicity. A complete description of all such functions was given above in Section 8.7.1 (see Theorem 8.7.10).

Thus, we have proved that there exist infinitely many different orthogonal scaling test functions. In contrast to the real setting, it turns out that all these functions generate the same MRA.

Theorem 8.7.12. ([5]) *There exists a unique MRA generated by an orthogonal scaling test function. This MRA coincides with the Haar MRA (described by Theorem 8.4.1) which is generated by the scaling function $\phi = \Omega(|\cdot|_p)$ (which is a solution of a "natural" refinement equation (8.1.5)).*

Thus in Sections 8.5 and 8.6 above *all p-adic compactly supported wavelet Haar bases were constructed.*

8.8. *p*-adic separable multidimensional MRA

Here we describe multidimensional wavelet bases constructed by means of a tensor product of one-dimensional MRAs. This standard approach for the construction of multivariate wavelets was suggested by Y. Meyer [187] (see, e.g., [192, §2.1]).

Let $\{V_j^{(\nu)}\}_{j\in\mathbb{Z}}$, $\nu = 1, \ldots, n$, be one-dimensional MRAs (see Section 8.3.1). We introduce subspaces V_j, $j \in \mathbb{Z}$, of $\mathcal{L}^2(\mathbb{Q}_p^n)$ by

$$V_j = \bigotimes_{\nu=1}^n V_j^{(\nu)} = \overline{\mathrm{span}\{F = f_1 \otimes \cdots \otimes f_n, \ f_\nu \in V_j^{(\nu)}\}}. \qquad (8.8.1)$$

Let $\phi^{(\nu)}$ be a refinable function of νth MRA $\{V_j^{(\nu)}\}_j$. Set

$$\Phi = \phi^{(1)} \otimes \cdots \otimes \phi^{(n)}. \qquad (8.8.2)$$

Since the system $\{\phi^{(\nu)}(\cdot - a)\}_{a_\nu \in I_p}$ is an orthonormal basis for $V_0^{(\nu)}$ (axiom (e) of Definition 8.1) for any $\nu = 1, \ldots, n$, it is clear that

$$V_0 = \overline{\mathrm{span}\{\Phi(\cdot - a) : a = (a_1, \ldots, a_n) \in I_p^n\}},$$

where $I_p^n = I_p \times \cdots \times I_p$ is the direct product of n sets I_p, and the system $\Phi(\cdot - a)$, $a \in I_p^n$, is an orthonormal basis for V_0. It follows from (8.8.1) and axiom (d) of Definition 8.1 that $f \in V_0$ if and only if $f(p^{-j}\cdot) \in V_j$ for all $j \in \mathbb{Z}$. Since axiom (a) from Definition 8.1 holds for any one-dimensional MRA $\{V_j^{(\nu)}\}_j$, it is easy to see that $\Phi(p^{-j} \cdot -a) \in V_{j+1}$ for any $a \in I_p^n$. Thus, $V_j \subset V_{j+1}$. It is not difficult to check that the axioms of completeness and separability for the spaces V_j hold. Thus we have the following statement.

Theorem 8.8.1. *Let $\{V_j^{(\nu)}\}_{j\in\mathbb{Z}}$, $\nu = 1, \ldots, n$, be KMAs in $\mathcal{L}^2(\mathbb{Q}_p)$. Then the subspaces V_j of $\mathcal{L}^2(\mathbb{Q}_p^n)$ defined by (8.8.1) satisfy the following properties:*
  (a) $V_j \subset V_{j+1}$ *for all $j \in \mathbb{Z}$;*
  (b) $\cup_{j\in\mathbb{Z}} V_j$ *is dense in $\mathcal{L}^2(\mathbb{Q}_p^n)$;*
  (c) $\cap_{j\in\mathbb{Z}} V_j = \{0\}$;

(d) $f(\cdot) \in V_j \Longleftrightarrow f(p^{-1}\cdot) \in V_{j+1}$ *for all* $j \in \mathbb{Z}$;

(e) *the system* $\{\Phi(x-a), a \in I_p^n\}$, *is an orthonormal basis for* V_0, *where* $\Phi \in V_0$ *is defined by* (8.8.2).

Similarly to Definition 8.1, the collection of spaces V_j, $j \in \mathbb{Z}$, which satisfies conditions (a)–(e) of Theorem 8.8.1 is called a *multiresolution analysis* in $\mathcal{L}^2(\mathbb{Q}_p^n)$, and the function Φ from axiom (e) is called *refinable*. It is clear that a multidimensional *refinable function* can be constructed by formula (8.8.2).

Next, we set

$$V_j = \overline{\operatorname{span}\{\Phi(p^{-j}\cdot -a) : a = (a_1, \ldots, a_n) \in I_p^n\}}, \quad j \in \mathbb{Z}, \quad (8.8.3)$$

and following the standard scheme (see, for example, [192, §2.1]), we define the wavelet spaces W_j as the orthogonal complement of V_j in V_{j+1}, i.e.,

$$W_j = V_{j+1} \ominus V_j, \quad j \in \mathbb{Z}.$$

Consequently,

$$\begin{aligned}
V_{j+1} &= \bigotimes_{\nu=1}^{n} V_{j+1}^{(\nu)} = \bigotimes_{\nu=1}^{n} \left(V_j^{(\nu)} \oplus W_j^{(\nu)}\right) \\
&= V_j \oplus \bigoplus_{e \subset \{1,\ldots,n\},\, e \neq \emptyset} \left(\bigotimes_{\nu \in e} W_j^{(\nu)}\right)\left(\bigotimes_{\mu \notin e} V_j^{(\mu)}\right).
\end{aligned}$$

So, the space W_j is a direct sum of $2^n - 1$ subspaces $W_{j,e}$, $e \subset \{1, \ldots, n\}$, $e \neq \emptyset$. Let $\psi^{(\nu)}$ be a wavelet function, i.e. a function whose shifts (with respect to $a \in I_p$) form an orthonormal basis for $W_0^{(\nu)}$. It is clear that the shifts (with respect to $a \in I_p^n$) of the function

$$\Psi_e = \left(\bigotimes_{\nu \in e} \psi^{(\nu)}\right)\left(\bigotimes_{\mu \notin e} \phi^{(\mu)}\right), \quad e \subset \{1, \ldots, n\}, \quad e \neq \emptyset, \quad (8.8.4)$$

form an orthonormal basis for $W_{0,e}$. So, we have

$$\mathcal{L}^2(\mathbb{Q}_p^n) = \bigoplus_{j \in \mathbb{Z}} W_j = \bigoplus_{j \in \mathbb{Z}} \left(\bigoplus_{e \subset \{1,\ldots,n\},\, e \neq \emptyset} W_{j,e}\right),$$

and the functions $p^{-nj/2}\Psi_e(p^j \cdot +a)$, $e \subset \{1, \ldots, n\}$, $e \neq \emptyset$, $j \in \mathbb{Z}$, $a \in I_p^n$, form an orthonormal basis for $\mathcal{L}^2(\mathbb{Q}_p^n)$.

8.9. Multidimensional p-adic Haar wavelet bases

8.9.1. Description of multidimensional 2-adic Haar MRA

Let us apply the above construction taking the 2-adic Haar MRA as νth one-dimensional multiresolution analysis $\{V_j^{(\nu)}\}_{j \in \mathbb{Z}}$, $\nu = 1, \ldots, n$.

To construct multivariate wavelet functions (8.8.4), we choose $\psi^{(0)}$ as a wavelet function for each one-dimensional MRA. Thus we have the following $2^n - 1$ multidimensional wavelet functions

$$\Psi_{\{1,\ldots,n\}}^{(0)}(x) = \psi^{(0)}(x_1)\psi^{(0)}(x_2)\cdots\psi^{(0)}(x_{n-1})\psi^{(0)}(x_n),$$

$$\Psi_{\{1,\ldots,n-1\}}^{(0)}(x) = \psi^{(0)}(x_1)\psi^{(0)}(x_2)\cdots\psi^{(0)}(x_{n-1})\phi(x_n),$$

$$\ldots\ldots\ldots\ldots\ldots\ldots\ldots\ldots\ldots\ldots\ldots\ldots\ldots\ldots\ldots$$

$$\Psi_{\{2,\ldots,n\}}^{(0)}(x) = \phi(x_1)\psi^{(0)}(x_2)\cdots\psi^{(0)}(x_{n-1})\psi^{(0)}(x_n),$$

$$\ldots\ldots\ldots\ldots\ldots\ldots\ldots\ldots\ldots\ldots\ldots\ldots\ldots\ldots\ldots$$

$$\ldots\ldots\ldots\ldots\ldots\ldots\ldots\ldots\ldots\ldots\ldots\ldots\ldots\ldots\ldots$$

$$\ldots\ldots\ldots\ldots\ldots\ldots\ldots\ldots\ldots\ldots\ldots\ldots\ldots\ldots\ldots$$

$$\Psi_{\{1\}}^{(0)}(x) = \psi^{(0)}(x_1)\phi(x_2)\cdots\phi(x_{n-1})\phi(x_n),$$

$$\ldots\ldots\ldots\ldots\ldots\ldots\ldots\ldots\ldots\ldots\ldots\ldots\ldots\ldots\ldots$$

$$\Psi_{\{n\}}^{(0)}(x) = \phi(x_1)\phi(x_2)\cdots\phi(x_{n-1})\psi^{(0)}(x_n).$$

According to (8.8.2), in this case the multidimensional *refinable function* is $\Phi(x) = \Omega(|x|_2)$, $x \in \mathbb{Q}_2^n$.

Let $e \subset \{1, \ldots, n\}$, $e \neq \emptyset$. Denote by $k_e = ((k_e)_1, \ldots, (k_e)_n)$ the vector whose coordinates are given by

$$(k_e)_\nu = \begin{cases} 1, & \nu \in e, \\ 0, & \nu \notin e, \end{cases} \quad \nu = 1, \ldots, n.$$

Since $\phi(x_\nu) = \Omega(|x_\nu|_2)$ and $\psi^{(0)}(x_\nu) = \chi_2(2^{-1}x_\nu)\Omega(|x_\nu|_2)$, $x_\nu \in \mathbb{Q}_2$, $\nu = 1, 2, \ldots, n$, (see (8.1.6) and (8.4.7)), it follows from (1.10.2) and (8.1.6) that the wavelet function $\Psi_e^{(0)}$ can be rewritten as

$$\Psi_e^{(0)}(x) = \chi_2(2^{-1}k_e \cdot x)\Omega(|x|_2), \quad x = (x_1, \ldots, x_n) \in \mathbb{Q}_2^n. \tag{8.9.1}$$

According to the above consideration, we have the following statement.

Theorem 8.9.1. *The system of functions*

$$\Psi_{e;ja}^{(0)}(x) = 2^{-nj/2}\Psi_e^{(0)}(2^j x - a)$$
$$= 2^{-nj/2}\chi_2(2^{-1}k_e \cdot (2^j x - a))\Omega(|2^j x - a|_2), \quad x \in \mathbb{Q}_2^n, \tag{8.9.2}$$

$e \subset \{1, \ldots, n\}$, $e \neq \emptyset$, $j \in \mathbb{Z}$, $a \in I_2^n$, *is an orthonormal basis for* $\mathcal{L}^2(\mathbb{Q}_2^n)$.

Now we construct multidimensional wavelet bases using different one-dimensional Haar wavelet bases (see Section 8.5.1). Namely, we apply the construction of Section 8.8 taking again the Haar MRA as νth one-dimensional multiresolution analysis $\{V_j^{(\nu)}\}_{j \in \mathbb{Z}}$, $\nu = 1, \ldots, n$, and choosing wavelet functions $\psi^{(s_\nu)}$ for construction of multivariate wavelet functions (8.8.4).

Let $s = (s_1, \ldots, s_n)$, where $s_\nu \in \mathbb{N}_0$, $\nu = 1, 2, \ldots, n$. We have the following $2^n - 1$ wavelet functions:

$$\Psi_{\{1,\ldots,n\}}^{(s)}(x) = \psi^{(s_1)}(x_1)\psi^{(s_2)}(x_2)\cdots\psi^{(s_{n-1})}(x_{n-1})\psi^{(s_n)}(x_n),$$
$$\Psi_{\{1,\ldots,n-1\}}^{(s)}(x) = \psi^{(s_1)}(x_1)\psi^{(s_2)}(x_2)\cdots\psi^{(s_{n-1})}(x_{n-1})\phi(x_n),$$
$$\cdots\cdots\cdots\cdots\cdots\cdots\cdots\cdots\cdots\cdots\cdots\cdots\cdots\cdots$$
$$\Psi_{\{2,\ldots,n\}}^{(s)}(x) = \phi(x_1)\psi^{(s_2)}(x_2)\cdots\psi^{(s_{n-1})}(x_{n-1})\psi^{(s_n)}(x_n),$$
$$\cdots\cdots\cdots\cdots\cdots\cdots\cdots\cdots\cdots\cdots\cdots\cdots\cdots\cdots$$
$$\cdots\cdots\cdots\cdots\cdots\cdots\cdots\cdots\cdots\cdots\cdots\cdots\cdots\cdots$$
$$\Psi_{\{1\}}^{(s)}(x) = \psi^{(s_1)}(x_1)\phi(x_2)\cdots\phi(x_{n-1})\phi(x_n),$$
$$\cdots\cdots\cdots\cdots\cdots\cdots\cdots\cdots\cdots\cdots\cdots\cdots\cdots\cdots$$
$$\Psi_{\{n\}}^{(s)}(x) = \phi(x_1)\phi(x_2)\cdots\phi(x_{n-1})\psi^{(s_n)}(x_n).$$

Set $\alpha_r^1 = \alpha_r$, where α_r is given by (8.5.5), $r = 0, 1, \ldots, 2^s - 1$, and $\alpha_0^0 = 1$, $\alpha_1^0 = \cdots \alpha_{2^s-1}^0 = 0$. Since $\phi(x_\nu) = \Omega(|x_\nu|_2)$, $x_\nu \in \mathbb{Q}_2$, and $\psi^{(s_\nu)}$ is given by (8.5.4), (8.5.5), $\nu = 1, 2, \ldots, n$, from (1.10.2), the wavelet functions $\Psi_e^{(s)}$ can be rewritten as

$$\Psi_e^{(s)}(x) = \sum_{r_1=0}^{2^{s_1}-1}\cdots\sum_{r_n=0}^{2^{s_n}-1} \alpha_{r_1}^{(k_e)_1}\cdots\alpha_{r_n}^{(k_e)_n}$$
$$\times \Psi_e^{(0)}\left(x - \left(\frac{r_1}{2^{s_1}}(k_e)_1, \ldots, \frac{r_n}{2^{s_n}}(k_e)_n\right)\right), \qquad (8.9.3)$$

$x \in \mathbb{Q}_2^n$, where $\Psi_e^{(0)}$ is defined by (8.9.1), $e \subset \{1, \ldots, n\}$, $e \neq \emptyset$.

According to the above consideration, we have the following theorem.

Theorem 8.9.2. *The system of functions*

$$\Psi_{e;ja}^{(s)}(x) = \sum_{r_1=0}^{2^{s_1}-1}\cdots\sum_{r_n=0}^{2^{s_n}-1} \alpha_{r_1}^{(k_e)_1}\cdots\alpha_{r_n}^{(k_e)_n}$$
$$\times \Psi_e^{(0)}\left(2^j x - a - \left(\frac{r_1}{2^{s_1}}(k_e)_1, \ldots, \frac{r_n}{2^{s_n}}(k_e)_n\right)\right), \qquad (8.9.4)$$

$x \in \mathbb{Q}_2^n$, $e \subset \{1, \ldots, n\}$, $e \neq \emptyset$, $j \in \mathbb{Z}$, $a \in I_2^n$, *forms an orthonormal basis for* $\mathcal{L}^2(\mathbb{Q}_2^n)$.

Since $\int_{\mathbb{Q}_2^n} \Psi_{e;ja}^{(s)}(x)\,d^2x = 0$, from (7.3.3), we have $\Psi_{e;ja}^{(s)} \in \Phi(\mathbb{Q}_2^n)$.

8.9.2. One multidimensional p-adic Haar basis

Now we construct the n-dimensional wavelet basis generated by the one-dimensional Kozyrev wavelet basis (8.1.3). As above, we construct $p^n - 1$ wavelet functions:

$$\theta_{k_1}(x_1)\theta_{k_2}(x_2)\cdots\theta_{k_{n-1}}(x_{n-1})\theta_{k_n}(x_n),$$
$$\theta_{k_1}(x_1)\theta_{k_2}(x_2)\cdots\theta_{k_{n-1}}(x_{n-1})\phi(x_n),$$
$$\cdots\cdots\cdots\cdots\cdots\cdots\cdots\cdots\cdots$$
$$\phi(x_1)\theta_{k_2}(x_2)\cdots\theta_{k_{n-1}}(x_{n-1})\theta_{k_n}(x_n),$$
$$\cdots\cdots\cdots\cdots\cdots\cdots\cdots\cdots\cdots \quad (8.9.5)$$
$$\theta_{k_1}(x_1)\phi(x_2)\cdots\phi(x_{n-1})\phi(x_n),$$
$$\cdots\cdots\cdots\cdots\cdots\cdots\cdots\cdots\cdots$$
$$\phi(x_1)\phi(x_2)\cdots\phi(x_{n-1})\theta_{k_n}(x_n),$$

where θ_{k_r} is given by (8.1.4) and is expressed in terms of the *refinable function* $\phi(x) = \Omega(|x|_p)$ by (8.4.3); $x_k \in \mathbb{Q}_p$; $k_r \in J_p = \{1, 2, \ldots, p-1\}$, $r = 1, 2, \ldots, n$. Let

$$J_{p0}^n = \{(k_1, \ldots, k_n) :$$
$$k_r = 0, 1, 2, \ldots, p-1; r = 1, 2, \ldots, n; k_1 + \cdots + k_n \neq 0\}. \quad (8.9.6)$$

In view of (1.10.2) and (8.1.4), the functions in (8.9.5) can be rewritten as

$$\Theta_k(x) = \chi_p(p^{-1}k \cdot x)\Omega(|x|_p), \quad x = (x_1, \ldots, x_n) \in \mathbb{Q}_p^n, \quad (8.9.7)$$

where $k = (k_1, \ldots, k_n) \in J_{p0}^n$.

According to the general scheme from Section 8.8, (8.9.7) is generating wavelet functions. Nevertheless, we prove this statement directly.

Theorem 8.9.3. *All dilations and shifts of the wavelet function (8.9.7) form an orthonormal (p-adic Haar wavelet basis) in $\mathcal{L}^2(\mathbb{Q}_p^n)$:*

$$\Theta_{k; ja}(x) = p^{-nj/2}\chi_p(p^{-1}k \cdot (p^j x - a))\Omega(|p^j x - a|_p), \quad (8.9.8)$$

where $x \in \mathbb{Q}_p^n$, $k = (k_1, \ldots, k_n) \in J_{p0}^n$, $j \in \mathbb{Z}$, $a \in I_p^n$.

Proof. The statement follows from the standard MRA scheme in Section 8.8 (see also, for example, [192, §2.1]). Nevertheless, we prove this statement directly.

According to [163] (or a particular case of Theorem 8.10.1), the wavelet functions (8.1.3) form an orthonormal basis in $\mathcal{L}^2(\mathbb{Q}_p)$, i.e.,

$$\left(\theta_{k_r; ja_r}(x_r), \theta_{k_r'; j'a_r'}(x_r)\right) = \delta_{k_r k_r'}\delta_{j'j}\delta_{a_r a_r'}.$$

Moreover, it is easy to calculate that

$$\left(\theta_{k_r;ja_r}(x_r),\phi(p^jx_r-a_r')\right)=\int_{\mathbb{Q}_p}p^{-j/2}\chi_p\left(p^{-1}k_r(p^jx_r-a_r)\right)$$
$$\times\Omega\left(|p^jx_r-a_r|_p\right)\Omega\left(|p^jx_r-a_r'|_p\right)dx_r=0,$$

$r=1,\ldots,n$. Thus the system of the functions (8.9.8) is orthonormal.

To prove the completeness of the system of functions (8.10.3), we repeat the corresponding proof [163] almost word for word. Since the system of the characteristic functions of the balls $B_k^n(0)$ is complete in $\mathcal{L}^2(\mathbb{Q}_p^n)$, it is sufficient to verify the Parseval identity for the characteristic function $\Omega(|x|_p)$.

According to Theorem 8.10.1 (see also [163]), we have

$$\left(\Omega(|x_r|_p),\theta_{k_r;ja_k}(x_r)\right)=\begin{cases}p^{-j/2}, & a=0,\ j\geq 1,\\ 0, & \text{otherwise},\end{cases}\quad r=1,\ldots,n.$$

Using this relation, we can calculate

$$\sum_{j\in\mathbb{Z},k\in J_{p0}^n,a\in I_p^n}\left|\left(\Omega(|x|_p),\Theta_{k;ja}(x)\right)\right|^2=\sum_{k\in J_{p0}^n}\sum_{j=1}^{\infty}p^{-nj}$$
$$=\left((p-1)^n+C_n^{n-1}(p-1)^{n-1}+\cdots+(p-1)\right)\frac{p^{-n}}{1-p^{-n}}=1$$
$$=\left|\left(\Omega(|x|_p),\Omega(|x|_p)\right)\right|^2,\quad x\in\mathbb{Q}_p^n,$$

where C_n^r are binomial coefficients.

Thus the system of functions (8.9.8) is an orthonormal basis in $\mathcal{L}^2(\mathbb{Q}_p^n)$. □

8.9.3. Multidimensional p-adic Haar wavelet bases

Now, using the scheme of Section 8.8, we construct multidimensional wavelet bases using the different one-dimensional Haar wavelet bases which were described in Section 8.6. Namely, taking the Haar MRA as νth one-dimensional multiresolution analysis $\{V_j^{(\nu)}\}_{j\in\mathbb{Z}}$, $\nu=1,\ldots,n$, and choosing arbitrary sets of wavelet functions $\{\psi_{\mu_\nu}:\mu_\nu=1,2,\ldots,p-1\}$, $\nu=1,2,\ldots,n$ (which were described by Theorem 8.6.1), we construct the multivariate wavelet functions (8.8.4).

First we construct the $p^n - 1$ wavelet functions:

$$\Psi_{\{1,\ldots,n\}}^{(\mu)}(x) = \psi_{\mu_1}(x_1)\psi_{\mu_2}(x_2)\cdots\psi_{\mu_{n-1}}(x_{n-1})\psi_{\mu_n}(x_n),$$

$$\Psi_{\{1,\ldots,n-1\}}^{(\mu)}(x) = \psi_{\mu_1}(x_1)\psi_{\mu_2}(x_2)\cdots\psi_{\mu_{n-1}}(x_{n-1})\phi(x_n),$$

$$\cdots\cdots\cdots\cdots\cdots\cdots\cdots\cdots\cdots$$

$$\Psi_{\{2,\ldots,n\}}^{(\mu)}(x) = \phi(x_1)\psi_{\mu_2}(x_2)\cdots\psi_{\mu_{n-1}}(x_{n-1})\psi_{\mu_n}(x_n), \qquad (8.9.9)$$

$$\cdots\cdots\cdots\cdots\cdots\cdots\cdots\cdots\cdots$$

$$\Psi_{\{1\}}^{(\mu)}(x) = \psi_{\mu_1}(x_1)\phi(x_2)\cdots\phi(x_{n-1})\phi(x_n),$$

$$\cdots\cdots\cdots\cdots\cdots\cdots\cdots\cdots\cdots$$

$$\Psi_{\{n\}}^{(\mu)}(x) = \phi(x_1)\phi(x_2)\cdots\phi(x_{n-1})\psi_{\mu_n}(x_n),$$

where $\mu = (\mu_1,\ldots,\mu_n)$, $x \in \mathbb{Q}_p^n$, and any set of one-dimensional wavelet functions $\{\psi_{\mu_\nu}(x_\nu) : \mu_\nu = 1, 2, \ldots, p - 1; \ x_\nu \in \mathbb{Q}_p\}$ is given by (8.6.2) and (8.6.3), $\nu = 1, 2, \ldots, n$. Just as above we denote the wavelet functions (8.9.9) by

$$\Psi_e^{(\mu)}(x), \quad e \subset \{1,\ldots,n\}, \quad e \neq \emptyset, \quad \mu = (\mu_1,\ldots,\mu_n). \qquad (8.9.10)$$

According to (8.8.2), here a multidimensional *refinable function* is

$$\Phi(x) = \Omega(|x|_p), \qquad x \in \mathbb{Q}_p^n. \qquad (8.9.11)$$

According to the above general consideration, we have the following statement.

Theorem 8.9.4. *All dilations and shifts of the wavelet functions* (8.9.10) *form an orthonormal complete (p-adic Haar wavelet basis) in* $\mathcal{L}^2(\mathbb{Q}_p^n)$:

$$p^{-nj/2}\Psi_e^{(\mu)}(p^j x - a), \qquad x \in \mathbb{Q}_p^n, \qquad (8.9.12)$$

where $e \subset \{1,\ldots,n\}$, $e \neq \emptyset$, $\mu = (\mu_1,\ldots,\mu_n)$, $\mu_\nu = 1, 2, \ldots, p - 1$, $\nu = 1, 2, \ldots, n$; $j \in \mathbb{Z}$; $a \in I_p^n$.

8.10. One non-Haar wavelet basis in $\mathcal{L}^2(\mathbb{Q}_p)$

Let

$$J_{p;m} = \{s = p^{-m}\big(s_0 + s_1 p + \cdots + s_{m-1}p^{m-1}\big) :$$

$$s_j = 0, 1, \ldots, p - 1; \ j = 0, 1, \ldots, m - 1; s_0 \neq 0\}, \qquad (8.10.1)$$

where $m \geq 1$ is a *fixed* positive integer.

Let us introduce the set of $(p - 1)p^{m-1}$ functions

$$\theta_s^{(m)}(x) = \chi_p(sx)\Omega(|x|_p), \qquad s \in J_{p;m}, \qquad x \in \mathbb{Q}_p, \qquad (8.10.2)$$

and the family of functions generated by their dilations and translations:

$$\theta^{(m)}_{s;ja}(x) = p^{-j/2}\chi_p(s(p^j x - a))\Omega(|p^j x - a|_p), \quad x \in \mathbb{Q}_p, \quad (8.10.3)$$

where $s \in J_{p;m}$, $j \in \mathbb{Z}$, $a \in I_p$, and $\Omega(t)$ is the characteristic function (4.2.2) of the segment $[0, 1]$.

Theorem 8.10.1. *The functions* (8.10.3) *form an orthonormal p-adic wavelet basis in* $\mathcal{L}^2(\mathbb{Q}_p)$.

Proof. Consider the scalar product (see Section 3.3)

$$\left(\theta^{(m)}_{s';j'a'}(x), \theta^{(m)}_{s;ja}(x)\right) = p^{-(j+j')/2}\int_{\mathbb{Q}_p}\chi_p\big(s'(p^{j'}x - a') - s(p^\gamma x - a)\big)$$

$$\times \Omega(|p^j x - a|_p)\Omega(|p^{j'}x - a'|_p)\,dx. \quad (8.10.4)$$

If $j \le j'$, according to formula (4.7.4)

$$\Omega(|p^j x - a|_p)\Omega(|p^{j'}x - a'|_p) = \Omega(|p^j x - a|_p)\Omega(|p^{j'-j}a - a'|_p), \quad (8.10.5)$$

(8.10.4) can be rewritten as

$$\left(\theta^{(m)}_{s';j'a'}(x), \theta^{(m)}_{s;ja}(x)\right)$$
$$= p^{-(j+j')/2}\Omega(|p^{j'-j}a - a'|_p)$$
$$\times \int_{\mathbb{Q}_p}\chi_p\big(s'(p^{j'}x - a') - s(p^j x - a)\big)\Omega(|p^j x - a|_p)\,dx. \quad (8.10.6)$$

Let $j < j'$. Making the change of variables $\xi = p^j x - a$ and taking into account (4.8.10), from (8.10.6) we obtain

$$\left(\theta^{(m)}_{s';j'a'}(x), \theta^{(m)}_{s;ja}(x)\right) = p^{-(j+j')/2}\chi_p\big(s'(p^{j'-j}a - a')\big)$$
$$\times \Omega(|p^{j'-j}a - a'|_p)\int_{\mathbb{Q}_p}\chi_p\big((p^{j'-j}s' - s)\xi\big)\Omega(|\xi|_p)\,d\xi$$
$$= p^{-(j+j')/2}\chi_p\big(s'(p^{j'-j}a - a')\big)$$
$$\times \Omega(|p^{j'-j}a - a'|_p)\Omega(|p^{j'-j}s' - s|_p). \quad (8.10.7)$$

Since

$$p^{j'-j}s' = p^{j'-j-m}\big(s'_0 + s'_1 p + \cdots + s'_{j-1}p^{m-1}\big),$$
$$s = p^{-m}\big(s_0 + s_1 p + \cdots + s_{j-1}p^{m-1}\big),$$

where $s'_0, s_0 \ne 0$, $j' - j \le 1$, it is clear that for the fractional part we have $\{p^{j'-j}s' - s\}_p \ne 0$. Thus $\Omega(|p^{j'-j}s' - s|_p) = 0$ and $\left(\theta^{(m)}_{s';j'a'}(x), \theta^{(m)}_{s;ja}(x)\right) = 0$.

Consequently, the scalar product $\left(\theta^{(m)}_{s';\,j'a'}(x),\,\theta^{(m)}_{s;\,ja}(x)\right) = 0$ can be nonzero only if $j = j'$. In this case (8.10.7) implies

$$\left(\theta^{(m)}_{s';\,ja'}(x),\,\theta^{(m)}_{s;\,ja}(x)\right) = p^{-j}\chi_p\big(s'(a - a')\big)\Omega\big(|a - a'|_p\big)\Omega\big(|s' - s|_p\big),$$
(8.10.8)

where $\Omega\big(|a - a'|_p\big) = \delta_{a'a}$, $\Omega\big(|s' - s|_p\big) = \delta_{s's}$, and $\delta_{s's}$, $\delta_{a'a}$ are the Kronecker symbols.

Since in view of (3.3.1) we have $\int_{\mathbb{Q}_p}\Omega\big(|p^j x - a|_p\big)\,dx = p^j$, formulas (8.10.7) and (8.10.8) imply that

$$\left(\theta^{(m)}_{s';\,j'a'}(x),\,\theta^{(m)}_{s;\,ja}(x)\right) = \delta_{s's}\delta_{j'j}\delta_{a'a}.$$
(8.10.9)

Thus the system of functions (8.10.3) is orthonormal.

To prove the completeness of the system of functions (8.10.3), we repeat the corresponding proof [163] almost word for word. Recall that the system of the characteristic functions of the balls $B_k(0)$ is complete in $\mathcal{L}^2(\mathbb{Q}_p)$. Consequently, taking into account that the system of functions $\{\theta^{(m)}_{s;\,ja}(x) : s \in J_{p;m}; j \in \mathbb{Z}, a \in I_p\}$ is invariant under dilations and translations, in order to prove that it is a complete system, it is sufficient to verify the Parseval identity for the characteristic function $\Omega(|x|_p)$.

If $0 \le j$, according to (8.10.5) and (4.8.10),

$$\big(\Omega(|x|_p),\,\theta^{(m)}_{s;\,ja}(x)\big) = p^{-j/2}\Omega\big(|-a|_p\big)\int_{\mathbb{Q}_p}\chi_p\big(s(p^j x - a)\big)\Omega\big(|x|_p\big)\,dx$$

$$= p^{-j/2}\chi_p\big(-sa\big)\Omega\big(|sp^j|_p\big)\Omega\big(|-a|_p\big)$$

$$= \begin{cases} 0, & a \ne 0, \\ 0, & a = 0, \quad j \le m - 1, \\ p^{-j/2}, & a = 0, \quad j \ge m. \end{cases}$$
(8.10.10)

If $0 > j$, according to (8.10.5) and (4.8.10),

$$\big(\Omega(|x|_p),\,\theta^{(m)}_{s;\,ja}(x)\big) = p^{-j/2}\Omega\big(|p^{-j}a|_p\big)\int_{\mathbb{Q}_p}\chi_p\big(s(p^j x - a)\big)\Omega\big(|p^j x - a|_p\big)\,dx$$

$$= p^{j/2}\Omega\big(|p^{-j}a|_p\big)\int_{\mathbb{Q}_p}\chi_p\big(s\xi\big)\Omega\big(|\xi|_p\big)\,d\xi$$

$$= p^{j/2}\Omega\big(|p^{-j}a|_p\big)\Omega\big(|s|_p\big) = 0.$$
(8.10.11)

Thus,

$$\sum_{s\in J_{p;m};\, j\in\mathbb{Z},\, a\in I_p} \left|\left(\Omega(|x|_p),\, \theta_{s;\,ja}^{(m)}(x)\right)\right|^2 = \sum_{j=m}^{\infty}\sum_{s\in J_{p;m}} p^{-j}$$

$$= p^{m-1}(p-1)\frac{p^{-m}}{1-p^{-1}} = 1 = \left|\left(\Omega(|x|_p),\, \Omega(|x|_p)\right)\right|^2.$$

Thus the system of functions (8.10.3) is an orthonormal basis in $\mathcal{L}^2(\mathbb{Q}_p)$.

Since elements of basis (8.10.3) can be obtained by dilatations and translations of the set of $(p-1)p^{m-1}$ functions (8.10.2), it is the p-adic wavelet basis. □

Theorem 8.10.2. *The wavelet basis (8.10.3) is non-Haar type for $m \geq 2$.*

Proof. Taking into account the decomposition formula (1.8.5), we represent the Kozyrev wavelet function (8.1.4) as

$$\theta_k(x) = \chi_p(p^{-1}kx)\Omega(|x|_p)$$

$$= \begin{cases} 0, & |x|_p \geq p, \\ e^{2\pi i k\frac{kr}{p}}, & x \in B_{-1}(r), \quad r = 1,\ldots,p-1, \\ 1, & x \in B_{-1}. \end{cases}$$

Thus the wavelet function $\theta_k(x)$ takes values in the set $\{e^{2\pi i \frac{kr}{p}} : r = 0, 1, \ldots, p-1\}$ of p elements on the balls $B_{-1}(r)$, $r = 0, 1, \ldots, p-1$, $k = 1, 2, \ldots, p-1$. The above formula is an analog of representation of the Kozyrev wavelet function $\theta_k(x)$ in terms of the Haar *refinable function* $\phi(x) = \Omega(|x|_p)$ by formula (8.4.5). Thus the wavelet basis (8.10.3) is the basis of the Haar type.

In contrast to the Kozyrev basis (8.10.3), the wavelet bases (8.10.3) for $m \geq 2$ are of a *non-Haar type*. Indeed, in the same way we consider the wavelet function $\theta_s^{(m)}(x) = \chi_p(sx)\Omega(|x|_p)$, $s \in J_{p;m}$. Let

$$B_0(0) = \cup_c B_{-m}(c) \cup B_{-m}(0)$$

be the *canonical covering* (1.8.4) of the ball $B_0(0)$ with p^m balls $B_{-m}(0)$ and $B_{-m}(c)$, where $c = c_r p^r + c_{r+1}p^{r+1} + \cdots + c_{m-1}p^{m-1}$, $r = 0, 1, 2, \ldots, m-1$, $0 \leq c_j \leq p-1$, $c_r \neq 0$; $m \geq 1$.

For $x \in B_{-m}(c)$, $s \in J_{p;m}$, we have $x = c + p^m(y_0 + y_1 p + y_2 p^2 + \cdots)$, $s = p^{-m}(s_0 + s_1 p + \cdots + s_{m-1}p^{m-1})$, $s_0 \neq 0$; $sx = sc + \xi$, $\xi \in \mathbb{Z}_p$; and $\{sx\}_p = \{sc\}_p = \{p^{r-m}(c_r + c_{r+1}p + \cdots + c_{m-1}p^{m-r-1})(s_0 + s_1 p + \cdots +$

$s_{m-1}p^{m-1})\}_p$, $r = 0, 1, 2, \ldots, m-1$ (see (1.6.4)). Thus,

$$\theta_s^{(m)}(x) = \chi_p(sx)\Omega(|x|_p)$$

$$= \begin{cases} 0, & |x|_p \geq p, \\ e^{2\pi i s a}, & x \in B_{-m}(c), \quad c = \sum_{l=r}^{m-1} c_l p^l, \quad (8.10.12) \\ 1, & x \in B_{-m}(0), \end{cases}$$

where $0 \leq c_j \leq p - 1$, $j = r, \ldots, m-1$, $c_r \neq 0$, $r = 0, 1, \ldots, m-1$; $s = p^{-m}(s_0 + s_1 p + \cdots + s_{m-1}p^{m-1})$, $0 \leq s_j \leq p - 1$, $j = 0, 1, \ldots, m-1$, $s_0 \neq 0$. Thus the wavelet function $\theta_s^{(m)}(x)$ takes values in the set $\{e^{2\pi i \{sc\}_p} : c = \sum_{l=r}^{m-1} c_l p^l, 0 \leq c_j \leq p - 1, j = r, \ldots, m-1, c_r \neq 0, r = 0, 1, \ldots, m-1\}$ of p^m elements on the balls $B_{-m}(b)$.

According to (8.10.12), the wavelet function (8.10.2) is represented in terms of the characteristic function of the unit ball $\phi(x) = \Omega(|x|_p)$ (which is a solution of the Haar *refinement equation* (8.1.5)), i.e., in the form

$$\theta_s^{(m)}(x) = \chi_p(sx)\Omega(|x|_p) = \sum_c h_c \phi\left(\frac{1}{p^m}x - \frac{c}{p^m}\right), \quad (8.10.13)$$

$x \in \mathbb{Q}_p$, where $h_0 = 1, h_c = e^{2\pi i s a}$, $c = c_r p^r + c_{r+1}p^{r+1} + \cdots + c_{m-1}p^{m-1}$, $r = 0, 1, \ldots, m-1, 0 \leq c_j \leq p - 1, c_r \neq 0$.

According to formulas (8.10.12) and (8.10.13), the wavelet function (8.10.2) *cannot be represented* in terms of the *Haar refinable function* $\phi(x) = \Omega(|x|_p)$ (which is a solution of the Haar *refinement equation* (8.1.5)), i.e., in the form

$$\theta_s^{(m)}(x) = \sum_{a \in I_p} \beta_a \phi(p^{-1}x - a), \quad \beta_a \in \mathbb{C}.$$

Thus the system of (8.10.3) is an orthonormal non-Haar wavelet basis for $m \geq 2$. $\qquad\square$

Remark 8.4. 1. In the case $m = 1$, i.e., for $s = p^{-1}k, k = 1, 2, \ldots, p - 1$, the functions (8.10.2) coincide with the Kozyrev wavelet functions (8.1.4), and the wavelet basis (8.10.3) coincides with the Kozyrev wavelet basis (8.1.3):

$$\theta_{s;ja}^{(1)}(x) \equiv \theta_{k;ja}(x) = p^{-j/2}\chi_p(p^{-1}k(p^j x - a))\Omega(|p^j x - a|_p), \quad x \in \mathbb{Q}_p,$$

$k = 1, 2, \ldots, p - 1; j \in \mathbb{Z}, a \in I_p$.

2. The non-Haar wavelet basis (8.10.3) can be constructed by the approach developed by J. J. Benedetto and R. L. Benedetto [50]. In their notation [50], $H^\perp$ is the ball B_0, and $A_1^* : \mathbb{Q}_p \to \mathbb{Q}_p$ is multiplication by p^{-1}, and $W = (A_1^*)^{m-1}H^\perp = B_{m-1}$. In addition, the "choice coset of representatives" $\mathcal{D}$ in [50] is the set (8.3.1) of shifts I_p, and the set $J_{p;m}$ given by (8.10.1) is precisely the set $\mathcal{D} \cap ((A_1^*W) \setminus W)$ that appears in equation [50, (4.1)]. The set

$J_{p;m}$ consists of $N = p^m - p^{m-1}$ elements, where N is the number of wavelet generators.

The algorithm of [50] starts with N sets $\Omega_{s,0}$ and N local translation functions T_s, one for each $s \in J_{p;m}$. In order to construct the wavelets (8.10.3) by using the algorithm of [50], we set

$$\Omega_{s,0} = B_0\big(p^{-m}(s_1 p + \cdots + s_{m-1} p^{m-1})\big)$$
$$= p^{-m}(s_1 p + \cdots + s_{m-1} p^{m-1}) + B_0 \subset B_{m-1},$$

i.e., we remove the term s_0 and add the set $B_0 = H^\perp$. We also define

$$T_s : B_{m-1} \to B_m \setminus B_{m-1} \quad \text{by} \quad T_s(w) = w + s_0 p^{-m},$$

so that T_s maps $\Omega_{s,0}$ to $B_0(s)$ by translation. It is easy to verify that this fits the requirements of [50]: each $\Omega_{s,0}$ is $(\tau, \mathcal{D})$-congruent to $H^\perp = B_0$, the union of all such sets contains a neighborhood of 0, each T_s has the form required by formula [50, (4.1)], and for each $s \neq s'$ in $J_{p;m}$, one of the two compatibility conditions [50, (4.2) or (4.3)] holds. Thus, by using the algorithm of [50], we can produce the wavelets (8.10.3).

Moreover, according to (8.12.6), any wavelet $\theta_s^{(m)}$ is the Fourier transform of the characteristic function of each ball $B_0(s)$, $s \in J_{p;m}$. We recall that the algorithm of [50] only allows the construction of wavelet functions whose Fourier transforms are the characteristic functions of some sets (see [50, Proposition 5.1.]).

Making the change of variables $\xi = p^j x - a$ and taking (4.8.10) into account, we obtain

$$\int_{\mathbb{Q}_p} \theta_{s;ja}^{(m)}(x)\, dx = p^{j/2} \int_{\mathbb{Q}_p} \chi_p(s\xi)\Omega\big(|\xi|_p\big)\, d\xi = p^{j/2}\Omega\big(|s|_p\big) = 0, \quad (8.10.14)$$

i.e., according to Lemma 7.3.3, the wavelet functions $\theta_{s;ja}^{(m)}(x)$ belong to the Lizorkin space $\Phi(\mathbb{Q}_p)$.

Corollary 8.10.3. *The functions*

$$\widetilde{\theta}_{s;ja}^{(m)}(\xi) = F[\theta_{s;ja}^{(m)}](\xi) = p^{j/2}\chi_p\big(p^{-j}a \cdot \xi\big)\Omega\big(|s + p^{-j}\xi|_p\big), \quad \xi \in \mathbb{Q}_p,$$

form an orthonormal basis in $\mathcal{L}^2(\mathbb{Q}_p)$, $j \in \mathbb{Z}$, $a \in I_p$, $s \in J_{p;m}$; $m \geq 1$ is a fixed positive integer.

The proof follows from Theorem 8.10.1, formula (8.12.5) (see below) and the Parseval–Steklov formula (5.3.5).

8.11. One infinite family of non-Haar wavelet bases in $\mathcal{L}^2(\mathbb{Q}_p)$

Now, using the proof scheme of Theorem 8.5.1, we construct infinitely many different non-Haar wavelet bases, which are distinct from the basis (8.10.3).

In what follows, we shall write the p-adic number $a = p^{-\gamma}(a_0 + a_1 p + \cdots + a_{\gamma-1} p^{\gamma-1}) \in I_p$, $a_j = 0, 1, \ldots, p-1$, $j = 0, 1, \ldots, \gamma - 1$, in the form $a = \frac{r}{p^\gamma}$, where $r = a_0 + a_1 p + \cdots + a_{\gamma-1} p^{\gamma-1}$.

Since the p-adic norm is non-Archimedean, it is easy to see that the wavelet functions (8.10.2) have the following property:

$$\theta_s^{(m)}(x \pm 1) = \chi_p(\pm s)\theta_s^{(m)}(x), \quad s \in J_{p;m}. \tag{8.11.1}$$

Theorem 8.11.1. *For every* $v = 0, 1, 2, \ldots$ *the functions*

$$\psi_s^{(m),\,v}(x) = \sum_{k=0}^{p^v-1} \alpha_{s;k}\theta_s^{(m)}\left(x - \frac{k}{p^v}\right), \quad s \in J_{p;m}, \tag{8.11.2}$$

are wavelet functions if and only if

$$\alpha_{s;k} = p^{-v} \sum_{r=0}^{p^v-1} \gamma_{s;r} e^{-2i\pi \frac{-s+r}{p^v}k}, \tag{8.11.3}$$

where wavelet functions $\theta_s^{(m)}$ *are given by (8.10.3),* $\gamma_{s;k} \in \mathbb{C}$ *is an arbitrary constant such that* $|\gamma_{s;k}| = 1$, $k = 0, 1, \ldots, p^v - 1$, $s \in J_{p;m}$.

Proof. Let $\psi_s^{(m),\,v}(x)$ be defined by (8.11.2), $s \in J_{p;m}$. According to Theorem 8.10.1, $\{\theta_s^{(m)}(\cdot - a), s \in J_{p;m}, a \in I_p\}$ is an orthonormal system. Hence, taking (8.11.1) into account, we see that $\psi_s^{(m),v}$ is orthogonal to $\psi_s^{(m),v}(x)(\cdot - a)$ whenever $a \in I_p$, $a \neq \frac{k}{p^v}$, $k = 0, 1, \ldots p^v - 1$; $s, s' \in J_{p;m}$ $(v = 1, 2, \ldots)$. Thus the system $\{\psi_s^{(m),\,v}(x - a), s \in J_{p;m}, a \in I_p\}$ is orthonormal if and only if the system consisting of the functions

$$\psi_s^{(m),\,v}\left(x - \frac{r}{p^v}\right) = \chi_p(-s)\alpha_{s;p^v-r}\theta_s^{(m)}(x) + \chi_p(-s)\alpha_{s;p^v-r+1}\theta_s^{(m)}\left(x - \frac{1}{p^v}\right)$$

$$+ \cdots + \chi_p(-s)\alpha_{s;p^v-1}\theta_s^{(m)}\left(x - \frac{r-1}{p^v}\right)$$

$$+ \alpha_{s;0}\theta_s^{(m)}\left(x - \frac{r}{p^v}\right) + \alpha_{s;1}\theta_s^{(m)}\left(x - \frac{r+1}{p^v}\right)$$

$$+ \cdots + \alpha_{s;p^v-1-r}\theta_s^{(m)}\left(x - \frac{p^v-1}{p^v}\right), \tag{8.11.4}$$

$r = 0, \ldots, p^\nu - 1, s \in J_{p;m}$, is orthonormal. Set

$$\Xi_s^0 = \left(\theta_s^{(m)}, \theta_s^{(m)}\left(\cdot - \frac{1}{p^\nu}\right), \ldots, \theta_s^{(m)}\left(\cdot - \frac{p^\nu - 1}{p^\nu}\right)\right)^T,$$

$$\Xi_s^\nu = \left(\psi_s^{(m), \nu}, \psi_s^{(m), \nu}\left(\cdot - \frac{1}{p^\nu}\right), \ldots, \psi_s^{(m), \nu}\left(\cdot - \frac{p^\nu - 1}{p^\nu}\right)\right)^T,$$

where T is the transposition operation. By (8.11.4), we have $\Xi_s^\nu = D_s \Xi_s^0$, where

$$D_s = \begin{pmatrix} \alpha_{s;0} & \alpha_{s;1} & \cdots & \alpha_{s;p^\nu-2} & \alpha_{s;p^\nu-1} \\ \chi_p(-s)\alpha_{s;p^\nu-1} & \alpha_{s;0} & \cdots & \alpha_{s;p^\nu-3} & \alpha_{s;p^\nu-2} \\ \chi_p(-s)\alpha_{s;p^\nu-2} & \chi_p(-s)\alpha_{s;p^\nu-1} & \cdots & \alpha_{s;p^\nu-4} & \alpha_{s;p^\nu-3} \\ \cdots\cdots\cdots\cdots\cdots\cdots\cdots\cdots\cdots\cdots\cdots\cdots \\ \chi_p(-s)\alpha_{s;2} & \chi_p(-s)\alpha_{s;3} & \cdots & \alpha_{s;0} & \alpha_{s;1} \\ \chi_p(-s)\alpha_{s;1} & \chi_p(-s)\alpha_{s;2} & \cdots & \chi_p(-s)\alpha_{s;p^\nu-1} & \alpha_{s;0} \end{pmatrix},$$

$$(8.11.5)$$

and $s \in J_{p;m}$. Due to orthonormality of $\{\psi_s^{(m), \nu}(x)(\cdot - a), s \in J_{p;m}, a \in I_p\}$, the coordinates of Ξ_s^ν form an orthonormal system if and only if the matrices D_s are unitary, $s \in J_{p;m}$.

Let $u_s = (\alpha_{s;0}, \alpha_{s;1}, \ldots, \alpha_{s;p^\nu-1})^T$ be a vector and let

$$A_s = \begin{pmatrix} 0 & 0 & \cdots & 0 & 0 & \chi_p(-s) \\ 1 & 0 & \cdots & 0 & 0 & 0 \\ 0 & 1 & \cdots & 0 & 0 & 0 \\ \cdots\cdots\cdots\cdots\cdots\cdots\cdots \\ 0 & 0 & \cdots & 1 & 0 & 0 \\ 0 & 0 & \cdots & 0 & 1 & 0 \end{pmatrix}$$

be a $p^\nu \times p^\nu$ matrix, $s \in J_{p;m}$. It is not difficult to see that

$$A_s^r u_\nu = \big(\chi_p(-s)\alpha_{s;p^\nu-r}, \chi_p(-s)\alpha_{s;p^\nu-r+1}, \ldots, \chi_p(-s)\alpha_{s;p^\nu-1},$$
$$\alpha_{s;0}, \alpha_{s;1}, \ldots, \alpha_{s;p^\nu-r-1}\big)^T,$$

where $r = 0, 1, \ldots, p^s - 1$, $s \in J_{p;m}$. Thus we have

$$D_s = \left(u_s, A_s u_s, \ldots, A_s^{p^\nu-1} u_s\right)^T.$$

Consequently, to describe all unitary matrices D_s, we should find all vectors $u_s = (\alpha_{s;0}, \alpha_{s;1}, \ldots, \alpha_{s;p^\nu-1})^T$ such that the system of vectors $\{A_s^r u_s, r = 0, \ldots, p^\nu - 1\}$ is orthonormal, $s \in J_{p;m}$. We already have such a vector $u_0 = (1, 0, \ldots, 0, 0)^T$ because the matrix

$$D_0 = \left(u_0, A u_0, \ldots, A^{p^\nu-1} u_0\right)^T$$

is the identity matrix.

Now we prove that the system $\{A_s^r u_s, r = 0, \ldots, p^\nu - 1\}$ is orthonormal if and only if $u_s = B_s u_0$, where B_s is a unitary matrix such that $A_s B_s = B_s A_s$, $s \in J_{p;m}$. Indeed, if $u_s = B_s u_0$, B_s is a unitary matrix such that $A_s B_s = B_s A_s$, then $A_s^r u_s = B_s A_s^r u_0, r = 0, 1, \ldots, p^\nu - 1, s \in J_{p;m}$. Since the system $\{A_s^r u_0, r = 0, 1, \ldots, p^\nu - 1\}$ is orthonormal and the matrix B_s is unitary, the vectors $A_s^r u_s$, $r = 0, 1, \ldots, p^\nu - 1$, are also orthonormal, $s \in J_{p;m}$. Conversely, if the system $A_s^r u_s$, $r = 0, 1, \ldots, p^\nu - 1$, is orthonormal, taking into account that $\{A_s^r u_0, r = 0, 1, \ldots, p^\nu - 1\}$ is also an orthonormal system, we conclude that there exists a unitary matrix B_s such that $A_s^r u_s = B_s(A_s^r u_0)$, $r = 0, 1, \ldots, p^\nu - 1$. Since $A_s^{p^\nu} u_s = \chi_p(-s) u_s$ and $A_s^{p^\nu} u_0 = \chi_p(-s) u_0$, we obtain additionally $A_s^{p^\nu} u_s = B_s A_s^{p^\nu} u_0$. It follows from the above relations that $(A_s B_s - B_s A_s)(A_s^r u_0) = 0$, $r = 0, 1, \ldots, p^\nu - 1$. Since the vectors $A_s^r u_0$, $r = 0, 1, \ldots, p^\nu - 1$, form a basis in the p^ν-dimensional space, we conclude that $A_s B_s = B_s A_s$, $s \in J_{p;m}$.

Thus all required unitary matrixes (8.11.5) are given by

$$D_s = \left(B_s u_0, B_s A_\nu u_0, \ldots, B_s A_s^{p^\nu - 1} u_0 \right)^T,$$

where $u_0 = (1, 0, \ldots, 0, 0)^T$ and B_s is a unitary matrix such that $A_s B_s = B_s A_s$, $s \in J_{p;m}$. It remains to describe all such matrices B_s.

It is easy to see that the eigenvalues of A_s and the corresponding normalized eigenvectors are respectively

$$\lambda_{s;r} = e^{2i\pi \frac{-s+r}{p^\nu}} \tag{8.11.6}$$

and

$$v_{s;r} = \left((v_{s;r})_0, \ldots, (v_{s;r})_{p^\nu - 1} \right)^T,$$

where

$$(v_{s;r})_l = p^{-\nu/2} e^{-2i\pi \frac{-s+r}{p^\nu} l}, \quad l = 0, 1, 2, \ldots, p^\nu - 1, \tag{8.11.7}$$

$r = 0, 1, \ldots, p^\nu - 1$, $s \in J_{p;m}$. Hence the matrix A_s can be represented as $A_s = C_s \tilde{A}_s C_s^{-1}$, where

$$\tilde{A}_s = \begin{pmatrix} \lambda_0 & 0 & \cdots & 0 \\ 0 & \lambda_1 & \cdots & 0 \\ \vdots & \vdots & \ddots & \vdots \\ 0 & 0 & \cdots & \lambda_{p^\nu - 1} \end{pmatrix}$$

is a diagonal matrix, and $C_s = \left(v_{s;0}, \ldots, v_{s;p^\nu - 1} \right)$ is a unitary matrix. It follows that the matrix $B_s = C_s \tilde{B}_s C_s^{-1}$ is unitary if and only if $\tilde{B}_s$ is unitary. On the other hand, $A_s B_s = B_s A_s$ if and only if $\tilde{A}_s \tilde{B}_s = \tilde{B}_s \tilde{A}_s$. Moreover, since according to (8.11.6) $\lambda_{s;k} \neq \lambda_{s;l}$ whenever $k \neq l$, all unitary matrices $\tilde{B}_s$ such that $\widetilde{A}_s \widetilde{B}_s = \widetilde{B}_s \widetilde{A}_s$ are given by

$$
\widetilde{B}_s = \begin{pmatrix} \gamma_{s;0} & 0 & \cdots & 0 \\ 0 & \gamma_{s;1} & \cdots & 0 \\ \vdots & \vdots & \ddots & \vdots \\ 0 & 0 & \cdots & \gamma_{s;p^\nu-1} \end{pmatrix},
$$

where $\gamma_{s;k} \in \mathbb{C}$, $|\gamma_{s;k}| = 1$, $k = 0, 1, \ldots, p^\nu - 1$, $s \in J_{p;m}$. Hence all unitary matrices B_s such that $A_s B_s = B_s A_s$ are given by $B_s = C_s \widetilde{B}_s C_s^{-1}$. Using (8.11.7), we can calculate

$$
\alpha_{s;k} = (B_s u_0)_k = (C_s \widetilde{B}_s C_s^{-1} u_0)_k
$$

$$
= \sum_{r=0}^{p^\nu-1} \gamma_{s;r}(v_{s;r})_k (\overline{v}_{s;r})_0 = p^{-\nu} \sum_{r=0}^{p^\nu-1} \gamma_{\nu;r} e^{-2i\pi \frac{-s+r}{p^\nu} k},
$$

where $\gamma_{s;k} \in \mathbb{C}$, $|\gamma_{s;k}| = 1$, $k = 0, 1, \ldots, p^\nu - 1,$, $s \in J_{p;m}$.

It remains to prove that

$$
\{p^{-j/2} \psi_s^{(m),\,\nu}(p^j x - a), s \in J_{p;m}, x \in \mathbb{Q}_p : j \in \mathbb{Z}, a \in I_p\} \qquad (8.11.8)
$$

is a basis for $\mathcal{L}^2(\mathbb{Q}_p)$ whenever $\psi_s^{(m),\,\nu}$ is defined by (8.11.2) and (8.11.3). Since according to Theorem 8.10.1,

$$
\{p^{-j/2} \theta_s^{(m)}(p^j x - a), s \in J_{p;m}, x \in \mathbb{Q}_p : j \in \mathbb{Z}, a \in I_p\}
$$

is a basis for $\mathcal{L}^2(\mathbb{Q}_p)$, it suffices to check that any basis function $p^{-j/2} \theta_s^{(m)}(p^j x - c)$, $c \in I_p$, can be decomposed with respect to the functions $p^{-j/2} \psi_s^{(m),\,\nu}(p^j x - a)$, $a \in I_p$; where $s \in J_{p;m}$, $j \in \mathbb{Z}$. Any $c \in I_p$, $c \neq 0$, can be represented in the form $c = \frac{r}{p^\nu} + b$, where $r = 0, 1, \ldots, p^\nu - 1$, $|b|_p \geq p^{\nu+1}$, $bp^\nu \in I_p$. Taking into account that $\Xi_s^0 = D_s^{-1} \Xi_s^\nu$, i.e.,

$$
\theta_s^{(m)}\left(p^j x - \frac{r}{p^\nu}\right) = \sum_{k=0}^{p^\nu-1} \beta_{s;k}^{(r)} \psi_s^{(m),\,\nu}\left(p^j x - \frac{k}{p^\nu}\right), \quad r = 0, 1, \ldots, p^\nu - 1,
$$

we have

$$
\theta_s^{(m)}\left(p^j x - c\right) = \theta_s^{(m)}\left(p^j x - \frac{r}{p^\nu} - b\right) = \sum_{k=0}^{p^\nu-1} \beta_{s;k}^{(r)} \psi_s^{(m),\,\nu}\left(p^j x - \frac{k}{p^\nu} - b\right),
$$

and $\frac{k}{p^\nu} + b \in I_p$, $k = 0, 1, \ldots, p^\nu - 1$ $(\nu = 1, 2, \ldots)$. Consequently, any function $f \in \mathcal{L}^2(\mathbb{Q}_p)$ can be decomposed with respect to the system of functions (8.11.8). $\qquad \square$

Thus, we have constructed an infinite family of non-Haar wavelet bases given by formulas (8.11.8), (8.11.2) and (8.11.3).

Remark 8.5. According to (8.11.8), (8.11.2), (8.11.3), (8.12.6) and (4.9.4), (4.8.10),

$$
F[\psi_s^{(m),\,\nu}(x-a)](\xi)
$$

$$
= \chi_p(a\xi)\Omega\big(|\xi+s|_p\big) \sum_{k=0}^{p^\nu-1} \alpha_{s;k} \chi_p\Big(\frac{k}{p^\nu}\xi\Big)
$$

$$
= \chi_p(a\xi)\Omega\big(|\xi+s|_p\big) p^{-\nu} \sum_{r=0}^{p^\nu-1} \gamma_{s;r} \sum_{k=0}^{p^\nu-1} e^{2i\pi \frac{\xi+s-r}{p^\nu}k}, \quad s \in J_{p;m}, \quad a \in I_p,
$$

where $\gamma_{s;r} \in \mathbb{C}, |\gamma_{s;r}| = 1; r = 0, 1, \dots, p^\nu - 1$. Thus the right-hand side of the last relation is not equal to zero only if $\xi = -s + \eta, \eta = \eta_0 + \eta_1 p + \cdots \in \mathbb{Z}_p$, i.e., $\xi \in B_0(-s)$. Hence, for $\xi \in B_0(-s)$ we have

$$
F[\psi_s^{(m),\,\nu}(x-a)](\xi) = \chi_p(a(-s+\eta))p^{-\nu} \sum_{r=0}^{p^\nu-1} \gamma_{s;r} \sum_{k=0}^{p^\nu-1} e^{2i\pi \frac{\tilde\eta-r}{p^\nu}k},
$$

where $\tilde\eta = \eta_0 + \eta_1 p + \cdots \eta_{\nu-1} p^{\nu-1}$. From Proposition 1.8.9, the ball $B_0(-s)$ is represented by the sum of p^ν disjoint balls $B_{-\nu}(-s+\tilde\eta)$. If $\xi \in B_{-\nu}(-s+\tilde\eta)$ then $\xi = -s + \tilde\eta + p^\nu\tilde\eta_0, \tilde\eta_0 \in \mathbb{Z}_p$. Thus, on any ball $B_{-\nu}(-s+\tilde\eta)$ we have

$$
F[\psi_s^{(m),\,\nu}(x-a)](\xi) = \chi_p(a(-s+\eta))\bigg(\gamma_{s;\tilde\eta} + p^{-\nu} \sum_{\substack{r=0 \\ r\neq\tilde\eta}}^{p^\nu-1} \gamma_{s;r} \sum_{k=0}^{p^\nu-1} e^{2i\pi \frac{\tilde\eta-r}{p^\nu}k}\bigg)
$$

$$
= \chi_p(a(-s+\eta))\bigg(\gamma_{s;\tilde\eta} + p^{-\nu} \sum_{\substack{r=0 \\ r\neq\tilde\eta}}^{p^\nu-1} \gamma_{s;r} \sum_{k=0}^{p^\nu-1} \frac{1-e^{2i\pi(\tilde\eta-r)}}{1-e^{2i\pi \frac{\tilde\eta-r}{p^\nu}}}\bigg)
$$

$$
= \chi_p(a(-s+\eta))\gamma_{s;\tilde\eta}, \quad \tilde\eta = \eta_0 + \eta_1 p + \cdots \eta_{\nu-1} p^{\nu-1},
$$

(8.11.9)

where $s \in J_{p;m}, a \in I_p$.

Let us recall that according to [50, Proposition 5.1.], *only* the functions whose Fourier transforms are the characteristic functions of some sets may be wavelet functions obtained by Benedettos' method. It is easy to see if in (8.11.9)

$$
\gamma_{s;\tilde\eta} = 1 \quad \forall \tilde\eta = \eta_0 + \eta_1 p + \cdots \eta_{\nu-1} p^{\nu-1}, \tag{8.11.10}
$$

where $\eta_j = 0, 1, \dots, p - 1$, $j = 0, 1, \dots, \nu - 1$, then for $a = 0$ the right-hand side of (8.11.9) *is a characteristic function* of the ball $B_0(-s)$. Consequently, in this case the wavelet basis generated by wavelet functions $\psi_s^{(m),\,\nu}$, $s \in J_{p;m}$, can be obtained by Benedettos' method [50]. If (8.11.10) does not hold, the right-hand side of (8.11.9) *is not a characteristic function of any set* for all shifts $a \in I_p$. Consequently, in this case the wavelet basis generated by wavelet functions $\psi_s^{(m), \nu}$, $s \in J_{p;m}$, *cannot be constructed* by the method of [50].

8.12. Multidimensional non-Haar p-adic wavelets

Since the one-dimensional wavelets (8.10.3) cannot be generated by standard p-adic MRA (see Section 8.3.1 above), we cannot construct the n-dimensional wavelet basis as the tensor products of the one-dimensional MRAs (see Section 8.8). In this case we introduce n-dimensional non-Haar wavelet functions as the n-direct product of the one-dimensional non-Haar wavelets (8.10.3):

$$\Theta_{s;\,ja}^{(m)\times}(x) = p^{-|j|/2}\theta_{s_1;\,j_1a_1}^{(m)\times}(x_1)\cdots\theta_{s_n;\,j_na_n}^{(m)\times}(x_n)$$

$$= p^{-|j|/2}\chi_p\big(s\cdot(\widehat{p^j}x - a)\big)\Omega\big(|\widehat{p^j}x - a|_p\big), \quad x \in \mathbb{Q}_p^n, \quad (8.12.1)$$

where

$$\widehat{p^j}x \overset{def}{=} (p^{j_1}x_1, \ldots, p^{j_n}x_n), \quad x = (x_1, \ldots, x_n) \in \mathbb{Q}_p^n, \quad (8.12.2)$$

is a multi-dilatation, $j = (j_1, \ldots, j_n) \in \mathbb{Z}^n$, $|j| = j_1 + \cdots + j_n$; $a = (a_1, \ldots, a_n) \in I_p^n$; $s = (s_1, \ldots, s_n) \in J_{p;m}^n$; $m = (m_1, \ldots, m_n)$, $m_l \geq 1$ is a *fixed* positive integer, $l = 1, 2, \ldots, n$. Here I_p^n, $J_{p;m}^n$ are the n-direct products of the corresponding sets (8.3.1) and (8.10.1).

In view of (1.10.2), Theorem 8.10.1 implies the following statement.

Theorem 8.12.1. *The non-Haar wavelet functions* (8.12.1) *form an orthonormal basis in* $\mathcal{L}^2(\mathbb{Q}_p^n)$.

Using (8.10.14) and (1.10.2), it is easy to verify that

$$\int_{\mathbb{Q}_p^n} \Theta_{s;\,ja}^{(m)\times}(x)\,d^n x = 0, \quad j \in \mathbb{Z}^n, \quad a \in I_p^n, \quad s \in J_{p;m}^n, \quad (8.12.3)$$

i.e., from Lemma 7.3.3, the wavelet functions $\Theta_{s;\,ja}^{(m)\times}(x)$ belong to the Lizorkin space $\Phi(\mathbb{Q}_p^n)$.

Corollary 8.12.2. *The n-direct products of the one-dimensional Kozyrev wavelets* (8.1.4)

$$\Theta_{k;\,ja}^{\times}(x) = p^{-|j|/2}\chi_p\big(p^{-1}k\cdot(\widehat{p^j}x - a)\big)\Omega\big(|\widehat{p^j}x - a|_p\big), \quad x \in \mathbb{Q}_p^n, \quad (8.12.4)$$

form the orthonormal basis in $\mathcal{L}^2(\mathbb{Q}_p^n)$, $k \in J_p^n$, $j \in \mathbb{Z}^n$, $a \in I_p^n$.

The proof follows from Theorem 8.12.1 if we set $m = 1$.

Corollary 8.12.3. *The functions*

$$\widetilde{\Theta}_{s;\,ja}^{(m)\times}(\xi) = F[\Theta_{s;\,ja}^{(m)\times}](\xi)$$
$$= p^{|j|/2}\chi_p\big(\widehat{p^{-j}a}\cdot\xi\big)\Omega\big(|s+\widehat{p^{-j}\xi}|_p\big), \quad \xi\in\mathbb{Q}_p^n, \quad (8.12.5)$$

form an orthonormal basis in $\mathcal{L}^2(\mathbb{Q}_p^n)$, $j\in\mathbb{Z}^n$; $a\in I_p^n$; $s\in J_{p;m}^n$; $m = (m_1,\ldots,m_n)$, $m_l \geq 1$ is a fixed positive integer, $l = 1, 2, \ldots, n$.

Proof. Consider the function $\Theta_s^{(m)\times}(x) = \chi_p(s\cdot x)\Omega(|x|_p)$ generated by the direct product of functions (8.10.2), $x\in\mathbb{Q}_p^n$, $s = (s_1,\ldots,s_n)\in J_{p;m}^n$, $s_k\in J_{p;m_k}$, $k = 1, 2, \ldots, n$. Using (1.10.2), (4.8.10) and (4.9.3), we have

$$F[\Theta_s^{(m)\times}(x)](\xi) = F\Big[\prod_{k=1}^{n}\chi_p(x_ks_k)\Omega(|x_k|_p)\Big](\xi)$$
$$= \prod_{k=1}^{n}F\Big[\Omega(|x_k|_p)\Big](|\xi_k + s_k|_p)$$
$$= \prod_{k=1}^{n}\Omega\big(|\xi_k + s_k|_p\big) = \Omega\big(|\xi + s|_p\big), \quad \xi\in\mathbb{Q}_p^n. \quad (8.12.6)$$

Here, from (1.10.2), $\Omega\big(|\xi + s|_p\big) = \Omega\big(|\xi_1 + s_1|_p\big)\times\cdots\times\Omega\big(|\xi_n + s_n|_p\big)$. According to (8.10.1), $|s_k|_p = p^{m_k}$ and $\Omega\big(|\xi_k + s_k|_p\big)\neq 0$ only if $\xi_k = -s_k + \eta_k$, where $\eta_k\in\mathbb{Z}_p$, $s_k\in J_{p;m_k}$, $k = 1, 2, \ldots, n$. This yields $\xi = -s + \eta$, where $\eta\in\mathbb{Z}_p^n$, $s\in J_{p;m}^n$, and from (1.10.1), $|\xi|_p = p^{\max\{m_1,\ldots,m_n\}}$.

In view of formulas (8.12.1), (8.12.6) and (4.9.3), we have

$$F[\Theta_{s;\,ja}^{(m)\times}(x)](\xi) = p^{-|j|/2}F[\Theta_s^{(m)\times}(\widehat{p^j}x - a)](\xi)$$
$$= p^{|j|/2}\chi_p\big(\widehat{p^{-j}a}\cdot\xi\big)\Omega\big(|s+\widehat{p^{-j}\xi}|_p\big),$$

i.e., (8.12.5).

Formula (8.12.5), the Parseval formula (5.3.5), and Theorem 8.12.1 imply the statement. $\qquad\qquad\qquad\qquad\qquad\qquad\qquad\qquad\qquad\square$

Similarly, we can construct n-dimensional non-Haar wavelet bases generated by the one-dimensional non-Haar wavelets (8.11.8) (as the n-direct product):

$$\Psi_{s;\,ja}^{(m),\,\nu\times}(x) = p^{-|j|/2}\psi_{s_1}^{(m_1),\,\nu}(p^{j_1}x_1 - a_1)\cdots\psi_{s_n}^{(m_n),\,\nu}(p^{j_n}x_n - a_n), \quad (8.12.7)$$

where $x\in\mathbb{Q}_p^n$, $\psi_{s_{j_k}}^{(m_{j_k}),\,\nu}$ is defined by (8.11.2) and (8.11.3), $j = (j_1,\ldots,j_n)\in\mathbb{Z}^n$; $|j| = j_1 + \cdots + j_n$; $a = (a_1,\ldots,a_n)\in I_p^n$; $s = (s_1,\ldots,s_n)\in J_{p;m}^n$; $m = (m_1,\ldots,m_n)$, $m_k \geq 1$ is a *fixed* positive integer, $k = 1, 2, \ldots, n$; $\nu = 1, 2, \ldots$.

8.13. The p-adic Shannon–Kotelnikov theorem

It is well known that the classical Shannon–Kotelnikov sampling theorem permits reconstructing a bandlimited function from its values on a set of equidistant points on the real line $\mathbb{R}$. The classical theorem is the following (see, for example, [111, Ch.5.1]): *Let $f \in L^2(\mathbb{R})$, and let $F[f](\xi) = 0$ for $|\xi| > M$. Then $f(t)$ can be reconstructed from its samples at the points $t_n = \frac{n}{2M}$, $n \in \mathbb{Z}$, by the interpolation formula*

$$f(t) = \sum_{n \in \mathbb{Z}} f(t_n) \frac{\sin\left(2\pi M(t - t_n)\right)}{2\pi M(t - t_n)}, \quad t \in \mathbb{R}, \tag{8.13.1}$$

where $\left\{ \frac{\sin\left(2\pi M(t - t_n)\right)}{2\pi M(t - t_n)} : n \in \mathbb{Z} \right\}$ *is the Shannon–Kotelnikov basis.* But it is well known that the above series reconstructing a signal converges *rather slowly*.

In this section we prove the multidimensional p-adic Shannon–Kotelnikov theorem (Theorem 8.13.2). Unlike the standard version of this theorem (8.13.1), in the p-adic case the series (8.13.3) reconstructing a signal has only *one term* at any point $x \in \mathbb{Q}_p$. Due to this fact, the p-adic Shannon–Kotelnikov theorem may give a better reconstruction of signals than its standard version.

Proposition 8.13.1. (See also [44].) *A family of functions*

$$p^{-jn/2} \chi_p(p^j a \cdot \xi), \quad a \in I_p^n, \quad \xi \in \mathbb{Q}_p^n, \tag{8.13.2}$$

form an orthonormal basis in $\mathcal{L}^2(B_j^n(0))$, where I_p is given by (8.3.1), and $B_j^n(0)$ is a ball in $\mathbb{Q}_p^n$ of radius p^j.

Proof. According to (4.8.10), for $a, b \in I_p^n$ we have

$$\int_{B_j} p^{-jn/2} \chi_p(p^j a \cdot \xi) p^{-jn/2} \chi_p(-p^j b \cdot \xi) d^n \xi = \Omega(|a - b|_p) = \delta_{a,b},$$

where $\delta_{a,b}$ is the Kronecker symbol ($\delta_{a,b} = 1$ for $a = b$ and $\delta_{a,b} = 0$ for $a \neq b$, $a, b \in I_p^n$). To prove the completeness of the system of functions (8.13.2), we repeat the corresponding parts of the proofs of Theorems 8.9.3 and 8.10.1. $\square$

Now we prove a p-adic version of Shannon–Kotelnikov theorem.

Theorem 8.13.2. *Let $f \in \mathcal{L}^2(\mathbb{Q}_p^n)$, and let $\operatorname{supp} F[f] \subset B_j^n(0)$. Then f can be reconstructed from its samples at the points $x_a = p^j a$, $a \in I_p^n$ by the formula*

$$f(x) = \sum_{a \in I_p^n} f(p^j a) \Omega(|p^{-j} x - a|_p), \quad x \in \mathbb{Q}_p^n, \tag{8.13.3}$$

where the series converges in $V_j \subset \mathcal{L}^2(\mathbb{Q}_p^n)$ and the space V_j is defined by (8.8.3) and (8.9.11). The family of functions

$$\Phi_{a,-j}(x) = p^{jn/2}\Omega(|p^{-j}x - a|_p), \quad a \in I_p^n, \quad \xi \in \mathbb{Q}_p^n, \qquad (8.13.4)$$

constitute the orthonormal Shannon–Kotelnikov basis in the space V_j. For any $x \in \mathbb{Q}_p^n$ the series giving (8.13.3) contains only one term.

Proof. According to Theorem 4.9.3, $F[\mathcal{L}^2(B_j^n)]$ is the space of locally constant functions with a parameter of constancy $\geq -j$. Expanding $F[f]$ in the Fourier series, we obtain

$$F[f](\xi) = \sum_{a \in I_p^n} c_a p^{-jn/2} \chi_p(p^j a \cdot \xi), \quad \xi \in B_j^n(0).$$

Moreover, according to the Parseval–Steklov theorem,

$$\int |F[f](\xi)|^2 d^n\xi = \int |f(x)|^2 d^n x = \sum_{a \in I_p^n} |c_a|^2.$$

Since $\operatorname{supp} F[f] \subset B_j^n(0)$, $F[f](\xi) = \sum_{a \in I_p^n} c_a p^{-jn/2} \chi_p(p^j a \cdot \xi) \Delta_j(\xi)$, where $\Delta_j(\xi)$ is the characteristic function of the ball $B_j^n(0)$, and consequently,

$$f(x) = \sum_{a \in I_p^n} c_a p^{-jn/2} F^{-1}[\chi_p(p^j a \cdot \xi)\Delta_j(\xi)](x)$$

$$= \sum_{a \in I_p^n} c_a p^{-jn/2} \int_{B_j^n(0)} \chi_p((p^j a - x)\xi) d^n\xi.$$

According to (1.10.2), (4.9.3) and (4.8.10), the latter integral satisfies

$$\int_{B_j^n} \chi_p((p^j a - x)\xi) d^n\xi = p^{jn}\Omega(p^j|p^j a - x|_p) = p^{jn}\Omega(|p^{-j}x - a|_p).$$

Consequently,

$$f(x) = \sum_{a \in I_p^n} c_a p^{jn/2}\Omega(|p^{-j}x - a|_p), \qquad (8.13.5)$$

where the equality holds in the sense of $\mathcal{L}^2(\mathbb{Q}_p)$. By the substitution of $x = bp^j$, $b \in I_p^n$, into equation (8.13.5), and taking into account that $\Omega(|b - a|_p) = \delta_{a,b}$, we find that $f(bp^j) = c_b p^{jn/2}$. Thus (8.13.5) implies (8.13.3).

Using a one-dimensional *refinable function* $\phi(x) = \Omega(|x|_p)$, $x \in \mathbb{Q}_p$, for the p-adic Haar MRA (which satisfies the *refinement equation* (8.1.5)), according to (8.8.2), we can construct a multidimensional *refinable function*

$\Phi(x) = \Omega(|x|_p), \; x \in \mathbb{Q}_p^n$. Thus

$$\Phi_{a,-j}(x) = p^{jn/2}\Omega(|p^{-j}x - a|_p) = \Phi(p^{-j}x - a), \quad a \in I_p^n, \quad \xi \in \mathbb{Q}_p^n.$$

Consequently, according to (8.8.3), $F\big[\mathcal{L}^2(B_j^n)\big] = V_j$.

Since $V_j = F\big[\mathcal{L}^2(B_j^n)\big]$, in view of Proposition 8.13.1, the *Shannon–Kotelnikov basis* (8.13.4) is an orthonormal basis in the space $V_j \subset \mathcal{L}^2(\mathbb{Q}_p^n)$.

The series (8.13.5) converges to the function $f(x)$ in $\mathcal{L}^2(\mathbb{Q}_p^n)$. Next, taking into account that $\Omega(|p^{-j}x - a|_p) \neq 0$ if and only if for the fractional part we have $\{p^{-j}x\}_p = a$, we can see that for any point $x \in \mathbb{Q}_p^n$ the series (8.13.5) contains only one term. $\qquad\square$

Corollary 8.13.3. (i) *A collection of spaces $V_j = F\big[\mathcal{L}^2(B_j^n)\big]$ constitutes the multidimensional p-adic Haar MRA.*

(ii) $f \in \mathcal{L}^2(\mathbb{Q}_p^n) \cap V_j \iff \operatorname{supp} F[f] \subset B_j^n(0), \quad j \in \mathbb{Z}$.

(iii) *For any $f \in V_j$ a representation (8.13.3) holds, $j \in \mathbb{Z}$.*

For the real case an analog of Corollary 8.13.3 was proved in [192, Theorem 1.4.1.].

Remark 8.6. In the real setting the function

$$\phi^S(t) = \frac{\sin(\pi t)}{\pi x}, \qquad t \in \mathbb{R},$$

used in representation (8.13.1) is a *refinable function* of the *Shannon–Kotelnikov MRA*. This MRA generates the corresponding system of wavelet functions (see [192, Theorem 1.4.3.]). Since

$$F\big[\phi^S\big](\xi) = \begin{cases} 1, & |\xi| \leq \pi, \\ 0, & |\xi| > \pi, \end{cases} = \chi_{[-\pi,\pi]}(\xi), \quad \xi \in \mathbb{R},$$

and for the real case (8.1.1), (8.1.2) the Haar refinable function given by (8.1.8) is $\phi^H(t) = \chi_{[0,1]}(t)$, then the real Shannon–Kotelnikov MRA is, in a sense, an "antipode" of the real Haar MRA. In contrast to this, in the p-adic setting, according to Theorem 8.13.2 and Corollary 8.13.3, the Shannon–Kotelnikov MRA coincides with the Haar MRA.

8.14. *p*-adic Lizorkin spaces and wavelets

Now we prove an analog of Lemma 4.3.1 for test functions from the Lizorkin space $\Phi(\mathbb{Q}_p^n)$.

Lemma 8.14.1. *Any function* $\phi \in \Phi(\mathbb{Q}_p^n)$ *can be represented in the form of a finite* sum

$$\phi(x) = \sum_{s \in J_{p;m}^n, j \in \mathbb{Z}^n, a \in I_p^n} c_{s;j,a} \Theta_{s;ja}^{(m)\times}(x), \quad x \in \mathbb{Q}_p^n, \tag{8.14.1}$$

where $c_{s;j,a}$ *are constants; and* $\Theta_{s;ja}^{(m)\times}(x)$ *are elements of the non-Haar wavelet basis* (8.12.1); $s = (s_1, \ldots, s_n) \in J_{p;m}^n$; $j = (j_1, \ldots, j_n) \in \mathbb{Z}^n$, $|j| = j_1 + \cdots + j_n$; $a = (a_1, \ldots, a_n) \in I_p^n$; $m = (m_1, \ldots, m_n)$, $m_l \geq 1$ *is a fixed positive integer,* $l = 1, 2, \ldots, n$.

Proof. Let us calculate the $\mathcal{L}^2$-scalar product $(\phi(x), \Theta_{s;ja}^{(m)\times}(x))$. Taking into account formula (8.12.5) and using the Parseval–Steklov theorem, we obtain

$$\begin{aligned} c_{s;j,a} &= (\phi(x), \Theta_{s;ja}^{(m)\times}(x)) = (F[\phi](\xi), F[\Theta_{s;ja}^{(m)\times}](\xi)) \\ &= (\psi(\xi), p^{|j|/2} \chi_p(\widehat{p^{-j}a} \cdot \xi) \Omega(|s + \widehat{p^{-j}\xi}|_p)), \end{aligned} \tag{8.14.2}$$

where $j \in \mathbb{Z}^n$, $a \in I_p^n$, $s \in J_{p;m}^n$. Here according to Definition 7.2 and Lemma 7.3.3, any function $\phi \in \Phi(\mathbb{Q}_p^n)$ belongs to one of the spaces $\mathcal{D}_N^l(\mathbb{Q}_p^n)$ and $\psi = F^{-1}[\phi] \in \Psi(\mathbb{Q}_p^n) \cap \mathcal{D}_{-l}^{-N}(\mathbb{Q}_p^n)$, supp $\psi \subset B_{-l}^n \setminus B_{-N}^n$.

From (8.14.2) it is possible that $c_{s;j,a} \neq 0$ if $p^{-N} \leq |\xi|_p \leq p^{-l}$ and $s + \widehat{p^{-j}\xi} = \eta \in \mathbb{Z}_p^n$. Since $\xi = \widehat{p^j}(\eta - s)$, we have $|\xi|_p = \max_{1 \leq k \leq n}(p^{-j_k}|s_k|_p)$ and $p^{-N} \leq |\xi|_p \leq p^{-l}$, where $p \leq |s_k|_p \leq p^{m_k}$, $1 \leq k \leq n$. Thus there is a finite quantity of indexes $j = (j_1, \ldots, j_n) \in \mathbb{Z}^n$ such that $c_{s;j,a} \neq 0$, and consequently the sum (8.14.1) is finite with respect to $j = (j_1, \ldots, j_n) \in \mathbb{Z}^n$.

Repeating the proof of Lemma 4.3.1 almost word for word, we observe that the sum (8.14.1) is also finite with respect to $a \in I_p^n$.

Thus equality (8.14.1) holds in the sense of $\mathcal{L}^2(\mathbb{Q}_p^n)$. Consequently, this equality holds in the usual sense. □

It is clear that in Lemma 8.14.1 we can use one of the non-Haar wavelet bases given by Theorem 8.11.1 and formulas (8.11.8), (8.11.2), (8.11.3) or their multidimensional generalization (8.12.7), instead of the wavelet basis (8.12.1).

Repeating the proof of Lemma 8.14.1 almost word for word, we obtain the corresponding assertion.

Lemma 8.14.2. *Any function* $\phi \in \Phi(\mathbb{Q}_p^n)$ *can be represented in the form of a finite* sum

$$\phi(x) = \sum_{k \in J_{p0}^n, j \in \mathbb{Z}, a \in I_p^n} c_{k;ja} \Theta_{k;ja}(x), \quad x \in \mathbb{Q}_p^n, \tag{8.14.3}$$

where $c_{k;ja}$ *are constants;* $\Theta_{k;ja}(x)$ *are elements of the Haar wavelet basis* (8.9.8), $k = (k_1, \ldots, k_n) \in J_{p0}^n$, $j \in \mathbb{Z}$, $a \in I_p^n$.

It is clear that in Lemma 8.14.2 we can use one of the Haar wavelet bases given by Theorems 8.9.2 and 8.9.4, instead of the wavelet basis $\Theta_{k;\,ja}(x)$.

Using standard results from the book [213] or repeating the reasoning of [149]–[151] almost word for word, we obtain the following assertion.

Proposition 8.14.3. *Any distribution $f \in \Phi'(\mathbb{Q}_p^n)$ can be realized in the form of an* infinite *sum of the form*

$$f(x) = \sum_{s\in J_{p;m}^n,\,j\in\mathbb{Z}^n,\,a\in I_p^n} d_{s;\,j,a}\Theta_{s;\,ja}^{(m)\times}(x), \quad x \in \mathbb{Q}_p^n, \tag{8.14.4}$$

where $d_{s;\,j,a}$ are constants; and $\Theta_{s;\,ja}^{(m)\times}(x)$ are elements of the non-Haar wavelet basis (8.12.1); $s = (s_1, \ldots, s_n) \in J_{p;m}^n$; $j = (j_1, \ldots, j_n) \in \mathbb{Z}^n$, $|j| = j_1 + \cdots + j_n$; $a = (a_1, \ldots, a_n) \in I_p^n$; $m = (m_1, \ldots, m_n)$, $m_l \geq 1$ is a fixed positive integer, $l = 1, 2, \ldots, n$.

Here any distribution $f \in \Phi'(\mathbb{Q}_p^n)$ is associated with the representation (8.14.4), where the coefficients

$$d_{s;\,j,a} \overset{def}{=} \langle f, \Theta_{s;\,ja}^{(m)\times}\rangle, \quad s \in J_{p;m}^n, \quad j \in \mathbb{Z}^n, \quad a \in I_p^n. \tag{8.14.5}$$

Conversely, taking into account Lemma 8.14.1 and orthonormality of the wavelet basis (8.12.1), any infinite sum (8.14.4) is associated with the distribution $f \in \Phi'(\mathbb{Q}_p^n)$ whose action on a test function $\phi \in \Phi(\mathbb{Q}_p^n)$ is defined as

$$\langle f, \phi \rangle = \sum_{s\in J_{p;m}^n,\,j\in\mathbb{Z}^n,\,a\in I_p^n} d_{s;\,j,a}\overline{c_{s;\,j,a}}, \tag{8.14.6}$$

where the sum is finite.

It is clear that in Proposition 8.14.3 instead of the basis (8.10.3) or its multidimensional generalization (8.12.1), we can use the bases (8.11.8), (8.11.2), (8.11.3) and their multidimensional generalizations (8.12.7), or the bases given by Theorems 8.9.2, 8.9.4.

In the paper [25], assertions of the type of Lemma 8.14.1 and Proposition 8.14.3 were stated for ultrametric Lizorkin spaces.

9

Pseudo-differential operators on the p-adic Lizorkin spaces

9.1. Introduction

For the p-adic analysis related to complex-valued functions which are defined on $\mathbb{Q}_p$, the operation of differentiation is *not defined*. As a result, a large number of p-adic models use pseudo-differential equations instead of differential equations. Thus, pseudo-differential operators (in particular, fractional operators) play an important role in p-adic mathematical physics.

The p-adic multidimensional fractional operator D^α was introduced by Taibleson [229] (see also [230]) in the space of distributions $\mathcal{D}'(\mathbb{Q}_p^n)$. The spectral theory of the one-dimensional version of this fractional operator was developed by Vladimirov in [238], in particular, explicit formulas for the eigenfunctions of this operator were constructed (see also [241]). The concept of pseudo-differential operators over the field $\mathbb{Q}_p^n$ was introduced by Vladimirov [237]. In [237] (see also [241]) Valdimirov constructed the spectral theory of the Schrödinger-type operator $D^\alpha + V(x)$, which was further developed by Kochubei in [153] and [154]. Explicit results in the theory of p-adic Schrödinger-type operators with point interactions were obtained by S. Kuzhel and S. Torba [169] (see Chapter 11), who carried further the previous investigations by Kochubei [157]. In [157] and [158], Kochubei studied pseudo-differential operators with symbols of the form $\mathcal{A}(\xi) = |f(\xi_1, \ldots, \xi_n)|_p^\alpha, \alpha > 0$, where $f(\xi_1, \ldots, \xi_n)$ is a quadratic form such that $f(\xi_1, \ldots, \xi_n) \neq 0$ when $|\xi_1|_p + \cdots |\xi_n|_p \neq 0$. In [256] and [257], Zuniga-Galindo considered pseudo-differential operators with symbols of the form $\mathcal{A}(\xi) = |f(\xi_1, \ldots, \xi_n)|_p^\alpha, \alpha > 0$, where $f(\xi_1, \ldots, \xi_n)$ is a non-constant polynomial. Some classes of p-adic pseudo-differential operators were studied by Kozyrev in [162], [164] and [165]. Pseudo-differential operators on general ultrametric spaces were studied by Kozyrev [162], and Khrennikov and Kozyrev [132].

In this chapter we study p-adic pseudo-differential operators on the Lizorkin spaces of distributions (these results are based on [21], [22], [142], [143], [144] and [225]).

A basic motivation for using Lizorkin spaces rather than the Bruhat–Schwartz space of distributions $\mathcal{D}'(\mathbb{Q}_p^n)$ is due to the fact that the latter space is not invariant under the fractional operator. Indeed, since the functions $|\xi_j|_p^\alpha$ and $|\xi_j|_p^{\alpha_j} F[\varphi](\xi)$, $\xi \in \mathbb{Q}^n$, are not locally constant on $\mathbb{Q}_p^n$, in general, one has

$$D_{x_j}^{\alpha_j}\varphi(x) = F^{-1}[|\xi_j|_p^{\alpha_j} F[\varphi](\xi)](x) \notin \mathcal{D}(\mathbb{Q}_p^n)$$

for $\varphi \in \mathcal{D}(\mathbb{Q}_p^n)$. Thus, the operation $D_{x_j}^{\alpha_j} f$ is well defined only for some distributions $f \in \mathcal{D}'(\mathbb{Q}_p^n)$. For example, in general, D^{-1} is *not defined* in the space of test functions $\mathcal{D}(\mathbb{Q}_p)$ [241, IX.2]. This fact restricts applications of fractional operators looked upon as acting in $\mathcal{D}'(\mathbb{Q}_p^n)$. On the other hand, as we will see below, the Lizorkin spaces of distributions introduced in Chapter 7 *are invariant under fractional operators, and, consequently, they constitute "natural" definition domains for them.*

In Section 9.2 two types of multidimensional fractional operators on the Lizorkin spaces of distributions are constructed. In Section 9.2.1, we introduce the Vladimirov multidimensional fractional operator $D_\times^\alpha$ on the space $\Phi_\times'(\mathbb{Q}_p^n)$ as the direct product of one-dimensional Vladimirov operators for all $\alpha \in \mathbb{C}^n$. In Section 9.2.2, we define the Taibleson multidimensional fractional operator D^α on the Lizorkin space of distributions $\Phi'(\mathbb{Q}_p^n)$ for all $\alpha \in \mathbb{C}$. The latter operator was introduced by Taibleson [229, §2], [230, III.4.] on the space $\mathcal{D}'(\mathbb{Q}_p^n)$ for $\alpha \in \mathbb{C}$, $\alpha \neq -n$. The Lizorkin space $\Phi_\times(\mathbb{Q}_p^n)$ is invariant under the Vladimirov fractional operator (Lemma 9.2.2), while the Lizorkin space $\Phi(\mathbb{Q}_p^n)$ is invariant under the Taibleson fractional operator (Lemma 9.2.5). These fractional operators form abelian groups on the corresponding Lizorkin spaces (see Theorems 9.2.3, 9.2.6). In Section 9.2.3, by analogy with the "$\mathbb{R}$-case" [210], [211], two types of p-adic Laplacians are discussed. Such types of p-adic Laplacians were introduced in [114].

In Section 9.3, a class of pseudo-differential operators (9.3.1) on the Lizorkin spaces is introduced. In particular, this class includes the Taibleson fractional operator (9.2.22), the Kochubei operators [157], [158], and the Zuniga-Galindo operators [256], [257], which were mentioned above. The Lizorkin spaces are *invariant* under pseudo-differential operators (9.3.1). The family of pseudo-differential operators (9.3.1) with symbols $\mathcal{A}(\xi) \neq 0$, $\xi \in \mathbb{Q}_p^n \setminus \{0\}$, forms an abelian group.

In Section 9.4, a spectral theory of pseudo-differential operators is developed. We recall that in [241, IX.4] the spectral theory of p-adic pseudo-differential operators was constructed and, in particular, a basis of eigenfunctions with compact support was found for the p-adic fractional operator (9.2.3). Another example of a basis of eigenfunctions for the fractional operator (9.2.3) was constructed in [156]. Later, in [163], the wavelet basis (8.1.3) was constructed whose elements are eigenfunctions of the fractional operator (9.2.3):

$$D^\alpha \theta_{k;ja}(x) = p^{\alpha(1-j)}\theta_{k;ja}(x), \quad x \in \mathbb{Q}_p, \quad \alpha > 0, \qquad (9.1.1)$$

where $k = 1, 2, \dots p - 1$, $j \in \mathbb{Z}$, $a \in I_p$. It is *typical* that the elements of the wavelet bases are eigenfunctions of p-adic pseudo-differential operators. It turned out that under a suitable condition the compactly supported p-adic wavelets constructed in Chapter 8 are eigenfunctions of the p-adic pseudo-differential operators (9.3.1). Thus *p-adic wavelet analysis is closely connected with the spectral analysis of pseudo-differential operators.* In contrast to the p-adic case, it is not typical for real pseudo-differential operators to have compactly supported eigenfunctions.

In Section 9.4.1, we derive the criteria for the multidimensional p-adic pseudo-differential operators (9.3.1) to have the wavelet functions (8.12.1), (8.12.4), (8.9.2), (8.9.4), (8.9.8), (8.9.12) as eigenfunctions. We also calculate the corresponding eigenvalues. In Section 9.4.2, we prove that all the wavelets mentioned above are eigenfunctions of the Taibleson fractional operator (9.2.22). In Section 9.4.3, we show that the p-adic pseudo-differential operators can have compactly supported eigenfunctions which are not wavelets.

The p-adic pseudo-differential operators studied here are important for some applications [115]–[117], [157], [241] as discussed in the Chapters 10, 12. In particular, fractional operators in the standard "$\mathbb{R}$-case", as well as in the p-adic case, are intensively used in mathematical physics [82], [83] [210], [211].

9.2. p-adic multidimensional fractional operators

9.2.1. The Vladimirov operator

Let us introduce a distribution from the space $\mathcal{D}'(\mathbb{Q}_p)$

$$f_\alpha(z) = \frac{|z|_p^{\alpha-1}}{\Gamma_p(\alpha)}, \quad \alpha \neq 0, 1 \quad \alpha \in \mathbb{C}, \quad z \in \mathbb{Q}_p, \qquad (9.2.1)$$

called the *Riesz kernel* [241, VIII.2.], where $|z|_p^{\alpha-1}$ is a homogeneous distribution of degree $\pi_\alpha(z) = |z|_p^{\alpha-1}$ defined by (6.2.2), the Γ-function $\Gamma_p(\alpha)$ being given by (6.2.6). The distribution $f_\alpha(z)$ is meromorphic on the whole complex plane α, and it has simple poles at the points $\alpha = 0, 1$. Following [241, VIII,

(2.20)], we define $f_0(\cdot)$ as a distribution from $\mathcal{D}'(\mathbb{Q}_p)$:

$$f_0(z) \overset{def}{=} \lim_{\alpha \to 0} f_\alpha(z) = \delta(z), \quad z \in \mathbb{Q}_p, \qquad (9.2.2)$$

where the limit is understood in the weak sense.

The one-dimensional fractional operator $D_x^\alpha : \varphi(x) \to D_x^\alpha \varphi(x)$ is defined on $\mathcal{D}(\mathbb{Q}_p)$ as a convolution operator:

$$\left(D_x^\alpha \varphi\right)(x) = f_{-\alpha}(x) * \varphi(x), \quad x \in \mathbb{Q}_p, \qquad \mathrm{Re}\,\alpha \neq -1. \qquad (9.2.3)$$

The operator D_x^α is called the operator of (fractional) differentiation of order α with respect to x, for $\mathrm{Re}\,\alpha > 0$; the operator of (fractional) integration of order α with respect to x, for $\mathrm{Re}\,\alpha < 0$, $\mathrm{Re}\,\alpha \neq -1$; for $\alpha = 0$, $D_x^0 \varphi(x) = \delta(x) * \varphi(x) = \varphi(x)$ is the identity operator.

For $\mathrm{Re}\,\alpha > 0$, the relation (9.2.3) can be rewritten equivalently (see [241, IX, (1.1)]) as

$$\left(D_x^\alpha \varphi\right)(x) = \frac{p^\alpha - 1}{1 - p^{-\alpha-1}} \int_{\mathbb{Q}_p} \frac{\varphi(x) - \varphi(\xi)}{|x - \xi|_p^{\alpha+1}} \, d\xi$$

$$= \int_{\mathbb{Q}_p} |\xi|_p^\alpha F[\varphi](\xi)\chi_p(-\xi x)\,d\xi, \quad x \in \mathbb{Q}_p. \qquad (9.2.4)$$

If $\mathrm{Re}\,\alpha < 0$, $\mathrm{Re}\,\alpha \neq -1$, the relation (9.2.3) implies (see [241, IX, (1.3), (1.4)]) that

$$\left(D_x^\alpha \varphi\right)(x) = \frac{1 - p^\alpha}{1 - p^{-\alpha-1}} \int_{\mathbb{Q}_p} |x - \xi|_p^{-\alpha}\varphi(\xi)\,d\xi$$

$$= \begin{cases} \displaystyle\int_{\mathbb{Q}_p} |\xi|_p^{\alpha_j} F[\varphi](\xi)\chi_p(-\xi x)\,d\xi, & \mathrm{Re}\,\alpha > -1, \\[2ex] \displaystyle\int_{\mathbb{Q}_p} |\xi|_p^\alpha \big(F[\varphi](\xi)\chi_p(-\xi x) - F[\varphi](0)\big)\,d\xi, & \mathrm{Re}\,\alpha < -1. \end{cases}$$

$$(9.2.5)$$

Following [241, IX, (2.3)], we define $f_1(\cdot)$ as a distribution from $\Phi'(\mathbb{Q}_p)$:

$$f_1(z) \overset{def}{=} \lim_{\alpha \to 1} f_\alpha(z) = -\frac{p-1}{\log p} \log |z|_p, \quad z \in \mathbb{Q}_p, \qquad (9.2.6)$$

where the limit is understood in the weak sense.

According to Definitions 6.1 and 6.2, if $\alpha \neq 1$ the Riesz kernel $f_\alpha(z)$ is a *homogeneous* distribution of degree $\alpha - 1$, and if $\alpha = 1$ then the Riesz kernel is an *associated homogeneous* distribution of degree 0 and order 1.

It is well known [241, VIII, (2.20), (3.8), (3.9)] that

$$f_\alpha(z) * f_\beta(z) = f_{\alpha+\beta}(z), \qquad \alpha, \beta, \ \alpha + \beta \neq 1, \qquad (9.2.7)$$

in the sense of the space $\mathcal{D}'(\mathbb{Q}_p)$. Formulas (9.2.7), (9.2.6) imply that

$$f_\alpha(z) * f_\beta(z) = f_{\alpha+\beta}(z), \qquad \alpha, \beta \in \mathbb{C}, \qquad (9.2.8)$$

in the sense of distributions from $\Phi'(\mathbb{Q}_p)$.

Let $\alpha = (\alpha_1, \ldots, \alpha_n) \in \mathbb{C}^n, \alpha_j \in \mathbb{C}, j = 1, 2, \ldots,$ and $|\alpha| = \alpha_1 + \cdots + \alpha_n$. We denote by

$$f_\alpha(x) = f_{\alpha_1}(x_1) \times \cdots \times f_{\alpha_n}(x_n), \qquad (9.2.9)$$

the *multi-Riesz kernel*, where the one-dimensional Riesz kernel $f_{\alpha_j}(x_j), j = 1, \ldots, n,$ is defined by (9.2.1), (9.2.2) and (9.2.6).

If $\alpha_j \neq 1, j = 1, 2, \ldots$ then the Riesz kernel

$$f_\alpha(x) = \frac{|x_1|_p^{\alpha_1 - 1}}{\Gamma_p(\alpha_1)} \times \cdots \times \frac{|x_n|_p^{\alpha_n - 1}}{\Gamma_p(\alpha_n)}$$

is a *homogeneous* distribution of degree $|\alpha| - n$ (see Definition 6.1(b)). If $\alpha_1 = \cdots = \alpha_k = 1, \ \alpha_{k+1}, \ldots, \alpha_n \neq 1$ then

$$f_\alpha(x) = (-1)^k \frac{(p-1)^k}{\log^k p} \log |x_1|_p \times \cdots \times \log |x_k|_p$$

$$\times \frac{|x_{k+1}|_p^{\alpha_{k+1} - 1}}{\Gamma_p(\alpha_{k+1})} \times \cdots \times \frac{|x_n|_p^{\alpha_n - 1}}{\Gamma_p(\alpha_n)}. \qquad (9.2.10)$$

Thus, if among all $\alpha_1, \ldots, \alpha_n$ there are k pieces which are equal to 1 and $n - k$ pieces which are different from 1 then the Riesz kernel $f_\alpha(x)$ is an *associated homogeneous* distribution of degree $|\alpha| - n$ and order k, $k = 1, \ldots, n$ (see Definition 6.2(b)).

For example, if $n = 2$ and $\alpha_1 = \alpha_2 = 1$ then we have $f_{(1,1)}(x_1, x_2) = \frac{(p-1)^2}{\log^2 p} \log |x_1|_p \log |x_2|_p, x = (x_1, x_2) \in \mathbb{Q}_p^2$ and

$$f_{(1,1)}(tx_1, tx_2) = \frac{(p-1)^2}{\log^2 p} \Big(\log |x_1|_p \log |x_2|_p$$

$$+ (\log |x_1|_p + \log |x_2|_p) \log |t|_p + \log^2 |t|_p \Big), \qquad t \in \mathbb{Q}_p^*.$$

Lemma 9.2.1. *The family of Riesz kernels* $\{f_\alpha(x) : \alpha \in \mathbb{C}^n\}$ *form an abelian group with respect to convolution (in the sense of the space of distributions* $\Phi'_\times(\mathbb{Q}_p^n)$*) which is isomorphic to the additive group* $\mathbb{C}^n$:

$$f_\alpha(x) * f_\beta(x) = f_{\alpha+\beta}(x), \qquad \alpha, \beta \in \mathbb{C}^n; \qquad x \in \mathbb{Q}_p^n. \qquad (9.2.11)$$

The proof of this lemma follows immediately from (9.2.9), (9.2.8).

We define the multidimensional Vladimirov operator $D_\times^\alpha$ on the Lizorkin space $\Phi_\times(\mathbb{Q}_p^n)$ by the convolution

$$\left(D_\times^\alpha \phi\right)(x) \overset{def}{=} f_{-\alpha}(x) * \phi(x)$$

$$= \langle f_{-\alpha_1}(x_1) \times \cdots \times f_{-\alpha_n}(x_n), \phi(x - \xi)\rangle, \quad x \in \mathbb{Q}_p^n, \quad (9.2.12)$$

where $\phi \in \Phi_\times(\mathbb{Q}_p^n)$, $\alpha = (\alpha_1, \ldots, \alpha_n) \in \mathbb{C}^n$. Here $D_\times^\alpha = D_{x_1}^{\alpha_1} \times \cdots \times D_{x_n}^{\alpha_n}$, where $D_{x_j}^{\alpha_j} = f_{-\alpha_j}(x_j)*$, $j = 1, 2, \ldots, n$.

Lemma 9.2.2. *The Lizorkin space of the first kind* $\Phi_\times(\mathbb{Q}_p^n)$ *is invariant under the Vladimirov fractional operator* $D_\times^\alpha$. *Moreover,*

$$D_\times^\alpha(\Phi_\times(\mathbb{Q}_p^n)) = \Phi_\times(\mathbb{Q}_p^n).$$

Proof. Taking into account the formula (6.2.8)

$$F[f_{\alpha_j}(x_j)](\xi) = |\xi_j|_p^{-\alpha_j}, \quad j = 1, \ldots, n \qquad (9.2.13)$$

and formulas (9.2.12) and (4.9.8), we see that

$$F[D_\times^\alpha \phi](\xi) = |\xi_1|_p^{\alpha_1} \times \cdots \times |\xi_n|_p^{\alpha_n} F[\phi](\xi), \quad \phi \in \Phi_\times(\mathbb{Q}_p^n).$$

Since according to Chapter 7, Section 7.3.1, $F[\phi](\xi) \in \Psi_\times(\mathbb{Q}_p^n)$ and $|\xi_1|_p^{\alpha_1} \times \cdots \times |\xi_n|_p^{\alpha_n} F[\phi](\xi) \in \Psi_\times(\mathbb{Q}_p^n)$ for any $\alpha = (\alpha_1, \ldots, \alpha_n) \in \mathbb{C}^n$, then $D_\times^\alpha \phi \in \Phi_\times(\mathbb{Q}_p^n)$, i.e., $D_\times^\alpha(\Phi_\times(\mathbb{Q}_p^n)) \subset \Phi_\times(\mathbb{Q}_p^n)$. Moreover, any function from $\Psi_\times(\mathbb{Q}_p^n)$ can be represented as $\psi(\xi) = |\xi_1|_p^{\alpha_1} \times \cdots \times |\xi_n|_p^{\alpha_n} \psi_1(\xi)$, $\psi_1 \in \Psi_\times(\mathbb{Q}_p^n)$. This implies that $D_\times^\alpha(\Phi_\times(\mathbb{Q}_p^n)) = \Phi_\times(\mathbb{Q}_p^n)$. $\square$

In view of (9.2.13) and (4.9.8), formula (9.2.12) can be rewritten as

$$\left(D_\times^\alpha \phi\right)(x) = F^{-1}\left[|\xi_1|_p^{\alpha_1} \times \cdots \times |\xi_n|_p^{\alpha_n} F[\phi](\xi)\right](x), \quad \phi \in \Phi_\times(\mathbb{Q}_p^n). \quad (9.2.14)$$

The operator $D_\times^\alpha = f_{-\alpha}(x)*$ is called the operator of fractional partial differentiation of order $|\alpha|$, for $\alpha_j > 0$, $j = 1, \ldots, n$; the operator of fractional partial integration of order $|\alpha|$, for $\alpha_j < 0$, $j = 1, \ldots, n$; for $\alpha_1 = \cdots = \alpha_n = 0$, $D_\times^0 = \delta(x)*$ is the identity operator.

According to (9.2.12) and (4.7.1), for a distribution $f \in \Phi_\times'(\mathbb{Q}_p^n)$ we define the Vladimirov fractional operator $D_\times^\alpha f$, $\alpha \in \mathbb{C}^n$ by the relation

$$\langle D_\times^\alpha f, \phi \rangle \overset{def}{=} \langle f, D_\times^\alpha \phi \rangle, \quad \forall \phi \in \Phi_\times(\mathbb{Q}_p^n). \qquad (9.2.15)$$

It follows from (9.2.15) and Lemma 9.2.2 that

$$D_\times^\alpha(\Phi_\times'(\mathbb{Q}_p^n)) = \Phi_\times'(\mathbb{Q}_p^n).$$

Theorem 9.2.3. *The family of operators* $\{D_\times^\alpha : \alpha \in \mathbb{C}^n\}$ *on the space of distributions* $\Phi'_\times(\mathbb{Q}_p^n)$ *forms an n-parametric abelian group: if* $f \in \Phi'_\times(\mathbb{Q}_p^n)$ *then*

$$D_\times^\alpha D_\times^\beta f = D_\times^\beta D_\times^\alpha f = D_\times^{\alpha+\beta} f,$$
$$D_\times^\alpha D_\times^{-\alpha} f = f, \qquad \alpha, \beta \in \mathbb{C}^n. \tag{9.2.16}$$

This assertion follows from formulas (9.2.11), (9.2.14) and (9.2.15), and Lemma 9.2.2.

Example 9.2.1. If $\alpha_j > 0$, $j = 1, 2, \ldots$ then the fractional integration formula for the delta function holds:

$$D_\times^{-\alpha}\delta(x) = \frac{|x_1|_p^{\alpha_1-1}}{\Gamma_p(\alpha_1)} \times \cdots \times \frac{|x_n|_p^{\alpha_n-1}}{\Gamma_p(\alpha_n)}.$$

9.2.2. The Taibleson operator

Let us introduce the distribution from $\mathcal{D}'(\mathbb{Q}_p^n)$

$$\kappa_\alpha(x) = \frac{|x|_p^{\alpha-n}}{\Gamma_p^{(n)}(\alpha)}, \qquad \alpha \neq 0, n, \qquad x \in \mathbb{Q}_p^n, \tag{9.2.17}$$

called the multidimensional *Riesz kernel* [229, §2], [230, III.4.], where the function $|x|_p$, $x \in \mathbb{Q}_p^n$ is given by (1.10.1). The Riesz kernel has a removable singularity at $\alpha = 0$ and according to [229, §2], [230, III.4.], [241, VIII.2], we have

$$\begin{aligned}
\langle \kappa_\alpha(x), \varphi(x) \rangle &= \frac{g_\alpha}{\Gamma_p^{(n)}(\alpha)} + \frac{1-p^{-n}}{(1-p^{-\alpha})\Gamma_p^{(n)}(\alpha)}\varphi(0) \\
&= g_\alpha \frac{1-p^{-\alpha}}{1-p^{\alpha-n}} + \frac{1-p^{-n}}{1-p^{\alpha-n}}\varphi(0), \qquad \varphi \in \mathcal{D}(\mathbb{Q}_p^n),
\end{aligned}$$

where g_α is an entire function in α. Passing to the limit $\alpha \to 0$ in the above relation, we obtain

$$\langle \kappa_0(x), \varphi(x) \rangle \overset{def}{=} \lim_{\alpha \to 0} \langle \kappa_\alpha(x), \varphi(x) \rangle = \varphi(0), \qquad \forall \varphi \in \mathcal{D}(\mathbb{Q}_p^n).$$

Thus we define $\kappa_0(\cdot)$ as a distribution from $\mathcal{D}'(\mathbb{Q}_p^n)$:

$$\kappa_0(x) \overset{def}{=} \lim_{\alpha \to 0} \kappa_\alpha(x) = \delta(x), \qquad x \in \mathbb{Q}_p^n. \tag{9.2.18}$$

Now, using (6.2.4) and (9.2.17), and taking into account (7.3.3), we define $\kappa_n(\cdot)$ as a distribution from the *Lizorkin space of distributions* $\Phi'(\mathbb{Q}_p^n)$:

$$
\begin{aligned}
\langle \kappa_n, \phi \rangle &\overset{def}{=} \lim_{\alpha \to n} \langle \kappa_\alpha, \phi \rangle = \lim_{\alpha \to n} \int_{\mathbb{Q}_p^n} \frac{|x|_p^{\alpha - n}}{\Gamma_p^{(n)}(\alpha)} \phi(x)\, d^n x \\
&= -\lim_{\beta \to 0} \left(1 - p^{-n-\beta}\right) \int_{\mathbb{Q}_p^n} \frac{|x|_p^\beta - 1}{p^\beta - 1} \phi(x)\, d^n x \\
&= -\frac{1 - p^{-n}}{\log p} \int_{\mathbb{Q}_p^n} \log |x|_p \phi(x)\, d^n x, \quad \forall \phi \in \Phi(\mathbb{Q}_p^n),
\end{aligned}
$$

where $|\alpha - n| \le 1$. The passage to the limit under the integral sign is justified by the Lebesgue-dominated convergence Theorem 3.2.4. Thus,

$$
\kappa_n(x) \overset{def}{=} \lim_{\alpha \to n} \kappa_\alpha(x) = -\frac{1 - p^{-n}}{\log p} \log |x|_p, \quad x \in \mathbb{Q}_p^n. \tag{9.2.19}
$$

Consequently, the Riesz kernel $\kappa_\alpha(x)$ is a well-defined distribution from the Lizorkin space $\Phi'(\mathbb{Q}_p^n)$ for all $\alpha \in \mathbb{C}$.

According to Definitions 6.1(b) and 6.2(b), if $\alpha \ne n$ then $\kappa_\alpha(x)$ is a *homogeneous* distribution of degree $\alpha - n$, and if $\alpha = n$ then $\kappa_\alpha(x)$ is an *associated homogeneous* distribution of degree 0 and order 1.

Lemma 9.2.4. *The family of Riesz kernels $\{\kappa_\alpha(x) : \alpha \in \mathbb{C}\}$ forms an abelian group with respect to a convolution (in the sense of the space of distributions $\Phi'(\mathbb{Q}_p^n)$) which is isomorphic to the additive group of complex numbers $\mathbb{C}$:*

$$
\kappa_\alpha(x) * \kappa_\beta(x) = \kappa_{\alpha+\beta}(x), \quad \alpha, \beta \in \mathbb{C}; \quad x \in \mathbb{Q}_p^n. \tag{9.2.20}
$$

Proof. With the help of (6.2.11), (9.2.17) and (9.2.18), we obtain the formula [229, (**)], [230, III, (4.6)], [241, VIII, (4.9), (4.10)]:

$$
\kappa_\alpha(x) * \kappa_\beta(x) = \kappa_{\alpha+\beta}(x), \quad \alpha, \beta, \alpha + \beta \ne n, \tag{9.2.21}
$$

in the sense of the space $\mathcal{D}'(\mathbb{Q}_p^n)$. Taking into account formula (9.2.19), it is easy to see that (9.2.20) folds in the sense of the Lizorkin space $\Phi'(\mathbb{Q}_p^n)$. $\square$

We define the multidimensional Taibleson operator D^α on the Lizorkin space $\Phi(\mathbb{Q}_p^n)$ as the convolution:

$$
\left(D^\alpha \phi\right)(x) \overset{def}{=} \kappa_{-\alpha}(x) * \phi(x) = \langle \kappa_{-\alpha}(x), \phi(x - \xi) \rangle, \quad x \in \mathbb{Q}_p^n, \tag{9.2.22}
$$

where $\phi \in \Phi(\mathbb{Q}_p^n)$, $\alpha \in \mathbb{C}$.

Lemma 9.2.5. *The Lizorkin space of the second kind $\Phi(\mathbb{Q}_p^n)$ is invariant under the Taibleson fractional operator D^α and $D^\alpha(\Phi(\mathbb{Q}_p^n)) = \Phi(\mathbb{Q}_p^n)$.*

Proof. The proof of Lemma 9.2.5 is carried out in the same way as the proof of Lemma 9.2.2. From formula (6.2.11), $F[\kappa_\alpha(x)](\xi) = |\xi|_p^{-\alpha}$. Consequently, using (4.9.8), we have

$$F[D^\alpha \phi](\xi) = |\xi|_p^\alpha F[\phi](\xi), \quad \phi \in \Phi(\mathbb{Q}_p^n).$$

According to Chapter 7, Section 7.3.2, $F[\phi](\xi)$, $|\xi|_p^\alpha F[\phi](\xi) \in \Psi(\mathbb{Q}_p^n)$, $\alpha \in \mathbb{C}$, and, consequently, $D^\alpha \phi \in \Phi(\mathbb{Q}_p^n)$. That is $D^\alpha(\Phi(\mathbb{Q}_p^n)) \subset \Phi(\mathbb{Q}_p^n)$. Since any function from $\Psi(\mathbb{Q}_p^n)$ can be represented as $\psi(\xi) = |\xi|_p^\alpha \psi_1(\xi)$, $\psi_1 \in \Psi(\mathbb{Q}_p^n)$, we have $D^\alpha(\Phi(\mathbb{Q}_p^n)) = \Phi(\mathbb{Q}_p^n)$. □

In view of (6.2.11) and (4.9.8), formula (9.2.22) can be represented in the form

$$\left(D^\alpha \phi\right)(x) = F^{-1}\left[|\xi|_p^\alpha F[\phi](\xi)\right](x), \quad \phi \in \Phi(\mathbb{Q}_p^n). \qquad (9.2.23)$$

According to (9.2.22) and (4.7.1), for $f \in \Phi'(\mathbb{Q}_p^n)$ we define the distribution $D^\alpha f$ by the relation

$$\langle D^\alpha f, \phi \rangle \overset{def}{=} \langle f, D^\alpha \phi \rangle, \quad \forall \phi \in \Phi(\mathbb{Q}_p^n). \qquad (9.2.24)$$

According to (9.2.24) and Lemma 9.2.5, we have

$$D^\alpha(\Phi'(\mathbb{Q}_p^n)) = \Phi'(\mathbb{Q}_p^n).$$

Theorem 9.2.6. *The family of operators $\{D^\alpha : \alpha \in \mathbb{C}\}$ on the space of distributions $\Phi'(\mathbb{Q}_p^n)$ forms a one-parametric abelian group: if $f \in \Phi'(\mathbb{Q}_p^n)$ then*

$$D^\alpha D^\beta f = D^\beta D^\alpha f = D^{\alpha+\beta} f,$$
$$D^\alpha D^{-\alpha} f = f, \quad \alpha, \beta \in \mathbb{C}. \qquad (9.2.25)$$

This assertion follows from formulas (9.2.20), (9.2.22) and (9.2.24), and Lemma 9.2.5.

Example 9.2.2. If $\alpha > 0$ then the fractional integration formula for the delta function holds:

$$D^{-\alpha}\delta(x) = \frac{|x|_p^{\alpha-n}}{\Gamma_p^{(n)}(\alpha)}.$$

9.2.3. *p*-adic Laplacians

By analogy with the "$\mathbb{R}$-case" [210], [211], and the *p*-adic case [114], [241, X.1, Example 2], using the fractional operators we can introduce the *p-adic Laplacians*.

The *Laplacian of the first kind* is the operator

$$-\widehat{\Delta} f(x) \overset{def}{=} \sum_{k=1}^{n} \left(D_{x_k}^2 f\right)(x), \quad f \in \Phi'(\mathbb{Q}_p^n)$$

with the symbol $-\sum_{k=1}^{n} |\xi_j|_p^2$, $\xi_k \in \mathbb{Q}_p, k = 1, 2, \ldots, n$; the *Laplacian of the second kind* is the operator

$$-\Delta f(x) \overset{def}{=} \left(D^2 f\right)(x), \quad f \in \Phi'(\mathbb{Q}_p^n).$$

with the symbol $-|\xi|_p^2$, $\xi \in \mathbb{Q}_p^n$. Moreover, we can define powers of the Laplacian by the formula

$$(-\Delta)^{\alpha/2} f(x) \overset{def}{=} \left(D^\alpha f\right)(x), \quad f \in \Phi'(\mathbb{Q}_p^n), \quad \alpha \in \mathbb{C}.$$

9.3. A class of pseudo-differential operators

9.3.1. Pseudo-differential operators

Let us consider a class of pseudo-differential operators in the Lizorkin space of test functions $\Phi(\mathbb{Q}_p^n)$

$$(A\phi)(x) = F^{-1}\left[\mathcal{A}(\xi) F[\phi](\xi)\right](x)$$
$$= \int_{\mathbb{Q}_p^n} \int_{\mathbb{Q}_p^n} \chi_p((y - x) \cdot \xi)\mathcal{A}(\xi)\phi(y) d^n\xi \, d^n y, \quad \phi \in \Phi(\mathbb{Q}_p^n), \quad (9.3.1)$$

with symbols $\mathcal{A}(\xi) \in \mathcal{E}(\mathbb{Q}_p^n \setminus \{0\})$.

This class includes as particular cases the Taibleson fractional operator ($\mathcal{A}(\xi) = |\xi|_p^\alpha$); the Kochubei operators with symbols of the form $\mathcal{A}(\xi) = |f(\xi_1, \ldots, \xi_n)|_p^\alpha$, $\alpha > 0$, where $f(\xi_1, \ldots, \xi_n)$ is a quadratic form such that $f(\xi_1, \ldots, \xi_n) \neq 0$ when $|\xi_1|_p + \cdots |\xi_n|_p \neq 0$ (see [157], [158]); the Zuniga–Galindo operators with symbols of the form $\mathcal{A}(\xi) = |f(\xi_1, \ldots, \xi_n)|_p^\alpha$, $\alpha > 0$, where $f(\xi_1, \ldots, \xi_n)$ is a non-constant polynomial (see [256], [257]).

We define a conjugate pseudo-differential operator A^T on $\Phi(\mathbb{Q}_p^n)$ by

$$(A^T\phi)(x) = F^{-1}[\mathcal{A}(-\xi)F[\phi](\xi)](x)$$
$$= \int_{\mathbb{Q}_p^n} \chi_p(-x \cdot \xi)\mathcal{A}(-\xi)F[\phi](\xi) d^n\xi. \quad (9.3.2)$$

Then the operator A in the Lizorkin space of distributions is defined in the usual way: for $f \in \Phi'(\mathbb{Q}_p^n)$ we have

$$\langle Af, \phi \rangle = \langle f, A^T\phi \rangle, \quad \forall \phi \in \Phi(\mathbb{Q}_p^n). \quad (9.3.3)$$

Taking into account the formula (4.9.2), from the last relation we have

$$Af = F^{-1}[\mathcal{A}\,F[f]], \qquad f \in \Phi'(\mathbb{Q}_p^n). \tag{9.3.4}$$

Lemma 9.3.1. *The Lizorkin spaces of the second kind* $\Phi(\mathbb{Q}_p^n)$ *and* $\Phi'(\mathbb{Q}_p^n)$ *are invariant under the operators* (9.3.1).

Proof. From Chapter 7, Section 7.3.2, both functions $F[\phi](\xi)$ and $\mathcal{A}(\xi)F[\phi](\xi)$ belong to $\Psi(\mathbb{Q}_p^n)$, and, consequently, $(A\phi)(\cdot) \in \Phi(\mathbb{Q}_p^n)$. Thus the pseudo-differential operators (9.3.1) are well defined and the Lizorkin space $\Phi(\mathbb{Q}_p^n)$ is invariant under them. Therefore, if $f \in \Phi'(\mathbb{Q}_p^n)$ then according to (9.3.4),

$$Af = F^{-1}[\mathcal{A}\,F[f]] \in \Phi'(\mathbb{Q}_p^n), \tag{9.3.5}$$

i.e., the Lizorkin space of distributions $\Phi'(\mathbb{Q}_p^n)$ is invariant under the pseudo-differential operators A. $\qquad\square$

Because of Lemma 9.3.1, definition (9.3.3) is correct.

If A, B are pseudo-differential operators with symbols $\mathcal{A}(\xi)$, $\mathcal{B}(\xi) \in \mathcal{E}(\mathbb{Q}_p^n \setminus \{0\})$, respectively, then the operator AB is well defined and represented by the formula

$$(AB)f = F^{-1}[\mathcal{A}\mathcal{B}\,F[f]] \in \Phi'(\mathbb{Q}_p^n).$$

If $\mathcal{A}(\xi) \neq 0$, $\xi \in \mathbb{Q}_p^n \setminus \{0\}$ then we define its inverse pseudo-differential operator by the formula

$$A^{-1}f = F^{-1}[\mathcal{A}^{-1}\,F[f]], \quad f \in \Phi'(\mathbb{Q}_p^n).$$

Proposition 9.3.2. *The set of pseudo-differential operators* (9.3.1) *with symbols* $\mathcal{A}(\xi) \neq 0$, $\xi \in \mathbb{Q}_p^n \setminus \{0\}$ *forms an abelian group.*

If the symbol $\mathcal{A}(\xi)$ of the operator A is an *associated homogeneous* function then the operator A is called an *associated homogeneous pseudo-differential operator*.

According to formula (9.2.23) and Definitions 6.1, 6.2, the operator D^α, $\alpha \neq -n$, is a *homogeneous* pseudo-differential operator of degree α with the symbol $\mathcal{A}(\xi) = |\xi|_p^\alpha$, and D^{-n} is a *associated homogeneous* pseudo-differential operator of degree $-n$ and order 1 with the symbol $\mathcal{A}(\xi) = P(|\xi|_p^{-n})$ (see (6.3.25)).

9.4. Spectral theory of pseudo-differential operators

9.4.1. Spectral analysis as wavelet analysis

Here we will develop the spectral theory of pseudo-differential operators (9.3.1) and derive criteria for them to have the wavelets constructed in Chapter 8 as eigenfunctions. According to Chapter 8, these wavelets belong to the Lizorkin space of test functions $\Phi(\mathbb{Q}_p^n)$, and the pseudo-differential operator (9.3.1) is well defined on them.

Proposition 9.4.1. *Let A be the pseudo-differential operator (9.3.1) with a symbol $\mathcal{A}(\xi)$ and $0 \neq z \in \mathbb{Q}_p^n$. Then the additive character $\chi_p(z \cdot x)$ is an eigenfunction of the operator A with the eigenvalue $\mathcal{A}(-z)$, i.e.,*

$$A\chi_p(z \cdot x) = \mathcal{A}(-z)\chi_p(z \cdot x).$$

Proof. Since $F[\chi_p(z \cdot x)] = \delta(\xi + z)$, $z \neq 0$, we have $\mathcal{A}(\xi)\delta(\xi + z) = \mathcal{A}(-z)\delta(\xi + z)$. Thus

$$\begin{aligned} A\chi_p(z \cdot x) &= F^{-1}[\mathcal{A}(\xi)F[\chi_p(z \cdot x)](\xi)](x) \\ &= \mathcal{A}(-z)F^{-1}[\delta(\xi + z)](x) = \mathcal{A}(-z)\chi_p(z \cdot x). \end{aligned}$$

$\square$

This proposition is an analog of the corresponding statement for the one-dimensional fractional operator [241, IX.1, Example 4].

First we consider the case of *non-Haar wavelet functions*.

Theorem 9.4.2. *Let A be a pseudo-differential operator (9.3.1) with a symbol $\mathcal{A}(\xi) \in \mathcal{E}(\mathbb{Q}_p^n \setminus \{0\})$; $j = (j_1, \ldots, j_n) \in \mathbb{Z}^n$; $a \in I_p^n$; $s \in J_{p;m}^n$; $m = (m_1, \ldots, m_n)$; let $m_l \geq 1$ be a fixed positive integer, $l = 1, 2, \ldots, n$. Then the n-dimensional non-Haar wavelet function (8.12.1)*

$$\Theta_{s;ja}^{(m)\times}(x) = p^{-|j|/2}\chi_p\big(s \cdot (\widehat{p^j}x - a)\big)\Omega\big(|\widehat{p^j}x - a|_p\big), \quad x \in \mathbb{Q}_p^n,$$

is an eigenfunction of the operator A if and only if

$$\mathcal{A}\big(\widehat{p^j}(-s + \eta)\big) = \mathcal{A}\big(-\widehat{p^j}s\big), \qquad \forall \eta \in \mathbb{Z}_p^n. \tag{9.4.1}$$

The corresponding eigenvalue is $\lambda = \mathcal{A}\big(-\widehat{p^j}s\big)$, i.e.,

$$A\Theta_{s;ja}^{(m)\times}(x) = \mathcal{A}(-\widehat{p^j}s)\Theta_{s;ja}^{(m)\times}(x).$$

Here the multi-dilatation is defined by (8.12.2), and I_p^n, $J_{p;m}^n$ are the n-direct products of the corresponding sets (8.3.1) and (8.10.1).

Proof. Let condition (9.4.1) be satisfied. Then (9.3.1) and the above formula (8.12.5) from Chapter 8 imply

$$A\Theta_{s;ja}^{(m)\times}(x) = F^{-1}\big[\mathcal{A}(\xi)F[\Theta_{s;ja}^{(m)\times}](\xi)\big](x)$$
$$= p^{|j|/2}F^{-1}\big[\mathcal{A}(\xi)\chi_p\big(\widehat{p^{-j}a}\cdot\xi\big)\Omega\big(|s+\widehat{p^{-j}\xi}|_p\big)\big](x).$$

(9.4.2)

Making the change of variables $\xi = \widehat{p^j}(\eta - s)$ and using (4.8.10), we obtain

$$A\Theta_{s;ja}^{(m)\times}(x) = p^{-|j|/2}\int_{\mathbb{Q}_p^n} \chi_p\big(-(\widehat{p^j}x - a)\cdot(\eta-s)\big)\mathcal{A}(\widehat{p^j}(\eta-s))\,\Omega(|\eta|_p)\,d^n\eta$$

$$= p^{-|j|/2}\mathcal{A}(-\widehat{p^j}s)\chi_p\big(s\cdot(\widehat{p^j}x - a)\big)\int_{B_0^n}\chi_p(-(\widehat{p^j}x - a)\cdot\eta)\,d^n\eta$$

$$= \mathcal{A}(-\widehat{p^j}s)\Theta_{s;ja}^{(m)}(x).$$

Consequently, $A\Theta_{s;ja}^{(m)\times}(x) = \lambda\Theta_{s;ja}^{(m)\times}(x)$, where $\lambda = \mathcal{A}(-\widehat{p^j}s)$.

Conversely, if $A\Theta_{s;ja}^{(m)\times}(x) = \lambda\Theta_{s;ja}^{(m)\times}(x)$, $\lambda \in \mathbb{C}$, taking the Fourier transform of both left and right hand sides of this identity and using (9.3.1), (8.12.5) and (9.4.2), we have

$$\big(\mathcal{A}(\xi) - \lambda\big)\chi_p\big(\widehat{p^{-j}a}\cdot\xi\big)\Omega\big(|s+\widehat{p^{-j}\xi}|_p\big) = 0, \qquad \xi \in \mathbb{Q}_p^n. \qquad (9.4.3)$$

If now $s + \widehat{p^{-j}\xi} = \eta$, $\eta \in \mathbb{Z}_p^n$, then $\xi = \widehat{p^j}(-s + \eta)$. Since $\chi_p\big(\widehat{p^{-j}a}\cdot\xi\big) \neq 0$, $\Omega\big(|s+\widehat{p^{-j}\xi}|_p\big) \neq 0$, it follows from (9.4.3) that $\lambda = \mathcal{A}\big(\widehat{p^j}(-s+\eta)\big)$ for any $\eta \in \mathbb{Z}_p^n$. Thus $\lambda = \mathcal{A}(-\widehat{p^j}s)$, and, consequently, (9.4.1) holds.

The proof of the theorem is complete. $\qquad\square$

Corollary 9.4.3. *Let A be a pseudo-differential operator* (9.3.1) *with a symbol* $\mathcal{A}(\xi) \in \mathcal{E}(\mathbb{Q}_p^n \setminus \{0\})$. *Then the n-dimensional wavelet function* (8.12.4)

$$\Theta_{k;ja}^{\times}(x) = p^{-|j|/2}\chi_p\big(p^{-1}k\cdot(\widehat{p^j}x - a)\big)\Omega\big(|\widehat{p^j}x - a|_p\big), \qquad x \in \mathbb{Q}_p^n,$$

is an eigenfunction of A if and only if

$$\mathcal{A}\big(\widehat{p^j}(-p^{-1}k + \eta)\big) = \mathcal{A}\big(-\widehat{p^{j-I}k}\big), \qquad \forall \eta \in \mathbb{Z}_p^n,$$

where $k \in J_p^n$, $j \in \mathbb{Z}^n$, $a \in I_p^n$, $I = (1, \dots, 1)$. *The corresponding eigenvalue is* $\lambda = \mathcal{A}\big(-\widehat{p^{j-I}j}\big)$, *i.e.,*

$$A\Theta_{k;ja}^{\times}(x) = \mathcal{A}(-\widehat{p^{j-I}k})\Theta_{k;ja}^{\times}(x).$$

If the symbol $\mathcal{A}(\xi)$ of a pseudo-differential operator A is a homogeneous distribution of degree π_β, then Corollary 9.4.3 coincides with [21, Corollary 1].

The representation (8.11.2) and Theorem 9.4.2 imply the following statement.

Theorem 9.4.4. *Let A be a pseudo-differential operator* (9.3.1) *with the symbol* $\mathcal{A}(\xi) \in \mathcal{E}(\mathbb{Q}_p^n \setminus \{0\})$. *Then the n-dimensional non-Haar wavelet function* (8.12.7)

$$\Psi_{s;\,ja}^{(m),\,\nu\times}(x) = p^{-|j|/2}\psi_s^{(m_1),\,\nu}(p^{j_1}x_1 - a_1)\cdots\psi_s^{(m_n),\,\nu}(p^{j_n}x_n - a_n), \quad x \in \mathbb{Q}_p^n,$$

is an eigenfunction of the operator A if and only if condition (9.4.1) *holds, where* $\psi_{s_{j_k}}^{(m_{j_k}),\,\nu}$ *is defined by* (8.11.2) *and* (8.11.3), $j = (j_1, \ldots, j_n) \in \mathbb{Z}^n$; $|j| = j_1 + \cdots + j_n$; $a = (a_1, \ldots, a_n) \in I_p^n$; $s = (s_1, \ldots, s_n) \in J_{p;m}^n$; $m = (m_1, \ldots, m_n)$; $m_k \geq 1$ *is a fixed positive integer,* $k = 1, 2, \ldots, n$; $\nu = 1, 2, \ldots$. *The corresponding eigenvalue is* $\lambda = \mathcal{A}(-\widehat{p^j s})$, *i.e.,*

$$A\Psi_{s;\,ja}^{(m),\,\nu\times}(x) = \mathcal{A}(-\widehat{p^j s})\Psi_{s;\,ja}^{(m),\,\nu\times}(x).$$

Now we consider the case of *Haar wavelet functions*.

Theorem 9.4.5. *Let A be a pseudo-differential operator* (9.3.1) *with a symbol* $\mathcal{A}(\xi) \in \mathcal{E}(\mathbb{Q}_p^n \setminus \{0\})$; *let* $k \in J_{p0}^n$, $j \in \mathbb{Z}$, $a \in I_p^n$, *where* J_{p0}^n *is defined by* (8.9.6). *Then the n-dimensional Haar wavelet function* (8.9.8)

$$\Theta_{k;\,ja}(x) = p^{-nj/2}\chi_p(p^{-1}k \cdot (p^j x - a))\Omega(|p^j x - a|_p), \quad x \in \mathbb{Q}_p^n,$$

is an eigenfunction of A if and only if

$$\mathcal{A}(p^j(-p^{-1}k + \eta)) = \mathcal{A}(-p^{j-1}k), \quad \forall \eta \in \mathbb{Z}_p^n, \qquad (9.4.4)$$

holds. The corresponding eigenvalue is $\lambda = \mathcal{A}(-p^{j-1}k)$, *i.e.,*

$$A\Theta_{k;\,ja}(x) = \mathcal{A}(-p^{j-1}k)\Theta_{k;\,ja}(x).$$

Proof. Let $\Theta_k(x) = \chi_p(p^{-1}k \cdot x)\Omega(|x|_p)$, $x \in \mathbb{Q}_p^n$, $k = (k_1, \ldots, k_n) \in J_{p0}^n$. Taking formulas (4.8.11), (1.10.2) and (4.9.3) into account, we obtain

$$F[\Theta_k(x)](\xi) = \prod_{r=1}^{n}\Omega(|p^{-1}k_r + \xi_r|_p) = \Omega(|p^{-1}k + \xi|_p), \quad \xi \in \mathbb{Q}_p^n.$$

$$(9.4.5)$$

Since $|p^{-1}k_r|_p = p$, then $\Omega(|\xi_r + p^{-1}k_r|_p) \neq 0$ only if $\xi_r = -p^{-1}k_r + \eta_r$, where $\eta_r \in \mathbb{Z}_p$, $r = 1, 2, \ldots, n$. Thus $\xi = -p^{-1}k + \eta$, $\eta \in \mathbb{Z}_p^n$, $k \in J_{p0}^n$, and $|\xi|_p = p$. According to (8.9.8), (9.4.5) and (4.9.3),

$$F[\Theta_{k;\,ja}(x)](\xi) = p^{-nj/2}F[\Theta_k(p^j x - a)](\xi)$$
$$= p^{nj/2}\chi_p(p^{-j}a \cdot \xi)\Omega(|p^{-1}k + p^{-j}\xi|_p). \quad (9.4.6)$$

If the condition (9.4.4) holds, using (9.3.1) and (9.4.6), we obtain

$$A\Theta_{k;ja}(x) = F^{-1}[\mathcal{A}(\xi)F[\Theta_{k;ja}](\xi)](x)$$
$$= p^{nj/2}F^{-1}[\mathcal{A}(\xi)\chi_p(p^{-j}a \cdot \xi)\Omega(|p^{-1}k + p^{-j}\xi|_p)](x). \quad (9.4.7)$$

Making the change of variables $\xi = p^j(\eta - p^{-1}k)$, we have $A\Theta_{k;ja}(x) = \lambda\Theta_{k;ja}(x)$, where $\lambda = \mathcal{A}(-p^{j-1}k)$.

Conversely, if $A\Theta_{k;ja}(x) = \lambda\Theta_{k;ja}(x)$, $\lambda \in \mathbb{C}$, then applying the inverse Fourier transform to the last equality and using formulas (9.4.6) and (9.4.7), we obtain $(\mathcal{A}(\xi) - \lambda)\Omega(|p^{-1}k + p^{-j}\xi|_p) = 0, \xi \in \mathbb{Q}_p^n$. If $p^{-1}k + p^{-j}\xi = \eta, \eta \in \mathbb{Z}_p^n$, then $\xi = p^j(-p^{-1}k + \eta)$ and $\lambda = \mathcal{A}(p^j(-p^{-1}k + \eta))$ for any $\eta \in \mathbb{Z}_p^n$. In particular, $\lambda = \mathcal{A}(-p^{j-1}j)$, and consequently (9.4.4) holds. $\quad\square$

Repeating the proof of Theorem 9.4.5 almost word for word, we obtain the following assertions.

Theorem 9.4.6. *Let A be a pseudo-differential operator* (9.3.1) *with a symbol* $\mathcal{A} \in \mathcal{E}(\mathbb{Q}_2^n \setminus \{0\})$, $e \subset \{1, \ldots, n\}$, $e \neq \emptyset$, $j \in \mathbb{Z}$, $a \in I_2^n$. *Then the n-dimensional Haar wavelet function* (8.9.2)

$$\Psi_{e;ja}^{(0)}(x) = 2^{-nj/2}\chi_2(2^{-1}k_e \cdot (2^j x - a))\Omega(|2^j x - a|_2), \quad x \in \mathbb{Q}_2^n,$$

is an eigenfunction of A if and only if

$$\mathcal{A}(2^j(-2^{-1}k_e + \eta)) = \mathcal{A}(-2^{j-1}k_e), \quad \forall \eta \in \mathbb{Z}_2^n. \quad (9.4.8)$$

The corresponding eigenvalue is $\lambda = \mathcal{A}(-2^{j-1}k_e)$, *i.e.,*

$$A\Psi_{e;ja}^{(0)} = \mathcal{A}(-2^{j-1}k_e)\Psi_{e;ja}^{(0)}.$$

Theorem 9.4.7. *Let A be a pseudo-differential operator* (9.3.1) *with a symbol* $\mathcal{A}(\xi) \in \mathcal{E}(\mathbb{Q}_p^n \setminus \{0\})$, $s = (s_1, \ldots, s_n)$, *where* $s_\nu \in \mathbb{N}_0$, $e \subset \{1, \ldots, n\}$, $e \neq \emptyset$, $j \in \mathbb{Z}$, $a \in I_2^n$. *Then the n-dimensional Haar wavelet function* $\Psi_{e;ja}^{(s)}$ *(given by* (8.9.4)*) is an eigenfunction of A if and only if* (9.4.8) *holds. The corresponding eigenvalue is* $\lambda = \mathcal{A}(-2^{j-1}k_e)$, *i.e.,*

$$A\Psi_{e;ja}^{(s)} = \mathcal{A}(-2^{j-1}k_e)\Psi_{e;ja}^{(s)}.$$

Proof. To prove the above theorem, we repeat the proof of Theorem 9.4.5 almost word for word. In particular, using the representations (8.9.4) and (8.5.4), we can see that if $A\Psi_{e;ja}^{(s)} = \lambda\Psi_{e;ja}^{(s)}$ then

$$(\mathcal{A}(\xi) - \lambda)2^{nj/2}\sum_{r_1=0}^{2^{s_1}-1}\cdots\sum_{r_n=0}^{2^{s_n}-1}\alpha_{r_1}^{(k_e)_1}\cdots\alpha_{r_n}^{(k_e)_n}$$

$$\times \chi_2\left(2^{-j}\left(a + \left(\frac{r_1}{2^{s_1}}(k_e)_1, \ldots, \frac{r_n}{2^{s_n}}(k_e)_n\right)\right) \cdot \xi\right)\Omega(|2^{-1}k_e + 2^{-j}\xi|_2) = 0,$$

$\xi \in \mathbb{Q}_2^n$. If $\xi = 2^j(-2^{-1}k_e + \eta)$, $\eta \in \mathbb{Z}_2^n$, then as above we have

$$\left(\mathcal{A}\big(2^j(-2^{-1}k_e + \eta)\big) - \lambda\right)2^{nj/2} \sum_{r_1=0}^{2^{s_1}-1} \cdots \sum_{r_n=0}^{2^{s_n}-1} \alpha_{r_1}^{(k_e)_1} \cdots \alpha_{r_n}^{(k_e)_n}$$

$$\times \chi_2\left(\left(a + \left(\frac{r_1}{2^{s_1}}(k_e)_1, \ldots, \frac{r_n}{2^{s_n}}(k_e)_n\right)\right) \cdot (-2^{-1}k_e + \eta)\right) = 0,$$

$\eta \in \mathbb{Z}_2^n$. Here

$$2^{nj/2} \sum_{r_1=0}^{2^{s_1}-1} \cdots \sum_{r_n=0}^{2^{s_n}-1} \alpha_{r_1}^{(k_e)_1} \cdots \alpha_{r_n}^{(k_e)_n}$$

$$\times \chi_2\left(\left(a + \left(\frac{r_1}{2^{s_1}}(k_e)_1, \ldots, \frac{r_n}{2^{s_n}}(k_e)_n\right)\right) \cdot (-2^{-1}k_e + \eta)\right) \neq 0,$$

because this term is the Fourier transform of the wavelet function $\Psi_{e;ja}^{(s)}$. This implies $\lambda = \mathcal{A}\big(2^j(-2^{-1}k_e + \eta)\big)$ for any $\eta \in \mathbb{Z}_2^n$. □

Let us consider the wavelet functions $p^{-nj/2}\Psi_e^{(\mu)}(p^j x - a)$, where $e \subset \{1, \ldots, n\}$, $e \neq \emptyset$, $\mu = (\mu_1, \ldots, \mu_n)$, $\mu_\nu = 1, 2, \ldots, p-1$, $\nu = 1, 2, \ldots, n$; $j \in \mathbb{Z}$; $a \in I_p^n$ (which are given by formulas (8.9.10) and (8.9.12)). Denote by $k_e = \big((k_e)_1, \ldots, (k_e)_n\big)$ the vector whose coordinates are given by

$$(k_e)_\nu = \begin{cases} \mu_\nu, & \nu \in e, \\ 0, & \nu \notin e, \end{cases} \quad \nu = 1, \ldots, n. \tag{9.4.9}$$

Using formulas (8.6.2) and (8.6.3) from Theorem 8.6.1 and repeating the first part of the proof of Theorem 9.4.5 almost word for word, it is easy to prove the following statement.

Theorem 9.4.8. *Let A be a pseudo-differential operator* (9.3.1) *with a symbol* $\mathcal{A}(\xi) \in \mathcal{E}(\mathbb{Q}_p^n \setminus \{0\})$ *and let* $k_e = \big((k_e)_1, \ldots, (k_e)_n\big)$, *where* $(k_e)_\nu$ *is defined by* (9.4.9). *If*

$$\mathcal{A}\big(p^j(-p^{-1}k_e + \eta)\big) = \mathcal{A}\big(-p^{j-1}k_e\big) = \lambda, \quad \forall k_e, \quad \forall \eta \in \mathbb{Z}_p^n, \tag{9.4.10}$$

then the n-dimensional Haar wavelet function $p^{-nj/2}\Psi_e^{(\mu)}(p^j x - a)$ *(given by* (8.9.10) *and* (8.9.12)) *is an eigenfunction of A:*

$$A\big(p^{-nj/2}\Psi_e^{(\mu)}(p^j x - a)\big) = \lambda p^{-nj/2}\Psi_e^{(\mu)}(p^j x - a).$$

9.4.2. Wavelets as eigenfunctions of the Taibleson fractional operator

As was mentioned above, the Taibleson fractional operator D^α has the symbol $\mathcal{A}(\xi) = |\xi|_p^\alpha$.

The symbol $\mathcal{A}(\xi) = |\xi|_p^\alpha$ of a fractional operator satisfies the condition (9.4.1):

$$\mathcal{A}\big(\widehat{p^j}(-s + \eta)\big) = |\widehat{p^j}(-s + \eta)|_p^\alpha = \Big(\max_{1 \leq r \leq n} \big(p^{-j_r}| - s_r|_p\big)\Big)^\alpha$$

$$= \mathcal{A}\big(-\widehat{p^j}s\big) = p^{\alpha \max_{1 \leq r \leq n}\{m_r - j_r\}}$$

for all $\eta \in \mathbb{Z}_p^n$; where $s \in J_{p;m}^n$; $j = (j_1, \ldots, j_n) \in \mathbb{Z}^n$; $m = (m_1, \ldots, m_n)$, $m_1, \ldots, m_n \geq 1$ are fixed positive integers. Thus according to Theorem 9.4.2, we have:

Corollary 9.4.9. *The n-dimensional p-adic wavelet (8.12.1) is an eigenfunction of the Taibleson fractional operator (9.2.22):*

$$D^\alpha \Theta_{s;\,ja}^{(m)\times}(x) = p^{\alpha \max_{1 \leq r \leq n}\{m_r - j_r\}} \Theta_{s;\,ja}^{(m)\times}(x), \quad x \in \mathbb{Q}_p^n, \quad \alpha \in \mathbb{C},$$

$$s \in J_{p;m}^n, \; j \in \mathbb{Z}^n, a \in I_p^n. \quad (9.4.11)$$

Corollary 9.4.10. *The n-dimensional p-adic wavelet (8.12.7) is an eigenfunction of the Taibleson fractional operator (9.2.22):*

$$D^\alpha \Psi_{s;\,ja}^{(m),\,\nu\times}(x) = p^{\alpha \max_{1 \leq r \leq n}\{m_r - j_r\}} \Psi_{s;\,ja}^{(m),\,\nu\times}(x), \quad \alpha \in \mathbb{C}, \quad x \in \mathbb{Q}_p^n,$$

$$s \in J_{p;m}^n, j \in \mathbb{Z}^n, a \in I_p^n, \nu = 1, 2, \ldots.$$

In particular, from Corollary 9.4.3, we have:

Corollary 9.4.11.

$$D^\alpha \Theta_{k;\,ja}^\times(x) = p^{\alpha(1 - \min_{1 \leq r \leq n} j_r)} \Theta_{k;\,ja}^\times(x), \quad \alpha \in \mathbb{C}, \quad x \in \mathbb{Q}_p^n,$$

$$k \in J_p^n, \; j \in \mathbb{Z}^n, a \in I_p^n \quad (9.4.12)$$

It is easy to verify that the symbol $\mathcal{A}(\xi) = |\xi|_p^\alpha$ satisfies the condition (9.4.4). Then from Theorem 9.4.5, we have:

Corollary 9.4.12. *The n-dimensional Haar wavelet (8.9.8) is an eigenfunction of the Taibleson fractional operator (9.2.22):*

$$D^\alpha \Theta_{k;\,ja} = p^{\alpha(1-j)} \Theta_{k;\,ja}(x), \quad \alpha \in \mathbb{C}, \quad x \in \mathbb{Q}_p^n,$$

$$j \in \mathbb{Z}, a \in I_p^n, \; k = (k_1, \ldots, k_n) \in J_{p0}^n. \quad (9.4.13)$$

Corollary 9.4.13. *Let $e \subset \{1, \ldots, n\}$, $e \neq \emptyset$, $j \in \mathbb{Z}$, $a \in I_2^n$. Then the n-dimensional Haar wavelet function $\Psi_{e;\,ja}^{(0)}$ (given by (8.9.4)) is an eigenfunction* of the Taibleson fractional operator (9.2.22):

$$D^\alpha \Psi_{e;ja}^{(0)} = 2^{\alpha(1-j)} \Psi_{e;ja}^{(0)}, \quad \alpha \in \mathbb{C}.$$

Proof. The symbol $\mathcal{A}(\xi) = |\xi|_2^\alpha$ of the fractional operator D^α satisfies condition (9.4.8):

$$
\begin{aligned}
\mathcal{A}\big(2^j(-2^{-1}k_e + \eta)\big) &= \big|2^j(-2^{-1}k_e + \eta)\big|_2^\alpha \\
&= 2^{-j\alpha}\Big(\max_{1 \le \nu \le n} \big| -2^{-1}(k_e)_\nu + \eta_\nu\big|_2\Big)^\alpha \\
&= 2^{\alpha(1-j)}\Big(\max_{1 \le \nu \le n} \big|2^j(-2^{-1}(k_e)_\nu)\big|_2\Big)^\alpha = \mathcal{A}(-2^{j-1}k_e)
\end{aligned}
$$

for all $\eta \in \mathbb{Z}_2^n$, $e \subset \{1, \ldots, n\}$, $e \ne \emptyset$. Here we take into account that $(k_e)_\nu = 0, 1$; $\nu = 1, 2, \ldots, n$; $(k_e)_1 + \cdots + (k_e)_n \ne 0$. Thus, by Theorem 9.4.6, $\Psi_{e;ja}^{(0)}$ is an eigenfunction and the corresponding eigenvalue is $\lambda = 2^{\alpha(1-j)}$. $\qquad\square$

Corollary 9.4.14. *Let $s = (s_1, \ldots, s_n)$, where $s_\nu \in \mathbb{N}_0$, $e \subset \{1, \ldots, n\}$, $e \ne \emptyset$, $j \in \mathbb{Z}$, $a \in I_2^n$. Then the n-dimensional Haar wavelet function $\Psi_{e;ja}^{(s)}$ is an eigenfunction of the Taibleson fractional operator (9.2.22):*

$$D^\alpha \Psi_{e;ja}^{(s)}(x) = 2^{\alpha(1-j)} \Psi_{e;ja}^{(s)}(x), \quad \alpha \in \mathbb{C}.$$

Similarly, from Theorem 9.4.8, we have:

Corollary 9.4.15. *The n-dimensional Haar wavelet (8.9.10), (8.9.12) is an eigenfunction of the Taibleson fractional operator (9.2.22):*

$$D^\alpha \big(p^{-nj/2}\Psi_e^{(\mu)}(p^j x - a)\big) = p^{\alpha(1-j)} p^{-nj/2}\Psi_e^{(\mu)}(p^j x - a), \quad \alpha \in \mathbb{C}.$$

9.4.3. Compactly supported eigenfunctions of pseudo-differential operators

Now we consider the pseudo-differential operator (9.3.1) with a symbol $\mathcal{A}(\xi) \in \mathcal{E}(\mathbb{Q}_2^n)$ in the space of test functions $\mathcal{D}(\mathbb{Q}_p^n)$. In contrast to the above results, the space $\mathcal{D}(\mathbb{Q}_p^n)$ is not invariant under this operator.

Theorem 9.4.16. *Let A be a pseudo-differential operator (9.3.1) with a symbol $\mathcal{A}(\xi) \in \mathcal{E}(\mathbb{Q}_p^n\})$. Then an element (8.13.4)*

$$\Phi_{ja}(x) = p^{-nj/2}\Omega\big(|p^j x - a|_p\big), \quad j \in \mathbb{Z}, \quad a \in I_p^n, \quad x \in \mathbb{Q}_p^n,$$

of Shannon–Kotelnikov basis is an eigenfunction of the pseudo-differential operator (9.3.1) if and only if

$$\mathcal{A}(p^j \xi) = \mathcal{A}(0), \quad \forall \eta \in \mathbb{Z}_p^n, \tag{9.4.14}$$

holds. The corresponding eigenvalue is $\lambda = \mathcal{A}(0)$, *i.e.,*

$$A\Phi_{ja}(x) = \mathcal{A}(0)\Phi_{ja}(x).$$

Proof. In view of (4.8.11), (1.10.2) and (4.9.3), we have

$$F[\Phi_{ja}(x)](\xi) = p^{nj/2}\chi_p(p^{-j}a \cdot \xi)\Omega(|p^{-j}\xi|_p), \quad \xi \in \mathbb{Q}_p^n. \tag{9.4.15}$$

If condition (9.4.14) holds, using (9.3.1) and (9.4.15), we obtain

$$
\begin{aligned}
(A\Phi_{ja})(x) &= F^{-1}\big[\mathcal{A}(\xi)F[\Phi_{ja}](\xi)\big](x) \\
&= p^{nj/2}F^{-1}\big[\mathcal{A}(\xi)\chi_p(p^{-j}a \cdot \xi)\Omega(|p^{-j}\xi|_p)\big](x). \tag{9.4.16}
\end{aligned}
$$

Making the change of variables $\xi = p^j\eta$, we calculate as in previous theorems

$$
\begin{aligned}
(A\Phi_{ja})(x) &= p^{-nj/2}\int_{\mathbb{Q}_p^n} \chi_p\big(-(p^jx - a) \cdot \eta\big)\mathcal{A}(p^j\eta)\,\Omega(|\eta|_p)\,d^n\eta \\
&= p^{-nj/2}\mathcal{A}(0)\int_{B_0^n} \chi_p(-(p^jx - a) \cdot \eta)\,d^n\eta = \mathcal{A}(0)\Phi_{ja}(x).
\end{aligned}
$$

Thus $(A\Phi_{ja})(x) = \lambda\Phi_{ja}(x)$, where $\lambda = \mathcal{A}(-p^{j-1}k)$.

Conversely, if $(A\Phi_{ja})(x) = \lambda\Phi_{ja}(x)$, $\lambda \in \mathbb{C}$, applying the inverse Fourier transform to the last equality and using formulas (9.4.15) and (9.4.16), we obtain $\big(\mathcal{A}(\xi) - \lambda\big)p^{nj/2}\chi_p(p^{-j}a \cdot \xi)\Omega(|p^{-j}\xi|_p) = 0$, $\xi \in \mathbb{Q}_p^n$. If $p^{-j}\xi = \eta$, $\eta \in \mathbb{Z}_p^n$, then $\xi = p^j\eta$ and $\lambda = \mathcal{A}(p^j\eta)$ for any $\eta \in \mathbb{Z}_p^n$. In particular, $\lambda = \mathcal{A}(0)$, and consequently (9.4.14) holds. □

Corollary 9.4.17. *Let A be a pseudo-differential operator* (9.3.1) *such that the condition* (9.4.14) *holds. Then the space*

$$V_{-j} = \overline{\text{span}\{\Phi(p^j \cdot -a) : a \in I_p^n\}}$$

(defined by (8.8.3)*), where* $\Phi(x) = \Omega(|x|_p)$, $x \in \mathbb{Q}_p^n$, *is a refinable function for the Haar MRA (see Section 8.9), and the space* $\mathcal{D}_N^j(\mathbb{Q}_p^n)$ *(introduced in Section 3.2) is invariant under the operator* (9.3.1).

Proof. The relation $AV_{-j} \subset V_{-j}$ follows from the definition of the space V_{-j}. Formulas (4.3.4) and (4.4.5) imply the inclusion $\mathcal{D}_N^j(\mathbb{Q}_p^n) \subset V_{-j}$. □

As a particular case we consider the pseudo-differential operator D_m^α with the symbol $\mathcal{A}_m(\xi) = |p^m + \xi|_p^\alpha$, $m \in \mathbb{Z}, \alpha \in \mathbb{C}, \xi \in \mathbb{Q}_p^n$:

$$(D_m^\alpha\varphi)(x) = F^{-1}\big[|p^m + \xi|_p^\alpha F[\varphi](\xi)\big](x), \quad \varphi \in \mathcal{D}(\mathbb{Q}_p^n). \tag{9.4.17}$$

Corollary 9.4.18. *An element* (8.13.4)

$$\Phi_{ja}(x) = \Omega(|p^jx - a|_p), \quad j \geq m + 1, \quad a \in I_p,$$

of Shannon–Kotelnikov basis is an eigenfunction of the operator D_m^α, Re $\alpha >$ -1. The spaces V_j for $j \leq -m - 1$ and the spaces $\mathcal{D}_N^j(\mathbb{Q}_p^n)$ for $j \geq m + 1$ are invariant under the operator (9.4.17).

Proof. It is clear that if $j \geq m + 1$ we have for the symbol of the operator (9.4.17):

$$\mathcal{A}_m(p^j \xi) = |p^m + p^j \xi|_p^\alpha = |p^m + p^j \xi|_p^\alpha = \mathcal{A}_m(0),$$

i.e., the condition (9.4.14) holds. This fact and Theorem 9.4.16 imply the statement.   □

10
Pseudo-differential equations

10.1. Introduction

In p-adic models pseudo-differential equations are intensively used. Stationary and non-stationary p-adic Shcrödinger equation were studied by Vladimirov and Volovich [243]. A p-adic stochastic equation was considered by Bikulov and Volovich [55]. A p-adic Einstein equation and quantum p-adic gravity were studied by Aref'eva, Dragovich, Frampton and Volovich [40], [41]. p-adic pseudo-differential equations are used to model basin-to-basin kinetics by Avetisov, Bikulov, Kozyrev and Osipov [44], [45], [164]. In the papers by Kozyrev [166] and Fischenko and Zelenov [93] ultrametric and p-adic nonlinear equations were used to model turbulence. A wide class of p-adic pseudo-differential equations was studied by Kochubei [157] (see also [159]). p-adic pseudo-differential equations were also studied in [205], [256]. Khrennikov and Kozyrev [133], [134] applied the wavelet analysis to study the Schrödinger equation on ultrametric spaces.

In this chapter we study linear and nonlinear pseudo-differential equations in the Lizorkin space of distributions of p-adics. Our results are based on the papers [21], [23], [143].

In Section 10.2 a solution of the pseudo-differential equation $Af = g$, $g \in \Phi'(\mathbb{Q}_p^n)$ is constructed. In Section 10.3 the Cauchy problems for linear evolutionary pseudo-differential equations of the first order in t (10.3.1), (10.3.3), (10.3.16), (10.3.17) are solved. By Theorem 10.3.4 and Corollary 10.3.5, respectively, the sufficient condition for solutions of the Cauchy problems (10.3.1) and (10.3.3) to stabilize as $t \to \infty$ are derived. The criterion of stabilization for the solution of the homogeneous Cauchy problem (10.3.3) was derived in [157, 4.3]. In Section 10.4 the Cauchy problems for the linear evolutionary pseudo-differential equation of the second order in t (10.4.1) are

solved. In Section 10.5 we solve the Cauchy problems for semi-linear evolutionary pseudo-differential equations (10.5.1) and (10.5.10). In all evolutionary
pseudo-differential equations $t \in \mathbb{R}$, $x \in \mathbb{Q}_p^n$. Equations (10.3.1) and (10.3.3)
are similar to classical parabolic equations. In particular, equation (10.3.3) is
the heat-type equation, and equations (10.3.17) and (10.5.10) are the linear and
non-linear Schrödinger-type equations, respectively.

Equations of such types are intensively used in applications (see, e.g., [43],
[44], [162], [157], [241]). In particular, in [44], the simplest p-adic pseudo-
differential heat-type equation $\frac{\partial u(x,t)}{\partial t} + D_x^\alpha u(x,t) = 0$ was used in the models
of interbasin kinetics of macromolecules. Here t is time (a real parameter), while
the p-adic parameter x describes the hierarchy of basins. Thus, the results of
this section allow significant advance in the theory of p-adic pseudo-differential
equations and can be used in applications.

To solve the Cauchy problems for p-adic evolutionary pseudo-differential
equations we develop the *"variable separation method"* (an analog of the
classical Fourier method) which is based on three important facts:

- for appropriate conditions wavelets constructed in Chapter 8 are eigenfunctions of pseudo-differential operators (9.3.1) constructed in Chapter 9;
- the Lizorkin space of distributions $\Phi'(\mathbb{Q}_p^n)$ is a natural domain of definition
  for pseudo-differential operators (9.3.1);
- any Lizorkin distribution can be realized as an infinite linear combination
  (8.14.4) of wavelets (see Chapter 8, Section 8.14).

Taking into account the above reasoning, one can see that it is natural to
seek solutions of the Cauchy problems for evolutionary pseudo-differential
equations in the space $\widetilde{\Phi}'(\mathbb{Q}_p^n \times \mathbb{R}_+)$ of distributions $f(x,t)$ such that $f(\cdot,t) \in
\Phi'(\mathbb{Q}_p^n)$ for any $t \geq 0$. According to Lemma 8.14.1, Proposition 8.14.3, and
formula (8.14.6), for any distribution $f \in \widetilde{\Phi}'(\mathbb{Q}_p^n \times \mathbb{R}_+)$ and any test function
$\phi \in \Phi(\mathbb{Q}_p^n)$ we have

$$\langle f(\cdot,t), \phi(\cdot,t) \rangle = \sum_{s \in J_{p;m}^n, j \in \mathbb{Z}^n, a \in I_p^n} \overline{\langle \phi(\cdot), \Theta_{s;ja}^{(m)\times}(\cdot) \rangle} \langle f(\cdot,t), \Theta_{s;ja}^{(m)\times}(\cdot) \rangle, \quad (10.1.1)$$

where the last sum is finite.

10.2. Simplest pseudo-differential equations

Let us consider the following simplest pseudo-differential equation:

$$Af = g, \qquad g \in \Phi'(\mathbb{Q}_p^n), \qquad (10.2.1)$$

where A is a pseudo-differential operator (9.3.1), $\Phi'(\mathbb{Q}_p^n)$ is the space of Lizorkin distributions introduced in Chapter 7, and f is the desired distribution.

Theorem 10.2.1. *If the symbol of a pseudo-differential operator A is such that* $\mathcal{A}(\xi) \neq 0, \xi \in \mathbb{Q}_p^n \setminus \{0\}$ *then equation* (10.2.1) *has the unique solution*

$$f(x) = F^{-1}\left[\frac{F[g](\xi)}{\mathcal{A}(\xi)}\right](x) = (A^{-1}g)(x) \in \Phi'(\mathbb{Q}_p^n).$$

Proof. Applying the Fourier transform to the left-hand and right-hand sides of the equation $Af = g$, in view of representation (9.3.5), we find that $\mathcal{A}(\xi)F[f](\xi) = F[g](\xi)$. Since according to Chapter 7, Section 7.3.2, $F[\Phi'(\mathbb{Q}_p^n)] = \Psi'(\mathbb{Q}_p^n)$, $F[\Psi'(\mathbb{Q}_p^n)] = \Phi'(\mathbb{Q}_p^n)$, and $\mathcal{A}(\xi)$ is a multiplier in $\Psi(\mathbb{Q}_p^n)$, we have $F[f](\xi) = \mathcal{A}^{-1}(\xi)F[g](\xi) \in \Psi'(\mathbb{Q}_p^n)$. Thus

$$f(x) = F^{-1}[\mathcal{A}^{-1}(\xi)F[g](\xi)](x) = (A^{-1}g)(x) \in \Phi'(\mathbb{Q}_p^n)$$

is a solution of the problem (10.2.1).

Now we study solutions of the homogeneous problem (10.2.2). Let $f \in \mathcal{D}'(\mathbb{Q}_p^n)$ and $Af = 0$, i.e., according to (9.3.3), $\langle Af, \phi \rangle = \langle f, A^T\phi \rangle = 0$, for all $\phi \in \Phi(\mathbb{Q}_p^n)$. Since $A^T(\Phi(\mathbb{Q}_p^n)) = \Phi(\mathbb{Q}_p^n)$, we have $\langle f, \phi \rangle = 0$, for all $\phi \in \Phi(\mathbb{Q}_p^n)$, and consequently $f \in \Phi^\perp$ (see Proposition 7.3.4). Thus the solutions of the homogeneous problem (10.2.1) are indistinguishable as elements of the space $\Phi'(\mathbb{Q}_p^n)$. $\square$

If $f_0(x)$ is a fundamental solution of the problem (10.2.1), i.e.,

$$Af_0(x) = \delta(x),$$

then

$$f_0(x) = F^{-1}\left[\mathcal{A}(\xi)^{-1}\right](x).$$

Using Theorem 10.2.1 and (4.9.8), we can represent a solution of the equation (10.2.1) as

$$f(x) = f_0(x) * g(x) \in \Phi'(\mathbb{Q}_p^n).$$

Let $P_N(z) = \sum_{k=0}^N a_k z^k$ be a polynomial, where $a_k \in \mathbb{C}$ are constants. Let us consider the equation

$$P_N(D_x^\alpha)f = g, \qquad g \in \Phi'(\mathbb{Q}_p^n), \tag{10.2.2}$$

where $\left(D_x^\alpha\right)^k \stackrel{def}{=} D_x^{\alpha k}$, $\alpha \in \mathbb{C}$ and f is the desired distribution.

Theorem 10.2.2. *If $P_N(z) \neq 0$ for all $z > 0$ then equation (10.2.2) has the unique solution*

$$f(x) = F^{-1}\left[\frac{F[g](\xi)}{P_N(|\xi|_p^\alpha)}\right](x) \in \Phi'(\mathbb{Q}_p^n). \qquad (10.2.3)$$

In particular, the unique solution of the equation

$$D_x^\alpha f = g, \qquad g \in \Phi'(\mathbb{Q}_p^n),$$

is given by the formula $f = D_x^{-\alpha} g \in \Phi'(\mathbb{Q}_p^n)$.

Proof. According to formulas (6.2.5)–(6.2.11) and (9.2.17)–(9.2.19),

$$F[\kappa_\alpha(x)] = |\xi|_p^{-\alpha}, \qquad \alpha \in \mathbb{C}$$

in $\Phi'(\mathbb{Q}_p^n)$. Consequently, applying the Fourier transform to the left-hand and right-hand sides of relation (10.2.2), we obtain (10.2.3). Here we must take into account the fact that $\frac{1}{P_N(|\xi|_p^\alpha)}$ is a multiplier in $\Psi(\mathbb{Q}_p^n)$. Thus (10.2.3) is the solution of the problem (10.2.2).

In view of the proof of Theorem 10.2.1, the homogeneous problem (10.2.2) has only a trivial solution. ☐

In a similar way we can prove the following theorem.

Theorem 10.2.3. *If $P_N(z) \neq 0$ for all $z > 0$ then the equation*

$$P_N(D_\times^\alpha) f = g, \qquad g \in \Phi'_\times(\mathbb{Q}_p^n),$$

$\alpha \in \mathbb{C}^n$ *has the unique solution*

$$f(x) = F^{-1}\left[\frac{F[g](\xi)}{P_N(|\xi_1|_p^{\alpha_1} \cdots |\xi_n|_p^{\alpha_n})}\right](x) \in \Phi'_\times(\mathbb{Q}_p^n).$$

In particular, the unique solution of the equation

$$D_\times^\alpha f = g, \qquad g \in \Phi'_\times(\mathbb{Q}_p^n),$$

is given by the formula $f = D_\times^{-\alpha} g \in \Phi'_\times(\mathbb{Q}_p^n)$.

10.3. Linear evolutionary pseudo-differential equations of the first order in time

Let us consider the Cauchy problem for the *linear evolutionary pseudo-differential equation of the first order in t*

$$
\begin{cases}
\frac{\partial u(x,t)}{\partial t} + A_x u(x,t) = f(x,t), & \text{in } \mathbb{Q}_p^n \times (0, \infty), \\
u(x,t) = u^0(x), & \text{in } \mathbb{Q}_p^n \times \{t = 0\},
\end{cases}
\tag{10.3.1}
$$

where $t \in \mathbb{R}$, $u^0 \in \Phi'(\mathbb{Q}_p^n)$ and

$$
A_x u(x,t) = F^{-1}\big[\mathcal{A}(\xi)\, F[u(\cdot, t)](\xi)\big](x)
\tag{10.3.2}
$$

is a pseudo-differential operator (9.3.1) (with respect to x) with symbols $\mathcal{A}(\xi) \in \mathcal{E}(\mathbb{Q}_p^n \setminus \{0\})$, $u(x,t) \in \widetilde{\Phi}'(\mathbb{Q}_p^n \times \mathbb{R}_+)$ is the desired distribution. In particular, we will consider the Cauchy problem

$$
\begin{cases}
\frac{\partial u(x,t)}{\partial t} + D_x^\alpha u(x,t) = f(x,t), & \text{in } \mathbb{Q}_p^n \times (0, \infty), \\
u(x,t) = u^0(x), & \text{in } \mathbb{Q}_p^n \times \{t = 0\},
\end{cases}
\tag{10.3.3}
$$

where $D_x^\alpha u(x,t) = F^{-1}\big[|\xi|_p^\alpha\, F[u(\cdot, t)](\xi)\big](x)$ is the Taibleson fractional operator (9.2.23) with respect to x, $\alpha \in \mathbb{C}$.

Theorem 10.3.1. *Suppose that $f(x,t)$ is a continuous function in t for all $x \in \mathbb{Q}_p^n$ and $f(x,t) \in \Phi(\mathbb{Q}_p^n)$ for all $t \geq 0$. Then the Cauchy problem (10.3.1) has the unique solution*

$$
u(x,t) = F^{-1}\Big[e^{-\mathcal{A}(\xi)t}\Big(F[u^0(\cdot)](\xi) + \int_0^t e^{-\mathcal{A}(\xi)\tau}\,\widehat{f}(\xi, \tau)\,d\tau\Big)\Big](x), \tag{10.3.4}
$$

where $\widehat{f}(\xi, t) = F[f](\xi, t)$.

Proof. Since $u(x,t)$ is a distribution such that $u(x,t) \in \Phi'(\mathbb{Q}_p^n)$ for any $t \geq 0$, the relation (10.3.1) is well defined. Applying the Fourier transform to (10.3.1), we obtain the equation

$$
\frac{\partial F[u(\cdot, t)](\xi)}{\partial t} + \mathcal{A}(\xi)\, F[u(\cdot, t)](\xi) = \widehat{f}(\xi, t),
$$

where according Section 7.3.2, $\widehat{f}(\xi, t) = F[f](\xi, t) \in \Psi(\mathbb{Q}_p^n)$ for all $t \geq 0$. Solving this equation, we obtain

$$
F[u(\cdot, t)](\xi) = e^{-\mathcal{A}(\xi)t}\Big(F[u(\cdot, 0)](\xi) + \int_0^t e^{-\mathcal{A}(\xi)\tau}\,\widehat{f}(\xi, \tau)\,d\tau\Big).
$$

Here because of our assumptions all terms are well defined. The last relation implies (10.3.4). □

Theorem 10.3.2. *Let a pseudo-differential operator A_x in (10.3.1) be such that its symbol $\mathcal{A}(\xi)$ satisfies condition (9.4.1),*

$$\mathcal{A}\big(\widehat{p^j}(-s+\eta)\big) = \mathcal{A}\big(-\widehat{p^j}s\big), \qquad \forall\, \eta \in \mathbb{Z}_p^n,$$

for any $j \in \mathbb{Z}^n$, $s \in J_{p;m}^n$. Let $f \in \widetilde{\Phi}'(\mathbb{Q}_p^n \times \mathbb{R}_+)$ and let $f(x,t)$ be continuous in t for all $x \in \mathbb{Q}_p^n$. Then the Cauchy problem (10.3.1) has in $\widetilde{\Phi}'(\mathbb{Q}_p^n \times \mathbb{R}_+)$ a unique solution

$$u(x,t) = \sum_{s \in J_{p;m}^n,\, j \in \mathbb{Z}^n,\, a \in I_p^n} \left(e^{-\mathcal{A}(-\widehat{p^j}s)t}\big\langle u^0, \Theta_{s;\,ja}^{(m)\times}\big\rangle \right.$$
$$\left. + \int_0^t e^{-\mathcal{A}(-\widehat{p^j}s)(t-\tau)}\big\langle f(\cdot,\tau), \Theta_{s;\,ja}^{(m)\times}(\cdot)\big\rangle d\tau \right) \Theta_{s;\,ja}^{(m)\times}(x), \quad (10.3.5)$$

where $\Theta_{s;\,ja}^{(m)\times}(x)$ are n-dimensional p-adic wavelets (8.12.1).

Proof. Since $f \in \widetilde{\Phi}'(\mathbb{Q}_p^n \times \mathbb{R}_+)$, according to the formula (8.14.4) from Proposition 8.14.3, this distribution is represented as an infinite sum

$$f(x,t) = \sum_{s \in J_{p;m}^n,\, j \in \mathbb{Z}^n,\, a \in I_p^n} d_{s;\,j,a}(t)\Theta_{s;\,ja}^{(m)\times}(x), \quad x \in \mathbb{Q}_p^n, \quad t \geq 0, \quad (10.3.6)$$

where due to (8.14.5)

$$d_{s;\,j,a}(t) = \langle f(\cdot,t), \Theta_{s;\,ja}^{(m)\times}(\cdot)\rangle, \quad s \in J_{p;m}^n, \quad j \in \mathbb{Z}^n, \quad a \in I_p^n. \quad (10.3.7)$$

Since we seek a solution $u(x,t)$ of the Cauchy problem (10.3.1) in the class $\widetilde{\Phi}'(\mathbb{Q}_p^n \times \mathbb{R}_+)$, in view of the above reasoning we will seek this solution in the form of an infinite sum

$$u(x,t) = \sum_{s \in J_{p;m}^n,\, j \in \mathbb{Z}^n,\, a \in I_p^n} \Lambda_{s;\,j,a}(t)\Theta_{s;\,ja}^{(m)\times}(x), \quad (10.3.8)$$

where $\Lambda_{s;\,j,a}(t)$ are the desired functions, $s \in J_{p;m}^n$, $j \in \mathbb{Z}^n$, $a \in I_p^n$.

Substituting (10.3.8) into (10.3.1), from Theorem 9.4.2, we obtain

$$\sum_{s \in J_{p;m}^n,\, j \in \mathbb{Z}^n,\, a \in I_p^n} \left(\Lambda_{s;\,j,a}'(t) + \mathcal{A}(-\widehat{p^j}s)\Lambda_{s;\,j,a}(t) - d_{s;\,j,a}(t) \right) \Theta_{s;\,ja}^{(m)\times}(x) = 0.$$

The last equation is understood in the weak sense, i.e.,

$$\sum_{s \in J_{p;m}^n,\, j \in \mathbb{Z}^n,\, a \in I_p^n} \bigg\langle \Big(\Lambda_{s;\,j,a}'(t) + \mathcal{A}(-\widehat{p^j}s)\Lambda_{s;\,j,a}(t)$$
$$- d_{s;\,j,a}(t)\Big)\Theta_{s;\,ja}^{(m)\times}(x), \phi(x)\bigg\rangle = 0, \quad (10.3.9)$$

for all $\phi \in \Phi(\mathbb{Q}_p^n)$. Since according to Lemma 8.14.1 any test function $\phi \in \Phi(\mathbb{Q}_p^n)$ is represented in the form of a *finite* sum (8.14.1), the equality (10.3.9) implies that

$$\Lambda'_{s;\,j,a}(t) + \mathcal{A}(-\widehat{p^j s})\Lambda_{s;\,j,a}(t) = d_{s;\,j,a}(t), \quad \forall s \in J_{p;m}^n, \ j \in \mathbb{Z}^n, \ a \in I_p^n,$$

for all $t \geq 0$. Solving this differential equation, we obtain

$$\Lambda_{s;\,j,a}(t) = e^{-\mathcal{A}(-\widehat{p^j s})t}\left(\Lambda_{s;\,j,a}(0) + \int_0^t e^{\mathcal{A}(-\widehat{p^j s})\tau} d_{s;\,j,a}(\tau)\,d\tau\right), \quad (10.3.10)$$

$s \in J_{p;m}^n, \ j \in \mathbb{Z}^n, a \in I_p^n.$

By substituting (10.3.10) into (10.3.8) we find a solution of the Cauchy problem (10.3.1) in the form

$$u(x,t) = \sum_{s \in J_{p;m}^n, j \in \mathbb{Z}^n, a \in I_p^n} e^{-\mathcal{A}(-\widehat{p^j s})t}\Big(\Lambda_{s;\,j,a}(0)$$

$$+ \int_0^t e^{\mathcal{A}(-\widehat{p^j s})\tau} d_{s;\,j,a}(\tau)\,d\tau\Big)\Theta_{s;\,ja}^{(m)\times}(x). \quad (10.3.11)$$

Setting $t = 0$, we find

$$u^0(x) = \sum_{s \in J_{p;m}^n, j \in \mathbb{Z}^n, a \in I_p^n} \Lambda_{s;\,j,a}(0)\Theta_{s;\,ja}^{(m)\times}(x),$$

where $u^0 \in \Phi'(\mathbb{Q}_p^n)$ and, according to (8.14.4), the coefficients $\Lambda_{s;\,j,a}(0)$ are uniquely determined by (8.14.5) as

$$\Lambda_{s;\,j,a}(0) = \langle u^0, \Theta_{s;\,ja}^{(m)\times}\rangle, \quad s \in J_{p;m}^n, \quad j \in \mathbb{Z}^n, \quad a \in I_p^n. \quad (10.3.12)$$

The relations (10.3.11), (10.3.12) and (10.3.7) imply (10.3.5). In view of (10.1.1), the sum (10.3.5) is finite on any test function from the Lizorkin space $\Phi(\mathbb{Q}_p^n)$.

The theorem is thus proved. $\qquad\square$

Theorem 10.3.2 and Corollary 9.4.9 imply the following assertion.

Corollary 10.3.3. *Let* $f \in \widetilde{\Phi}'(\mathbb{Q}_p^n \times \mathbb{R}_+)$ *and let* $f(x,t)$ *be continuous in t for all* $x \in \mathbb{Q}_p^n$. *Then the Cauchy problem* (10.3.3) *has in* $\widetilde{\Phi}'(\mathbb{Q}_p^n \times \mathbb{R}_+)$ *a unique solution*

$$u(x,t) = \sum_{s \in J_{p;m}^n, j \in \mathbb{Z}^n, a \in I_p^n} \Big(e^{-p^{\alpha \max_{1 \leq r \leq n}\{m_r - j_r\}}t}\langle u^0, \Theta_{s;\,ja}^{(m)\times}\rangle$$

$$+ \int_0^t e^{-p^{\alpha \max_{1 \leq r \leq n}\{m_r - j_r\}}(t-\tau)}\langle f(\cdot,\tau), \Theta_{s;\,ja}^{(m)\times}(\cdot)\rangle\,d\tau\Big)\Theta_{s;\,ja}^{(m)\times}(x),$$

$$(10.3.13)$$

where $\Theta_{s;\,ja}^{(m)\times}(x)$ *are n-dimensional p-adic wavelets* (8.12.1).

For the case of homogeneous Cauchy problems (10.3.1) and (10.3.3), when $f(x, t) = 0$, the solutions (10.3.5) and (10.3.13) describe diffusion processes in the space $\mathbb{Q}_p^n$.

Example 10.3.1. Now we consider the one-dimensional homogeneous Cauchy problem (10.3.3) for the initial data

$$u^0(x) = \Omega(|x|_p) = \begin{cases} 1, & |x|_p \leq 1, \\ 0, & |x|_p > 1. \end{cases}$$

Since $f(x, t) = 0$, substituting (8.10.3), (8.10.10) and (8.10.11) into formula (10.3.13), we obtain a solution of this Cauchy problem:

$$u(x, t) = \sum_{s \in J_{p;m}} \sum_{j=m}^{\infty} p^{-j} e^{-p^{\alpha(m-j)}t} \chi_p(sp^j x) \Omega(|p^j x|_p).$$

Theorem 10.3.4. *Suppose that the symbol $\mathcal{A}(\xi)$ of a pseudo-differential operator A_x in (10.3.1) satisfies the condition*

$$\mathcal{A}(\widehat{p^j}(-s + \eta)) = \mathcal{A}(-\widehat{p^j s}) > 0, \qquad \forall \eta \in \mathbb{Z}_p^n,$$

for any $j \in \mathbb{Z}^n$, $s \in J_{p;m}^n$. Suppose that $f \in \widetilde{\Phi}'(\mathbb{Q}_p^n \times \mathbb{R}_+)$ and $f(x, t)$ is continuous in t for all $x \in \mathbb{Q}_p^n$. Suppose also that, for any $s \in J_{p;m}^n$, $j \in \mathbb{Z}^n$, $a \in I_p^n$

$$\lim_{t \to \infty} \langle f(\cdot, t), \Theta_{s;\,ja}^{(m)\times}(\cdot) \rangle = c_{s;\,ja} (= const). \tag{10.3.14}$$

Then the solution (10.3.5) of the Cauchy problem (10.3.1) is stabilized as $t \to \infty$:

$$\lim_{t \to \infty} u(x, t) = g(x) = \sum_{s \in J_{p;m}^n, j \in \mathbb{Z}^n, a \in I_p^n} \frac{c_{s;\,ja}}{\mathcal{A}(-\widehat{p^j s})} \Theta_{s;\,ja}^{(m)\times}(x) \in \Phi'(\mathbb{Q}_p^n), \tag{10.3.15}$$

for every $x \in \mathbb{Q}_p^n$.

Proof. Taking into account that sum (10.3.5) is finite on any test function $\phi \in \Phi(\mathbb{Q}_p^n)$ and using the L'Hospital rule, it is easy to find that (10.3.15) holds for all $\phi \in \Phi(\mathbb{Q}_p^n)$. □

Corollary 10.3.5. *Let $f \in \widetilde{\Phi}'(\mathbb{Q}_p^n \times \mathbb{R}_+)$ and let $f(x, t)$ be continuous in t for all $x \in \mathbb{Q}_p^n$. Suppose that (10.3.14) holds. Then the solution (10.3.13) of the Cauchy problem (10.3.3) is stabilized as $t \to \infty$, i.e., (10.3.15) holds, where $\mathcal{A}(-\widehat{p^j s}) = p^{\alpha \max_{1 \leq r \leq n}\{m_r - j_r\}}, \alpha \in \mathbb{R}$.*

(b) Now we consider the Cauchy problem

$$\begin{cases} i\frac{\partial u(x,t)}{\partial t} - A_x u(x, t) = f(x, t), & \text{in } \mathbb{Q}_p^n \times (0, \infty), \\ u(x, t) = u^0(x), & \text{in } \mathbb{Q}_p^n \times \{t = 0\}, \end{cases} \tag{10.3.16}$$

where $u^0 \in \Phi'(\mathbb{Q}_p^n)$ and a pseudo-differential operator A_x is given by (10.3.2). In particular, we have the Cauchy problem

$$
\begin{cases}
i\frac{\partial u(x,t)}{\partial t} - D_x^\alpha u(x,t) = f(x,t), & \text{in} \quad \mathbb{Q}_p^n \times (0, \infty), \\
u(x,t) = u^0(x), & \text{in} \quad \mathbb{Q}_p^n \times \{t = 0\},
\end{cases}
\tag{10.3.17}
$$

where D_x^α is the Taibleson fractional operator (9.2.23) with respect to x, $\alpha \in \mathbb{C}$.

Using the above results, we can construct a solution of the Cauchy problems (10.3.16) and (10.3.17).

Theorem 10.3.6. *Suppose that the symbol $\mathcal{A}(\xi)$ of a pseudo-differential operator A_x in (10.3.16) satisfies condition (9.4.1)*

$$
\mathcal{A}\big(\widehat{p^j}(-s + \eta)\big) = \mathcal{A}\big(-\widehat{p^j}s\big), \qquad \forall \eta \in \mathbb{Z}_p^n,
$$

for any $j \in \mathbb{Z}^n$, $s \in J_{p;m}^n$. Let $f \in \widetilde{\Phi}'(\mathbb{Q}_p^n \times \mathbb{R}_+)$ and let $f(x,t)$ be continuous in t for all $x \in \mathbb{Q}_p^n$. Then the Cauchy problem (10.3.16) has in $\widetilde{\Phi}'(\mathbb{Q}_p^n \times \mathbb{R}_+)$ a unique solution

$$
\begin{aligned}
u(x,t) = \sum_{s \in J_{p;m}^n, j \in \mathbb{Z}^n, a \in I_p^n} \bigg(& e^{-i\mathcal{A}(-\widehat{p^j}s)t} \big\langle u^0, \Theta_{s;ja}^{(m)\times} \big\rangle \\
& - i \int_0^t e^{-i\mathcal{A}(-\widehat{p^j}s)(t-\tau)} \big\langle f(\cdot, \tau), \Theta_{s;ja}^{(m)\times}(\cdot) \big\rangle \, d\tau \bigg) \Theta_{s;ja}^{(m)\times}(x), \quad (10.3.18)
\end{aligned}
$$

where $\Theta_{s;ja}^{(m)\times}(x)$ are n-dimensional p-adic wavelets (8.12.1).

Corollary 10.3.7. *Let $f \in \widetilde{\Phi}'(\mathbb{Q}_p^n \times \mathbb{R}_+)$ and let $f(x,t)$ be continuous in t for all $x \in \mathbb{Q}_p^n$. Then the Cauchy problem (10.3.17) has in $\widetilde{\Phi}'(\mathbb{Q}_p^n \times \mathbb{R}_+)$ a unique solution*

$$
\begin{aligned}
u(x,t) = \sum_{s \in J_{p;m}^n, j \in \mathbb{Z}^n, a \in I_p^n} \bigg(& e^{-ip^{\alpha \max_{1 \le r \le n} \{m_r - j_r\}} t} \big\langle u^0, \Theta_{s;ja}^{(m)\times} \big\rangle \\
& - i \int_0^t e^{-ip^{\alpha \max_{1 \le r \le n} \{m_r - j_r\}}(t-\tau)} \big\langle f(\cdot, \tau), \Theta_{s;ja}^{(m)\times}(\cdot) \big\rangle \, d\tau \bigg) \Theta_{s;ja}^{(m)\times}(x),
\end{aligned}
$$

$$\tag{10.3.19}$$

where $\Theta_{s;ja}^{(m)\times}(x)$ are n-dimensional p-adic wavelets (8.12.1).

10.4. Linear evolutionary pseudo-differential equations of the second order in time

Let us consider the Cauchy problem for a linear evolutionary pseudo-differential equation of the second order in t

$$\begin{cases} \frac{\partial^2 u}{\partial t^2} + A_{1x}\frac{\partial u}{\partial t} + A_{2x}u(x,t) + u(x,t) = f(x,t), & \text{in } \mathbb{Q}_p^n \times (0,\infty), \\ u(x,t) = u^0(x), \quad \frac{\partial u(x,t)}{\partial t} = u^1(x), & \text{in } \mathbb{Q}_p^n \times \{t=0\}, \end{cases}$$

$$(10.4.1)$$

where $t \in \mathbb{R}$, $u^0, u^1(x) \in \Phi'(\mathbb{Q}_p^n)$, pseudo-differential operators A_{1x} and A_{2x} are given by (10.3.2) and their symbols $\mathcal{A}_1(\xi)$, $\mathcal{A}_2(\xi) \in \mathcal{E}(\mathbb{Q}_p^n \setminus \{0\})$, $u(x,t) \in \widetilde{\Phi}'(\mathbb{Q}_p^n \times \mathbb{R}_+)$ is the desired distribution.

Theorem 10.4.1. *Suppose that the symbol $\mathcal{A}_l(\xi)$ of the pseudo-differential operator A_{lx} in (10.4.1) satisfies condition (9.4.1)*

$$\mathcal{A}_l\big(\widehat{p^j}(-s+\eta)\big) = \mathcal{A}_l\big(-\widehat{p^j}s\big), \qquad \forall \eta \in \mathbb{Z}_p^n, \quad l=1,2,$$

for any $j \in \mathbb{Z}^n$, $s \in J_{p;m}^n$. Let $f \in \widetilde{\Phi}'(\mathbb{Q}_p^n \times \mathbb{R}_+)$ and let $f(x,t)$ be continuous in t for all $x \in \mathbb{Q}_p^n$. Then the Cauchy problem (10.4.1) has a unique solution

$$u(x,t) = \sum_{s \in J_{p;m}^n, j \in \mathbb{Z}^n, a \in I_p^n} \Lambda_{s;j,a}(t)\Theta_{s;ja}^{(m)\times}(x) \in \widetilde{\Phi}'(\mathbb{Q}_p^n \times \mathbb{R}_+), \quad (10.4.2)$$

where $\Theta_{s;ja}^{(m)\times}(x)$ are n-dimensional p-adic wavelets (8.12.1). Here, if
(a) $\big(\mathcal{A}_1(-\widehat{p^j}s)\big)^2 \neq 4\big(\mathcal{A}_2(-\widehat{p^j}s)+1\big)$ then

$$\Lambda_{s;j,a}(t) = \frac{1}{k_+ - k_-}\bigg(\int_0^t \Big(e^{k_+(t-\tau)} - e^{k_-(t-\tau)} \Big) d_{s;j,a}(\tau)\,d\tau$$
$$+ \big(\langle u^1, \Theta_{s;ja}^{(m)\times}\rangle - k_-\langle u^0, \Theta_{s;ja}^{(m)\times}\rangle\big)e^{k_+ t}$$
$$- \big(\langle u^1, \Theta_{s;ja}^{(m)\times}\rangle - k_+\langle u^0, \Theta_{s;ja}^{(m)\times}\rangle\big)e^{k_- t}\bigg), \qquad (10.4.3)$$

where $k_\pm = \frac{-\mathcal{A}_1(-\widehat{p^j}s) \pm \sqrt{(\mathcal{A}_1(-\widehat{p^j}s))^2 - 4(\mathcal{A}_2(-\widehat{p^j}s)+1)}}{2}$;
(b) $\big(\mathcal{A}_1(-\widehat{p^j}s)\big)^2 = 4\big(\mathcal{A}_2(-\widehat{p^j}s)+1\big)$ then

$$\Lambda_{s;j,a}(t) = \int_0^t e^{k(t-\tau)}(t-\tau)d_{s;j,a}(\tau)\,d\tau$$
$$+ e^{kt}\Big(\langle u^0, \Theta_{s;ja}^{(m)\times}\rangle + \big(\langle u^1, \Theta_{s;ja}^{(m)\times}\rangle - k\langle u^0, \Theta_{s;ja}^{(m)\times}\rangle\big)\Big), \quad (10.4.4)$$

where $k = \frac{-A_1(-\widehat{p^j s})}{2}$; $d_{s;\,j,a}(t)$ are coefficients of the distribution $f(x,t)$ in representation (10.3.6); $s \in J^n_{p;m}$, $j \in \mathbb{Z}^n$, $a \in I^n_p$.

Proof. Since $f \in \widetilde{\Phi}'(\mathbb{Q}^n_p \times \mathbb{R}_+)$, according to (8.14.3), this distribution is represented in the form (10.3.6). As in Theorem 10.3.2, we will seek a solution of the Cauchy problem (10.4.1) in the form (10.3.8). Substituting (10.3.8) into equation (10.4.1), from Theorem 10.2.3, we obtain

$$\sum_{s \in J^n_{p;m},\, j \in \mathbb{Z}^n,\, a \in I^n_p} \left(\Lambda''_{s;\,j,a}(t) + A_1(-\widehat{p^j s})\Lambda'_{s;\,j,a}(t) \right.$$

$$\left. + \left(A_2(-\widehat{p^j s}) + 1 \right) \Lambda_{s;\,j,a}(t) - d_{s;\,j,a}(t) \right) \Theta^{(m)\times}_{s;\,ja}(x) = 0,$$

where $d_{s;\,j,a}(t)$ is given by (10.3.7). Next, similarly to the proof of Theorem 10.3.2, the last relation implies that $\Lambda_{s;\,j,a}(t)$ satisfies the differential equation

$$\Lambda''_{s;\,j,a}(t) + A_1(-\widehat{p^j s})\Lambda'_{s;\,j,a}(t) + \left(A_2(-\widehat{p^j s}) + 1 \right) \Lambda_{s;\,j,a}(t) = d_{s;\,j,a}(t),$$

$$(10.4.5)$$

for all $t \geq 0$, $s \in J^n_{p;m}$, $j \in \mathbb{Z}^n$, $a \in I^n_p$. Since $d_{s;\,j,a}(t)$ is continuous, for given $\Lambda_{s;\,j,a}(0)$ and $\frac{d\Lambda_{s;\,j,a}(0)}{dt}$ the differential equation (10.4.5) has a unique solution. This solution can be calculated explicitly.

The characteristic equation corresponding to the homogeneous differential equation (10.4.5) is the following:

$$k^2 + kA_1(-\widehat{p^j s}) + \left(A_2(-\widehat{p^j s}) + 1 \right) = 0. \qquad (10.4.6)$$

(a) Let $\left(A_1(-\widehat{p^j s}) \right)^2 \neq 4\left(A_2(-\widehat{p^j s}) + 1 \right)$. In this case, from (10.4.6) we have $k_{\pm} = \frac{-A_1(-\widehat{p^j s}) \pm \sqrt{(A_1(-\widehat{p^j s}))^2 - 4(A_2(-\widehat{p^j s})+1)}}{2}$. Hence, according to the well-known results from the theory of ordinary differential equations, a solution of (10.4.5) is

$$\Lambda_{s;\,j,a}(t) = \frac{1}{k_+ - k_-} \left(\int_0^t \left(e^{k_+(t-\tau)} - e^{k_-(t-\tau)} \right) d_{s;\,j,a}(\tau)\, d\tau + \left(\Lambda'_{s;\,j,a}(0) \right. \right.$$

$$\left. \left. - k_-\Lambda_{s;\,j,a}(0) \right) e^{k_+ t} - \left(\Lambda'_{s;\,j,a}(0) - k_+\Lambda_{s;\,j,a}(0) \right) e^{k_- t} \right).$$

$$(10.4.7)$$

(b) Let $\left(A_1(-\widehat{p^j s}) \right)^2 = 4\left(A_2(-\widehat{p^j s}) + 1 \right)$ then from (10.4.6) we have $k = \frac{-A_1(-\widehat{p^j s})}{2}$. Hence, according to the well-known results from the theory of ordinary differential equations, a solution of (10.4.5) is

$$\Lambda_{s;\,j,a}(t) = \int_0^t e^{k(t-\tau)}(t-\tau)d_{s;\,j,a}(\tau)\,d\tau$$

$$+ e^{kt}\Big(\Lambda_{s;\,j,a}(0) + \big(\Lambda'_{s;\,j,a}(0) - k\Lambda_{s;\,j,a}(0)\big)t\Big). \quad (10.4.8)$$

By substituting the solution (10.4.7) (or (10.4.8)) of the differential equation (10.4.5) into (10.3.8) and taking into account formulas (10.3.7) and

$$\Lambda_{s;\,j,a}(0) = \big\langle u^0, \Theta^{(m)\times}_{s;\,ja}\big\rangle, \qquad \Lambda'_{s;\,j,a}(0) = \big\langle u^1, \Theta^{(m)\times}_{s;\,ja}\big\rangle,$$

$s \in J^n_{p;m}$, $j \in \mathbb{Z}^n$, $a \in I^n_p$, we find a unique solution of the Cauchy problem (10.4.1) in the form (10.3.8), where the coefficients $\Lambda_{s;\,j,a}(t)$ are given by (10.4.3) or (10.4.4).

In view of (10.1.1), the sum (10.4.2) is finite on any test function from the Lizorkin space $\Phi(\mathbb{Q}^n_p)$.

The theorem is proved. $\qquad\qquad\qquad\qquad\qquad\qquad\qquad\qquad\qquad\square$

In particular, if A_{1x} and A_{2x} are fractional operators, Theorem 10.4.1 gives a solution of the Cauchy problem

$$\begin{cases} \frac{\partial^2 u}{\partial t^2} + D_x^{\alpha_1}\frac{\partial u}{\partial t} + D_x^{\alpha_2}u(x,t) + u(x,t) = f(x,t), & \text{in} \quad \mathbb{Q}^n_p \times (0,\infty), \\[2mm] u(x,t) = u^0(x), \quad \frac{\partial u(x,t)}{\partial t} = u^1(x), & \text{in} \quad \mathbb{Q}^n_p \times \{t = 0\}, \end{cases}$$
$$(10.4.9)$$

A particular case of the problem (10.4.9) was solved in the paper [63].

Remark 10.1. The variable separation method developed for solving the above Cauchy problems can be used for solving the Cauchy problems for the linear pseudo-differential equations of order m in t:

$$\sum_{j=0}^m A_{jx}\frac{\partial^j u(x,t)}{\partial t^j} + u(x,t) = f(x,t),$$

where pseudo-differential operators A_{jx} are given by (10.3.2), $j = 0, 1, \ldots, m$, $u(x,t) \in \widetilde{\Phi}'(\mathbb{Q}^n_p \times \mathbb{R}_+)$ is the desired distribution. A particular case of the above equation is the following:

$$\sum_{j=0}^m D_x^{\alpha_j}\frac{\partial^j u(x,t)}{\partial t^j} + u(x,t) = f(x,t),$$

where $D_x^{\alpha_j}$ is the Taibleson fractional operator (9.2.23) with respect to x, $\alpha_j \in \mathbb{C}$, $j = 0, 1, \ldots, m$.

10.5. Semi-linear evolutionary pseudo-differential equations

(a) Consider the Cauchy problem for the semi-linear pseudo-differential equation

$$\begin{cases} \frac{\partial u(x,t)}{\partial t} + A_x u(x,t) + u(x,t)|u(x,t)|^{2k} = 0, & \text{in } \mathbb{Q}_p^n \times (0, \infty), \\ u(x,t) = u^0(x), & \text{in } \mathbb{Q}_p^n \times \{t = 0\}, \end{cases}$$
(10.5.1)

where a pseudo-differential operator A_x is given by (10.3.2), $k \in \mathbb{N}$, $u(x,t) \in \tilde{\Phi}'(\mathbb{Q}_p^n \times \mathbb{R}_+)$ is the desired distribution.

According to Proposition 8.14.3, the distribution $u(x,t)$ can be realized as an infinite sum of the form

$$u(x,t) = \sum_{s \in J_{p;m}^n, j \in \mathbb{Z}^n, a \in I_p^n} \Lambda_{s;j,a}(t) \Theta_{s;ja}^{(m)\times}(x),$$
(10.5.2)

where $\Lambda_{s;j,a}(t)$ are the desired functions, $\Theta_{s;ja}^{(m)\times}(x)$ are elements of the wavelet basis (8.12.1). We will solve the Cauchy problem in a particular class of distributions $u(x,t)$ such that in the representation (10.5.2)

$$\widehat{p^{j'-j}}a - a' \notin \mathbb{Z}_p^n, \quad \text{if } j_l < j_l', \quad l = 1, \dots, n.$$
(10.5.3)

In view of (8.10.5), in this case all the sets $\{x \in \mathbb{Q}_p^n : |\widehat{p^j}x - a|_p \leq 1\}$, $\{x \in \mathbb{Q}_p^n : |\widehat{p^{j'}}x - a'|_p \leq 1\}$ are disjoint.

Theorem 10.5.1. *Let a pseudo-differential operator A_x in (10.5.1) be such that its symbol $\mathcal{A}(\xi)$ satisfies condition (9.4.1)*

$$\mathcal{A}(\widehat{p^j}(-s+\eta)) = \mathcal{A}(-\widehat{p^j}s), \quad \forall \eta \in \mathbb{Z}_p^n,$$

for any $j \in \mathbb{Z}^n$, $s \in J_{p;m}^n$. Then in the above-mentioned class of distributions (10.5.2) and (10.5.3) the Cauchy problem (10.5.1) has the unique solution

$$u(x,t)$$

$$= \sum_{\substack{s \in J_{p;m}^n, \\ j \in \mathbb{Z}^n, \\ a \in I_p^n}} \left(\frac{\text{Re}\mathcal{A}(-\widehat{p^j}s)}{\text{Re}\mathcal{A}(-\widehat{p^j}s) + |\langle u^0, \Theta_{s;ja}^{(m)\times} \rangle|^{2k} p^{-k|j|}(1 - e^{-2k\text{Re}\mathcal{A}(-\widehat{p^j}s)t})} \right)^{1/2k}$$

$$\times \langle u^0, \Theta_{s;ja}^{(m)\times} \rangle e^{-\mathcal{A}(-\widehat{p^j}s)t} \Theta_{s;ja}^{(m)\times}(x),$$
(10.5.4)

for $t \geq 0$, where $\Theta_{s;ja}^{(m)\times}(x)$ are n-dimensional p-adic wavelets (8.12.1). Moreover, this formula is applicable for the case $\mathcal{A} \equiv 0$.

Proof. Since $|\chi_p(s \cdot (p^j x - a))| = 1$, taking into account formulas (10.5.2), (10.5.3), (8.12.1), we obtain a formal series

$$|u(x,t)|^2 = \sum_{s \in J_{p;m}^n, j \in \mathbb{Z}^n, a \in I_p^n} |\Lambda_{s;j,a}(t)|^2 p^{-|j|} \Omega(|\widehat{p^j}x - a|_p)$$

and

$$u(x,t)|u(x,t)|^{2k}$$
$$= \sum_{s \in J_{p;m}^n, j \in \mathbb{Z}^n, a \in I_p^n} |\Lambda_{s;j,a}(t)|^{2k} \Lambda_{s;j,a}(t) p^{-k|j|} \Theta_{s;ja}^{(m)\times}(x) \in \widetilde{\Phi}'(\mathbb{Q}_p^n \times \mathbb{R}_+),$$

$$(10.5.5)$$

where the indexes in the above sums satisfy the condition (10.5.3).

Substituting (10.5.5) and (10.5.2) into (10.5.1), in view of of Theorem 9.4.2, we find that

$$\sum_{j \in \mathbb{Z}, k \in J_{p0}^n, a \in I_p^n} \left(\Lambda_{s;j,a}'(t) + \mathcal{A}(-\widehat{p^j}s) \Lambda_{s;j,a}(t) \right.$$
$$\left. + p^{-k|j|} |\Lambda_{s;j,a}(t)|^{2k} \Lambda_{s;j,a}(t) \right) \Theta_{s;ja}^{(m)\times}(x) = 0, \qquad (10.5.6)$$

where the latter equation is understood in the weak sense. Since, according to Lemma 8.14.1, any test function $\phi \in \Phi(\mathbb{Q}_p^n)$ is represented in the form of a *finite* sum (8.14.1), the equality (10.5.6) implies that

$$\Lambda_{s;j,a}'(t) + \mathcal{A}(-\widehat{p^j}s) \Lambda_{s;j,a}(t) + p^{-k|j|} |\Lambda_{s;j,a}(t)|^{2k} \Lambda_{s;j,a}(t) = 0,$$
$$(10.5.7)$$

for all $s \in J_{p;m}^n$, $j \in \mathbb{Z}^n$, $a \in I_p^n$, and for all $t \geq 0$.

Substituting $\Lambda_{s;j,a}(t) = R_{s;j,a}(t) e^{i\alpha_{s;j,a}(t)}$ and $\mathcal{A}(-\widehat{p^j}s) = \mathrm{Re}\mathcal{A}(-\widehat{p^j}s) + i\mathrm{Im}\mathcal{A}(-\widehat{p^j}s)$ into (10.5.7), we obtain the system of differential equations

$$R_{s;j,a}'(t) + \mathrm{Re}\mathcal{A}(-\widehat{p^j}s) R_{s;j,a}(t) + p^{-k|j|} R_{s;j,a}^{2k+1}(t) = 0,$$

$$\alpha_{s;j,a}'(t) + \mathrm{Im}\mathcal{A}(-\widehat{p^j}s) = 0,$$

$s \in J_{p;m}^n$, $j \in \mathbb{Z}^n$, $a \in I_p^n$, $t \geq 0$. Integrating the last system, we find $\alpha_{s;j,a}(t) = -\mathrm{Im}\mathcal{A}(-\widehat{p^j}s)t + \alpha_{s;j,a}(0)$ and

$$\frac{R_{s;j,a}^{2k}(t)}{\mathrm{Re}\mathcal{A}(-\widehat{p^j}s) + p^{-k|j|} R_{s;j,a}^{2k}(t)} = E_{s;j,a} e^{-2k\mathrm{Re}\mathcal{A}(-\widehat{p^j}s)t},$$

i.e.,

$$\Lambda_{s;j,a}(t) = \left(\frac{E_{s;j,a} \mathrm{Re}\mathcal{A}(-\widehat{p^j}s)}{1 - E_{s;j,a} p^{-k|j|} e^{-2k\mathrm{Re}\mathcal{A}(-\widehat{p^j}s)t}} \right)^{1/2k} e^{-\mathcal{A}(-\widehat{p^j}s)t} e^{i\alpha_{s;j,a}(0)},$$

$$(10.5.8)$$

Substituting (10.5.8) into (10.5.2), we find a solution of the problem (10.5.1)

$$u(x,t) = \sum_{\substack{s \in J^n_{p;m}, \\ j \in \mathbb{Z}^n, \\ a \in I^n_p}} \left(\frac{E_{s;j,a} \mathrm{Re}\mathcal{A}(-\widehat{p^j}s)}{1 - E_{s;j,a} p^{-k|j|} e^{-2k\mathrm{Re}\mathcal{A}(-\widehat{p^j}s)t}} \right)^{1/2k}$$

$$\times e^{-\mathcal{A}(-\widehat{p^j}s)t} e^{i\alpha_{s;j,a}(0)} \Theta^{(m)\times}_{s;ja}(x), \qquad (10.5.9)$$

$x \in \mathbb{Q}^n_p$, $t \geq 0$. Setting in (10.5.9) $t = 0$, we obtain that

$$u^0(x) = \sum_{s \in J^n_{p;m}, j \in \mathbb{Z}^n, a \in I^n_p} \left(\frac{E_{s;j,a} \mathrm{Re}\mathcal{A}(-\widehat{p^j}s)}{1 - E_{s;j,a} p^{-k|j|}} \right)^{1/2k} e^{i\alpha_{s;j,a}(0)} \Theta^{(m)\times}_{s;ja}(x),$$

where $u^0 \in \Phi'(\mathbb{Q}^n_p)$. Hence, according to (8.14.4) , the coefficients $E_{s;j,a}$ are uniquely determined by (8.14.5) as

$$\left(\frac{E_{s;j,a} \mathrm{Re}\mathcal{A}(-\widehat{p^j}s)}{1 - E_{s;j,a} p^{-k|j|}} \right)^{1/2k} e^{i\alpha_{s;j,a}(0)} = \langle u^0, \Theta^{(m)\times}_{s;ja} \rangle.$$

The latter equation implies that

$$E_{s;j,a} = \frac{|\langle u^0(x), \Theta^{(m)\times}_{s;ja} \rangle|^{2k}}{\mathrm{Re}\mathcal{A}(-\widehat{p^j}s) + p^{-k|j|}|\langle u^0, \Theta^{(m)\times}_{s;ja} \rangle|^{2k}}.$$

Substituting $E_{s;j,a}$ into (10.5.9), we obtain (10.5.4). In view of formula (10.1.1), the sum (10.5.4) is finite on any test function from the Lizorkin space $\Phi(\mathbb{Q}^n_p)$.

Now by passing to the limit as $\mathcal{A} \to 0$ in formula (10.5.4), one can easily see that this formula (10.5.4) is applicable for the case $\mathcal{A} \equiv 0$.

The theorem is thus proved. $\qquad\qquad\square$

(b) Now we consider the Cauchy problem

$$\begin{cases} i\frac{\partial u(x,t)}{\partial t} - A_x u(x,t) + u(x,t)|u(x,t)|^{2k} = 0, & \text{in} \quad \mathbb{Q}^n_p \times (0, \infty), \\ u(x,t) = u^0(x), & \text{in} \quad \mathbb{Q}^n_p \times \{t = 0\}, \end{cases}$$
$$(10.5.10)$$

where a pseudo-differential operator A_x is given by (10.3.2), $k \in \mathbb{N}$, $u(x,t) \in \widetilde{\Phi}'(\mathbb{Q}^n_p \times \mathbb{R}_+)$ is the desired distribution.

Theorem 10.5.2. *Let a pseudo-differential operator A_x in equation (10.5.10) be such that its symbol $\mathcal{A}(\xi)$ satisfies the condition (9.4.1):*

$$\mathcal{A}(\widehat{p^j}(-s + \eta)) = \mathcal{A}(-\widehat{p^j}s), \qquad \forall \eta \in \mathbb{Z}^n_p,$$

for any $j \in \mathbb{Z}^n$, $s \in J^n_{p;m}$. Then in the above-mentioned class of distributions

(10.5.2), (10.5.3) *the Cauchy problem* (10.5.10) *has a unique solution*

$$u(x,t) = \sum_{s\in J^n_{p;m}, j\in\mathbb{Z}^n, a\in I^n_p} e^{ip^{-k|j|}\frac{|\langle u^0,\Theta^{(m)\times}_{s;ja}\rangle|^{2k}}{2k\mathrm{Im}\mathcal{A}(-\widehat{p^j}s)}(e^{2k\mathrm{Im}\mathcal{A}(-\widehat{p^j}s)t}-1)}$$

$$\times\langle u^0,\Theta^{(m)\times}_{s;ja}\rangle e^{-i\mathcal{A}(-\widehat{p^j}s)t}\Theta^{(m)\times}_{s;ja}(x), \qquad (10.5.11)$$

for $t \geq 0$, *where* $\Theta^{(m)\times}_{s;ja}(x)$ *are n-dimensional p-adic wavelets* (8.12.1). *Moreover, the passage to the limit in* (10.5.11) *as* $\mathrm{Im}\mathcal{A}\to 0$ *gives exactly solution for the case* $\mathrm{Im}\mathcal{A}=0$.

Proof. We will seek a solution of the problem in the form (10.5.2) and (10.5.3). Substituting (10.5.5) and (10.5.2) into (10.5.10), in view of Theorem 9.4.2, we find that

$$i\Lambda'_{s;j,a}(t) - \mathcal{A}(-\widehat{p^j}s)\Lambda_{s;j,a}(t) + p^{-k|j|}|\Lambda_{s;j,a}(t)|^{2k}\Lambda_{s;j,a}(t) = 0, \quad (10.5.12)$$

$s \in J^n_{p;m}$, $j \in \mathbb{Z}^n$, $a \in I^n_p$, $t \geq 0$. Setting $\mathcal{A}(-\widehat{p^j}s) = \mathrm{Re}\mathcal{A}(-\widehat{p^j}s) + i\mathrm{Im}\mathcal{A}(-\widehat{p^j}s)$, and $\Lambda_{s;j,a}(t) = R_{s;j,a}(t)e^{i\alpha_{s;j,a}(t)}$ in (10.5.12), we obtain

$$R'_{s;j,a}(t) - \mathrm{Im}\mathcal{A}(-\widehat{p^j}s)R_{s;j,a}(t) = 0,$$

$$\alpha'_{s;j,a}(t) + \mathrm{Re}\mathcal{A}(-\widehat{p^j}s) - p^{-k|j|}R^{2k}_{s;j,a}(t) = 0,$$

$s \in J^n_{p;m}, j \in \mathbb{Z}^n, a \in I^n_p$. Solving this system, we find

$$R_{s;j,a}(t) = R_{s;j,a}(0)e^{\mathrm{Im}\mathcal{A}(-\widehat{p^j}s)t},$$

$$\alpha_{s;j,a}(t) = -\mathrm{Re}\mathcal{A}(-\widehat{p^j}s)t + p^{-k|j|}\frac{R^{2k}_{s;j,a}(0)}{2k\mathrm{Im}\mathcal{A}(-\widehat{p^j}s)}e^{2k\mathrm{Im}\mathcal{A}(-\widehat{p^j}s)t} + \alpha_{s;j,a}(0),$$

and

$$\Lambda_{s;j,a}(t) = R_{s;j,a}(0)e^{i\alpha_{s;j,a}(0)}e^{-i\mathcal{A}(-\widehat{p^j}s)t}$$

$$\times e^{ip^{-k|j|}\frac{R^{2k}_{s;j,a}(0)}{2k\mathrm{Im}\mathcal{A}(-\widehat{p^j}s)}e^{2k\mathrm{Im}\mathcal{A}(-\widehat{p^j}s)t}}, \qquad (10.5.13)$$

$s \in J^n_{p;m}, j \in \mathbb{Z}^n, a \in I^n_p$. We have

$$\Lambda_{s;j,a}(0) = \langle u^0, \Theta^{(m)\times}_{s;ja}\rangle = R_{s;j,a}(0)e^{i\left(\alpha_{s;j,a}(0)+p^{-k|j|}\frac{R^{2k}_{s;j,a}(0)}{2k\mathrm{Im}\mathcal{A}(-\widehat{p^j}s)}\right)}.$$

The last relation, (10.5.13) and (10.5.2) imply (10.5.11). $\square$

A particular solution of the problem (10.5.10) for $A_x = D^\alpha_x$ and $k = 1$ was first obtained by S. V. Kozyrev.

According to (10.5.4) and (10.5.11), *if the initial data* $u^0(x)$ *of the problem* (10.5.1) *or* (10.5.10) *have a support in the ball* $B^n_N \subset \mathbb{Q}^n_p$, *then the support of a solution* $u(x,t)$ *with respect to x belongs to this ball for all* $t > 0$. *This effect of localization of a solution* for p-adic Schrödinger equation was first described in [132].

11

A p-adic Schrödinger-type operator with point interactions

11.1. Introduction

In this chapter a p-adic Schrödinger-type operator $D^\alpha + V_Y$ is studied, where D^α ($\alpha > 0$) is the operator of fractional differentiation (which was studied in Chapter 9) and $V_Y = \sum_{i,j=1}^n b_{ij} \langle \delta_{x_j}, \cdot \rangle \delta_{x_i}$ ($b_{ij} \in \mathbb{C}$) is a singular potential containing the Dirac delta functions δ_x concentrated on a set of points $Y = \{x_1, \ldots, x_n\}$ of the field of p-adic numbers $\mathbb{Q}_p$. It is shown that such a problem is well-posed for $\alpha > 1/2$ and the singular perturbation V_Y is form-bounded for $\alpha > 1$. In the latter case, the spectral analysis of η-self-adjoint operator realizations of $D^\alpha + V_Y$ in $\mathcal{L}_2(\mathbb{Q}_p)$ is carried out.

The results of this chapter are based on the papers by S. Albeverio, S. Kuzhel, S. Torba [31] and S. Kuzhel and S. Torba [169] (see also [102, Example 5.6]) and continue the investigation of p-adic Schrödinger-type operators with point interactions started by A. Kochubei [157]. We recall that the concept of a p-adic Schrödinger-type operator was first introduced and studied by V. S. Vladimirov and I. V. Volovich [241].

In "usual" mathematical physics Schrödinger operators with point interactions are well-studied; they are used in quantum mechanics to obtain Hamiltonians describing realistic physical systems having the important property of being exactly solvable, i.e., such that all eigenfunctions, spectrum, and scattering matrix can be calculated [9], [27].

Since we deal with the mapping $\mathbb{Q}_p \to \mathbb{C}$, i.e., complex-valued functions defined on $\mathbb{Q}_p$ are considered, the operation of differentiation *is not defined* and the operator of fractional differentiation D^α of order α ($\alpha > 0$) takes over the role of differentiation [157], [243], [241]. In particular, p-adic Schrödinger-type operators with potentials $V(x) : \mathbb{Q}_p \to \mathbb{C}$ are defined as $D^\alpha + V(x)$.

One of the remarkable features of the p-adic theory of distribution is that any distribution $f \in \mathcal{D}'(\mathbb{Q}_p)$ with point support $\operatorname{supp} f = \{x\}$ coincides with the Dirac delta function at the point x multiplied by a constant $c \in \mathbb{C}$, i.e., $f = c\delta_x$. For this reason, it is natural to consider the expression $D^\alpha + V_Y$ where the singular potential $V_Y = \sum_{i,j=1}^{n} b_{ij} \langle \cdot, \delta_{x_j} \rangle \delta_{x_i}$ ($b_{ij} \in \mathbb{C}$) contains the Dirac delta functions δ_x concentrated on points x_r of the set $Y = \{x_1, \ldots, x_n\} \subset \mathbb{Q}_p$ as a p-adic analogue of the Schrödinger operator with point interactions. Since D^α is a p-adic pseudo-differential operator the expression $D^\alpha + V_Y$ gives an example of pseudo-differential operators with point interactions. In the "usual" (Archimedean) theory, expressions of this (and more general) type are studied in [28].

Obviously, the domain of definition $\mathcal{D}(D^\alpha)$ of the unperturbed operator D^α need not contain functions continuous on $\mathbb{Q}_p$ and, in general, it may happen that the singular potential V_Y is not well-defined on $\mathcal{D}(D^\alpha)$.

In Section 11.2, we discuss the problem of characterizing $\mathcal{D}(D^\alpha)$ and studying in detail the solutions of the equation $D^\alpha - \lambda I = \delta_x$.

In Sections 11.3–11.9, we deal with the spectral analysis of operator realizations of $D^\alpha + V_Y$ ($\alpha > 1$) in $\mathcal{L}_2(\mathbb{Q}_p)$. We do not restrict ourselves only to the self-adjoint case and consider also η-self-adjoint operators. The investigation of such operators is motivated by the intensive development of pseudo-Hermitian ($\mathcal{PT}$-symmetric) quantum mechanics in the last few years [48], [100], [190], [231], [255].

Among the self-adjoint extensions of the symmetric operator A_{sym} associated with $D^\alpha + V_Y$ ($\alpha > 1$), we pay special attention to the Friedrichs extension A_F. Since A_F is the "hard" extension of A_{sym} (see [37] for the terminology) and the singular potential V_Y is form-bounded, the hypothesis that the discrete spectrum of A_F depends on the geometrical structure of Y looks likely. In this way we discuss the connection between the minimal distance $p^{\gamma_{min}}$ between elements of Y and an infinite sequence of points of the discrete spectrum (type-1 part of the discrete spectrum).

We will use the following notation: $\mathcal{D}(A)$ and $\ker A$ denote the domain and the null-space of a linear operator A, respectively. $A \upharpoonright_X$ denotes the restriction of A onto a set X.

11.2. The equation $D^\alpha - \lambda I = \delta_x$

As mentioned above, the operator of differentiation is not defined in $\mathcal{L}_2(\mathbb{Q}_p)$, and it is replaced by the operator of fractional differentiation $D^\alpha, \alpha > 0$ (9.2.4). It is easy to see that $D^\alpha f$ is well defined for all $f \in \mathcal{D}(\mathbb{Q}_p)$, but the element $D^\alpha f$ need not necessarily belong to $\mathcal{D}(\mathbb{Q}_p)$ (since the function $|\xi|_p^\alpha$ is not locally constant); however $D^\alpha f \in L_2(\mathbb{Q}_p)$ [157]. Since $\mathcal{D}(\mathbb{Q}_p)$ is not invariant with respect to D^α we cannot define D^α on the whole space $\mathcal{D}'(\mathbb{Q}_p)$. For a distribution $f \in \mathcal{D}'(\mathbb{Q}_p)$ the operator D^α is well defined only if the right-hand side of (9.2.4) exists. To overcome this obstacle, we need to consider this operator in the Lizorkin spaces instead of $\mathcal{D}(\mathbb{Q}_p)$ (see Chapter 9).

In what follows we will consider D^α, $\alpha > 0$, as an unbounded operator in $\mathcal{L}_2(\mathbb{Q}_p)$. In this case, the domain of definition $\mathcal{D}(D^\alpha)$ consists of the functions $f \in \mathcal{L}_2(\mathbb{Q}_p)$ such that $|\xi|_p^\alpha F[f](\xi) \in \mathcal{L}_2(\mathbb{Q}_p)$. Since D^α is unitarily equivalent to the operator of multiplication by $|\xi|_p^\alpha$, this operator is positive self-adjoint in $\mathcal{L}_2(\mathbb{Q}_p)$ and its spectrum consists of eigenvalues $\lambda_m = p^{\alpha m}$ ($m \in \mathbb{Z}$) of infinite multiplicity and with accumulation point $\lambda = 0$.

According to Kozyrev's paper [163], the orthonormal p-adic wavelet basis in $\mathcal{L}_2(\mathbb{Q}_p)$ (8.1.3),

$$\theta_{k;\,ja}(x) = p^{-j/2}\chi_p\big(p^{-1}k(p^j x - a)\big)\Omega\big(|p^j x - a|_p\big), \quad x \in \mathbb{Q}_p, \quad (11.2.1)$$

coincides with the set of eigenfunctions of D^α, and according to (9.1.1),

$$D^\alpha \theta_{k;\,ja}(x) = p^{\alpha(1-j)}\theta_{k;\,ja}(x), \quad x \in \mathbb{Q}_p, \quad \alpha > 0, \quad (11.2.2)$$

where $k = 1, 2, \ldots p - 1$, $j \in \mathbb{Z}$, $a \in I_p = \mathbb{Q}_p/\mathbb{Z}_p$, and $\Omega(t)$ is the characteristic function (4.2.2) of the segment $[0, 1]$ (for details, see Section 9.4).

In view of (11.2.2) the Kozyrev p-adic wavelet basis (11.2.1) does not depend on the choice of α and it provides a convenient framework for the investigation of D^α. In particular, analyzing the expansion of any element $u \in \mathcal{D}(D^\alpha)$ with respect to (11.2.1), it is not hard to establish the uniform convergence of the corresponding series for $\alpha > 1/2$. This fact and the property of eigenfunctions $\theta_{k;\,ja}(x)$ to be continuous on $\mathbb{Q}_p$ imply the following statement.

Proposition 11.2.1. ([168]) *The domain $\mathcal{D}(D^\alpha)$ consists of functions which are continuous on $\mathbb{Q}_p$ if and only if $\alpha > 1/2$.*

Let us consider the equation

$$(D^\alpha - \lambda I)h = \delta_{x_r}, \quad \lambda \in \mathbb{C}, \quad x_r \in \mathbb{Q}_p, \quad \alpha > 0, \quad (11.2.3)$$

where $D^\alpha : \mathcal{L}_2(\mathbb{Q}_p) \to \mathcal{D}'(\mathbb{Q}_p)$ is understood in the distribution sense.

It follows from [157, Lemma 3.7] that equation (11.2.3) has no solutions belonging to $\mathcal{L}_2(\mathbb{Q}_p)$ for $\alpha \leq 1/2$.

Theorem 11.2.2. *The following statements are valid:*

1. *If $\alpha > 1/2$, then (11.2.3) has a unique solution $h = h_{r,\lambda} \in \mathcal{L}_2(\mathbb{Q}_p)$ if and only if $\lambda \neq p^{\alpha m}$, where m runs $\mathbb{Z} \cup \{-\infty\}$.*

2. *If $\alpha > 1$ and $\lambda \neq p^{\alpha m}$ ($\forall m \in \mathbb{Z} \cup \{-\infty\}$), then $h_{r,\lambda} \in \mathcal{D}(D^{\alpha/2})$.*

Proof. First of all we note that any function $u \in \mathcal{D}(D^\alpha)$ can be expanded in a uniformly convergent series with respect to the complex-conjugate p-adic wavelet basis

$$\{\overline{\theta_{k;\,ja}}(x) : k = 1, 2, \ldots p - 1;\; j \in \mathbb{Z};\; a \in I_p\}.$$

This means (since $\{\overline{\theta_{k;\,ja}}\}$ are continuous functions on $\mathbb{Q}_p$) that

$$u(x_r) = \sum_{j=-\infty}^{\infty} \sum_{k=1}^{p-1} \sum_{a\in I_p} (u, \overline{\theta_{k;\,ja}})\overline{\theta_{k;\,ja}}(x_r) \quad \text{for} \quad x_r \in \mathbb{Q}_p,$$

where $(f, g) = \int_{\mathbb{Q}_p} f(x)\overline{g(x)}dx$ is the scalar product in $\mathcal{L}_2(\mathbb{Q}_p)$.

Obviously, $\overline{\theta_{k;\,ja}}(x_r) \neq 0 \iff |p^j x_r - a|_p \leq 1$. Here $a \in I_p$ and hence $|a|_p > 1$ for $a \neq 0$. It follows from the strong triangle inequality that $|p^j x_r - a|_p \leq 1 \iff a = \{p^j x_r\}_p$. But then, recalling (11.2.1), we obtain

$$\overline{\theta_{k;\,ja}}(x_r) = \begin{cases} 0, & a \neq \{p^j x_r\}_p, \\ p^{-j/2}\chi_p\big(-p^{-1}k(p^j x_r - a)\big), & a = \{p^j x_r\}_p \end{cases} \quad (11.2.4)$$

Therefore,

$$\langle \delta_{x_r}, u \rangle = u(x_r)$$

$$= \sum_{j=-\infty}^{\infty} \sum_{k=1}^{p-1} p^{-j/2}\chi_p\big(-p^{-1}k(p^j x_r - \{p^j x_r\}_p)\big)\big(u, \overline{\theta_{k;\,j\{p^j x_r\}_p}}\big)$$

$$= \sum_{j=-\infty}^{\infty} \sum_{k=1}^{p-1} p^{-j/2}\chi_p\big(-p^{-1}k(p^j x_r - \{p^j x_r\}_p)\big)\langle\theta_{k;\,j\{p^j x_r\}_p}, u\rangle.$$

$$(11.2.5)$$

Since $\mathcal{D}(\mathbb{Q}_p) \subset \mathcal{D}(D^\alpha)$ the equality (11.2.5) yields that

$$\delta_{x_r} = \sum_{j=-\infty}^{\infty} \sum_{k=1}^{p-1} p^{-j/2}\chi_p\big(-p^{-1}k(p^j x_r - \{p^j x_r\}_p)\big)\theta_{k;\,j\{p^j x_r\}_p}, \quad (11.2.6)$$

where the series converges in $\mathcal{D}'(\mathbb{Q}_p)$.

Suppose that a function $h \in \mathcal{L}_2(\mathbb{Q}_p)$ is represented as a convergent series in $\mathcal{L}_2(\mathbb{Q}_p)$:

$$h(x) = \sum_{j=-\infty}^{\infty} \sum_{k=1}^{p-1} \sum_{a\in I_p} c_{jka}\theta_{k;\,ja}(x).$$

Applying the operator $D^\alpha - \lambda I$ termwise, we get the series

$$(D^\alpha - \lambda I)h = \sum_{j=-\infty}^{\infty} \sum_{k=1}^{p-1} \sum_{a \in I_p} c_{jka}\big(p^{\alpha(1-j)} - \lambda\big)\theta_{k;\,ja}, \qquad (11.2.7)$$

converging in $\mathcal{D}'(\mathbb{Q}_p)$ (since $D^\alpha \mathcal{D}(\mathbb{Q}_p) \subset \mathcal{L}_2(\mathbb{Q}_p)$). Since (11.2.6) and (11.2.7) are the right and left hand sides of the identity (11.2.3), the comparison of these formulas gives

$$c_{jka} = \begin{cases} 0, & a \neq \{p^j x_r\}_p \\ \dfrac{p^{-j/2}\chi_p\big(-p^{-1}k(p^j x_r - \{p^j x_r\}_p)\big)}{p^{\alpha(1-j)} - \lambda}, & a = \{p^j x_r\}_p \end{cases}$$

Thus

$$h_{r,\lambda}(x) = \sum_{j=-\infty}^{\infty} \sum_{k=1}^{p-1} \frac{p^{-j/2}\chi_p\big(-p^{-1}k(p^j x_r - \{p^j x_r\}_p)\big)}{p^{\alpha(1-j)} - \lambda}\theta_{k;\,j\{p^j x_r\}_p}(x)$$

$$(11.2.8)$$

is the unique solution of (11.2.3).

Since the functions $\theta_{k;\,j\{p^j x_r\}_p}(x)$ in (11.2.8) are elements of the orthonormal basis (11.2.1) in $\mathcal{L}_2(\mathbb{Q}_p)$, the function $h_{r,\lambda}(x)$ belongs to $\mathcal{L}_2(\mathbb{Q}_p)$ if and only if

$$(p-1) \sum_{j=-\infty}^{\infty} \frac{p^{-j}}{\big(p^{\alpha(1-j)} - \lambda\big)^2} < \infty.$$

This inequality holds if and only if $\lambda \neq p^{\alpha m}$ ($\forall m \in \mathbb{Z} \cup \{-\infty\}$). Assertion 1 is proved.

Let $\alpha > 1$. Applying the operator $D^{\alpha/2}$ to the function (11.2.8) and taking (11.2.1) and (11.2.2) into account, we obtain

$$D^{\alpha/2}h_{r,\lambda}$$

$$= \sum_{j=1}^{\infty} \sum_{k=1}^{p-1} \frac{p^{-j/2}\chi_p\big(-p^{-1}k(p^j x_r - \{p^j x_r\}_p)\big)}{p^{\alpha(1-j)} - \lambda}\, p^{\frac{\alpha}{2}(1-j)}\theta_{k;\,j\{p^j x_r\}_p}(x)$$

$$+ \sum_{j=-\infty}^{0} \sum_{k=1}^{p-1} \frac{p^{-j/2}\chi_p\big(-p^{-1}k(p^j x_r - \{p^j x_r\}_p)\big)}{p^{\alpha(1-j)} - \lambda}\, p^{\frac{\alpha}{2}(1-j)}\theta_{k;\,j\{p^j x_r\}_p}(x)$$

It is easy to see that $h_{r,\lambda} \in \mathcal{D}(D^{\alpha/2})$ if and only if the above series converge in $\mathcal{L}_2(\mathbb{Q}_p)$. If these series converge then their sum coincides with $D^{\alpha/2}h_r$. For the general term of the first series we have

$$\left| \frac{p^{-j/2}p^{\frac{\alpha}{2}(1-j)}\chi_p\big(-p^{-1}k(p^j x_r - \{p^j x_r\}_p)\big)}{p^{\alpha(1-j)} - \lambda} \right|^2 \leq Cp^{-(\alpha+1)j}, \qquad j \geq 1$$

(since $\lambda \neq p^{\alpha m}$, $\forall m \in \mathbb{Z} \cup \{-\infty\}$) which implies its convergence in $L_2(\mathbb{Q}_p)$ for $\alpha > 1$. Similarly, the general term of the second series can be estimated from above by $Cp^{(\alpha-1)j}$ ($j \leq 0$), which implies its convergence in $\mathcal{L}_2(\mathbb{Q}_p)$ for $\alpha > 1$.

Theorem 11.2.2 is thus proved. $\qquad\square$

Let us study the solutions $h_{r,\lambda}(x)$ of (11.2.3) in more detail for $\alpha > 1$. To do this we consider the family of functions $M_{p^\gamma}(\lambda)$ ($\gamma \in \mathbb{Z} \cup \{-\infty\}$) represented by the series

$$M_{p^\gamma}(\lambda) = \frac{p-1}{p} \sum_{N=-\infty}^{-\gamma} \frac{p^N}{p^{\alpha N} - \lambda} - \frac{p^{-\gamma}}{p^{\alpha(1-\gamma)} - \lambda}, \qquad (11.2.9)$$

$$M_{p^{-\infty}}(\lambda) \stackrel{def}{=} M_0(\lambda) = \frac{p-1}{p} \sum_{N=-\infty}^{\infty} \frac{p^N}{p^{\alpha N} - \lambda}, \qquad \gamma \in \mathbb{Z}. \quad (11.2.10)$$

Obviously, $M_0(\lambda)$ is differentiable for $\lambda \in \mathbb{C} \setminus \{p^{\alpha N} : \forall N \in \mathbb{Z} \cup \{-\infty\}\}$ and $M_0'(\lambda) = \frac{p-1}{p} \sum_{N=-\infty}^{\infty} \frac{p^N}{(p^{\alpha N}-\lambda)^2}$.

Proposition 11.2.3. *Let $\alpha > 1$ and $\lambda \neq p^{\alpha N}$ ($\forall N \in \mathbb{Z} \cup \{-\infty\}$). Then*

$$h_{r,\lambda}(x) = \begin{cases} M_0(\lambda) & \text{if } x = x_r \\ M_{p^\gamma}(\lambda) & \text{if } |x - x_r|_p = p^\gamma \end{cases}, \qquad \|h_{r,\lambda}\|^2 = M_0'(\lambda).$$

Proof. If $\alpha > 1$ and $\lambda \neq p^{\alpha N}$ ($\forall N \in \mathbb{Z} \cup \{-\infty\}$), then $h_{r,\lambda} \in \mathcal{D}(D^{\alpha/2})$, where $\alpha/2 > 1/2$, and hence the series (11.2.8) point-wise converges to $h_{r,\lambda}(x)$.

Employing (11.2.1) and (11.2.10), we immediately deduce from (11.2.8) that $h_{r,\lambda}(x_r) = M_0(\lambda)$, $\|h_{r,\lambda}\|^2 = M_0'(\lambda)$, and

$$h_{r,\lambda}(x) = \sum_{N=-\infty}^{\infty} \sum_{k=1}^{p-1} \frac{p^{-N} \chi_p(p^{N-1}k(x-x_r))}{p^{\alpha(1-N)} - \lambda} \cdot \Omega\big(\big|p^N x - \{p^N x_r\}_p\big|_p\big)$$

$$(11.2.11)$$

for $x \neq x_r$.

The expression (11.2.11) can be simplified with the use of the following arguments: 1. It follows from the strong triangle inequality and the definitions (1.6.4) of $\{\cdot\}_p$ and $\Omega(\cdot)$ that $\Omega\big(\big|p^N x - \{p^N x_r\}_p\big|_p\big) = \Omega(|p^N(x - x_r)|_p)$ and

$$\Omega\big(\big|p^N x - \{p^N x_r\}_p\big|_p\big) = 0 \Leftrightarrow |p^N(x-x_r)|_p > 1 \Leftrightarrow |x - x_r|_p > p^N.$$

If $x \neq x_r$, then $|x - x_r|_p = p^\gamma$ for some $\gamma \in \mathbb{Z}$. Therefore, the terms of (11.2.11) with indexes $N < \gamma$ are equal to zero.

2. Since $|p^{N-1}k(x - x_r)|_p = |p^{N-1}|_p |k|_p |x - x_r|_p = p^{\gamma+1-N}$ the fractional part $\{p^{N-1}k(x - x_r)\}_p$ is equal to zero for $N \geq \gamma + 1$. Hence, $\chi_p(p^{N-1}k(x - x_r)) \equiv 1$ where $N \geq \gamma + 1$.

3. Set for brevity $y = p^{N-1}(x - x_r)$ and consider the case where $N = \gamma$. Then $|y|_p = p$ and hence $\{y\}_p = p^{-1}y_0$, where $y_0 \in \{1, \ldots, p - 1\}$ is the first term in the canonical presentation of a p-adic number y (see (1.6.2)). Since p is a prime number, it is easy to verify that the set of numbers $\{ky\}_p$ $(k = 1 \ldots p - 1)$ coincides with the set $p^{-1}k$ $(k = 1 \ldots p - 1)$ modulo p. This means that

$$\sum_{k=1}^{p-1} \chi_p(p^{\gamma-1}k(x - x_r)) = \sum_{k=1}^{p-1} \chi_p(ky) = \sum_{k=1}^{p-1} \exp\left(k\frac{2\pi i}{p}\right) = -1$$

(the latter equality holds because $\sum_{k=1}^{p} \exp ki\omega = 0$ for $\omega = \frac{2\pi}{p}$).

Statements 1–3 allow us to rewrite (11.2.11) as follows:

$$h_{k,\lambda}(x) = (p - 1) \sum_{N=\gamma+1}^{\infty} \frac{p^{-N}}{p^{\alpha(1-N)} - \lambda} - \frac{p^{-\gamma}}{p^{\alpha(1-\gamma)} - \lambda} = M_{p^\gamma}(\lambda).$$

Proposition 11.2.3 is proved.   □

By Proposition 11.2.3, $h_{r,\lambda}(x)$ is a "radial" function which takes exactly one value $M_{p^\gamma}(\lambda)$ for all points x of the sphere $S_\gamma(x_r) = \{x \in \mathbb{Q}_p : |x - x_r|_p = p^\gamma\}$. Such a property of the solution $h_{r,\lambda}(x)$ of equation (11.2.3) is related to the property of the distribution δ to be homogeneous of degree $|x|_p^{-1}$ (see Definition 6.1 and Theorem 6.2.1).

In conclusion, we single out the properties of the functions $M_{p^\gamma}(\lambda)$ and $M_0(\lambda)$ which will be useful for the spectral analysis in the next sections.

Lemma 11.2.4. *Let $\alpha > 1$ and let $M_{p^\gamma}(\lambda)$ and $M_0(\lambda)$ be defined by (11.2.9) and (11.2.10). Then:*

1. The function $M_0(\lambda)$ is continuous and monotonically increasing on each interval $(-\infty, 0)$, $(p^{\alpha N}, p^{\alpha(N+1)})$ $(\forall N \in \mathbb{Z})$. Furthermore, $M_0(\lambda)$ maps $(-\infty, 0)$ onto $(0, \infty)$ and maps $(p^{\alpha N}, p^{\alpha(N+1)})$ onto $(-\infty, \infty)$.

2. The function $M_{p^\gamma}(\lambda)$ is continuous and monotonically increasing (decreasing) on $(-\infty, 0)$ (on $(p^{\alpha(1-\gamma)}, \infty)$). Furthermore, $M_{p^\gamma}(\lambda)$ maps $(-\infty, 0)$ onto $(0, \infty)$ and maps $(p^{\alpha(1-\gamma)}, \infty)$ onto $(0, \infty)$.

The proof of Lemma 11.2.4 is quite elementary and is based on a simple analysis of the series (11.2.9) and (11.2.10). In particular, rewriting the definition of $M_{p^\gamma}(\lambda)$ as

$$
\begin{aligned}
M_{p^\gamma}(\lambda) &= \sum_{N=-\infty}^{-\gamma} \frac{p^N}{p^{\alpha N} - \lambda} - \sum_{N=-\infty}^{-\gamma+1} \frac{p^{N-1}}{p^{\alpha N} - \lambda} \\
&= \sum_{N=-\infty}^{-\gamma} \frac{p^N}{p^{\alpha N} - \lambda} - \sum_{N=-\infty}^{-\gamma} \frac{p^N}{p^{\alpha(N+1)} - \lambda} \\
&= \sum_{N=-\infty}^{-\gamma} \frac{p^N(p^{\alpha(N+1)} - p^{\alpha N})}{(p^{\alpha N} - \lambda)(p^{\alpha(N+1)} - \lambda)}
\end{aligned}
$$

we easily establish assertion 2.

In the following sections we are going to study finite rank point perturbations of D^α determined by the expression

$$
D^\alpha + V_Y, \qquad V_Y = \sum_{i,j=1}^{n} b_{ij} \langle \delta_{x_j}, \cdot \rangle \delta_{x_i}, \tag{11.2.12}
$$

where $b_{ij} \in \mathbb{C}$, $Y = \{x_1, \ldots, x_n\}$.

Since $\delta_{x_j} \notin \mathcal{L}_2(\mathbb{Q}_p)$ expression (11.2.12) does not determine an operator in $\mathcal{L}_2(\mathbb{Q}_p)$. Moreover, in contrast to the standard theory of point interactions [9], the potential V_Y is not defined on the domain of definition $\mathcal{D}(D^\alpha)$ of the unperturbed operator D^α for $\alpha \leq 1/2$ (Proposition 11.2.1). For this reason we will assume $\alpha > 1/2$.

11.3. Definition of operator realizations of $D^\alpha + V$ in $\mathcal{L}_2(\mathbb{Q}_p)$

Let $\mathfrak{H}_2 \subset \mathfrak{H}_1 \subset L_2(\mathbb{Q}_p) \subset \mathfrak{H}_{-1} \subset \mathfrak{H}_{-2}$ be the standard scale of Hilbert spaces (A-scale) associated with the positive self-adjoint operator $A = D^\alpha$ in $\mathcal{L}_2(\mathbb{Q}_p)$. Here $\mathfrak{H}_s = \mathcal{D}(A^{s/2})$, $s = 1, 2$ with the norm $\|u\|_s = \|(D^\alpha + I)^{s/2}u\|$ and $\mathfrak{H}_{-s}$ is the completion of $\mathcal{L}_2(\mathbb{Q}_p)$ with respect to the norm $\|u\|_{-s}$. Naturally, $\mathfrak{H}_s$ and $\mathfrak{H}_{-s}$ are dual, and the inner product in $\mathcal{L}_2(\mathbb{Q}_p)$ is extended to the pairing

$$
\langle \phi, u \rangle = ((D^\alpha + I)^{s/2}u, (D^\alpha + I)^{-s/2}\phi), \qquad u \in \mathfrak{H}_s, \quad \phi \in \mathfrak{H}_{-s}
$$

(for details, see [27]).

By virtue of Proposition 11.2.3, the solutions $h_{r,\lambda}$ of (11.2.3) satisfy the relation $\overline{h_{r,\lambda}} = h_{r,\bar\lambda}$. Taking this into account and using (11.2.5) and (11.2.8) we get

$$
\langle \delta_{x_r}, u \rangle = u(x_r) = ((D^\alpha - \bar\lambda I)u, h_{r,\lambda})_{\mathcal{L}_2(\mathbb{Q}_p)}, \tag{11.3.1}
$$

$u \in \mathcal{D}(D^\alpha)$, $x_r \in \mathbb{Q}_p$, for any complex $\lambda \neq p^{\alpha m}$ ($\forall m \in \mathbb{Z} \cup \{-\infty\}$). Hence, $\delta_{x_r} \in \mathfrak{H}_{-2}$.

In order to give a meaning to (11.2.12) as an operator acting in $\mathcal{L}_2(\mathbb{Q}_p)$, we consider the positive symmetric operator A_{sym} defined by

$$A_{\text{sym}} = D^\alpha \restriction_\mathcal{D},$$
$$\mathcal{D} = \{u \in \mathcal{D}(D^\alpha) : u(x_1) = \ldots = u(x_n) = 0\}, \qquad (11.3.2)$$

where $\alpha > 1/2$.

It follows from (11.3.1) that A_{sym} is a closed densely defined symmetric operator in $\mathcal{L}_2(\mathbb{Q}_p)$ and the linear span of $\{h_{r,\lambda}\}_{r=1}^n$ coincides with $\ker(A_{\text{sym}}^* - \lambda I)$. It is convenient to present the domain of the adjoint $\mathcal{D}(A_{\text{sym}}^*)$ as $\mathcal{D}(A_{\text{sym}}^*) = \mathcal{D}(D^\alpha) \dotplus \mathcal{H}$, where $\mathcal{H} = \ker(A_{\text{sym}}^* + I)$. Then

$$A_{\text{sym}}^* f = A_{\text{sym}}^*(u + h) = D^\alpha u - h, \quad \forall f = u + h \in \mathcal{D}(A_{\text{sym}}^*) \qquad (11.3.3)$$

($u \in \mathcal{D}(D^\alpha)$, $h \in \mathcal{H}$).

In the additive singular perturbation theory, the algorithm of the determination of operator realizations of $D^\alpha + V_Y$ is well known [27] and is based on the construction of some extension (regularization) $A_{\text{reg}} := D^\alpha + V_{Y\text{reg}}$ of (11.2.12) onto the domain $\mathcal{D}(A_{\text{sym}}^*) = \mathcal{D}(D^\alpha) \dotplus \mathcal{H}$.

The $\mathcal{L}_2(\mathbb{Q}_p)$ part

$$\widetilde{A} = A_{\text{reg}} \restriction_{\mathcal{D}(\widetilde{A})},$$
$$\mathcal{D}(\widetilde{A}) = \{f \in \mathcal{D}(A_{\text{sym}}^*) : A_{\text{reg}} f \in \mathcal{L}_2(\mathbb{Q}_p)\} \qquad (11.3.4)$$

of the regularization A_{reg} is called the *operator realization* of $D^\alpha + V_Y$ in $\mathcal{L}_2(\mathbb{Q}_p)$.

Since the action of D^α on elements of $\mathcal{H}$ is defined by (11.2.3) the regularization A_{reg} depends on the definition of $V_{Y\text{reg}}$.

If $\alpha > 1$, Theorem 11.2.2 gives that $\delta_{x_r} \in \mathfrak{H}_{-1}$. Hence, the singular potential $V_Y = \sum_{i,j=1}^n b_{ij} \langle \delta_{x_j}, \cdot \rangle \delta_{x_i}$ is form bounded [9]. In this case, the set $\mathcal{D}(A_{\text{sym}}^*) \subset \mathfrak{H}_1$ consists of continuous functions on $\mathbb{Q}_p$ (in view of Proposition 11.2.1 and Theorem 11.2.2) and the δ_{x_r} are uniquely determined on elements $f \in \mathcal{D}(A_{\text{sym}}^*)$ by the formula (cf. (11.3.1))

$$\langle \delta_{x_r}, f \rangle = ((D^\alpha + I)^{1/2} f, (D^\alpha + I)^{1/2} h_{r,-1})_{\mathcal{L}_2(\mathbb{Q}_p)} = f(x_r). \quad (11.3.5)$$

Thus the regularization $A_{Y\text{reg}}$ is uniquely defined for $\alpha > 1$ and formula (11.3.4) provides a unique operator realization of (11.2.12) in $\mathcal{L}_2(\mathbb{Q}_p)$ corresponding to a fixed singular potential V_Y.

If $1/2 < \alpha \leq 1$, then the delta functions δ_{x_r} form an $\mathfrak{H}_{-1}$-independent system (since the linear span of $\{\delta_{x_r}\}_1^n$ does not intersect with $\mathfrak{H}_{-1}$) and $V_{Y\text{reg}}$ is not uniquely defined on $\mathcal{D}(A_{\text{sym}}^*)$ (see [169] for a detailed discussion of this part).

11.4. Description of operator realizations

Let η be an invertible bounded self-adjoint operator in $L_2(\mathbb{Q}_p)$.

An operator A is called *η-self-adjoint* in $\mathcal{L}_2(\mathbb{Q}_p)$ if $A^* = \eta A \eta^{-1}$, where A^* denotes the adjoint of A [46]. Obviously, self-adjoint operators are η-self-adjoint ones for $\eta = I$. In that case we will use the simpler terminology "self-adjoint" instead of "I-self-adjoint".

Our goal is to describe η-self-adjoint operator realizations of $D^\alpha + V_Y$ in $\mathcal{L}_2(\mathbb{Q}_p)$ for $\alpha > 1$.

Since the solutions $h_r \overset{def}{=} h_{r,-1}$ ($1 \leq r \leq n$) of (11.2.3) form a basis of $\mathcal{H}$ any function $f \in \mathcal{D}(A_{\text{sym}}^*) = \mathcal{D}(D^\alpha) \dotplus \mathcal{H}$ admits a decomposition $f = u + \sum_{r=1}^n c_r h_r$ ($u \in \mathcal{D}(D^\alpha)$, $c_r \in \mathbb{C}$). Using such a presentation we define the linear mappings $\Gamma_i : \mathcal{D}(A_{\text{sym}}^*) \to \mathbb{C}^n$ ($i = 0, 1$):

$$\Gamma_0 f = \begin{pmatrix} f(x_1) \\ \vdots \\ f(x_n) \end{pmatrix}, \quad \Gamma_1 f = -\begin{pmatrix} c_1 \\ \vdots \\ c_n \end{pmatrix}, \quad (11.4.1)$$

for all $f = u + \sum_{k=1}^n c_k h_k \in \mathcal{D}(A_{\text{sym}}^*)$.

In what follows we assume that

$$D^\alpha \eta = \eta D^\alpha \quad \text{and} \quad \eta : \mathcal{H} \to \mathcal{H}. \quad (11.4.2)$$

By the second relation in (11.4.2), the action of η on elements of $\mathcal{H}$ can be described by the matrix $\mathcal{Y} = \|y_{ij}\|_{i,j=1}^n$ where entries y_{ij} are determined by the relations $\eta h_j = \sum_{i=1}^n y_{ij} h_i$ ($1 \leq j \leq n$). In general, the basis $\{h_i\}_{i=1}^n$ of $\mathcal{H}$ is not orthogonal and the matrix $\mathcal{Y}$ is not Hermitian ($\mathcal{Y} \neq \overline{\mathcal{Y}}^T$).

Theorem 11.4.1. ([169]) *Let $\alpha > 1$ and let $\widetilde{A}$ be the operator realization of $D^\alpha + V_Y$ defined by (11.3.4). Then $\widetilde{A}$ coincides with the operator*

$$A_\mathcal{B} = A_{\text{sym}}^* \upharpoonright_{\mathcal{D}(A_\mathcal{B})},$$
$$\mathcal{D}(A_\mathcal{B}) = \{f \in \mathcal{D}(A_{\text{sym}}^*) : \mathcal{B}\Gamma_0 f = \Gamma_1 f\}, \quad (11.4.3)$$

where $\mathcal{B} = \|b_{ij}\|_{i,j=1}^n$ is the coefficient matrix of the potential V_Y.

The operator $A_\mathcal{B}$ is self-adjoint if and only if the matrix $\mathcal{B}$ is Hermitian.

If η satisfies (11.4.2), then $A_\mathcal{B}$ is η-self-adjoint if and only if the matrix $\mathcal{Y}\mathcal{B}$ is Hermitian.

Example 1. $\mathcal{P}$-*self-adjoint realizations.*

Let $Y = \{x_1, x_2\}$, where $x_2 = -x_1$, and let $\eta = \mathcal{P}$ be the space parity operator $\mathcal{P}f(x) = f(-x)$ in $L_2(\mathbb{Q}_p)$. It follows from Proposition 11.2.3 that $\mathcal{P}h_1 = h_2$ and $\mathcal{P}h_2 = h_1$. Hence, the corresponding matrix $\mathcal{Y}$ has the form $\mathcal{Y} = \begin{pmatrix} 0 & 1 \\ 1 & 0 \end{pmatrix}$ and $\mathcal{P}$ satisfies (11.4.2).

By Theorem 11.4.1 the formula (11.4.3) determines $\mathcal{P}$-self-adjoint realizations $A_{\mathcal{B}}$ of $D^\alpha + V_Y$ if and only if the entries b_{ij} of the matrix $\mathcal{B} = \|b_{ij}\|_{i,j=1}^2$ satisfy the relations $b_{12}, b_{21} \in \mathbb{R}$, $b_{11} = \bar{b}_{22}$.

Under such conditions imposed on b_{ij} the corresponding singular potential V_Y is not symmetric in the standard sense (except for the case $b_{ij} \in \mathbb{R}$, $b_{11} = b_{22}$, $b_{12} = b_{21}$) but satisfies the condition of $\mathcal{P}$-symmetry $\mathcal{P}V_Y^* = V_Y\mathcal{P}$, where the adjoint V_Y^* is determined by the relation $\langle V_Y u, v \rangle = \langle u, V_Y^* v \rangle$ $(u, v \in \mathcal{D}(D^\alpha))$. Assuming formally that $\mathcal{T}V_Y = V_Y^*\mathcal{T}$, where $\mathcal{T}$ is the complex conjugation operator $\mathcal{T}f(x) = \overline{f(x)}$, we can rewrite the condition of $\mathcal{P}$-symmetry as follows: $\mathcal{P}\mathcal{T}V_Y = V_Y\mathcal{P}\mathcal{T}$. This means that the expression $D^\alpha + V_Y$ is $\mathcal{P}\mathcal{T}$-symmetric (since $\mathcal{P}\mathcal{T}D^\alpha = D^\alpha\mathcal{P}\mathcal{T}$). Thus the $\mathcal{P}$-self-adjoint operators $A_{\mathcal{B}}$ described above are operator realizations of the $\mathcal{P}\mathcal{T}$-symmetric expression $D^\alpha + V_Y$ in $\mathcal{L}_2(\mathbb{Q}_p)$.

11.5. Spectral properties

As a rule, the spectral properties of finite rank perturbations are described in terms of a Nevanlinna function (Krein–Langer Q-function) appearing as a parameter in a Krein-type resolvent formula relating the resolvents of perturbed and unperturbed operators [27], [74], [201]. The choice of a resolvent formula has to be motivated by simple links with the parameters of the perturbations.

Denote by $\mathcal{L}$ and $\mathcal{L}_Y$ the closed subspaces of $\mathcal{L}_2(\mathbb{Q}_p)$ spanned by the p-adic wavelets $\theta_{k;ja}(x)$ $(k = 1, 2, \ldots, p-1, j \in \mathbb{Z})$ with $a \neq \{p^N x_i\}_p$ $(\forall x_i \in Y)$ and $a = \{p^N x_i\}_p$ $(\exists x_i \in Y)$, respectively (see (11.2.1)). Obviously, $\mathcal{L} \oplus \mathcal{L}_Y = \mathcal{L}_2(\mathbb{Q}_p)$. The relations (11.2.2), (11.2.4), and (11.3.2) imply that the subspaces $\mathcal{L}$ and $\mathcal{L}_Y$ reduce the operators D^α and A_{sym}. Furthermore $A_{\mathrm{sym}} = D^\alpha \upharpoonright_{\mathcal{L}} \oplus A_{\mathrm{sym}} \upharpoonright_{\mathcal{L}_Y}$.

Let $A_{\mathcal{B}}$ be the operator realization of $D^\alpha + V$ defined by (11.4.3). Then $A_{\mathcal{B}} = D^\alpha \upharpoonright_{\mathcal{L}} \oplus A_{\mathcal{B}} \upharpoonright_{\mathcal{L}_Y}$. Therefore, the spectrum of $A_{\mathcal{B}}$ consists of eigenvalues $\lambda = p^{\alpha N}$ $(\forall N \in \mathbb{Z})$ of infinite multiplicity and with accumulation point $\lambda = 0$.

To describe eigenvalues of finite multiplicity we consider the matrix

$$M(\lambda) = \left\| M_{|x_i - x_j|_p}(\lambda) \right\|_{i,j=1}^n, \quad \forall \lambda \in \mathbb{C} \setminus \{p^{\alpha N} : \forall N \in \mathbb{Z} \cup \{-\infty\}\},$$

$$(11.5.1)$$

where the functions $M_{|x_i-x_j|_p}(\lambda)$ $(|x_i - x_j|_p = p^{\gamma(x_i,x_j)})$ are defined by (11.2.9) and (11.2.10).

Theorem 11.5.1. *Let the matrix $\mathcal{B}$ in* (11.4.3) *be invertible. Then a point $\lambda \in \mathbb{C} \setminus \{p^{\alpha N} : \forall N \in \mathbb{Z} \cup \{-\infty\}\}$ is an eigenvalue of finite multiplicity of $A_{\mathcal{B}}$ if and only if $\det[M(\lambda) + \mathcal{B}^{-1}] = 0$. In this case, the (geometric) multiplicity of λ is $n - r$, where r is the rank of $M(\lambda) + \mathcal{B}^{-1}$.*

If $\det[M(\lambda) + \mathcal{B}^{-1}] \neq 0$, then $\lambda \in \rho(A_{\mathcal{B}})$ and the corresponding Krein resolvent formula has the form

$$(A_{\mathcal{B}} - \lambda I)^{-1} = (D^\alpha - \lambda I)^{-1} - (h_{1,\lambda}, \ldots, h_{n,\lambda})[M(\lambda)$$

$$+ \mathcal{B}^{-1}]^{-1} \begin{pmatrix} (\cdot, h_{1,\bar{\lambda}}) \\ \vdots \\ (\cdot, h_{n,\bar{\lambda}}) \end{pmatrix}. \tag{11.5.2}$$

Proof. It is easy to see from (11.2.8) that $h_{r,\lambda} = u + h_{r,-1}$, where $u \in \mathcal{D}(D^\alpha)$. This relation and (11.4.1) give

$$\Gamma_1 h_{r,\lambda} = (0, \ldots, \underbrace{-1}_{r\,\text{th}}, \ldots, 0)^T. \tag{11.5.3}$$

On the other hand, in view of Proposition 11.2.3 and (11.4.1),

$$\Gamma_0 h_{r,\lambda} = (M_{|x_r-x_1|_p}(\lambda), \ldots, \underbrace{M_0(\lambda)}_{r\,\text{th}}, \ldots, M_{|x_r-x_n|_p}(\lambda))^T. \tag{11.5.4}$$

It is clear that λ is an eigenvalue of finite multiplicity of $A_{\mathcal{B}}$ if and only if $\lambda \neq p^{\alpha N}$ ($\forall N \in \mathbb{Z} \cup \{-\infty\}$) and there exists a nontrivial element $f_\lambda \in \ker(A^*_{\text{sym}} - \lambda I) \cap \mathcal{D}(A_{\mathcal{B}})$. Representing f_λ as $f_\lambda = \sum_{r=1}^n c_r h_{r,\lambda}$, using (11.5.1), (11.5.3) and (11.5.4), and keeping in mind that $\mathcal{D}(A_{\mathcal{B}}) = \ker(\Gamma_0 - \mathcal{B}^{-1}\Gamma_1)$, we rewrite the latter condition as follows: $[M(\lambda) + \mathcal{B}^{-1}](c_1, \ldots, c_n)^T = 0$. Therefore, λ is an eigenvalue if and only if this matrix equation has a nontrivial solution. Obviously, the (geometric) multiplicity of λ is $n - r$, where r is the rank of $M(\lambda) + \mathcal{B}^{-1}$.

The resolvent formula (11.5.2) can be established by a direct verification with the help of (11.3.1), (11.5.3), and (11.5.4). Theorem 11.5.1 is proved. $\square$

Remark. It is easy to see that the triple $(\mathbb{C}^n, -\Gamma_1, \Gamma_0)$, where Γ_i are defined by (11.4.1), is a boundary value space (BVS) of A_{sym} and the matrix $M(\lambda)$ is the corresponding Weyl–Titchmarsh function of A_{sym} [75]. From this point of view, Theorem 11.5.1 is a direct consequence of the general BVS theory. However, we prefer not to employ the general constructions in cases where the required results can be established in a more direct way.

11.6. The case of η-self-adjoint operator realizations

One of the principal motivations for the study of η-self-adjoint operators in framework of quantum mechanics is the observation that some of them have real spectrum (like self-adjoint operators) and, therefore, they can be used as an alternative to standard Hamiltonians to explain experimental data [189].

Since an arbitrary η-self-adjoint operator A is self-adjoint with respect to the indefinite metric $[f, g] \stackrel{def}{=} (\eta f, g)(f, g \in \mathcal{L}_2(\mathbb{Q}_p))$, one can attempt to develop a consistent quantum theory for η-self-adjoint Hamiltonians with real spectrum. However, in this case, we encounter the difficulty of dealing with the Hilbert space $\mathcal{L}_2(\mathbb{Q}_p)$ equipped with the indefinite metric $[\cdot, \cdot]$. One of the natural ways of overcoming this problem consists of constructing a certain physical symmetry $\mathcal{C}$ for A (see, e.g., [48], [49], [190]).

By analogy with [48], we will say that an η-self-adjoint operator A acting in $\mathcal{L}_2(\mathbb{Q}_p)$ possesses the property of $\mathcal{C}$-*symmetry* if there exists a bounded linear operator $\mathcal{C}$ in $\mathcal{L}_2(\mathbb{Q}_p)$ such that the following conditions are satisfied:

(i) $A\mathcal{C} = \mathcal{C}A$;

(ii) $\mathcal{C}^2 = I$;

(iii) the sesquilinear form $(f, g)_\mathcal{C} \stackrel{def}{=} [\mathcal{C}f, g]$ $(\forall f, g \in \mathcal{L}_2(\mathbb{Q}_p))$ determines an inner product in $\mathcal{L}_2(\mathbb{Q}_p)$ that is equivalent to the initial one.

The existence of a $\mathcal{C}$-symmetry for an η-self-adjoint operator A ensures unitarity of the dynamics generated by A in the norm $\| \cdot \|_\mathcal{C}^2 = (\cdot, \cdot)_\mathcal{C}$.

In ordinary quantum theory, it is crucial that any state vector can be expressed as a linear combination of the eigenstates of the Hamiltonian. For this reason, it is natural to assume that every physically acceptable η-self-adjoint operator must admit an unconditional basis composed of its eigenvectors or at least of its root vectors (see [231] for a detailed discussion of this point).

Theorem 11.6.1. *Let $A_\mathcal{B}$ be the η-self-adjoint operator realization of $D^\alpha + V$ defined by* (11.4.3). *Then the following statements are equivalent:*

(i) *$A_\mathcal{B}$ possesses the property of $\mathcal{C}$-symmetry;*

(ii) *the spectrum $\sigma(A_\mathcal{B})$ is real and there exists a Riesz basis of $\mathcal{L}_2(\mathbb{Q}_p)$ composed of the eigenfunctions of $A_\mathcal{B}$.*

Proof. It is known that the property of $\mathcal{C}$-symmetry for η-self-adjoint operators is equivalent to their similarity to self-adjoint ones ([29], [190]). Hence, if $A_\mathcal{B}$ possesses $\mathcal{C}$-symmetry, then there exists an invertible bounded operator Z such that

$$A_\mathcal{B} = ZHZ^{-1}, \tag{11.6.1}$$

where H is a self-adjoint operator in $\mathcal{L}_2(\mathbb{Q}_p)$. So the spectrum of A_B lies on the real axis. Furthermore, it follows from Theorem 11.5.1 that $\sigma(A_B)$ has no more than a countable set of points of condensation. Obviously, this property holds for the spectrum of the self-adjoint operator H. Using Lemma 4.2.7 in [46], we immediately derive the existence of an orthonormal basis of $\mathcal{L}_2(\mathbb{Q}_p)$ composed of the eigenfunctions of H. To complete the proof of the implication (i) $\Rightarrow$ (ii) it is sufficient to use (11.6.1).

Let us verify that (ii) $\Rightarrow$ (i). Indeed, if $\{f_i\}_1^\infty$ is a Riesz basis composed of the eigenfunctions of A_B (i.e., $A_B f_i = \lambda_i f_i$, $\lambda_i \in \mathbb{R}$), then $f_i = Z e_i$, where $\{e_i\}_1^\infty$ is an orthonormal basis of $\mathcal{L}_2(\mathbb{Q}_p)$ and Z is an invertible bounded operator. This means that (11.6.1) holds for a self-adjoint operator H defined by the relations $H e_i = \lambda_i e_i$. Theorem 11.6.1 is proved. $\qquad\square$

The next statement is a direct consequence of Theorem 11.6.1.

Corollary 11.6.2. *An arbitrary self-adjoint operator realization A_B of $D^\alpha + V$ possesses a complete set of eigenfunctions in $\mathcal{L}_2(\mathbb{Q}_p)$.* $\qquad\square$

In conclusion we note that the spectral properties of η-self-adjoint operators can have rather unexpected features. In particular, the standard one-dimensional Schrödinger operator with a certain kind of $\mathcal{PT}$-symmetric zero-range potentials gives examples of $\mathcal{P}$-self-adjoint operators in $L_2(\mathbb{R})$ whose spectra coincide with $\mathbb{C}$ [30].

11.7. The Friedrichs extension

Let A_F be the Friedrichs extension of the symmetric operator A_{sym} defined by (11.3.2). The standard arguments of the extension theory lead to the conclusion (for details, see [169]) that $A_F = D^\alpha$ when $1/2 < \alpha \le 1$ and

$$A_F = A_{\mathrm{sym}}^* \upharpoonright_{\mathcal{D}(A_F)},$$
$$\mathcal{D}(A_F) = \{f(x) \in \mathcal{D}(A_{\mathrm{sym}}^*) : f(x_1) = \ldots = f(x_n) = 0\}$$

when $\alpha > 1$. In the latter case, $\mathcal{D}(A_F) = \ker \Gamma_0$ and the operator A_F can formally be described by (11.4.3) with $\mathcal{B} = \infty$.

Obviously, the essential spectrum of A_F consists of the eigenvalues $\lambda = p^{\alpha N}$ ($N \in \mathbb{Z}$) of infinite multiplicity, and with accumulation point $\lambda = 0$.

Let $\alpha > 1$. Repeating step by step the proof of Theorem 11.5.1 and taking the relation $\mathcal{D}(A_F) = \ker \Gamma_0$ into account, we conclude that the discrete spectrum $\sigma_{\mathrm{dis}}(A_F)$ coincides with the set of solutions λ of the equation $\det M(\lambda) = 0$.

The obtained relation allows us to establish some connections between $\sigma_{\text{dis}}(A_F)$ and the geometrical characteristics of the set Y. To illustrate this fact we consider the two points case $Y = \{x_1, x_2\}$.

Indeed, $\lambda \in \sigma_{\text{dis}}(A_F) \iff$

$$0 = \det \left\| M_{|x_i - x_j|_p}(\lambda) \right\|_{i,j=1}^2 = (M_0(\lambda) - M_{p^\gamma}(\lambda))(M_0(\lambda) + M_{p^\gamma}(\lambda)),$$

where $p^\gamma = |x_1 - x_2|_p$. Therefore, the discrete spectrum is determined by the equations $M_0(\lambda) - M_{p^\gamma}(\lambda) = 0$ and $M_0(\lambda) + M_{p^\gamma}(\lambda) = 0$.

In view of (11.2.9) and (11.2.10),

$$M_0(\lambda) - M_{p^\gamma}(\lambda) = \frac{p-1}{p} \sum_{N=-\gamma+2}^{\infty} \frac{p^N}{p^{\alpha N} - \lambda} + \frac{p^{1-\gamma}}{p^{\alpha(1-\gamma)} - \lambda}. \qquad (11.7.1)$$

A simple analysis of (11.7.1) shows that the function $M_0(\lambda) - M_{p^\gamma}(\lambda)$ is monotonically increasing on the intervals $(-\infty, p^{\alpha(1-\gamma)})$ and $(p^{\alpha N}, p^{\alpha(N+1)})$, $\forall N \geq -\gamma + 1$, and it maps $(-\infty, p^{\alpha(1-\gamma)})$ onto $(0, \infty)$ and maps $(p^{\alpha N}, p^{\alpha(N+1)})$ onto $(-\infty, \infty)$. This means that the set of solutions of $M_0(\lambda) - M_{p^\gamma}(\lambda) = 0$ coincides with the infinite series of numbers $\lambda = \lambda_N^-$, $N \geq -\gamma + 1$, each of which is situated in the interval $(p^{\alpha N}, p^{\alpha(N+1)})$. We will call the series of numbers $\{\lambda_N^-\}_{N=-\gamma+1}^{\infty}$ *the type-1 part* of the discrete spectrum of A_F. So the type-1 part σ_{dis}^- of $\sigma_{\text{dis}}(A_F)$ consists of solutions of the equation $M_0(\lambda) - M_{p^\gamma}(\lambda) = 0$.

By virtue of (11.2.9) and (11.2.10),

$$M_0(\lambda) + M_{p^\gamma}(\lambda) = 2\frac{p-1}{p} \sum_{N=-\infty}^{-\gamma} \frac{p^N}{p^{\alpha N} - \lambda}$$
$$+ \frac{p-2}{p} \frac{p^{1-\gamma}}{p^{\alpha(1-\gamma)} - \lambda} + \frac{p-1}{p} \sum_{N=-\gamma+2}^{\infty} \frac{p^N}{p^{\alpha N} - \lambda}.$$

Analyzing this relation, it is easy to see that there exists exactly one solution $\lambda = \lambda_N^+$ of $M_0(\lambda) + M_{p^\gamma}(\lambda) = 0$ lying inside an interval $(p^{\alpha N}, p^{\alpha(N+1)})$ $\forall N \in \mathbb{Z}$. We will call the infinite series of numbers $\{\lambda_N^+\}_{-\infty}^{\infty}$ *the type-2 part* σ_{dis}^+ of the discrete spectrum $\sigma_{\text{dis}}(A_F)$.

Obviously, $\sigma_{\text{dis}}^- \cup \sigma_{\text{dis}}^+ = \sigma_{\text{dis}}(A_F)$. Let $N \geq -\gamma + 1$ and let $\lambda_N^\pm \in \sigma_{\text{dis}}^\pm$ be the corresponding discrete spectrum points in $(p^{\alpha N}, p^{\alpha(N+1)})$. It follows from Lemma 11.2.4 that $\lambda_N^+ < \lambda_N^-$. Therefore, $\sigma_{\text{dis}}^- \cap \sigma_{\text{dis}}^+ = \emptyset$.

Thus the discrete spectrum $\sigma_{\text{dis}}(A_F)$ consists of an infinite series of eigenvalues of multiplicity one, which are disposed as follows: an interval

$(p^{\alpha N}, p^{\alpha(N+1)})$ contains exactly one eigenvalue λ_N^+ if $N < -\gamma$ (type-2 only) and exactly two eigenvalues $\lambda_N^+ < \lambda_N^-$ if $N \geq -\gamma + 1$ (type-1 and type-2).

The obtained description shows that the type-1 part σ_{dis}^- of $\sigma_{\text{dis}}(A_F)$ uniquely determines the distance $|x_1 - x_2|_p$.

In the general case $Y = \{x_1, \ldots, x_n\}$, the discrete spectrum $\sigma_{\text{dis}}(A_F)$ also contains the type-1 part. Indeed, denote by $p^{\gamma_{\min}}$ the minimal distance between the points of Y. Without loss of generality we may assume that $|x_1 - x_2|_p = p^{\gamma_{\min}}$. Then, by the strong triangle inequality, $|x_j - x_1|_p = |x_j - x_2|_p = p^{\gamma_j} \geq p^{\gamma_{\min}}$ for any point $x_j \in Y$ $(j \neq 1, 2)$. This means that the first two rows (columns) of the matrix $M(\lambda)$ (see (11.5.1)) differ from each other by the first two terms only. Subtracting the second row from the first, we get

$$\det M(\lambda) = (M_0(\lambda) - M_{p^{\gamma_{\min}}}(\lambda))$$

$$\times \begin{vmatrix} 1 & -1 & 0 & \ldots & 0 \\ M_{p^{\gamma_{\min}}}(\lambda) & M_0(\lambda) & M_{p^{\gamma_3}}(\lambda) & \ldots & M_{p^{\gamma_n}}(\lambda) \\ M_{p^{\gamma_3}}(\lambda) & M_{p^{\gamma_3}}(\lambda) & \ddots & & \\ \vdots & \vdots & & \ddots & \\ M_{p^{\gamma_n}}(\lambda) & M_{p^{\gamma_n}}(\lambda) & & & \ddots \end{vmatrix}.$$

Thus the type-1 part σ_{dis}^- of the discrete spectrum always exists and it characterizes the minimal distance $p^{\gamma_{\min}}$ between elements of Y.

11.8. Two points interaction

11.8.1. Invariance with respect to the change of points of interaction

Let $Y = \{x_1, x_2\}$ and let the symmetric potential $V_Y = \sum_{i,j=1}^{2} b_{ij} \langle \delta_{x_j}, \cdot \rangle \delta_{x_i}$ be invariant under the change $x_1 \leftrightarrow x_2$. This means that $b_{ij} \in \mathbb{R}$ and $b_{11} = b_{22}$, $b_{12} = b_{21}$. In this case, the inverse $\mathcal{B}^{-1}$ of the coefficient matrix $\mathcal{B}$ has the form $\mathcal{B}^{-1} = \begin{pmatrix} a & b \\ b & a \end{pmatrix}$, where $a = b_{11}/\Delta$, $b = -b_{12}/\Delta$, and $\Delta = b_{11}^2 - b_{12}^2 \neq 0$. (We omit the case $b_{11} = b_{12} = b_{21} = b_{22}$.)

The operator $A_{\mathcal{B}}$ is self-adjoint in $L_2(\mathbb{Q}_p)$ and (by Theorem 11.5.1)

$$\lambda \in \sigma_{\text{dis}}(A_{\mathcal{B}}) \iff (M_0(\lambda) - M_{p^{\gamma}}(\lambda) + a - b)(M_0(\lambda) + M_{p^{\gamma}}(\lambda) + a + b)$$
$$= 0,$$

where $p^\gamma = |x_1 - x_2|_p$. Thus, the description of $\sigma_{\text{dis}}(A_{\mathcal{B}})$ is similar to the description of $\sigma_{\text{dis}}(A_F)$ and we can define some analogs of the type-1

$$\sigma_{\text{dis}}^-(A_{\mathcal{B}}) \overset{def}{=} \{\lambda \in \mathbb{R} \setminus \sigma(D^\alpha) : M_0(\lambda) - M_{p^\gamma}(\lambda) + a - b = 0\}$$

and the type-2

$$\sigma_{\text{dis}}^+(A_{\mathcal{B}}) \overset{def}{=} \{\lambda \in \mathbb{R} \setminus \sigma(D^\alpha) : M_0(\lambda) + M_{p^\gamma}(\lambda) + a + b = 0\}$$

parts of the discrete spectrum $\sigma_{\text{dis}}(A_{\mathcal{B}})$.

By analogy with the Friedrichs extension case (see formula (11.7.1)), $\sigma_{\text{dis}}^-(A_{\mathcal{B}})$ contains an infinite series of eigenvalues λ_N^- lying in the intervals $(p^{\alpha N}, p^{\alpha(N+1)})$ $\forall N \geq -\gamma + 1$. However, in contrast to the Friedrichs case, the interval $(-\infty, p^{\alpha(-\gamma+1)})$ contains an additional (unique) point $\lambda^- \in \sigma_{\text{dis}}^-(A_{\mathcal{B}})$ if and only if

$$0 < b - a \quad \text{and} \quad b - a \neq [M_0(\lambda) - M_{p^\gamma}(\lambda)]|_{\lambda = p^{\alpha m}}, \quad -\infty \leq m \leq -\gamma,$$

where the difference $[M_0(\lambda) - M_{p^\gamma}(\lambda)]|_{\lambda = p^{\alpha m}}$ is determined by formula (11.7.1). In particular, $\lambda^- < 0 \iff 0 < b - a < M_0(0) - M_{p^\gamma}(0) = p^{(1-\alpha)(-\gamma+1)}$.

The type-2 part $\sigma_{\text{dis}}^+(A_{\mathcal{B}})$ contains an infinite series of eigenvalues λ_N^+ lying in the intervals $(p^{\alpha N}, p^{\alpha(N+1)})$ $\forall N \in \mathbb{Z}$ covering the positive semi-axis. An additional (unique) negative point $\lambda^+ \in \sigma_{\text{dis}}^+(A_{\mathcal{B}})$ arises $\iff b + a < 0$.

Obviously $\sigma_{\text{dis}}^-(A_{\mathcal{B}}) \cup \sigma_{\text{dis}}^+(A_{\mathcal{B}}) = \sigma_{\text{dis}}(A_{\mathcal{B}})$ but $\sigma_{\text{dis}}^-(A_{\mathcal{B}})$ and $\sigma_{\text{dis}}^+(A_{\mathcal{B}})$ need not be disjoint.

11.8.2. Examples of $\mathcal{P}$-self-adjoint realizations

Consider the case $Y = \{x_1, x_2\}$ where $x_2 = -x_1$ and let $A_{\mathcal{B}}$ be the $\mathcal{P}$-self-adjoint realizations of $D^\alpha + V_Y$ described in Example 1 in Section 11.4. We restrict ourselves to the case where the inverse $\mathcal{B}^{-1}$ of the coefficient matrix $\mathcal{B}$ has the form $\mathcal{B}^{-1} = \begin{pmatrix} -ia & b \\ -b & ia \end{pmatrix}$ $(a, b \in \mathbb{R})$.

The operator $A_{\mathcal{B}}$ is $\mathcal{P}$-self-adjoint in $\mathcal{L}_2(\mathbb{Q}_p)$ and λ is an eigenvalue of finite multiplicity of $A_{\mathcal{B}}$ if and only if

$$(M_0(\lambda) - M_{p^\gamma}(\lambda))(M_0(\lambda) + M_{p^\gamma}(\lambda)) + a^2 + b^2 = 0 \quad (p^\gamma = |2x_1|_p).$$

Using the properties of $M_0(\lambda) - M_{p^\gamma}(\lambda)$ and $M_0(\lambda) + M_{p^\gamma}(\lambda)$ presented in Section 11.7, it is easy to describe the real eigenvalues of $A_{\mathcal{B}}$. Precisely:

(i) The negative semi-axis $\mathbb{R}_- = (-\infty, 0)$ belongs to $\rho(A_{\mathcal{B}})$.

(ii) If $N < -\gamma$, then the interval $(p^{\alpha N}, p^{\alpha(N+1)})$ contains an eigenvalue λ_N of $A_{\mathcal{B}}$ such that $p^{\alpha N} < \lambda_N < \lambda_N^+$, where λ_N^+ is the corresponding type-2 discrete spectrum point of the Friedrichs extension A_F.

(iii) If $N \geq -\gamma + 1$, then eigenvalues of A_B may appear only in the sub-interval $(\lambda_N^+, \lambda_N^-) \subset (p^{\alpha N}, p^{\alpha(N+1)})$, where λ_N^- is the type-1 point of $\sigma_{\text{dis}}(A_F)$. The existence of such eigenvalues in $(\lambda_N^+, \lambda_N^-)$ can be guaranteed by the decreasing of a and b (for fixed $N \geq -\gamma + 1$).

11.9. One point interaction

Without loss of generality we will assume $x_1 = 0$. Then the general expression (11.2.12) takes the form $D^\alpha + b\langle \delta_0, \cdot \rangle \delta_0$ ($b \in \mathbb{R} \cup \infty$) and the corresponding self-adjoint operator realizations A_b in $L_2(\mathbb{Q}_p)$ are defined by the formula

$$A_b f = A_b(u + \beta h_{1,-1}) = D^\alpha u - \beta h_{1,-1}, \qquad (11.9.1)$$

where the parameter $\beta = \beta(u, b) \in \mathbb{C}$ is uniquely determined by the relation $bu(0) = -\beta\big(1 + b M_0(-1)\big)$. The operators A_b are self-adjoint extensions of the symmetric operator $A_{\text{sym}} = D^\alpha \restriction_{\mathcal{D}}, \mathcal{D} = \{u \in \mathcal{D}(D^\alpha) : u(0) = 0\}$.

In our case, the subspace $\mathcal{L}_Y$ (which is defined in Section 11.5) is the closed linear span of $\theta_{k;j0}(x)$ ($j \in \mathbb{Z}, k = 1, \ldots, p - 1$) and $A_b = D^\alpha \restriction_{\mathcal{L}} \oplus A_b \restriction_{\mathcal{L}_Y}$. The operator $A_b \restriction_{\mathcal{L}_Y}$ is a self-adjoint extension of $A_{\text{sym}} \restriction_{\mathcal{L}_Y}$ and the points $p^{\alpha(1-j)}$ are eigenvalues of multiplicity $p - 2$ of the symmetric operator $A_{\text{sym}} \restriction_{\mathcal{L}_Y}$. The orthonormal basis $\{\widetilde{\theta}_{k;j0}(x)\}_{k=1}^{p-2}$ of the corresponding subspace $\ker(A_{\text{sym}} \restriction_{\mathcal{L}_Y} - p^{\alpha(1-j)}I)$ can be chosen as follows:

$$\widetilde{\theta}_{k;j0}(x) = \left(\frac{k}{k+1}\right)^{1/2} \left(\theta_{k+1;j0}(x) - \frac{1}{k}\sum_{i=1}^{k} \theta_{i;j0}(x)\right). \qquad (11.9.2)$$

The decomposition $A_b = D^\alpha \restriction_{\mathcal{L}} \oplus A_b \restriction_{\mathcal{L}_Y}$, Lemma 11.2.4, and Theorem 11.5.1 allow us to describe in detail the spectral properties of A_b ($b \neq 0$). Precisely:

(i) The operator A_b is positive $\iff b > 0$. Otherwise ($b < 0$), the unique solution of the equation $M_0(\lambda) = -1/b$ on the semi-axis $(-\infty, 0)$ gives a negative eigenvalue λ_b^- of multiplicity one. The corresponding normalized eigenfunction has the form

$$\phi_b^-(x) = \frac{h_{1,\lambda_b^-}(x)}{\sqrt{M_0'(\lambda_b^-)}}$$

$$= \frac{1}{\sqrt{M_0'(\lambda_b^-)}} \sum_{m=-\infty}^{\infty} \sum_{k=1}^{p-1} \frac{p^{-m/2}}{p^{\alpha(1-m)} - \lambda_b^-} \theta_{k;m0}(x). \qquad (11.9.3)$$

(ii) The positive part of the discrete spectrum of A_b consists of an infinite series of points λ_{jb} of multiplicity one, each of which is the unique solution of $M_0(\lambda) = -1/b$ in the interval $(p^{\alpha j}, p^{\alpha(j+1)})$ ($j \in \mathbb{Z}$). The corresponding normalized eigenfunction is (cf. (11.9.3))

$$\phi_{jb}(x) = \frac{1}{\sqrt{M_0'(\lambda_{jb})}} \sum_{m=-\infty}^{\infty} \sum_{k=1}^{p-1} \frac{p^{-m/2}}{p^{\alpha(1-m)} - \lambda_{jb}} \theta_{k;m0}(x). \qquad (11.9.4)$$

(iii) The points $p^{\alpha(1-j)}$ are eigenvalues of infinite multiplicity of A_b. The orthonormal basis of the corresponding subspace $\ker(A_b - p^{\alpha(1-j)}I)$ can be chosen as follows:

$$\theta_{k;ja}(x) \quad (1 \le k \le p-1, \quad a \neq 0), \qquad \widetilde{\theta}_{k;j0} \quad (1 \le k \le p-2),$$

where $\theta_{k;ja}(x)$ and $\widetilde{\theta}_{k;j0}$ are defined by (11.2.1) and (11.9.2), respectively.

(iv) The coefficient b of the singular perturbation $b \langle \delta_0, \cdot \rangle \delta_0$ is uniquely recovered by any point of the discrete spectrum and

$$\sigma_{\text{dis}}(A_{b_1}) \cap \sigma_{\text{dis}}(A_{b_2}) = \emptyset \quad (b_1 \neq b_2); \qquad \bigcup_{b \in \mathbb{R}} \sigma_{\text{dis}}(A_b) = \mathbb{R} \setminus \sigma(D^\alpha).$$

Combining properties (i)–(iii) with Corollary 11.6.2 we immediately establish the following statement.

Proposition 11.9.1. *The set of eigenfunctions of A_b*

$$
\begin{aligned}
&\theta_{k;ja}(x) \quad (j \in \mathbb{Z}, \quad 1 \le k \le p-1, \quad a \neq 0), \\
&\widetilde{\theta}_{k;j0}(x) \quad (j \in \mathbb{Z}, \quad 1 \le k \le p-2), \\
&\phi_{jb}(x) \quad (j \in \mathbb{Z}), \\
&\phi_b^-(x) \quad (\text{for the case } b < 0 \text{ only})
\end{aligned}
\qquad (11.9.5)
$$

forms an orthonormal basis of $\mathcal{L}_2(\mathbb{Q}_p)$. □

The Krein spectral shift $\xi_b(\lambda) = \frac{1}{\pi} \arg\left(1 + bM_0(\lambda + i0)\right)$ is easily calculated

$$\xi_b(\lambda) = \begin{cases} 0 & \text{if } \lambda \in (-\infty, \lambda_-) \bigcup \left[\bigcup_{-\infty}^{\infty} (\lambda_{jb}, p^{\alpha(j+1)}) \right] \\ 1 & \text{if } \lambda \in (\lambda_-, 0) \bigcup \left[\bigcup_{-\infty}^{\infty} (p^{\alpha j}, \lambda_{jb}) \right] \end{cases}$$

(the interval $(\lambda_-, 0)$ is omitted for $b > 0$). Therefore [227], the difference of the spectral projectors $P_\lambda(A_b) - P_\lambda(D^\alpha)$ ($P_\lambda \overset{\text{def}}{=} P_{(-\infty, \lambda)}$) is trace class and $\text{Tr}[P_\lambda(A_b) - P_\lambda(D^\alpha)] = 0$ for all $\lambda \in \ker \xi_b(\lambda)$.

Let us consider the transformation of dilation $Uf(x) = p^{-1/2} f(px)$. Obviously, U is an unitary operator in $\mathcal{L}_2(\mathbb{Q}_p)$ and the p-adic wavelet basis (11.2.1) $\{\theta_{k;ja}(x) : k = 1, 2, \ldots, p-1; j \in \mathbb{Z}; a \in I_p\}$ is invariant with respect to the dilation

$$U\theta_{k;ja}(x) = \theta_{k;j+1a}(x). \tag{11.9.6}$$

Furthermore, in view of (11.2.2)

$$U^m D^\alpha = p^{\alpha m} D^\alpha U^m, \qquad m \in \mathbb{Z}. \tag{11.9.7}$$

In this sense the operator D^α is $p^{\alpha m}$-homogeneous with respect to the one parameter family $\mathfrak{U} = \{U^m\}_{m \in \mathbb{Z}}$ of unitary operators [27], [103].

Proposition 11.9.2. *Among the self-adjoint operators A_b described by* (11.9.1) *there are only two $p^{\alpha m}$-homogeneous operators with respect to the family $\mathfrak{U}$. One of them, $A_0 = D^\alpha$, is the Krein–von Neumann extension of A_{sym}; the other one coincides with the Friedrichs extension $A_\infty = A_F$.*

An orthonormal basis of $\mathcal{L}_2(\mathbb{Q}_p)$ composed of the eigenfunctions of A_b and invariant with respect to the dilation U exists if and only if $b = 0$ or $b = \infty$.

Proof. The first part of the proposition is a direct consequence of [103, Section 4.4].

The p-adic wavelet basis $\{\theta_{k;ja}(x)\}$ is an example of an orthonormal basis composed of the eigenfunctions of A_0 and invariant with respect to U.

Let us show that the orthonormal basis of eigenfunctions of A_∞ defined by (11.9.5) is also invariant with respect to U. Indeed, relations (11.9.2) and (11.9.6) yield $U\widetilde{\theta}_{k;j0}(x) = \widetilde{\theta}_{k;j+10}(x)$.

It follows from (11.2.10) that

$$p^{\alpha-1} M_0(p^\alpha \lambda) = M_0(\lambda). \tag{11.9.8}$$

Using (11.9.8) and recalling that $\lambda_{j\infty}$ is the solution of $M_0(\lambda) = 0$ in the interval $(p^{\alpha j}, p^{\alpha(j+1)})$, we derive the recurrence relation $\lambda_{(j+1)\infty} = p^\alpha \lambda_{j\infty}$. The obtained relation and formulas (11.9.4) and (11.9.6) imply $U\phi_{j\infty}(x) = \phi_{(j-1)\infty}(x)$. Hence, the basis (11.9.5) is invariant with respect to U for $b = \infty$.

Let $\mathcal{M}$ be an arbitrary orthonormal basis composed of the eigenfunctions of A_b ($b \in \mathbb{R} \setminus \{0\}$). Since $\lambda_{jb} \in (p^{\alpha j}, p^{\alpha(j+1)})$ is an eigenvalue of A_b of multiplicity one, the corresponding eigenfunction $\phi_{jb}(x)$ belongs to $\mathcal{M}$. Assuming that $\mathcal{M}$ is invariant with respect to U, we get $A_b U\phi_{jb} = \mu U\phi_{jb}$, where $\mu \in \sigma(A_b)$. To find μ we note that the $p^{\alpha m}$-homogeneity of A_0 and A_∞ with respect to $\mathfrak{U}$ implies that A_{sym} and A_{sym}^* also are $p^{\alpha m}$-homogeneous with respect to U. Therefore,

$$\begin{aligned}
\lambda_{jb} U\phi_{jb} &= U A_b \phi_{jb} = U A_{\text{sym}}^* \phi_{jb} \\
&= p^\alpha A_{\text{sym}}^* U\phi_{jb} = p^\alpha A_b U\phi_{jb} = p^\alpha \mu U\phi_{jb}.
\end{aligned}$$

Thus $\mu = p^{-\alpha}\lambda_{jb}$. Obviously, $\mu \in (p^{\alpha(j-1)}, p^{\alpha j})$ and μ is the solution of $M_0(\lambda) = -1/b$ (since μ is an eigenvalue of A_b). Employing (11.9.8) for $\lambda = \mu$, we arrive at the contradiction

$$-1/b = M_0(\mu) = p^{\alpha-1}M_0(p^\alpha\mu) = p^{\alpha-1}M_0(\lambda_{Nb}) = -p^{\alpha-1}/b$$

which completes the proof of Proposition 11.9.2. $\square$

12

Distributional asymptotics and p-adic Tauberian theorems

12.1. Introduction

Tauberian theorems is a generic name used to indicate results connecting the asymptotic behavior of a function (distribution) at zero with the asymptotic behavior of its Fourier, Laplace or other integral transforms at infinity; the inverse theorems are usually called *abelian*. In the real setting *Tauberian theorems* have numerous applications, in particular, in mathematical physics (for example, see Drozzinov and Zavyalov [80], [81], Korevaar [160], Nikolić-Despotović, Pilipović [191], Vladimirov, Drozzinov and Zavyalov [240], Yakymiv [248] and the references cited therein). Multidimensional Tauberian theorems for distributions are treated in the fundamental book [240]. Some of them are connected with the fractional operator. In [240], as a rule, theorems of this type are proved for distributions whose supports belong to a cone in $\mathbb{R}^n$ (semi-axis for $n = 1$). This is related to the fact that such distributions form a convolution algebra. In this case the kernel of the fractional operator is a distribution whose support belongs to the cone in $\mathbb{R}^n$ or a semi-axis for $n = 1$ [240, §2.8.]

p-adic analogs of *Tauberian theorems* do not seem to have been discussed so far except for [140], [141], [21]. In this chapter, we present a first study of them based on the above papers.

In the beginning, in Sections 12.2 and 12.3, we introduce the notion of the p-adic distributional asymptotics [140], [141]. In Section 12.2, the definition of *distributional (stabilized) asymptotic estimate* at infinity is introduced. In Section 12.3, we extend the notion of regular variation introduced by I. Karamata [110] and studied by Seneta [216] to the p-adic case. Next we introduce the definitions of *quasi-asymptotics* at infinity and at zero adapted to the case of p-adic distributions (for the corresponding real case definition see [80], [240]).

230

In Sections 12.4 and 12.5, using definitions and results of Sections 12.2 and 12.3, multidimensional p-adic Tauberian type theorems (Theorems 12.4.4–12.4.7, and Theorems 12.5.1–12.5.6, Corollary 12.5.2) for p-adic distributions are proved. Theorem 12.5.1 and Corollary 12.5.2 are related to the Fourier transform and hold for distributions belonging to $\mathcal{D}'(\mathbb{Q}_p^n)$. Theorems 12.5.3–12.5.5 are related to fractional operators and hold for distributions belonging to the Lizorkin spaces $\Phi'_\times(\mathbb{Q}_p^n)$ and $\Phi'(\mathbb{Q}_p^n)$. Theorem 12.5.6 is related to the pseudo-differential operator (9.3.1) in the Lizorkin space $\Phi'(\mathbb{Q}_p^n)$. Taking into account the fact that the kernels of fractional operators are defined on the whole space $\mathbb{Q}_p^n$ (by virtue of the p-adic field nature), the Tauberian theorems proved here are not direct analogs of the Tauberian theorems for the usual generalized functions discussed in [240].

12.2. Distributional asymptotics

Let us recall the usual (in a real case) definition of the *distribution asymptotic estimate at infinity*. According to [57, 6.1], we say that a distribution $f \in \mathcal{D}(\mathbb{R}^n)$ has the *distribution asymptotic estimate at infinity* g if there exist constants R and C such that the following relation holds:

$$|\langle f, \varphi \rangle| \leq C \int_{\mathbb{R}^n} |g(x)| |\varphi(x)| \, d^n x, \quad \forall \, \varphi \in \mathcal{D}(\mathbb{R}^n \setminus \{x : |x| > R\}).$$

Now we adapt this definition for the p-adic case. In contrast to the above definition our adaptation allows us to obtain the estimate for the Fourier transform (see Theorem 12.4.6 below).

Definition 12.1. We say that a distribution $f \in \mathcal{D}'(\mathbb{Q}_p^n)$ has the *distributional (stabilized) asymptotic estimate* $g \in \mathcal{D}'(\mathbb{Q}_p^n)$ at infinity if there exists a ball $B_\gamma^n = \{x : |x|_p \leq p^\gamma\}$ such that

$$\langle f, \varphi \rangle = \langle g, \varphi \rangle, \quad \forall \, \varphi \in \mathcal{D}(\mathbb{Q}_p^n \setminus B_\gamma^n). \quad (12.2.1)$$

We shall write it as follows

$$f(x) \overset{\mathcal{D}'}{\approx} g(x), \quad |x|_p \to \infty.$$

12.3. p-adic distributional quasi-asymptotics

We recall some facts from our papers [140], [141], where we introduced the notion of *quasi-asymptotics* [80], [240] adapted to the p-adic case.

Definition 12.2. ([140], [141]) A continuous complex-valued function $\rho(z)$ on the multiplicative group $\mathbb{Q}_p^\times$ such that for any $z \in \mathbb{Q}_p^\times$ the limit

$$\lim_{|t|_p \to \infty} \frac{\rho(tz)}{\rho(t)} = C(z)$$

exists is called an *automodel* (or *regular varying*) function, where the convergence is uniform with respect to z for any compact $K \subset \mathbb{Q}_p^\times$.

It is easy to see that the function $C(z)$ is continuous and satisfies the functional equation

$$C(ab) = C(a)C(b), \qquad a, b \in \mathbb{Q}_p^\times.$$

According to Lemma 2.3.7, the solution of this equation is a multiplicative character π_α of the field $\mathbb{Q}_p$ defined by (2.3.12) and (2.3.13), i.e.,

$$C(z) = |z|_p^{\alpha-1} \pi_1(z), \quad z \in \mathbb{Q}_p^\times, \quad \alpha \in \mathbb{C}. \qquad (12.3.1)$$

In this case we say that the *automodel function* ρ *has degree* π_α. In particular, if $\pi_\alpha(z) = |z|_p^{\alpha-1}$ we say that the *automodel* function has degree $\alpha - 1$.

If an *automodel* function $\rho(t)$, $t \in \mathbb{Q}_p^\times$, has degree π_α then the *automodel* function $|t|_p^\beta \rho(t)$ has degree $\pi_\alpha(t)\pi_0^{-\beta}(t) = \pi_1(t)|t|_p^{\alpha+\beta}$, where $\pi_0(t) = |t|_p^{-1}$.

For example, the functions $|t|_p^{\alpha-1}\pi_1(t)$ and $|t|_p^{\alpha-1}\pi_1(t)\log_p^m |t|_p$ constitute an *automodel* of degree π_α.

Definition 12.3. ([140], [141]) Let $f \in \mathcal{D}'(\mathbb{Q}_p^n)$. If there exists an *automodel* function $\rho(t)$, $t \in \mathbb{Q}_p^\times$, of degree π_α such that

$$\frac{f(tx)}{\rho(t)} \to g(x) \not\equiv 0, \quad |t|_p \to \infty, \quad \text{in} \quad \mathcal{D}'(\mathbb{Q}_p^n)$$

then we say that the distribution f has the *quasi-asymptotics* g of degree π_α at *infinity* with respect to $\rho(t)$, and write

$$f(x) \overset{\mathcal{D}'}{\sim} g(x), \quad |x|_p \to \infty \ (\rho(t)).$$

If for any α we have

$$\frac{f(tx)}{|t|_p^\alpha} \to 0, \quad |t|_p \to \infty, \quad \text{in} \quad \mathcal{D}'(\mathbb{Q}_p^n),$$

then we say that the distribution f has a *quasi-asymptotics* of degree $-\infty$ at infinity and write $f(x) \overset{\mathcal{D}'}{\sim} 0$, $|x|_p \to \infty$.

Since $|t|_p \to \infty$ if and only if $|xt|_p = \max_{1 \le j \le n} |tx_j|_p \to \infty$ for any $x \ne 0 \in \mathbb{Q}_p^n$, our definition is natural.

Lemma 12.3.1. ([140], [141]) *Let* $f \in \mathcal{D}'(\mathbb{Q}_p^n)$. *If* $f(x) \overset{\mathcal{D}'}{\sim} g(x) \neq 0$, *as* $|x|_p \to$ ∞ *with respect to the* automodel *function* $\rho(t)$ *of degree* π_α, *then* g *is a homogeneous distribution of degree* π_α *(with respect to Definition 6.1 (b)).*

Proof. This lemma is proved by repeating the corresponding assertion from the book [240] practically word for word. Let $a \in \mathbb{Q}_p^\times$. In view of Definition 12.2 and (12.3.1), we obtain

$$\langle g(a\cdot), \varphi(\cdot) \rangle = \lim_{|t|_p \to \infty} \left\langle \frac{f(ta\cdot)}{\rho(t)}, \varphi(\cdot) \right\rangle$$

$$= \pi_\alpha(a) \lim_{|t|_p \to \infty} \left\langle \frac{f(ta\cdot)}{\rho(ta)}, \varphi(\cdot) \right\rangle = \pi_\alpha(a) \langle g, \varphi \rangle,$$

for all $a \in \mathbb{Q}_p^\times$, $\varphi \in \mathcal{D}(\mathbb{Q}_p^n)$. Thus $g(ax) = \pi_\alpha(a) g(x)$ for all $a \in \mathbb{Q}_p^\times$. $\square$

For $n = 1$, it follows from Theorem 6.2.1 (describing all one-dimensional homogeneous distributions) and Lemma 12.3.1 that if $f \in \mathcal{D}'(\mathbb{Q}_p)$ has the quasi-asymptotics of degree π_α at infinity then

$$f(x) \overset{\mathcal{D}'}{\sim} g(x) = \begin{cases} C\pi_\alpha(x), & \pi_\alpha \neq \pi_0 = |x|_p^{-1}, \\ C\delta(x), & \pi_\alpha = \pi_0 = |x|_p^{-1}, \end{cases} \quad |x|_p \to \infty, \quad (12.3.2)$$

where C is a constant, and the distribution π_α is defined by (6.2.2).

Definition 12.4. ([140], [141]) Let $f \in \mathcal{D}'(\mathbb{Q}_p^n)$. If there exists an *automodel* function $\rho(t)$, $t \in \mathbb{Q}_p^\times$ of degree π_α such that

$$\frac{f\left(\frac{x}{t}\right)}{\rho(t)} \to g(x) \neq 0, \quad |t|_p \to \infty, \quad \text{in} \quad \mathcal{D}'(\mathbb{Q}_p^n),$$

then we say that the distribution f has the *quasi-asymptotics* g of degree $(\pi_\alpha)^{-1}$ *at zero* with respect to $\rho(t)$, and write

$$f(x) \overset{\mathcal{D}'}{\sim} g(x), \quad |x|_p \to 0 \ (\rho(t)).$$

If for any α we have

$$\frac{f\left(\frac{x}{t}\right)}{|t|_p^\alpha} \to 0, \quad |t|_p \to \infty, \quad \text{in} \quad \mathcal{D}'(\mathbb{Q}_p^n),$$

then we say that the distribution f has a *quasi-asymptotics of degree* $-\infty$ at zero, and write $f(x) \overset{\mathcal{D}'}{\sim} 0$, $|x|_p \to 0$.

Example 12.3.1. Let $f_m \in \mathcal{D}'(\mathbb{Q}_p)$, $m \in \mathbb{N}$, be a *quasi-associated homogeneous* distribution of degree $\pi_\alpha(x)$ and order m defined by (6.3.2) and (6.3.5). In view of Definition 6.2, we have the asymptotic formulas:

$$f_m(tx) = \pi_1(t)|t|_p^{\alpha-1} f_m(x)$$
$$+ \sum_{j=1}^{m} \pi_1(t)|t|_p^{\alpha-1} \log_p^j |t|_p f_{m-j}(x), \qquad |t|_p \to \infty,$$

$$f_m\left(\frac{x}{t}\right) = \pi_1^{-1}(t)|t|_p^{-\alpha+1} f_m(x)$$
$$+ \sum_{j=1}^{m} (-1)^j \pi_1^{-1}(t)|t|_p^{-\alpha+1} \log_p^j |t|_p f_{m-j}(x), \quad |t|_p \to \infty.$$

Here the coefficients of the *leading term* of both asymptotics are homogeneous distributions f_0 and $(-1)^m f_0$ of degree π_α defined by the relation given in Definition 6.2. For the real case, this type of distributional asymptotics is studied in [87].

According to the latter relations and Definitions 12.3 and 12.4, we can easily see that

$$f_m(x) \overset{\mathcal{D}'}{\sim} f_0(x), \qquad |x|_p \to \infty \quad \left(|t|_p^{\alpha-1} \pi_1(t) \log_p^m |t|_p\right),$$
$$f_m(x) \overset{\mathcal{D}'}{\sim} (-1)^m f_0(x), \quad |x|_p \to 0 \quad \left(|t|_p^{-\alpha+1} \pi_1^{-1}(t) \log_p^m |t|_p\right).$$

Example 12.3.2. Consider the quasi-associated homogeneous distribution $P\left(|x|_p^{-1}\right) \in \mathcal{D}'(\mathbb{Q}_p)$ of degree $\pi_\alpha(x) = |x|_p^{-1}$ and order 1 defined by (6.3.5). Since

$$P\left(|tx|_p^{-1}\right) = |t|_p^{-1} P\left(|x|_p^{-1}\right) + |t|_p^{-1} \log_p |t|_p \left(1 - p^{-1}\right) \delta(x),$$

for all $t \in \mathbb{Q}_p^\times$, $x \in \mathbb{Q}_p$, using the latter relation, Definition 12.3, and setting $\rho(t) = |t|_p^{-1} \log_p |t|_p$, we obtain

$$\lim_{|t|_p \to \infty} \left\langle \frac{P\left(|t \cdot |_p^{-1}\right)}{|t|_p^{-1} \log_p |t|_p}, \varphi(\cdot) \right\rangle = \left(1 - p^{-1}\right) \langle \delta, \varphi \rangle, \quad \forall \, \varphi \in \mathcal{D}(\mathbb{Q}_p).$$

Thus the distribution $P\left(|x|_p^{-1}\right)$ has a quasi-asymptotics $\delta(x)$ of degree -1, i.e.,

$$P\left(|x|_p^{-1}\right) \overset{\mathcal{D}'}{\sim} \left(1 - p^{-1}\right) \delta(x), \quad |x|_p \to \infty \quad \left(|t|_p^{-1} \log_p |t|_p\right).$$

12.4. Tauberian theorems with respect to asymptotics

We intend to prove Tauberian type theorems with respect to asymptotics.

It is known that in the general case if $\varphi \in \mathcal{D}(\mathbb{Q}_p)$ then the function $D_x^{\alpha_j}\varphi(x) \in \mathcal{E}(\mathbb{Q}_p)$ is *not compactly supported*. Thus the operation $D_x^{\alpha_j}f(x)$, $f \in \mathcal{D}'(\mathbb{Q}_p)$, is well defined only if the convolution $f_{-\alpha_j}(x) * f(x)$ exists ($\alpha \neq -1$) [241, IX]. Moreover, from (9.2.4) and (9.2.5) it is easy to see that

$$\left(D_x^{\alpha}\varphi\right)(x) = O\left(|x|_p^{-\alpha_j-1}\right), \quad x \in \mathbb{Q}_p, \quad |x|_p \to \infty. \tag{12.4.1}$$

This estimate is the direct consequence of the fact that the functions $|\xi|_p^{\alpha}$ and $|\xi|_p^{\alpha}F[\varphi](\xi)$ are *not locally constant in a zero neighborhood*.

Since for $\alpha > -1$ we have $|\xi|_p^{\alpha}F[\varphi](\xi)\chi_p(-\xi x) = O(|\xi|_p^{\alpha})$, $|\xi|_p \to 0$, in view of (12.4.1), we can see that on the background of these properties of the fractional operator is a *prototype of a p-adic analog* of the Tauberian type theorem (see Lemmas 12.4.1, 12.4.2, and Theorems 12.4.4–12.4.7).

The above mentioned situation with the fractional operator is reflected by the following lemmas.

Lemma 12.4.1. *Let* $\varphi \in \mathcal{D}'_N(\mathbb{Q}_p)$ *and* $\psi(t) = F\left[|x|_p^{\alpha}\varphi(x)\right](t)$, $\mathrm{Re}\,\alpha > -1$. *Then*

$$\psi(t) = \begin{cases} \in \mathcal{D}_{-l}^{-N}(\mathbb{Q}_p), & |t|_p \leq p^{-l}, \\ \varphi(0)\Gamma_p(\alpha+1)\frac{1}{|t|_p^{\alpha+1}}, & |t|_p > p^{-l}, \end{cases} \tag{12.4.2}$$

where $\Gamma_p(\alpha)$ *is given by formula* (6.2.6).

Proof. Since $\mathrm{Re}\,\alpha > -1$ the integral $\psi(t)$ converges absolutely. We rewrite it as the sum $\psi(t) = \psi_1(t) + \psi_2(t)$, where

$$\begin{aligned} \psi_1(t) &= \int_{B_l} \chi_p(tx)|x|_p^{\alpha}\varphi(x)\,dx, \\ \psi_2(t) &= \int_{\mathbb{Q}_p \setminus B_l} \chi_p(tx)|x|_p^{\alpha}\varphi(x)\,dx. \end{aligned} \tag{12.4.3}$$

If $x \in \mathbb{Q}_p \setminus B_l$ the function $|x|_p^{\alpha}$ has a parameter of constancy larger than or equal to l, i.e., $|x|_p^{\alpha}\varphi(x) \in \mathcal{D}_N^l$. Hence according to (4.8.5),

$$\psi_2(t) = F\left[|x|_p^{\alpha}(1 - \Delta_l(x))\varphi(x)\right](t) \in \mathcal{D}_{-l}^{-N}, \tag{12.4.4}$$

i.e. $\psi_2(t) = 0$ if $|t|_p > p^{-l}$.

Since $\varphi \in \mathcal{D}'_N(\mathbb{Q}_p)$, the function ψ_1 can be rewritten as

$$\psi_1(t) = \int_{B_l} \chi_p(tx)|x|_p^{\alpha}\,\varphi(x)\,dx = \varphi(0)\int_{B_l} \chi_p(tx)|x|_p^{\alpha}\,dx.$$

Next, according to [241, VII.2, Example 9] and (6.2.6), for Re $\alpha > -1$ we have

$$
F\big[|x|_p^\alpha \Delta_l(x)\big](t) = \int_{B_l} \chi_p(tx)|x|_p^\alpha \, dx
$$

$$
= \frac{1 - p^{-1}}{1 - p^{-(\alpha+1)}} p^{l(\alpha+1)} \Delta_{-l}(t) + \frac{\Gamma_p(\alpha + 1)}{|t|_p^{\alpha+1}} \big(1 - \Delta_{-l}(t)\big).
$$

$$
(12.4.5)
$$

To complete the proof of the lemma, it remains to use (12.4.3)–(12.4.5). $\square$

Lemma 12.4.2. *Let* $\varphi \in \mathcal{D}_N^l(\mathbb{Q}_p)$, $\psi(t) = F\big[|x|_p^\alpha \log_p^m |x|_p \varphi(x)\big](t)$, Re $\alpha > -1$, $m = 1, 2, \ldots$. *Then*

$$
\psi(t) = \begin{cases} \in \mathcal{D}_{-l}^{-N}(\mathbb{Q}_p), & |t|_p \le p^{-l}, \\ \varphi(0) \sum_{k=0}^m C_m^k \log_p^{m-k} e \frac{d^{m-k}\Gamma_p(\alpha+1)}{d\alpha^{m-k}} \frac{\log_p^k |t|_p}{|t|_p^{\alpha+1}}, & |t|_p > p^{-l}. \end{cases} \quad (12.4.6)
$$

Proof. Since Re $\alpha > 0$, by differentiating the identity (12.4.5) with respect to α, we derive the following identity:

$$
F\big[|x|_p^\alpha \log_p^m |x|_p \Delta_l(x)\big](t)
$$

$$
= \int_{B_l} \chi_p(tx)|x|_p^\alpha \log_p^m |x|_p \, dx
$$

$$
= \big(1 - p^{-1}\big) \log_p^m e \frac{d^m}{d\alpha^m} \left(\frac{p^{l(\alpha+1)}}{1 - p^{-(\alpha+1)}} \right) \Delta_{-l}(t)
$$

$$
+ \big(1 - \Delta_{-l}(t)\big) \sum_{k=0}^m C_m^k \log_p^{m-k} e \frac{d^{m-k}\Gamma_p(\alpha + 1)}{d\alpha^{m-k}} \frac{\log_p^k |t|_p}{|t|_p^{\alpha+1}}, \quad (12.4.7)
$$

where C_m^k are binomial coefficients and $\Gamma_p(\alpha)$ is given by formula (6.2.6).

Next, repeating the constructions of Lemma 12.4.1 practically word for word, we obtain the proof of Lemma 12.4.2. $\square$

Thus, the functions ψ in Lemmas 12.4.1 and 12.4.2 are *not compactly supported*. These lemmas are a direct consequence of the fact that $|x|_p^\alpha$ is *not a locally constant* function in a zero neighborhood.

The simplest Tauberian theorem is the following:

Corollary 12.4.3. *If* $\varphi \in \mathcal{D}(\mathbb{Q}_p)$, Re $\alpha > -1$, *then*

$$
\psi(x) = F\big[|\xi|_p^\alpha \log_p^m |x|_p \varphi(\xi)\big](x)
$$

$$
= O\big(|x|_p^{-\alpha-1} \log_p^m |x|_p\big), \quad |x|_p \to \infty, x, \xi \in \mathbb{Q}_p.
$$

Using Lemma 12.4.1, we can prove the following Tauberian type theorems.

Theorem 12.4.4. *Let $f : U \to \mathbb{Q}_p$ be a locally analytic function (see Definition 2.2) (where $U \subset \mathbb{Q}_p^n$ is a convex open subset containing 0) such that $f(\xi) = \xi_1^{m_1} \cdots \xi_n^{m_n} f_0(\xi)$, $f_0(0) \neq 0$ for ξ in some neighborhood of zero, and $f(\xi) \neq 0$, $\xi \in U \setminus \{0\}$. Then we have for the Fourier transform of $|f(\xi)|_p$ the relation*

$$g(x) = \int_U \chi_p(x\xi)|f(\xi)|_p \, d\xi = O\big(|x_1|_p^{-m_1-1} \cdots |x_n|_p^{-m_n-1}\big), \quad |x|_p \to \infty.$$

Proof. Let us cover the ball $B^n(0)$ by a finite union of disjoint balls $B_\rho^n(\xi_{(j)})$ with centers at points $\xi_{(j)} \in B^n(0)$ such that

$$|f_0(\xi)|_p = |f_0(\xi_{(j)})|_p = p^{r_j}, \qquad \xi \in B_\rho^n(\xi_j)$$

(see the proof of Theorem 3.4.1). Thus $|f_0(\xi)|_p$ is a locally constant function on $B^n(0)$. It follows from the same argument that $|f(\xi)|_p$ is a locally constant function on $U \setminus B^n(0)$. Consequently, applying Corollary 12.4.3, we obtain the proof of the theorem. $\qquad\square$

Corollary 12.4.5. *Let $f : U \to \mathbb{Q}_p$ be a locally analytic function on a convex open subset $U \subset \mathbb{Q}_p$ containing 0 such that $f^{(k)}(0) = 0$, $k = 0, 1, \ldots, m$, $f^{(m+1)}(0) \neq 0$, $f(\xi) \neq 0, \xi \in U \setminus \{0\}$. Then we have the following relation for the Fourier transform of $|f(\xi)|_p$:*

$$g(x) = \int_U \chi_p(x\xi)|f(\xi)|_p \, d\xi = O\big(|x|_p^{-m-1}\big), \quad |x|_p \to \infty.$$

Proof. It is sufficient to realize that under the assumption of the corollary there is a ball $B(0)$ such that $f(\xi) = \xi^m f_0(\xi)$, where $f_0(\xi) \neq 0$, for any $\xi \in B(0)$. Then the statement follows from Theorem 12.4.4. $\qquad\square$

Theorem 12.4.6. *Let $f \in \mathcal{D}'(\mathbb{Q}_p^n)$ and*

$$f \stackrel{\mathcal{D}'}{\approx} \prod_{j=1}^n |x_j|_p^{-(\alpha_j+1)} \log_p^{m_j} |x_j|_p, \quad |x|_p \to \infty, \tag{12.4.8}$$

(see Definition 12.1), where $\operatorname{Re} \alpha_j > -1, m_j \in \mathbb{N}_0, j = 1, \ldots, n$. Then $g(\xi) = F\big[f(x)\big](\xi) \in \mathcal{E}(\mathbb{Q}_p^n \setminus \{0\})$ and

$$g(\xi) = O\Big(1 + \prod_{j=1}^n \big(1 + \sum_{s=0}^{m_j} |\xi_j|_p^{\alpha_j} \log_p^s |\xi_j|_p\big)\Big), \quad |\xi|_p \to 0. \tag{12.4.9}$$

Proof. In view of relation (12.4.8), there exists a ball B_γ^n such that

$$\langle f, \varphi \rangle = \int_{\mathbb{Q}_p^n \setminus B_\gamma^n} \prod_{j=1}^n |x_j|_p^{-(\alpha_j+1)} \log_p^{m_j} |x_j|_p \varphi(x) \, d^n x, \tag{12.4.10}$$

for all $\varphi \in \mathcal{D}(\mathbb{Q}_p^n \setminus B_\gamma^n)$.

Let $N \geq \gamma$. According to Theorem 4.4.5 on "piecewise sewing", $f = f_1 + f_2$, where $f_1 = \Delta_N(x)f$, $f_2 = (1 - \Delta_N(x))f$. Consequently,

$$\langle F[f], \varphi \rangle = \left\langle (\Delta_N(x) + (1 - \Delta_N(x)))f, F[\varphi] \right\rangle$$

$$= \left\langle f(x), \Delta_N(x) \int \chi_p(x\xi)\varphi(\xi)\,d^n\xi \right\rangle$$

$$+ \left\langle f(x), (1 - \Delta_N(x)) \int \chi_p(x\xi)\varphi(\xi)\,d^n\xi \right\rangle$$

$$= \int \left(\left\langle f_1(x), \Delta_N(x)\chi_p(x\xi) \right\rangle \right.$$

$$\left. + \left\langle f_2(x), (1 - \Delta_N(x))\chi_p(x\xi) \right\rangle \right) \varphi(\xi)\,d^n\xi,$$

for all $\varphi \in \mathcal{D}(\mathbb{Q}_p^n)$. Thus,

$$F[f] = \left\langle f_1(x), \Delta_N(x)\chi_p(x\xi) \right\rangle + \left\langle f_2(x), (1 - \Delta_N(x))\chi_p(x\xi) \right\rangle.$$

Since the distribution $f_1 = \Delta_N(x)f$ has a compact support in B_N^n, according to Theorem 4.9.3, its Fourier transform

$$g_1(\xi) = F[f_1(x)](\xi) = \langle f_1(x), \Delta_N(x)\chi_p(x\xi) \rangle \in \mathcal{E}(\mathbb{Q}_p^n)$$

is a locally constant function with a parameter of constancy larger than or equal to $-N$.

Let $g_2(\xi) = \left\langle f_2(x), (1 - \Delta_N(x))\chi_p(x\xi) \right\rangle$. Since $(1 - \Delta_N(x))\chi_p(x\xi)$ is a test function having the support on $\mathbb{Q}_p^n \setminus B_N^n$, using (12.4.10), we obtain

$$g_2(\xi) = \int_{\mathbb{Q}_p^n \setminus B_N^n} \prod_{j=1}^{n} |x_j|_p^{-(\alpha_j+1)} \log_p^{m_j} |x_j|_p \chi_p(x\xi)\,d^n x.$$

It is easy to calculate, by (12.4.5), that

$$F\left[|x_j|_p^{-(\alpha_j+1)}(1 - \Delta_N(x_j))\right](\xi_j)$$

$$= \int_{\mathbb{Q}_p \setminus B_N} |x_j|_p^{-(\alpha_j+1)} \chi_p(x_j\xi_j)\,dx_j$$

$$= \frac{1 - p^{-(\alpha_j+1)}}{1 - p^{\alpha_j}} \left(|\xi_j|_p^{\alpha_j} - \frac{1 - p^{-1}}{1 - p^{-(\alpha_j+1)}} p^{-N\alpha_j} \right) \Delta_{-N}(\xi_j), \quad (12.4.11)$$

for $j = 1, \ldots, n$. By differentiation of (12.4.11) with respect to α_j we obtain the relation

$$F\big[|x_j|_p^{-(\alpha_j+1)} \log_p^{m_j} |x_j|_p \big(1 - \Delta_N(x_j)\big)\big](\xi_j)$$

$$= \int_{\mathbb{Q}_p \backslash B_N} |x_j|_p^{-(\alpha_j+1)} \log_p^{m_j} |x_j|_p \chi_p(x_j \xi_j)\, dx_j$$

$$= \Big(\sum_{s=0}^{m_j} C_{m_j}^s A_{m_j-s}(\alpha_j)|\xi_j|_p^{\alpha_j} \log_p^s |\xi_j|_p$$

$$- (1 - p^{-1})B(\alpha_j)\Big)\Delta_{-N}(\xi_j), \qquad (12.4.12)$$

where $C_{m_j}^s$ are the binomial coefficients,

$$A_s(\alpha_j) = \log_p^{m-s} e \frac{d^s}{d\alpha_j^s}\Big(\frac{1 - p^{-(\alpha_j+1)}}{1 - p^{\alpha_j}}\Big),$$

$$B(\alpha_j) = \log_p^m e \frac{d^m}{d\alpha_j^m}\Big(\frac{p^{-N\alpha_j}}{1 - p^{-(\alpha_j+1)}}\Big),$$

and $s = 0, 1, \ldots, m_j$, $j = 1, \ldots, n$.

Since $g_1(\xi)$ is a locally constant function, $g_1(\xi) = g_1(0)$ for all $\xi \in B_{-N}^n$. Next, with the help of formulas (12.4.11) and (12.4.12), we obtain

$$|g(\xi)| = |g_1(\xi) + g_2(\xi)| = |g_1(0) + g_2(\xi)|$$

$$\leq C\Big(1 + \prod_{j=1}^{n}\Big(1 + \sum_{s=0}^{m_j} |\xi_j|_p^{\alpha_j} \log_p^s |\xi_j|_p\Big)\Big),$$

for all $\xi \in B_{-N}^n \backslash \{0\}$, $-N \leq -\gamma$, i.e., (12.4.9) holds. The proof of the theorem is complete. $\qquad \square$

Theorem 12.4.7. *Consider the following properties:*

(i) $f \in \mathcal{D}'(\mathbb{Q}_p)$;

(ii) $f \overset{\mathcal{D}'}{\approx} |x|_p^{-(\alpha+1)}$, $|x|_p \to \infty$, Re $\alpha > -1$, *i.e., there exists a ball B_γ such that* $\langle f, \varphi \rangle = \int |x|_p^{-(\alpha+1)}\varphi(x)\, dx$, *for all* $\varphi \in \mathcal{D}(\mathbb{Q}_p^n \backslash B_\gamma^n)$;

(iii) $\langle f, \Delta_\gamma \rangle = -\frac{1-p^{-1}}{1-p^\alpha} p^{-\gamma\alpha}$.

For them to hold it is necessary and sufficient that

$$g(\xi) = F\big[f(x)\big](\xi) = \begin{cases} \in \mathcal{E}(\mathbb{Q}_p), & \xi \in \mathbb{Q}_p \backslash B_{-\gamma}, \\ \frac{1-p^{-(\alpha+1)}}{1-p^\alpha}|\xi|_p^\alpha, & \xi \in B_{-\gamma}. \end{cases}$$

To prove this assertion, it suffices to repeat the proof of Theorem 12.4.6 almost word for word. Moreover, we must take into account formula (12.4.11) and the fact that $g_1(0) = \langle f(x), \Delta_\gamma(x) \rangle$ (see [241, IX, Lemma 1]).

12.5. Tauberian theorems with respect to quasi-asymptotics

Now we prove some Tauberian theorems with respect to the distributional quasi-asymptotics introduced in Section 12.3.

Theorem 12.5.1. ([141]) *A distribution* $f \in \mathcal{D}'(\mathbb{Q}_p^n)$ *has a quasi-asymptotics of degree* π_α *at infinity with respect to the automodel function* $\rho(t)$, $t \in \mathbb{Q}_p^\times$, *i.e.,*

$$f(x) \overset{\mathcal{D}'}{\sim} g(x), \quad |x|_p \to \infty \; (\rho(t)),$$

if and only if its Fourier transform has a quasi-asymptotics at zero of degree $\pi_\alpha^{-1}\pi_0^n = \pi_{\alpha+n}^{-1}$ *with respect to the automodel function* $|t|_p^n \rho(t)$, *i.e.,*

$$F[f(x)](\xi) \overset{\mathcal{D}'}{\sim} F[g(x)](\xi), \quad |\xi|_p \to 0 \; \left(|t|_p^n \rho(t)\right).$$

Proof. Let us prove the necessity. Let $f(x) \overset{\mathcal{D}'}{\sim} g(x)$, $|x|_p \to \infty \; (\rho(t))$, i.e.,

$$\lim_{|t|_p \to \infty} \left\langle \frac{f(tx)}{\rho(t)}, \varphi(x) \right\rangle = \langle g(\cdot), \varphi(\cdot) \rangle, \quad \forall \, \varphi \in \mathcal{D}(\mathbb{Q}_p^n), \qquad (12.5.1)$$

where $\rho(t)$ is an automodel function of degree π_α. In view of formula (4.9.3), $F[f(x)](\frac{\xi}{t}) = |t|_p^n F[f(tx)](\xi)$, $x, \xi \in \mathbb{Q}_p^n, t \in \mathbb{Q}_p^\times$, we have

$$\left\langle F[f(x)]\left(\frac{\xi}{t}\right), \varphi(\xi) \right\rangle$$
$$= |t|_p^n \langle F[f(tx)](\xi), \varphi(\xi) \rangle = |t|_p^n \langle f(tx), F[\varphi(\xi)](x) \rangle,$$

$\varphi \in \mathcal{D}(\mathbb{Q}_p^n)$. Hence, taking into account relation (12.5.1), we obtain

$$\lim_{|t|_p \to \infty} \left\langle \frac{F[f(x)](\frac{\xi}{t})}{|t|_p^n \rho(t)}, \varphi(\xi) \right\rangle = \lim_{|t|_p \to \infty} \left\langle \frac{f(tx)}{\rho(t)}, F[\varphi(\xi)](x) \right\rangle$$
$$= \langle g(x), F[\varphi(\xi)](x) \rangle = \langle F[g(x)](\xi), \varphi(\xi) \rangle,$$

for all $\varphi \in \mathcal{D}(\mathbb{Q}_p^n)$, i.e., the distribution $F[f(x)](\xi)$ has the quasi-asymptotics $F[g(x)](\xi)$ of degree $\pi_{\alpha+n}^{-1}$ at zero with respect to $|t|_p^n \rho(t)$.

The sufficiency can be proved similarly.   □

   For $n = 1$ Theorem 12.5.1, Lemma 12.3.1, and formula (9.2.13) imply the following corollary.

Corollary 12.5.2. *A distribution* $f \in \mathcal{D}'(\mathbb{Q}_p)$ *has a quasi-asymptotics of degree* $\pi_\alpha(x)$ *at infinity, i.e.,*

$$f(x) \overset{\mathcal{D}'}{\sim} g(x) = \begin{cases} C|x|_p^{\alpha-1}\pi_1(x), & \pi_\alpha \neq \pi_0 = |x|_p^{-1}, \\ C\delta(x), & \pi_\alpha = \pi_0 = |x|_p^{-1}, \end{cases} \quad |x|_p \to \infty,$$

$$(12.5.2)$$

if and only if its Fourier transform $F[f]$ *has a quasi-asymptotics of degree* $\pi_{\alpha+1}^{-1}(\xi)$ *at zero, i.e.,*

$$F[f(x)](\xi) \overset{\mathcal{D}'}{\sim} F[g(x)](\xi)$$

$$= \begin{cases} C\Gamma_p(\pi_\alpha)|\xi|_p^{-\alpha}\pi_1^{-1}(\xi), & \pi_\alpha \neq \pi_0 = |x|_p^{-1}, \\ C, & \pi_\alpha = \pi_0 = |x|_p^{-1}, \end{cases} \quad |\xi|_p \to 0,$$

where the distribution $\pi_\alpha(x) = |x|_p^{\alpha-1}\pi_1(x)$ *is given by (6.2.2).*

Theorem 12.5.3. *Let* $f \in \Phi_\times'(\mathbb{Q}_p^n)$. *Then*

$$f(x) \overset{\Phi_\times'}{\sim} g(x), \quad |x|_p \to \infty \quad (\rho(t)),$$

if and only if

$$D_\times^\beta f(x) \overset{\Phi_\times'}{\sim} D_\times^\beta g(x), \quad |x|_p \to \infty \quad (|t|_p^{-\beta|}\rho(t)),$$

where $\beta = (\beta_1, \dots, \beta_n) \in \mathbb{C}^n$, $|\beta| = \beta_1 + \cdots + \beta_n$.

Proof. 1. Let $\beta_j \neq -1$, $j = 1, 2, \dots$. In this case the Riesz kernel $f_{-\beta}(x)$ is a *homogeneous* distribution of degree $|-\beta| - n$. According to Lemma 9.2.2 and formulas (9.2.9), (9.2.12) and (9.2.15), we have

$$\langle (D_\times^\beta f)(tx), \phi(x) \rangle = \langle (f * f_{-\beta})(tx), \phi(x) \rangle$$

$$= |t|_p^{-n}\langle f(x), \langle f_{-\beta}(y), \phi\left(\frac{x+y}{t}\right)\rangle\rangle$$

$$= |t|_p^n\langle f(tx), \langle f_{-\beta}(ty), \phi(x+y)\rangle\rangle$$

$$= |t|_p^{-\beta|}\langle f(tx), \langle f_{-\beta}(y), \phi(x+y)\rangle\rangle$$

$$= |t|_p^{-\beta|}\langle f(tx), (D_\times^\beta\phi)(x)\rangle,$$

for all $\phi \in \Phi_\times(\mathbb{Q}_p^n)$. Thus

$$\left\langle \frac{(D_\times^\beta f)(tx)}{|t|_p^{-\beta|}\rho(t)}, \phi(x) \right\rangle = \left\langle \frac{f(tx)}{\rho(t)}, (D_\times^\beta\phi)(x) \right\rangle.$$

Next, passing to the limit in the above relation, as $|t|_p \to \infty$, we obtain

$$\lim_{|t|_p \to \infty} \left\langle \frac{(D_\times^\beta f)(tx)}{|t|_p^{-|\beta|} \rho(t)}, \phi(x) \right\rangle = \lim_{|t|_p \to \infty} \left\langle \frac{f(tx)}{\rho(t)}, (D_\times^\beta \phi)(x) \right\rangle.$$

Thus $\lim_{|t|_p \to \infty} \frac{(D_\times^\beta f)(tx)}{|t|_p^{-|\beta|} \rho(t)} = D_\times^\beta g(x)$ in $\Phi'_\times(\mathbb{Q}_p^n)$ if and only if

$$\lim_{|t|_p \to \infty} \frac{f(tx)}{\rho(t)} = g(x)$$

in $\Phi'_\times(\mathbb{Q}_p^n)$. Thus this case of the theorem is proved.

2. Consider the case where among numbers $\beta_1, \ldots, \beta_n$ there are k numbers equal to -1 and $n - k$ numbers $\neq -1$. In this case the Riesz kernel $f_{-\beta}(x)$ is a *quasi-associated homogeneous* distribution of degree $|-\beta| - n$ and order k, $k = 1, \ldots, n$. Let $\beta_1 = \cdots = \beta_k = -1$, $\beta_{k+1}, \ldots, \beta_n \neq -1$. Then according to (9.2.10),

$$f_{-\beta}(ty)$$

$$= |t|_p^{|-\beta|-n} (-1)^k \frac{(p-1)^k}{\log^k p} (\log |y_1|_p + \log |t|_p)$$

$$\times \cdots \times (\log |y_k|_p + \log |t|_p) \times \frac{|y_{k+1}|_p^{-\beta_{k+1}-1}}{\Gamma_p(-\beta_{k+1})} \times \cdots \times \frac{|y_n|_p^{-\beta_n-1}}{\Gamma_p(-\beta_n)}$$

$$= |t|_p^{|-\beta|-n} f_{-\beta}(y)$$

$$+ |t|_p^{|-\beta|-n} (-1)^k \frac{(p-1)^k}{\log^k p} \frac{|y_{k+1}|_p^{-\beta_{k+1}-1}}{\Gamma_p(-\beta_{k+1})} \times \cdots \times \frac{|y_n|_p^{-\beta_n-1}}{\Gamma_p(-\beta_n)}$$

$$\times \left(\left(\log |y_2|_p \times \cdots \times \log |y_k|_p + \cdots + \log |y_1|_p \times \cdots \times \log |y_{k-1}|_p \right) \right.$$

$$\left. \times \log |t|_p + \cdots + \left(\log |y_1|_p + \cdots + \log |y_k|_p \right) \log^{k-1} |t|_p + \log^k |t|_p \right).$$

$$(12.5.3)$$

It is easy to verify that since the Lizorkin space $\Phi_\times(\mathbb{Q}_p^n)$ has the characterization (7.3.1), we have

$$\langle f_{-\beta}(ty), \phi(x+y) \rangle = |t|_p^{|-\beta|-n} \langle f_{-\beta}(y), \phi(x+y) \rangle$$

$$= |t|_p^{|-\beta|-n} (D_\times^\beta \phi)(x), \quad \phi \in \Phi_\times(\mathbb{Q}_p^n). \quad (12.5.4)$$

For example, taking (7.3.1) into account, we obtain

$$\left\langle \times_{j=2}^k \log |x_j - y_j|_p \times \times_{i=k+1}^n |x_i - y_i|_p^{-\beta_i-1}, \int_{\mathbb{Q}_p} \phi(y_1, y_2, \ldots, y_n)\, dy_1 \right\rangle = 0,$$

for all $\phi \in \Phi_\times(\mathbb{Q}_p^n)$. In a similar way, we can prove that all terms in (12.5.3), with the exception of $|t|_p^{|-\beta|-n} f_{-\beta}(y)$, *do not give any contribution to the* functional $\langle f_{-\beta}(ty), \phi(x+y)\rangle$, where $\beta_1 = \cdots = \beta_k = -1$, $\beta_{k+1}, \dots, \beta_n \neq -1$. Thus repeating the above calculations almost word for word and using (12.5.4), we prove this case of the theorem. $\qquad\square$

Theorem 12.5.4. *Let $f \in \Phi'(\mathbb{Q}_p^n)$. Then*

$$f(x) \overset{\Phi'}{\sim} g(x), \quad |x|_p \to \infty \quad (\rho(t)),$$

if and only if

$$D^\beta f(x) \overset{\Phi'}{\sim} D^\beta g(x), \quad |x|_p \to \infty \quad (|t|_p^{-\beta}\rho(t)),$$

where $\beta \in \mathbb{C}$.

Proof. Let $\beta \neq -n$, $j = 1, 2, \dots$. Since the Riesz kernel $\kappa_{-\beta}(x)$ is a *homogeneous* distribution of degree $-\beta - n$, according to Lemma 9.2.5 and formulas (9.2.17), (9.2.22) and (9.2.24), we have

$$
\begin{aligned}
\langle (D^\beta f)(tx), \phi(x)\rangle &= \langle (f * \kappa_{-\beta})(tx), \phi(x)\rangle \\
&= |t|_p^{-n}\Big\langle f(x), \Big\langle \kappa_{-\beta}(y), \phi\Big(\frac{x+y}{t}\Big)\Big\rangle\Big\rangle \\
&= |t|_p^{n}\langle f(tx), \langle \kappa_{-\beta}(ty), \phi(x+y)\rangle\rangle \\
&= |t|_p^{-\beta}\langle f(tx), \langle \kappa_{-\beta}(y), \phi(x+y)\rangle\rangle, \\
&= |t|_p^{-\beta}\langle f(tx), (D^\beta\phi)(x)\rangle,
\end{aligned}
$$

for all $\phi \in \Phi(\mathbb{Q}_p^n)$.

Passing to the limit in the above relation, as $|t|_p \to \infty$, we obtain

$$\lim_{|t|_p\to\infty}\Big\langle \frac{(D^\beta f)(tx)}{|t|_p^{-\beta}\rho(t)}, \phi(x)\Big\rangle = \lim_{|t|_p\to\infty}\Big\langle \frac{f(tx)}{\rho(t)}, (D^\beta\phi)(x)\Big\rangle.$$

Thus this case of the theorem is proved.

Let $\beta = -n$. In this case the Riesz kernel $\kappa_n(x)$ is an *associated homogeneous* distribution of degree 0 and order 1. According to (9.2.19), we have

$$\kappa_n(ty) = -\frac{1-p^{-n}}{\log p}\log|y|_p - \frac{1-p^{-n}}{\log p}\log|t|_p.$$

In view of (7.3.3),

$$
\begin{aligned}
\langle \kappa_n(ty), \phi(x+y)\rangle &= \langle \kappa_n(y), \phi(x+y)\rangle \\
&- \frac{1-p^{-n}}{\log p}\log|t|_p\langle 1, \phi(x+y)\rangle = (D^{-n}\phi)(x), \quad \phi \in \Phi(\mathbb{Q}_p^n).
\end{aligned}
$$

Thus repeating the above calculations almost word for word and using the latter relation, we prove this case of the theorem.   □

Theorem 12.5.5. *A distribution $f \in \Phi'(\mathbb{Q}_p)$ has a quasi asymptotics at infinity with respect to an automodel function ρ of degree π_α if and only if there exists a positive integer $N > -\alpha + 1$ such that*

$$\lim_{|x|_p \to \infty} \frac{D^{-N} f(x)}{|x|_p^N \rho(x)} = C \neq 0,$$

i.e., the (fractional) primitive $D^{-N} f(x)$ of order N has an asymptotics at infinity of degree $\pi_{\alpha+N}$ (understood in the usual sense), where C ia a constant.

Proof. By setting $\beta = -N$, $N > -\alpha + 1$ in Theorem 12.5.3, we obtain that relation (12.5.2) holds if and only if

$$D^{-N} f(x) \overset{\Phi'}{\sim} D^{-N} g(x) = C \begin{cases} D^{-N}\left(|x|_p^{\alpha-1} \pi_1(x)\right), & \pi_\alpha \neq \pi_0, \\ D^{-N}\left(\delta(x)\right), & \pi_\alpha = \pi_0, \end{cases} \qquad (12.5.5)$$

as $|x|_p \to \infty$ $\left(|t|_p^N \rho(t)\right)$, where $\pi_0 = |x|_p^{-1}$.

If $\pi_\alpha \neq \pi_0 = |x|_p^{-1}$, with the help of formulas (6.2.9), (6.2.10) and (9.2.8), we find that

$$D^{-N} g(x) = C \frac{|x|_p^{N-1}}{\Gamma_p(N)} * \left(|x|_p^{\alpha-1} \pi_1(x)\right)$$

$$= C \frac{\Gamma_p(\pi_\alpha)}{\Gamma_p(\pi_{\alpha+N})} |x|_p^{\alpha+N-1} \pi_1(x), \qquad (12.5.6)$$

where the Γ-functions are given by (6.2.7) and (6.2.6). If $\pi_\alpha = \pi_0 = |x|_p^{-1}$ then

$$D^{-N} g(x) = C \frac{|x|_p^{N-1}}{\Gamma_p(N)} * \delta(x) = C \frac{|x|_p^{N-1}}{\Gamma_p(N)}. \qquad (12.5.7)$$

Formulas (12.5.5), (12.5.6) and (12.5.7) imply that

$$\lim_{|t|_p \to \infty} \left\langle \frac{(D^{-N} f)(tx)}{|t|_p^N \rho(t)}, \phi(x) \right\rangle = C \frac{\Gamma_p(\pi_\alpha)}{\Gamma_p(\pi_{\alpha+N})} \left\langle |x|_p^{\alpha+N-1} \pi_1(x), \phi(x) \right\rangle,$$

for all $\phi \in \Phi(\mathbb{Q}_p)$. Since $\alpha + N - 1 > 0$, we have

$$\lim_{|t|_p \to \infty} \frac{(D^{-N} f)(tx)}{|t|_p^N \rho(t)} = C \frac{\Gamma_p(\pi_\alpha)}{\Gamma_p(\pi_{\alpha+N})} |x|_p^{\alpha+N-1} \pi_1(x). \qquad (12.5.8)$$

By using Definition 12.2 and formula (12.3.1), relation (12.5.8) can be rewritten in the form

$$A = C \frac{\Gamma_p(\pi_\alpha)}{\Gamma_p(\pi_{\alpha+N})} = \lim_{|t|_p \to \infty} \frac{(D^{-N} f)(tx)}{|t|_p^N \rho(t) |x|_p^{\alpha+N-1} \pi_1(x)}$$

$$= \lim_{|tx|_p \to \infty} \frac{(D^{-N} f)(tx)}{|tx|_p^N \rho(tx)} \lim_{|t|_p \to \infty} \frac{\rho(tx)}{|x|_p^{\alpha-1} \pi_1(x)\rho(t)}$$

$$= \lim_{|y|_p \to \infty} \frac{(D^{-N} f)(y)}{|y|_p^N \rho(y)}.$$

$\qquad\qquad\qquad\qquad\qquad\qquad\qquad\qquad\qquad\qquad\qquad\qquad\qquad\qquad$ □

Theorem 12.5.6. *Let* $\mathcal{A}(\xi) \in \mathcal{E}(\mathbb{Q}_p^n \setminus \{0\})$ *be the symbol of a* homogeneous *pseudo-differential operator A of degree π_β, which is given by (9.3.1) and $f \in \Phi'(\mathbb{Q}_p^n)$. Then*

$$f(x) \overset{\Phi'}{\sim} g(x), \quad |x|_p \to \infty \quad \big(\rho(t)\big)$$

if and only if

$$(Af)(x) \overset{\Phi'}{\sim} (Ag)(x), \quad |x|_p \to \infty \quad \big(\pi_\beta^{-1}(t)\rho(t)\big).$$

Proof. According to Lemma 9.3.1, the Lizorkin space $\Phi(\mathbb{Q}_p^n)$ is invariant under the pseudo-differential operator A. Consequently, due to formulas (9.3.5), (9.3.2), and (4.9.3), (6.2.1), we have

$$\langle (Af)(tx), \phi(x) \rangle = |t|_p^{-n} \Big\langle f(x), A^T \phi\Big(\frac{x}{t}\Big) \Big\rangle$$

$$= |t|_p^{-n} \Big\langle f(x), F^{-1}\Big[\mathcal{A}(-\xi) F[\phi\Big(\frac{x}{t}\Big)](\xi)\Big](x) \Big\rangle$$

$$= \frac{1}{\pi_\beta(t)} \big\langle f(x), F^{-1}\big[\mathcal{A}(-t\xi) F[\phi(x)](t\xi)\big](x) \big\rangle$$

$$= \frac{|t|_p^{-n}}{\pi_\beta(t)} \Big\langle f(x), F^{-1}\big[\mathcal{A}(-\xi) F[\phi(x)](\xi)\big]\Big(\frac{x}{t}\Big) \Big\rangle$$

$$= \frac{1}{\pi_\beta(t)} \big\langle f(tx), F^{-1}\big[\mathcal{A}(-\xi) F[\phi(x)](\xi)\big](x) \big\rangle,$$

for all $\phi \in \Phi(\mathbb{Q}_p^n)$.

Passing to the limit in the above relation, as $|t|_p \to \infty$, we obtain

$$\lim_{|t|_p \to \infty} \left\langle \frac{(Af)(tx)}{\pi_\beta^{-1}(t)\rho(t)}, \phi(x) \right\rangle = \lim_{|t|_p \to \infty} \left\langle \frac{f(tx)}{\rho(t)}, \left(A^T \phi\right)(x) \right\rangle,$$

i.e., in view of (9.3.2),

$$\lim_{|t|_p \to \infty} \frac{(Af)(tx)}{\pi_\beta^{-1}(t)\rho(t)} = Ag(x) \qquad \text{in} \quad \Phi'(\mathbb{Q}_p^n)$$

if and only if $\lim_{|t|_p \to \infty} \frac{f(tx)}{\rho(t)} = g(x)$ in $\Phi'(\mathbb{Q}_p^n)$. Thus the theorem is proved. $\quad\square$

13

Asymptotics of the *p*-adic singular Fourier integrals

13.1. Introduction

We recall that in the usual real setting one studies the important class of *oscillatory* (sometimes also called *oscillating*) integrals

$$\int_{\mathbb{R}^n} e^{itf(x)} \varphi(x) \, d^n x.$$

Such integrals often arise in applied and mathematical physics. The classical problem related to *oscillatory integrals* is to investigate their asymptotic behavior when the parameter t tends to infinity [42], [92], [106, 7.8]. In the *p*-adic setting oscillatory integrals are studied in [107], [104], [258].

The *Fourier integrals* are particular cases of *oscillatory integrals*. There are many problems where solutions are obtained as Fourier integrals which cannot be evaluated exactly. Nevertheless, these solutions are no less important because it is often possible to study the asymptotic behavior of such integrals [109, 9]. The problem of the asymptotic behavior of *singular Fourier integrals* is related to the well-known Erdélyi lemma (see [84], [85]). In the one-dimensional case this lemma describes the asymptotics of the Fourier transform of functions $f(x)$ defined on $\mathbb{R}$ and having singularities of the type

$$x_{\pm}^{\alpha-1} \log^m x_{\pm} \varphi(x), \quad m = 0, 1, 2, \ldots, \quad \mathrm{Re}\,\alpha > 0, \quad x \in \mathbb{R},$$

where $\varphi(x)$ is sufficiently smooth [92, Ch.III, §1]. Here $x_{\pm}^{\alpha-1} \log^m x_{\pm}$ belongs to the class of *quasi-associated homogeneous distributions* from $\mathcal{D}'(\mathbb{R})$ described in Chapter 6. There are multidimensional generalizations of this lemma [92, Ch.III], [42, Ch.II, §7], [252].

The latter problem has not been studied so far for the *p*-adic case. In this chapter we shall perform such a study. These results are based on [145]. In Section 13.2, for comparison, results on asymptotics of singular

Fourier integrals for the real case are quoted. Next, in Sections 13.4 and 13.5, we prove Theorems 13.4.1, 13.4.2 and 13.5.1 which describe the asymptotic behavior of the simplest p-adic singular Fourier integrals

$$J_{\pi_\alpha, m; \varphi}(t) = \langle f_{\pi_\alpha; m}(x) \chi_p(xt), \varphi(x) \rangle, \quad |t|_p \to \infty, \qquad (13.1.1)$$

where $f_{\pi_\alpha; m}(x) \in \mathcal{D}'(\mathbb{Q}_p)$ is a *quasi-associated homogeneous distribution* of degree $\pi_\alpha(x) = |x|_p^{\alpha-1} \pi_1(x)$ and order m (the class of p-*adic quasi-associated homogeneous distributions* from $\mathcal{D}'(\mathbb{Q}_p)$ was studied above in Chapter 6), $\pi_\alpha(x)$, $\pi_1(x)$, and $\chi_p(x)$ are multiplicative, resp. normed multiplicative, resp. additive characters of the field $\mathbb{Q}_p$ of p-adic numbers. $\varphi(x) \in \mathcal{D}(\mathbb{Q}_p)$ is a test function, $m = 0, 1, 2, \ldots, \alpha \in \mathbb{C}$. In Section 13.6, by Corollary 13.6.1 a p-adic version of the well-known Erdélyi lemma is given. This lemma is a direct consequence of Theorems 13.4.1 and 13.5.1 for the case Re $\alpha > 0$. To prove Theorems 13.4.1, 13.4.2 and 13.5.1, a new technique of constructing p-adic weak asymptotics is developed.

The asymptotic relations (13.4.1)–(13.4.4) are p-adic analogs of relations (13.2.1), (13.2.2) for the corresponding real case. However, the p-adic asymptotic relations (13.4.1)–(13.4.4), (13.4.15)–(13.4.18), and (13.5.1)–(13.5.4), in contrast to the real case (13.2.1), (13.2.2), have a *specific stabilization property*. Namely, these relations are *exact equalities* for sufficiently big $|t|_p > s(\varphi)$, where $s(\varphi)$ is the stabilization parameter (see Definition 13.3 and Remark 13.1). The stabilization parameter $s(\varphi)$ depends on the *parameter of constancy* of the function φ (see (4.3.1)) and the rank of the character $\pi_1(x)$ (see (2.3.14)). Asymptotics of this type will be called *stable* asymptotical expansions (see Definitions 13.1–13.3 below). This p-adic phenomenon is quite different from asymptotic properties in the real setting. This unexpected property of *stabilization* of the p-adic asymptotics also arises in another context (see Theorem 14.6.1 in Chapter 14 and Propositions C.1–C.9 in Appendix C), where some weak asymptotics are calculated.

The *asymptotic stabilization property* is similar to another p-adic phenomenon which is described by Proposition 2.2.1: *if* $\lim_{n \to \infty} x_n = x$, $x_n, x \in \mathbb{Q}_p$, $|x|_p \neq 0$, *then the sequence of norms* $\{ |x_n|_p : n \in \mathbb{N} \}$ *must stabilize for sufficiently large* n. It may well be that stabilization is a *typical property of* p-*adic asymptotics*.

Theorems which give asymptotic expansions of singular Fourier integrals are the abelian-type theorems (see [240]). Indeed, according to (13.1.1), we study the asymptotic behavior of the singular Fourier integrals

$$J_{\pi_\alpha, m; \varphi}(t) = F[g(x)](t),$$

where the functions

$$g(x) = |x|_p^{\alpha-1} \log_p^m |x|_p \pi_1(x)\varphi(x)$$

admit the estimate $g(x) = O(|x|_p^{\alpha-1} \log_p^m |x|_p)$, $|x|_p \to 0$. If $\alpha \neq 0$, according to Theorems 13.4.1 and 13.5.1, we have

$$J_{\pi_\alpha,m;\varphi}(t) = O\left(|t|_p^{-\alpha} \log_p^m |t|_p\right), \quad |t|_p \to \infty.$$

This statement on the connection between the asymptotic behavior of $g(x)$ and $J_{\pi_\alpha,m;\varphi}(t)$ is a typical abelian-type theorem.

Since the asymptotic relations for the Fourier transform of associated homogeneous distributions from $\mathcal{D}'(\mathbb{R})$ have many applications (see, for example, [61], [226], [252]), we hope that their p-adic versions may also be useful in p-adic mathematical physics.

13.2. Asymptotics of singular Fourier integrals for the real case

As was mentioned in Section 13.1, asymptotic relations for the Fourier transform of distributions from $\mathcal{D}'(\mathbb{R})$,

$$(x \pm i0)^{\alpha-1} \log^m(x \pm i0)\varphi(x), \quad m = 0, 1, 2 \ldots, \quad \alpha \in \mathbb{C},$$

where $\varphi \in \mathcal{D}(\mathbb{R})$ and $(x \pm i0)^{\alpha-1} \log^m(x \pm i0)$ are *associated homogeneous distributions* of degree $\alpha - 1$ and order m, have many applications (see, for example, [61], [226], [252]). These asymptotics were constructed in [60], [61], [58] (see also [92, Ch.III,§1.6.,§8]). In particular, in these papers the following asymptotic relations were derived:

$$(x \pm i0)^{\alpha-1} \log^m(x \pm i0)e^{itx}$$

$$\approx \delta(x)2\pi \sum_{k=0}^{m} \frac{d^{m-k}}{d\alpha^{m-k}} \left(\frac{e^{\pm i \frac{\pi(\alpha-1)}{2}}}{\Gamma(-\alpha+1)}\right) \frac{\log^k |t|}{|t|^\alpha}, \quad t \to \mp\infty, \quad (13.2.1)$$

$$(x \pm i0)^{\alpha-1} \log^m(x \pm i0)e^{itx} = o(|t|^{-N}), \quad t \to \pm\infty, \quad (13.2.2)$$

for any $N \in \mathbb{N}$, where $\alpha \notin \mathbb{N}$. Some particular cases of these formulas were studied in [109, 9]. In [200, Appendix 0] (see also [109, 9.1.]), the Riemann–Lebesgue type lemma for the case of distributions was proved.

13.3. *p*-adic distributional asymptotic expansions

Let us introduce a definition of a distributional asymptotics [59] adapted to the case of $\mathbb{Q}_p^n$.

Definition 13.1. A sequence $\psi_k(t)$ of continuous complex-valued functions on the multiplicative group $\mathbb{Q}_p^\times$ is called an *asymptotic sequence, as* $|t|_p \to \infty$, if

$$\psi_{k+1}(t) = o(\psi_k(t)), \qquad |t|_p \to \infty$$

for all $k = 1, 2, \ldots$.

Definition 13.2. Let $f(\cdot, t) \in \mathcal{D}'(\mathbb{Q}_p^n)$ be a distribution depending on t as a parameter, and $C_k \in \mathcal{D}'(\mathbb{Q}_p^n)$ be distributions, $k = 1, 2, \ldots$. We say that the relation

$$f(x, t) \approx \sum_{k=1}^{\infty} C_k(x)\psi_k(t), \quad |t|_p \to \infty, \quad x \in \mathbb{Q}_p^n, \qquad (13.3.1)$$

is an *asymptotic expansion of the distribution* $f(x, t)$, as $|t|_p \to \infty$, with respect to an asymptotic sequence $\{\psi_k(t)\}$ if

$$\langle f(\cdot, t), \varphi(\cdot) \rangle \approx \sum_{k=1}^{\infty} \langle C_k(\cdot), \varphi(\cdot) \rangle \psi_k(t), \quad |t|_p \to \infty, \qquad \forall \varphi \in \mathcal{D}(\mathbb{Q}_p^n).$$

$$(13.3.2)$$

This means that for any test function $\varphi \in \mathcal{D}(\mathbb{Q}_p^n)$ and for any N we have

$$\langle f(\cdot, t), \varphi(\cdot) \rangle - \sum_{k=1}^{N} \langle C_k(\cdot), \varphi(\cdot) \rangle \psi_k(t) = o(\psi_N(t)), \quad |t|_p \to \infty.$$

Definition 13.3. Suppose that a distribution $f(\cdot, t) \in \mathcal{D}'(\mathbb{Q}_p^n)$ depending on t as a parameter has the asymptotic expansion (13.3.1). If for any test function $\varphi \in \mathcal{D}(\mathbb{Q}_p^n)$ there exists a number $s(\varphi)$ depending on φ such that for all $|t|_p > s(\varphi)$ the relation (13.3.2) is an *exact equality*, we say that the *asymptotic expansion* (13.3.1) is *stable* and write

$$f(x, t) = \sum_{k=1}^{\infty} C_k(x)\psi_k(t), \quad |t|_p \to \infty. \qquad (13.3.3)$$

The number $s(\varphi)$ is called the *stabilization parameter of the asymptotic expansion* (13.3.1).

13.4. Asymptotics of singular Fourier integrals $(\pi_1(x) \equiv 1)$

13.4.1. The case $f_{\pi_\alpha;m}(x) = |x|_p^{\alpha-1} \log_p^m |x|_p$, $\alpha \neq 0, m = 0, 1, 2, \ldots$

Theorem 13.4.1. *Let* $\varphi \in \mathcal{D}_N^l(\mathbb{Q}_p)$. *Then the functional* $J_{\pi_\alpha, m; \varphi}(t)$ *has the following asymptotic behavior:*
(a) *If* $\alpha \neq 0$, $m = 0$ *then*

$$J_{\pi_\alpha, 0; \varphi}(t) = \left\langle |x|_p^{\alpha-1} \chi_p(xt), \varphi(x) \right\rangle$$
$$= \varphi(0) \frac{\Gamma_p(\alpha)}{|t|_p^\alpha}, \quad |t|_p > p^{-l}, \qquad (13.4.1)$$

where the Γ-*function* $\Gamma_p(\alpha)$ *is given by* (6.2.6), *i.e., in the weak sense*

$$|x|_p^{\alpha-1} \chi_p(xt) = \delta(x) \frac{\Gamma_p(\alpha)}{|t|_p^\alpha}, \quad |t|_p \to \infty. \qquad (13.4.2)$$

(b) *If* $\alpha \neq 0$, $m = 1, 2, \ldots$ *then*

$$J_{\pi_\alpha, m; \varphi}(t) = \left\langle |x|_p^{\alpha-1} \log_p^m |x|_p \chi_p(xt), \varphi(x) \right\rangle$$
$$= \varphi(0) \sum_{k=0}^{m} C_m^k \log_p^k e \frac{d^k \Gamma_p(\alpha)}{d\alpha^k} \frac{\log_p^{m-k} |t|_p}{|t|_p^\alpha}, \quad |t|_p > p^{-l}, \ (13.4.3)$$

i.e., in the weak sense

$$|x|_p^{\alpha-1} \log_p^m |x|_p \chi_p(xt)$$
$$= \delta(x) \sum_{k=0}^{m} C_m^k \log_p^k e \frac{d^k \Gamma_p(\alpha)}{d\alpha^k} \frac{\log_p^{m-k} |t|_p}{|t|_p^\alpha}, \quad |t|_p \to \infty, \quad (13.4.4)$$

with respect to the asymptotic sequence

$$\{|t|_p^{-\alpha} \log_p^{m-k} |t|_p : k = 0, 1, \ldots, m\}.$$

Thus for any $\varphi \in \mathcal{D}(\mathbb{Q}_p)$, *the relations* (13.4.1) *and* (13.4.3) *are exact equalities for sufficiently large* $|t|_p > p^{-l}$, *i.e., these asymptotic expansions are stable with the stabilization parameter* $s(\varphi) = p^{-l}$.

Proof. 1. Let Re $\alpha > 0$. In this case $|x|_p^{\alpha-1} \log_p^m |x|_p \varphi(x) \in \mathcal{L}^1(\mathbb{Q}_p)$, and the integral

$$J_{\pi_\alpha, m; \varphi}(t) = \left\langle |x|_p^{\alpha-1} \log_p^m |x|_p \chi_p(xt), \varphi(x) \right\rangle$$
$$= \int_{\mathbb{Q}_p} |x|_p^{\alpha-1} \log_p^m |x|_p \chi_p(xt) \varphi(x) \, dx$$

converges absolutely. Hence, according to the Riemann–Lebesgue Theorem 5.2.1, $J_{\pi_\alpha,m;\varphi}(t) \to 0$, as $|t|_p \to \infty$. More precisely, since $\varphi(x) \in \mathcal{D}_N^l(\mathbb{Q}_p)$ then, in view of Lemmas 12.4.1 and 12.4.2,

$$J_{\pi_\alpha,m;\varphi}(t) = \frac{\varphi(0)}{|t|_p^\alpha} \sum_{k=0}^m C_m^k \log_p^k e \frac{d^k \Gamma_p(\alpha)}{d\alpha^k} \log_p^{m-k} |t|_p, \quad \forall |t|_p > p^{-l}.$$

(13.4.5)

Thus relations (13.4.1)–(13.4.4) hold.

2. Let Re $\alpha < 0$. In this case we define the functional $J_{\pi_\alpha,m;\varphi}(t)$ by an analytic continuation with respect to α. According to (6.2.2), (6.2.3), (6.3.2) and (6.3.3):

$$
\begin{aligned}
J_{\pi_\alpha,m;\varphi}(t) &= \left\langle |x|_p^{\alpha-1} \log_p^m |x|_p \chi_p(xt), \varphi(x) \right\rangle \\
&= \int_{B_0} |x|_p^{\alpha-1} \log_p^m |x|_p \chi_p(xt) \big(\varphi(x) - \varphi(0)\big) \, dx \\
&\quad + \int_{\mathbb{Q}_p \setminus B_0} |x|_p^{\alpha-1} \log_p^m |x|_p \chi_p(xt) \varphi(x) \, dx \\
&\quad + \varphi(0) \int_{B_0} |x|_p^{\alpha-1} \log_p^m |x|_p \chi_p(xt) \, dx,
\end{aligned}
$$

(13.4.6)

for all $\varphi \in \mathcal{D}(\mathbb{Q}_p)$, where the latter integral in (13.4.6) is defined by means of an analytic continuation with respect to α.

Since $\varphi \in \mathcal{D}_N^l(\mathbb{Q}_p)$, it is *natural* to rewrite the functional (13.4.6) as the following sum:

$$J_{\pi_\alpha,m;\varphi}(t) = J_{\pi_\alpha,m;\varphi}^1(t) + J_{\pi_\alpha,m;\varphi}^2(t) + \varphi(0) J_{\pi_\alpha,m}^0(t),$$

(13.4.7)

where

$$J_{\pi_\alpha,m;\varphi}^1(t) = \int_{B_l} |x|_p^{\alpha-1} \log_p^m |x|_p \chi_p(xt) \big(\varphi(x) - \varphi(0)\big) \, dx, \quad (13.4.8)$$

$$J_{\pi_\alpha,m;\varphi}^2(t) = \int_{\mathbb{Q}_p \setminus B_l} |x|_p^{\alpha-1} \log_p^m |x|_p \chi_p(xt) \varphi(x) \, dx, \quad (13.4.9)$$

$$J_{\pi_\alpha,m}^0(t) = \int_{B_l} |x|_p^{\alpha-1} \log_p^m |x|_p \chi_p(xt) \, dx. \quad (13.4.10)$$

The integral (13.4.10) is defined by means of an analytic continuation with respect to α.

For Re $\alpha > 0$ and $m = 0$, according to (12.4.5), the integral (13.4.10) is equal to

$$J^0_{\pi_\alpha,0}(t) = F\big[|x|_p^{\alpha-1}\Delta_l(x)\big](t) = \int_{B_l} \chi_p(tx)|x|_p^{\alpha-1}\,dx$$

$$= \frac{1-p^{-1}}{1-p^{-\alpha}} p^{\alpha l}\Delta_{-l}(t) + \frac{\Gamma_p(\alpha)}{|t|_p^\alpha}\big(1 - \Delta_{-l}(t)\big). \quad (13.4.11)$$

For any $\alpha \neq \alpha_j = \frac{2\pi i}{\ln p}j$, $j \in \mathbb{Z}$, we define $J^0_{\pi_\alpha,0}(t)$ by means of an analytic continuation with respect to α.

Differentiating relation (13.4.11) with respect to α, we obtain

$$J^0_{\pi_\alpha,m}(t) = F\big[|x|_p^{\alpha-1}\log_p^m|x|_p\Delta_l(x)\big](t)$$

$$= \int_{B_l} \chi_p(tx)|x|_p^{\alpha-1}\log_p^m|x|_p\,dx = \log_p^m e \frac{d^m}{d\alpha^m}J^0_{\pi_\alpha,0}(t)$$

$$= \Delta_{-l}(t)\big(1-p^{-1}\big)\frac{d^m}{d\alpha^m}\Big(\frac{p^{\alpha l}}{1-p^{-\alpha}}\Big)\log_p^m e$$

$$+ \big(1-\Delta_{-l}(t)\big)\frac{1}{|t|_p^\alpha}\sum_{k=0}^m C_m^k \log_p^k e \frac{d^k\Gamma_p(\alpha)}{d\alpha^k}\log_p^{m-k}|t|_p. \quad (13.4.12)$$

We point out that by using formulas (3.3.1)–(3.3.4), the relation (13.4.12) can be derived explicitly.

According to (13.4.12), we have

$$J^0_{\pi_\alpha,m}(t) = \frac{1}{|t|_p^\alpha}\sum_{k=0}^m C_m^k \frac{d^k\Gamma_p(\alpha)}{d\alpha^k}\log_p^{m-k}|t|_p, \quad |t|_p > p^{-l}. \quad (13.4.13)$$

Since $\varphi \in \mathcal{D}_N^l(\mathbb{Q}_p)$, it is clear that the following relations hold:

$$|x|_p^{\alpha-1}\log_p^m|x|_p\big(\varphi(x)-\varphi(0)\big)\Delta_l(x) = 0,$$

$$|x|_p^{\alpha-1}\log_p^m|x|_p\varphi(x)\big(1-\Delta_l(x)\big) \in \mathcal{D}_N^l(\mathbb{Q}_p).$$

Thus taking Fourier transforms, according to (4.8.5), we have

$$J^1_{\pi_\alpha,m;\varphi}(t) = \int_{B_l}|x|_p^{\alpha-1}\log_p^m|x|_p\chi_p(xt)\big(\varphi(x)-\varphi(0)\big)\,dx = 0,$$

$$J^2_{\pi_\alpha,m;\varphi}(t) = \int_{\mathbb{Q}_p\setminus B_l}|x|_p^{\alpha-1}\log_p^m|x|_p\chi_p(xt)\varphi(x)\,dx = 0,$$

(13.4.14)

for all $|t|_p > p^{-l}$. Thus for Re $\alpha < 0$ relations (13.4.7), (13.4.14) and (13.4.13) imply (13.4.1)–(13.4.4). $\qquad\square$

13.4.2. The case $f_{\pi_0;m}(x) = P\Big(\dfrac{\log_p^m |x|_p}{|x|_p}\Big)$, $m = 0, 1, 2, \ldots$

Theorem 13.4.2. (a) *If* $\alpha = 0$, $m = 0$ *then*

$$
\begin{aligned}
J_{\pi_0,0;\varphi}(t) &= \Big\langle P\Big(\frac{1}{|x|_p}\Big)\chi_p(xt),\, \varphi(x)\Big\rangle \\
&= \varphi(0)\Big(-\frac{1}{p} - \Big(1 - \frac{1}{p}\Big)\log_p\Big(\frac{|t|_p}{p^{-l}}\Big)\Big), \quad |t|_p > p^{-l}, \quad (13.4.15)
\end{aligned}
$$

i.e., in the weak sense

$$
P\Big(\frac{1}{|x|_p}\Big)\chi_p(xt) = \delta(x)\Big(-\frac{1}{p} - \Big(1 - \frac{1}{p}\Big)\log_p\Big(\frac{|t|_p}{p^{-l}}\Big)\Big), \quad |t|_p \to \infty.
$$
$$(13.4.16)$$

(b) *If* $\alpha = 0$, $m = 1, 2, \ldots$ *then*

$$
\begin{aligned}
J_{\pi_0,m;\varphi}(t) &= \Big\langle P\Big(\frac{\log_p^m |x|_p}{|x|_p}\Big)\chi_p(xt),\, \varphi(x)\Big\rangle \\
&= \varphi(0)\Big\{ \frac{1}{p}(-1)^{m+1}(\log_p |t|_p - 1)^m \\
&\quad + \Big(1 - \frac{1}{p}\Big)\frac{1}{m+1}\big((-1)^{m+1}\big(\log_p^{m+1}|t|_p - \log_p^{m+1} p^{-l}\big)\big) \\
&\quad - (-1)^m C_{m+1}^1 \mathbf{B}_1\big(\log_p^m |t|_p - \log_p^m p^{-l}\big) \\
&\quad + \sum_{r=2}^{m}(-1)^{m+1-r} C_{m+1}^r \mathbf{B}_r \\
&\quad\quad \times \big(\log_p^{m+1-r}|t|_p - \log_p^{m+1-r} p^{-l}\big)\Big\}, \quad |t|_p > p^{-l},
\end{aligned}
$$
$$(13.4.17)$$

where the Bernoulli numbers $\mathbf{B}_r$, $r = 0, 1, \ldots, m$, *are defined by the relation* (B.1), *i.e., in the weak sense,*

$$
\begin{aligned}
P\Big(&\frac{\log_p^m |x|_p}{|x|_p}\Big)\chi_p(xt) \\
&= \delta(x)\Big\{ -\frac{1}{p}\big(-\log_p |t|_p + 1\big)^m \\
&\quad + \Big(1 - \frac{1}{p}\Big)\frac{1}{m+1}\big((-1)^{m+1}\big(\log_p^{m+1}|t|_p - \log_p^{m+1} p^{-l}\big)\big) \\
&\quad - (-1)^m C_{m+1}^1 \mathbf{B}_1\big(\log_p^m |t|_p - \log_p^m p^{-l}\big)
\end{aligned}
$$

$$+ \sum_{r=2}^{m} (-1)^{m+1-r} C_{m+1}^r \mathbf{B}_r$$

$$\times \left(\log_p^{m+1-r} |t|_p - \log_p^{m+1-r} p^{-l} \right) \bigg) \bigg\}, \quad |t|_p \to \infty, \quad (13.4.18)$$

with respect to the asymptotic sequence

$$\{ \log_p^{m+1-k} |t|_p : k = 0, 1, \ldots, m+1 \}.$$

Thus for any $\varphi \in \mathcal{D}(\mathbb{Q}_p)$, relations (13.4.15) and (13.4.17) are exact equalities for sufficiently large $|t|_p > p^{-l}$, i.e., these asymptotic expansions are stable with the stabilization parameter $s(\varphi) = p^{-l}$.

Proof. Let $\alpha = 0$, and $m = 0, 1, 2, \ldots$. According to (6.3.5), we have

$$J_{\pi_0,m;\varphi}(t) = \left\langle P\left(\frac{\log_p^m |x|_p}{|x|_p} \right) \chi_p(xt), \varphi(x) \right\rangle$$

$$= \int_{B_0} \frac{\log_p^m |x|_p}{|x|_p} \chi_p(xt) \big(\varphi(x) - \varphi(0) \big)\, dx$$

$$+ \int_{\mathbb{Q}_p \setminus B_0} \frac{\log_p^m |x|_p}{|x|_p} \chi_p(xt) \varphi(x)\, dx, \quad (13.4.19)$$

for all $\varphi \in \mathcal{D}(\mathbb{Q}_p)$.

Since $\varphi \in \mathcal{D}_N^l(\mathbb{Q}_p)$, it is *natural* to rewrite the functional $J_{\pi_0,m;\varphi}(t)$ as a sum of integrals:

$$J_{\pi_0,m;\varphi}(t) = J_{\pi_0,m;\varphi}^1(t) + J_{\pi_0,m;\varphi}^2(t) + \varphi(0) J_{\pi_0,m}^0(t), \quad (13.4.20)$$

where

$$J_{\pi_0,m;\varphi}^1(t) = \int_{B_l} \frac{\log_p^m |x|_p}{|x|_p} \chi_p(xt) \big(\varphi(x) - \varphi(0) \big)\, dx, \quad (13.4.21)$$

$$J_{\pi_0,m;\varphi}^2(t) = \int_{\mathbb{Q}_p \setminus B_l} \frac{\log_p^m |x|_p}{|x|_p} \chi_p(xt) \varphi(x)\, dx, \quad (13.4.22)$$

$$J_{\pi_0,m}^0(t) = \int_{B_l} \frac{\log_p^m |x|_p}{|x|_p} \big(\chi_p(xt) - 1 \big)\, dx. \quad (13.4.23)$$

Since $\varphi \in \mathcal{D}_N^l(\mathbb{Q}_p)$, it is clear that

$$\frac{\log_p^m |x|_p}{|x|_p} \big(\varphi(x) - \varphi(0) \big) \Delta_l(x) = 0,$$

$$\frac{\log_p^m |x|_p}{|x|_p} \varphi(x) \big(1 - \Delta_l(x) \big) \in \mathcal{D}_N^l(\mathbb{Q}_p).$$

Thus as above, according to (4.8.5), for (13.4.21) and (13.4.22) we have

$$J^1_{\pi_0,m;\varphi}(t) = J^2_{\pi_0,m;\varphi}(t) = 0, \quad \forall \, |t|_p > p^{-l}. \qquad (13.4.24)$$

Let us calculate the integral (13.4.23). Suppose that $|t|_p = p^M$, $M > -l$.
We start with the case $m = 0$. Taking into account that $-M + 1 \le l$, according to formulas (3.3.1)–(3.3.4), we have

$$
\begin{aligned}
J^0_{\pi_0,0}(t) &= \int_{B_l} \frac{\chi_p(xt) - 1}{|x|_p} \, dx = \sum_{\gamma=-\infty}^{l} p^{-\gamma} \int_{S_\gamma} (\chi_p(xt) - 1) \, dx \\
&= -p^{-(-M+1)} p^{-M+1-1} - \sum_{\gamma=-M+1}^{l} p^{-\gamma} \left(1 - \frac{1}{p}\right) p^\gamma \\
&= -\frac{1}{p} - \left(1 - \frac{1}{p}\right)(l + M) = -\frac{1}{p} - \left(1 - \frac{1}{p}\right) \log_p \left(\frac{|t|_p}{p^{-l}}\right).
\end{aligned}
$$

$$(13.4.25)$$

The relations (13.4.24) and (13.4.25) imply that

$$J_{\pi_0,0;\varphi}(t) = \varphi(0)\left(-\frac{1}{p} - \left(1 - \frac{1}{p}\right) \log_p \left(\frac{|t|_p}{p^{-l}}\right)\right), \quad |t|_p > p^{-l}. \quad (13.4.26)$$

Note that the latter relation can also be proved if we use the representation of the functional (13.4.19) in the form of the convolution (4.7.1) $J_{\pi_0,m;\varphi}(t) = F\big[P\big(\frac{1}{|x|_p}\big)\big](t) * F[\varphi(x)](t)$, and the formula form of Example 6.4.1:

$$F\big[(1 - p^{-1}) \log_p |x|_p\big](t) = -P\Big(\frac{1}{|t|_p}\Big) - p^{-1}\delta(t).$$

In the case $m = 1, 2, \ldots$, for $|t|_p = p^M$, $M > -l$, using formulas (3.3.1)–(3.3.4), we obtain

$$
\begin{aligned}
J^0_{\pi_0,m}(t) &= \int_{B_l} \frac{\log_p^m |x|_p}{|x|_p} (\chi_p(xt) - 1) \, dx = \sum_{\gamma=-\infty}^{l} p^{-\gamma} \gamma^m \int_{S_\gamma} (\chi_p(xt) - 1) \, dx \\
&= -p^{-(-M+1)}(-M + 1)^m p^{-M+1-1} - \sum_{\gamma=-M+1}^{l} p^{-\gamma} \gamma^m \left(1 - \frac{1}{p}\right) p^\gamma \\
&= -\frac{1}{p}(-M + 1)^m - \left(1 - \frac{1}{p}\right) \sum_{\gamma=-M+1}^{l} \gamma^m.
\end{aligned}
$$

$$(13.4.27)$$

Next, using equality (B.2) from Appendix B and formula (B.3), the relation (13.4.27) can easily be transformed into the form

$$
J^0_{\pi_0,m}(t) = -\frac{1}{p}(-M+1)^m - \left(1 - \frac{1}{p}\right)(S_m(l) - S_m(-M))
$$

$$
= -\frac{1}{p}(-M+1)^m + \left(1 - \frac{1}{p}\right)\frac{1}{m+1}\left((-M)^{m+1} - l^{m+1}\right.
$$

$$
- C^1_{m+1}\mathbf{B}_1\left((-M)^m - l^m\right) + \sum_{r=2}^{m} C^r_{m+1}\mathbf{B}_r\left((-M)^{m+1-r} - l^{m+1-r}\right)\bigg)
$$

$$
= -\frac{1}{p}(-1)^m(\log_p |t|_p - 1)^m
$$

$$
+ \left(1 - \frac{1}{p}\right)\frac{1}{m+1}\left((-1)^{m+1}\left(\log_p^{m+1} |t|_p - \log_p^{m+1} p^{-l}\right)\right.
$$

$$
- (-1)^m C^1_{m+1}\mathbf{B}_1\left(\log_p^m |t|_p - \log_p^m p^{-l}\right)
$$

$$
+ \sum_{r=2}^{m}(-1)^{m+1-r} C^r_{m+1}\mathbf{B}_r\left(\log_p^{m+1-r} |t|_p - \log_p^{m+1-r} p^{-l}\right)\bigg),
$$

$$
(13.4.28)
$$

where the Bernoulli numbers $\mathbf{B}_r, r = 0, 1, \dots, m$ are defined by formula (B.1), and the polynomial $\mathbf{S}_m(\gamma_0)$ is given by formula (B.2) from Appendix B.

The relations (13.4.20), (13.4.24) and (13.4.28) imply

$$
J_{\pi_0,m;\varphi}(t) = \varphi(0)\bigg\{\frac{1}{p}(-1)^{m+1}(\log_p |t|_p - 1)^m
$$

$$
+ \left(1 - \frac{1}{p}\right)\frac{1}{m+1}\left((-1)^{m+1}\left(\log_p^{m+1} |t|_p - \log_p^{m+1} p^{-l}\right)\right.
$$

$$
- (-1)^m C^1_{m+1}\mathbf{B}_1\left(\log_p^m |t|_p - \log_p^m p^{-l}\right)
$$

$$
+ \sum_{r=2}^{m}(-1)^{m+1-r} C^r_{m+1}\mathbf{B}_r\left(\log_p^{m+1-r} |t|_p - \log_p^{m+1-r} p^{-l}\right)\bigg)\bigg\},
$$

$$
(13.4.29)
$$

for all $|t|_p > p^{-l}$.

Thus the relations (13.4.17) and (13.4.18) hold. $\square$

The following corollary is immediate:

Corollary 13.4.3. *For $\alpha = 1$, since the Fourier transform is an isomorphism on $\mathcal{D}(\mathbb{Q}_p)$, the formulas (13.4.1) and (6.2.6) imply the statement (4.8.5) of Lemma 4.8.3.*

Indeed, in view of (6.2.6), $\Gamma_p(1) = 0$, and consequently for $\varphi \in \mathcal{D}_N^l(\mathbb{Q}_p)$ we have $F[\varphi] = 0$ for $|t|_p > p^{-l}$.

Remark 13.1. The asymptotic expansion (13.4.4) can be represented in the form

$$|x|_p^{\alpha-1} \log_p^m |x|_p \chi_p(xt)$$

$$= \delta(x) \frac{\log_p^m |t|_p}{|t|_p^\alpha} \left(\sum_{k=0}^N A_k(\alpha) \log_p^{-k} |t|_p + o\left(\log_p^{-N} |t|_p \right) \right), \quad |t|_p \to \infty,$$

$$(13.4.30)$$

where $A_k(\alpha)$ is an explicit computable constant, $k = 0, 1, \ldots$. Here the *stabilization property* is expressed by the following assertion: *for $N \geq m$ and for large enough $|t|_p$ the remainder $o\left(\log_p^{-N} |t|_p \right)$ disappears and the asymptotic expansion becomes an exact equality.*

The same remark is also true for the case of the asymptotic expansions (13.4.18).

Remark 13.2. Since $\varphi \in \mathcal{D}_N^l(\mathbb{Q}_p)$, to calculate the asymptotics of the functionals $J_{\pi_\alpha, m; \varphi}(t)$ and $J_{\pi_0, m; \varphi}(t)$ (for the case $\pi_1(x) \equiv 1, \alpha \neq 0$), it is *natural* to represent these functionals as the sums of the integrals (13.4.7) and (13.4.20), respectively. However, we can represent these functionals as the sums of integrals

$$\tilde{J}_{\pi_\alpha, m; \varphi}^1(t) = \int_{B_{l_0}} |x|_p^{\alpha-1}(x) \log_p^m |x|_p \chi_p(xt)\big(\varphi(x) - \varphi(0)\big) \, dx, \quad (13.4.31)$$

$$\tilde{J}_{\pi_\alpha, m; \varphi}^2(t) = \int_{\mathbb{Q}_p \backslash B_{l_0}} |x|_p^{\alpha-1} \log_p^m |x|_p \chi_p(xt)\varphi(x) \, dx, \quad (13.4.32)$$

$$\tilde{J}_{\pi_\alpha, m}^0(t) = \int_{B_{l_0}} |x|_p^{\alpha-1} \log_p^m |x|_p \chi_p(xt) \, dx, \quad (13.4.33)$$

where $l_0 \in \mathbb{Z}$. For example, we can choose $l_0 = 0$, as in the standard representations (13.4.6) and (13.4.19). In this case

$$|x|_p^{\alpha-1} \log_p^m |x|_p \big(\varphi(x) - \varphi(0)\big)\Delta_{l_0}(x) \in \mathcal{D}_N^{\min(l, l_0)}(\mathbb{Q}_p),$$

$$|x|_p^{\alpha-1} \log_p^m |x|_p \varphi(x)\big(1 - \Delta_{l_0}(x)\big) \in \mathcal{D}_N^{\min(l, l_0)}(\mathbb{Q}_p),$$

and, as above, according to (4.8.5), $\tilde{J}_{\pi_\alpha, m; \varphi}^1(t) = \tilde{J}_{\pi_\alpha, m; \varphi}^2(t) = 0$ for all $|t|_p > p^{\max(-l, -l_0)}$. Thus repeating the above calculations almost word for word, we obtain the asymptotic relations from Theorem 13.4.1. However, in this case the stabilization parameter is equal to $s(\varphi) = p^{\max(-l, -l_0)}$.

The same remark is also true for the case of Theorem 13.4.2.

13.5. Asymptotics of singular Fourier integrals ($\pi_1(x) \not\equiv 1$)

Let us consider the case $f_{\pi_\alpha;m}(x) = |x|_p^{\alpha-1}\pi_\alpha(x)\log_p^m |x|_p$, $m = 0, 1, 2, \ldots$.

Theorem 13.5.1. *Let $\varphi \in \mathcal{D}_N^l(\mathbb{Q}_p)$, and let $k_0 > 0$ be the rank of the character $\pi_1(x)$. Then the functional $J_{\pi_\alpha,m;\varphi}(t)$ has the following asymptotic behavior:*
(a) If $\alpha \neq 0$ and $m = 0$ then

$$J_{\pi_\alpha,0;\varphi}(t) = \left\langle |x|_p^{\alpha-1}\pi_1(x)\chi_p(xt), \varphi(x) \right\rangle$$

$$= \varphi(0)\frac{\Gamma_p(\pi_\alpha)}{|t|_p^\alpha \pi_1(t)}, \quad |t|_p > p^{-l+k_0}, \quad (13.5.1)$$

for all $\varphi \in \mathcal{D}_N^l(\mathbb{Q}_p)$, where the Γ-function $\Gamma_p(\pi_\alpha)$ is given by (6.2.7), i.e., in the weak sense

$$|x|_p^{\alpha-1}\pi_1(x)\chi_p(xt) = \delta(x)\frac{\Gamma_p(\pi_\alpha)}{|t|_p^\alpha \pi_1(t)}, \quad |t|_p \to \infty. \quad (13.5.2)$$

(b) If $\alpha \neq 0$ and $m = 1, 2, \ldots$ then

$$J_{\pi_\alpha,m;\varphi}(t) = \left\langle |x|_p^{\alpha-1}\pi_1(x)\log_p^m |x|_p \chi_p(xt), \varphi(x) \right\rangle$$

$$= \varphi(0)\sum_{k=0}^m C_m^k \log_p^k e \frac{d^k \Gamma_p(\pi_\alpha)}{d\alpha^k} \frac{\log_p^{m-k} |t|_p}{|t|_p^\alpha \pi_1(t)}, \quad |t|_p > p^{-l+k_0},$$

$$(13.5.3)$$

for all $\varphi \in \mathcal{D}_N^l(\mathbb{Q}_p)$, i.e., in the weak sense

$$|x|_p^{\alpha-1}\pi_1(x)\log_p^m |x|_p \chi_p(xt)$$

$$= \delta(x)\sum_{k=0}^m C_m^k \log_p^k e \frac{d^k \Gamma_p(\pi_\alpha)}{d\alpha^k} \frac{\log_p^{m-k} |t|_p}{|t|_p^\alpha \pi_1(t)}, \quad |t|_p \to \infty, \quad (13.5.4)$$

with respect to the asymptotic sequence

$$\{\pi_{\alpha+1}^{-1}(t)\log_p^{m-k} |t|_p : k = 0, 1, \ldots, m\}.$$

Thus for any $\varphi \in \mathcal{D}(\mathbb{Q}_p)$, the relations (13.5.1) and (13.5.3) are exact equalities for sufficiently large $|t|_p > p^{-l+k_0}$, i.e., these asymptotic expansions are stable with the stabilization parameter $s(\varphi) = p^{-l+k_0}$.

Proof. Let $m = 0$. By formulas (4.9.5) and (4.9.8), the functional $J_{\pi_\alpha,0;\varphi}(t)$ can be rewritten as a convolution (4.7.1):

$$J_{\pi_\alpha,0;\varphi}(t) = \left\langle |x|_p^{\alpha-1}\pi_1(x)\chi_p(xt), \varphi(x) \right\rangle$$

$$= F[|x|_p^{\alpha-1}\pi_1(x)\varphi(x)](t) = F[|x|_p^{\alpha-1}\pi_1(x)](t) * \psi(t), \quad (13.5.5)$$

where $\psi(\xi) = F[\varphi(x)](\xi)$ and according to (6.2.8)

$$F[|x|_p^{\alpha-1}\pi_1(x)](t) = \Gamma_p(\pi_\alpha)|t|_p^{-\alpha}\pi_1^{-1}(t). \qquad (13.5.6)$$

Since $\varphi \in \mathcal{D}_N^l(\mathbb{Q}_p)$ then, in view of (4.8.5), $\psi \in \mathcal{D}_{-l}^{-N}(\mathbb{Q}_p)$. If $|t|_p > p^{-l}$, according to (13.5.6) and (4.7.7), relation (13.5.5) can be rewritten as

$$J_{\pi_\alpha,0;\varphi}(t) = \Gamma_p(\pi_\alpha)\int_{\mathbb{Q}_p} |t - \xi|_p^{-\alpha}\pi_1^{-1}(t - \xi)\psi(\xi)\,d\xi. \qquad (13.5.7)$$

Since $|t|_p > p^{-l}$ and $\xi \in B_{-l}$ the latter integral is well defined for any α. Moreover, we have $|t - \xi|_p = |t|_p$ for $|t|_p > p^{-l}$, $\xi \in B_{-l}$. Thus the relation (13.5.7) can be transformed into the form

$$J_{\pi_\alpha,0;\varphi}(t) = \Gamma_p(\pi_\alpha)|t|_p^{-\alpha}\pi_1^{-1}(t)\Psi(t), \quad |t|_p > p^{-l}, \qquad (13.5.8)$$

where

$$\Psi(t) = \int_{B_{-l}} \pi_1^{-1}\left(1 - \frac{\xi}{t}\right)\psi(\xi)\,d\xi. \qquad (13.5.9)$$

Let k_0 be the rank of the character $\pi_1(x) \not\equiv 1$. In is clear that if $|t|_p > p^{-l+k_0}$ then the inequality $\left|\frac{\xi}{t}\right|_p \le p^{-k_0}$ holds for all $\xi \in B_{-l}$. Thus in view of (2.3.14), we see that $\pi_1^{-1}\left(1 - \frac{\xi}{t}\right) \equiv 1$ for all $\xi \in B_{-l}$ and $|t|_p > p^{-l+k_0}$. Next, applying an analog of the Lebesgue theorem to the limiting passage under the sign of an integral (Theorem 3.2.4) to (13.5.9), and taking into account that $|\pi_1(x)| = 1$ and

$$\int_{B_{-l}} \psi(\xi)\,d\xi = \int_{\mathbb{Q}_p} \psi(\xi)\,d\xi = \varphi(0),$$

we see that the formulas (13.5.8) and (13.5.9) imply the relations (13.5.1) and (13.5.2) for all $|t|_p > p^{-l+k_0}$.

If $m = 1, 2, \ldots$, differentiating (13.5.8) with respect to α, we obtain

$$J_{\pi_\alpha,m;\varphi}(t) = \log_p^m e\,\frac{d^m}{d\alpha^m}J_{\pi_\alpha,0;\varphi}(t)$$

$$= \sum_{k=0}^m C_m^k \log_p^k e\,\frac{d^k\Gamma_p(\pi_\alpha)}{d\alpha^k}|t|_p^{-\alpha}\log_p^{m-k}|t|_p\pi_1^{-1}(t)\Psi(t), \quad |t|_p > p^{-l}.$$

$$(13.5.10)$$

Just as above, since $\Psi(t) = \varphi(0)$ for $|t|_p > p^{-l+k_0}$, relation (13.5.10) implies (13.5.3) and (13.5.4). $\qquad \square$

The analogs of Remarks 13.1 and 13.2 are also true for the case of Theorem 13.5.1.

13.6. *p*-adic version of the Erdélyi lemma

Theorems 13.4.1 and 13.5.1 for Re $\alpha > 0$ imply the following *p*-adic version of the well-known Erdélyi lemma.

Corollary 13.6.1. *Let k_0 be the rank of the character π_1, and $\varphi \in \mathcal{D}_N^l(\mathbb{Q}_p)$. Then for* Re $\alpha > 0$, *$m = 0, 1, 2, \ldots$, we have*

$$\int_{\mathbb{Q}_p} |x|_p^{\alpha-1} \pi_1(x) \log_p^m |x|_p \chi_p(xt) \varphi(x)\, dx$$

$$= \varphi(0) \sum_{k=0}^m C_m^k \log_p^k e \frac{d^k \Gamma_p(\pi_\alpha)}{d\alpha^k} \frac{\log_p^{m-k} |t|_p}{|t|_p^\alpha \pi_1(t)}, \quad |t|_p > p^{-l+k_0}. \quad (13.6.1)$$

Moreover, for any $\varphi \in \mathcal{D}(\mathbb{Q}_p)$, the latter relation is a stable asymptotic expansion.

14

Nonlinear theories of p-adic generalized functions

14.1. Introduction

In standard mathematical physics (in the real setting) there are problems which require the definition of products of distributions (generalized functions) [66], [68], [86], [97], [194]. Such problems appear in quantum mechanics [53], [9], [27], quantum field theory, some problems of gas dynamics, elasticity theory, and also in the description of, e.g., shock waves, δ-shock waves, and typhoons. In the framework of the approaches connected to problems of multiplications of distributions, a theory of singular solutions of nonlinear equations has been developed [8], [68]–[71], [149], [150], [195], [220]. Solving problems of this kind requires the development of special analytical methods, the construction of algebras containing the space of distributions, and the development of a technique for constructing singular asymptotics. As a result, the demand arises for a construction of a *nonlinear theory of generalized functions*. Besides, the development of nonlinear theories of distributions is of great interest in itself.

Since p-adic mathematical physics is a relatively young science, p-adic analogs of the above mentioned problems have not been studied so far (to the best of our knowledge). The problems of p-adic analysis related to the theory of p-adic distributions which have been solved up to now are of the *linear* type. To deal with *nonlinear singular problems* one needs some *additional technique* similar to that developed in the *usual* real mathematical physics mentioned above. From the aforesaid it seems natural to introduce the p-adic analogs of *the usual* (the real case) nonlinear theories of generalized functions, which we construct in this chapter. These results are based on the papers [18], [19], [20]. We also intensively use some algebraic constructions related to distributions from [149], [150], [70, Section 1–2].

At first, in Section 14.2, we recall some standard (in the real case) results on nonlinear theories of generalized functions and algebras of distributions and give a brief survey of this area.

In Sections 14.3 and 14.4, by using some ideas of Livchak, Egorov, and Colombeau, we construct the p-adic algebra of generalized functions $\mathcal{G}_p(\mathbb{Q}_p^n)$, called the Colombeau–Egorov algebra. The Colombeau–Egorov algebra $\mathcal{G}_p(\mathbb{Q}_p^n)$ is associative and commutative (see Section 14.3.2). $\mathcal{G}_p(\mathbb{Q}_p^n)$ is also an *associative convolution algebra* (see Section 14.4.5). In the algebra $\mathcal{G}_p(\mathbb{Q}_p^n)$ the product of p-adic distributions from $\mathcal{D}'(\mathbb{Q}_p^n)$ can be defined, which, in the general case, presents a generalized function from $\mathcal{G}_p(\mathbb{Q}_p^n)$. As in the real theories, there is a linear injective embedding $\mathcal{D}'(\mathbb{Q}_p^n) \subset \mathcal{G}_p(\mathbb{Q}_p^n)$, which, however, is *not unique*. This is known to be a general shortcoming of the Colombeau approach. The space $\mathcal{D}(\mathbb{Q}_p^n)$ is a subalgebra in the algebra $\mathcal{G}_p(\mathbb{Q}_p^n)$ (Theorem 14.3.3). In the p-adic Colombeau–Egorov theory we can define arbitrary continuous functions of generalized functions from $\mathcal{G}_p(\mathbb{Q}_p^n)$ (Section 14.4.1). In Section 14.5, in the p-adic Colombeau–Egorov theory the operation of fractional differentiation and fractional integration are introduced in two ways: by the Vladimirov–Taibleson fractional operators (instead of the usual operation of differentiation and integration in the case of real theories). According to Theorems 14.5.1 and 14.5.2, the families of fractional operators form abelian groups on $\mathcal{G}_p(\mathbb{Q}_p^n)$.

In Section 14.6, we construct the p-adic algebra of *asymptotic distributions* $\mathcal{A}^*$ generated by the linear span $\mathcal{A}$ of the set of p-adic *quasi associated homogeneous distributions* (*QAHD*) (which were introduced and studied in Chapter 6, Sections 6.2–6.5). The algebra $\mathcal{A}^*$ is an analog of algebras (in the real case), constructed in the papers [68], [220], [149], [150] (see also Section 14.2.2). $\mathcal{A}^*$ is also an *associative convolution algebra*. Each element of the algebra $\mathcal{A}^*$ (called an *asymptotic distribution*) is a product of QAHDs and is represented as a *vector-valued Schwartz distribution* (which is finite from the left and from the right) (14.6.7):

$$f^*(x) = \big(f_{(\lambda_m, n)}(x)\big), \quad \lambda_m \in \mathbb{R}, \quad m, n \in \mathbb{N}_0,$$

where $\lambda_0 = 0$, $f_{(\lambda_0, 0)} \in \mathcal{A}$, $f_{(\lambda_m, n)} = c_{mn} \delta(x)$; $\lambda_m \in \mathbb{R}$ is an increasing sequence, c_{mn} are constants; $1 \leq m \leq M$, $0 \leq n \leq N$. Each distribution $f(x)$ from the subspace $\mathcal{A}$ can be identified with the vector-valued distribution $f^*(x) = \big(f_{(\lambda_m, n)}(x)\big)$, where $f_{0,0}(x) = f(x)$, while all the other components are equal to zero, i.e., $\mathcal{A} \subset \mathcal{A}^*$. Elements of the algebra $\mathcal{A}^*$ can be represented in the form of sums (14.6.8). The vector-valued distribution $f^* \in \mathcal{A}^*$ is a linear continuous functional over the space $\mathcal{D}_*$ of test vector-valued functions

$$\varphi_*(x) = \big(\varphi_{(\beta_m, n)}(x)\big), \quad \varphi_{(\beta_m, n)} \in \mathcal{D}(\mathbb{Q}_p), \quad \beta_m \in \mathbb{R},$$

where $\beta_m \in \mathbb{R}$ is an increasing sequence, $m, n \in \mathbb{N}_0$. On $\mathcal{D}_*$ a locally convex topology can be introduced analogously to the case of the real algebra of asymptotic distributions [149]–[151]. However, here the problem of introducing this type of topology is not further discussed. The construction of the algebra $\mathcal{A}^*$ is based on the main Theorem 14.6.1, which gives weak asymptotic expansions of products of regularizations of QAHD. The proof of the main theorem (see Appendix C) uses a new technique of constructing p-adic weak asymptotics. Theorem 14.6.1 is a p-adic analog of Theorem 14.2.1 from Section 14.2. According to Section 14.7, the algebra of asymptotic distributions $\mathcal{A}^*$ can be embedded into the Colombeau–Egorov algebra $\mathcal{G}_p(\mathbb{Q}_p)$ as a subalgebra. The latter fact is an analog of results [219], [222] for the real Colombeau algebra.

Unlike the Colombeau approach, where, in the general case, a product of distributions is not a distribution, *asymptotic distributions* from the algebra $\mathcal{A}^*$ are Schwartz vector-valued distributions. Therefore, these results show the way to the construction of a *nonlinear theory of distributions* of the same type as the standard Schwartz theory of distributions.

14.2. Nonlinear theories of distributions (the real case)

14.2.1. Colombeau type algebras

Let us recall standard (in the real setting) results which serve as a prototype for our p-adic investigations.

The first nonlinear "sequential" theory of distributions was constructed in papers by Ya. B. Livchak [177]–[179]. These excellent papers are full of deep ideas. Some ideas used by Livchak can also be found in the papers by C. Schmieden and D. Laugwitz [215], and D. Laugwitz [172], [173]. In order to construct an algebra of generalized functions Ya. B. Livchak used the following. Let $\mathcal{M}$ be the set of all sequences $\{f_k(x)\}$ of real-valued measurable functions on $\mathbb{R}^n$. A sequence $\{f_k(x)\} \in \mathcal{M}$ is called an *almost zero* if there exists k_0 such that $f_k(x) = 0$ for all $k \geq k_0$. Let $T(\mathcal{M})$ be the set of all almost zero sequences. The *Livchak algebra of generalized functions* [179, §8] is defined as the quotient algebra

$$\mathcal{L} = \mathcal{M}/T(\mathcal{M}).$$

In Ya. B. Livchak's papers the point value characterization of generalized functions was introduced, which is an analog of the point value characterization on the set of *nonstandard numbers* introduced later by M. Oberguggenberger and M. Kunzingerin in [196] (see also [97]).

A particular case of the Livchak's algebra was later introduced by Yu. V. Egorov in [86] by the following definition. Let Ω be a domain in $\mathbb{R}^n$. We will call two sequences $\{f_k\}$ and $\{g_k\}$ $(f_k, g_k \in C^\infty(\Omega))$ *equivalent* if for each compact subset $K \subset \Omega$ there exists $N > 0$ such that $f_k(x) = g_k(x)$ for all $k \geq N$, $x \in K$. The *Egorov algebra of generalized functions* $\widehat{\mathcal{G}}(\Omega)$ is introduced as the set of the equivalence classes in this case.

The nonlinear "continuous" theory of distributions developed in the works of J. Colombeau [66] is widely used and finds extensive application in mathematical physics [54], [66], [86], [97], [113], [193], [194], [195]. Let $\mathbb{N}_0 = 0 \cup \mathbb{N}$. For any $q \in \mathbb{N}_0$ we define subsets of test functions $\mathcal{A}_0(\Omega) = \{\chi \in \mathcal{D}(\Omega) : \int \chi(x)\,dx = 1\}$,

$$\mathcal{A}_q(\Omega) = \{\chi(x) \in \mathcal{A}_0(\Omega) : \int x^\alpha \chi(x)\,dx = 0, \quad 1 \leq |\alpha| \leq q\},$$

where $\Omega \subset \mathbb{R}^n$, $q = 1, 2, \ldots$. Obviously, $\mathcal{D}(\Omega) \supset \mathcal{A}_0(\Omega) \supset \mathcal{A}_1(\Omega) \supset \mathcal{A}_2(\Omega) \supset \cdots \mathcal{A}_q(\Omega) \supset \cdots$ and $\mathcal{A}_q(\Omega) \neq \emptyset$ for all $q \in \mathbb{N}_0$.

Denote by $\mathcal{E}(\Omega)$ the algebra with differentiation of all mappings $f : \mathcal{A}_0(\Omega) \to C^\infty(\mathbb{R}^n)$. The elements of this algebra are families $f(\psi, x)$, where $\psi \in \mathcal{A}_q(\Omega)$ is considered as a parameter.

Let us define a family of functions $\psi_\varepsilon(x) = \frac{1}{\varepsilon^n} \psi\left(\frac{x}{\varepsilon}\right)$, $\varepsilon > 0$, where $\psi \in \mathcal{A}_0(\Omega)$. Let $\mathcal{E}_M(\Omega)$ be the subalgebra with differentiation of *moderate elements* $f(\psi, x)$ from $\mathcal{E}(\Omega)$, i.e., such that for any compact $K \subset \Omega$ and any $\alpha \in \mathbb{N}_0^n$ there exists $N \in \mathbb{N}$ such that for all $\psi \in \mathcal{A}_N(\Omega)$

$$\sup_{x \in K} |\partial_x^\alpha f(\psi_\varepsilon, x)| \leq M\varepsilon^{-N}, \quad \forall \varepsilon \in (0, \varepsilon_0),$$

where $M > 0$ is a constant, $\varepsilon_0 > 0$ is a certain number. We define an ideal $\mathcal{N}(\Omega)$ in the algebra $\mathcal{E}_M(\Omega)$ as the set of all elements $f(\psi, x) \in \mathcal{E}(\Omega)$ such that for any compact $K \subset \Omega$ and any $\alpha \in \mathbb{N}_0^n$ there exists $N \in \mathbb{N}_0$ such that for all $\psi \in \mathcal{A}_q(\Omega)$, where $q \geq N$, the following inequality holds:

$$\sup_{x \in K} |\partial_x^\alpha f(\psi_\varepsilon, x)| \leq M\varepsilon^{q-N}, \quad \forall \varepsilon \in (0, \varepsilon_0),$$

where $M > 0$ and $\varepsilon_0 > 0$.

The Colombeau algebra is defined as the quotient algebra

$$\mathcal{G}(\Omega) = \mathcal{E}_M(\Omega)/\mathcal{N}(\Omega)$$

and its elements are called *generalized Colombeau functions*.

Note that in the Colombeau, Livchak, and Egorov algebras products of distributions, in the general case, are not distributions. The elements of these algebras have no explicit functional interpretation, which restricts their application.

14.2.2. Algebras of asymptotic distributions

There is another approach to the problem of multiplication of distributions associated with the names of Ya. B. Livchak [179], Li Bang-He [176] and V. K. Ivanov [108]. The main idea of this approach is that the product $f(x)g(x)$ of distributions $f(x)$ and $g(x)$ from a certain class is defined as a weak asymptotic expansion of the product $f(x, \varepsilon)g(x, \varepsilon)$ as $\varepsilon \to +0$, where $f(x, \varepsilon)$ and $g(x, \varepsilon)$ are regularizations of the distributions $f(x)$ and $g(x)$, respectively, and $\varepsilon > 0$ is the approximation parameter. The development of this approach in [217]–[220] led to the construction of the associative and commutative algebra E_1^* of *asymptotic distributions*, generated by a linear span E_1 of distributions $\delta^{(m-1)}(x - c_k)$, $P\left(\frac{1}{(x-c_k)^m}\right)$, $m = 1, 2, \ldots, c_k \in \mathbb{R}, k = 1, \ldots, s$.

In the framework of the same approach, in [68], the associative and commutative algebra E^* of *asymptotic distributions*, generated by the linear span of *quasi associated homogeneous distributions* (*QAHD*) (which were studied in Chapter 6 and Sections A.2–A.7) was constructed. Now we recall the scheme of these constructions [68].

Let us introduce the linear space $E \subset \mathcal{D}'(\mathbb{R})$ generated by the linear combinations of quasi associated homogeneous distributions (*QAHD*s): $(x - c_k \pm i0)^\lambda \log^p(x - c_k \pm i0)$, $\lambda \in \mathbb{R}$, $p = 0, 1, 2, \ldots$; $c_k \in \mathbb{R}$, $k = 1, \ldots, s$ (see Chapter 6, Section A.3). Let us also introduce the space of harmonic regularizations of distributions from E:

$$h = \text{span}\{(x - c_k \pm i\varepsilon)^\lambda \log^p(x - c_k \pm i\varepsilon) :$$
$$\lambda \in \mathbb{R}, p = 0, 1, 2, \ldots; c_k \in \mathbb{R}, k = 1, \ldots, s; \varepsilon > 0\},$$

where $z = x + i\varepsilon$, $\bar{z} = x - i\varepsilon$. The elements of this space belong to the Vladimirov algebras [236].

Let us further introduce an *associative and commutative algebra* h^* *with unity and without zero divisors* generated by the space h. The elements $f^*(x, \varepsilon)$ of this algebra are finite sums of finite products of functions $f(x, \varepsilon) \in h$.

The following assertion holds.

Theorem 14.2.1. ([68, Corollary from Theorem 4.3.]) *Let* $f^*(x, \varepsilon) \in h^*$. *For all* $\varphi \in \mathcal{D}(\mathbb{R})$ *and all* $M \geq 0$, $N = 0, 1, \ldots, n_0(M)$:

$$\left\{ \langle f^*(x, \varepsilon), \varphi(x) \rangle - \sum_{m=0}^{M} \sum_{n=0}^{N} \langle f_{\alpha_m, n}(x), \varphi(x) \rangle \varepsilon^{\alpha_m} \log^{n_0(m)-n} \varepsilon \right\}$$
$$= o\left(\varepsilon^{\alpha_M} \log^{n_0(M)-N} \varepsilon\right), \quad \varepsilon \to +0,$$

where $\alpha_m \in \mathbb{R}$ *is an increasing sequence*, $\alpha_0 = 0$, $n_0(m) \in \mathbb{N}_0$, *the coefficients* $f_{\alpha_m, n}(x)$ *are distributions from the space* E. *Thus, each element* $f^*(x, \varepsilon)$ *of the* *algebra h^* has a unique representation in the form of a weak asymptotic series (finite from the left)*

$$f^*(x, \varepsilon) = \sum_{m=0}^{\infty} \sum_{n=0}^{n_0(m)} f_{\alpha_m, n}(x) \varepsilon^{\alpha_m} \log^{n_0(m)-n} \varepsilon, \quad \varepsilon \to +0, \qquad (14.2.1)$$

In particular, for any element $f^*(x, \varepsilon) \in h \subset h^*$ its representation (14.2.1) is a "Taylor type" expansion.

Let A^* be the linear mapping which associates each element $f^*(x, \varepsilon)$ from the algebra h^* with the unique vector-valued distribution:

$$f^*(x) = \big(f_{(\alpha_m, n)}(x) \big), \quad \alpha_m \in \mathbb{R}, \quad n \in \mathbb{Z}_+, \qquad (14.2.2)$$

where the components of the vector are $f_{(\alpha_m, n)}(x) \in E$, $\alpha_m \in \mathbb{R}$ is an increasing sequence, $n = 0, 1, 2, \ldots, n_0(m)$, $m \in \mathbb{N}_0$. The mapping A^* is defined in the following way. A^* associates to each element $f^*(x, \varepsilon) \in h^* \setminus h$ the unique vector-valued distribution (14.2.2), whose components $f_{\alpha_m, n}(x)$ are the corresponding coefficients of the asymptotics (14.2.1). By A^* each element $f^*(x, \varepsilon) \in h \subset h^*$ is associated with the unique vector-valued distribution (14.2.2) whose components $f_{\alpha_m, n}(x)$ are equal to zero except for the component $f_{0, n_0(0)}(x) = \lim_{\varepsilon \to +0} f^*(x, \varepsilon) \in E$. According to (14.2.2), any vector-valued distribution $f^*(x)$ is finite from the left.

The image $f^* = A^*[f^*(\cdot, \varepsilon)]$, $f^*(\cdot, \varepsilon) \in h^*$, is called an *asymptotic distribution*. By

$$E^* = A^*[h^*] = \{ f^* = A^*[f^*(\cdot, \varepsilon)] : f^*(\cdot, \varepsilon) \in h^* \}$$

we denote the *linear space of all asymptotic distributions*. Each distribution $f(x)$ from the subspace E can be identified with the vector-valued distribution $f^*(x) = \big(f_{(\alpha_m, n)}(x) \big)$, where $f_{0, n_0(0)}(x) = f(x)$, while all the other components are equal to zero, i.e., there is an embedding $E \subset E^*$.

As in [220], [67], [149]–[151], we can consider vector-valued distributions $f^*(x)$ from E^* as continuous linear vector-functionals on the space of test vector-valued functions (finite from the left) $\mathcal{D}_*$:

$$\varphi_*(x) = \big(\varphi_{\beta_m, n}(x) \big), \quad \varphi_{\beta_m, n}(x) \in \mathcal{D}(\mathbb{R}), \qquad (14.2.3)$$

where $\beta_m \in \mathbb{R}$ is an increasing sequence, $n \in \mathbb{N}_0$. Here the action of the vector-valued distribution f^* on a test vector-valued function φ_* is defined as

$$\langle f^*, \varphi_* \rangle = \sum_{m=0}^{\infty} \sum_{n=0}^{\infty} \langle f_{\alpha_m, n}, \varphi_{-\alpha_m, -n} \rangle,$$

where the latter sum contains a finite number of terms. Using results of the papers [149]–[151], the space $\mathcal{D}_*$ can be equipped with the sequentially complete topology.

From Theorem 14.2.1, and taking into account the construction of the mapping A^*, the asymptotic distribution (14.2.2) can be identified with the corresponding formal series

$$f^*(x) = \sum_{m=0}^{\infty} \sum_{n=0}^{n_0(m)} f_{\alpha_m,n}(x)\varepsilon^{\alpha_m}\log^{n_0-n}\varepsilon, \qquad (14.2.4)$$

where $f_{\alpha_m,n}(x) \in E$, $\alpha_m \in \mathbb{R}$ is an increasing sequence, and $n = 0, 1, 2, \ldots, n_0(m)$, $m, n_0(m) \in \mathbb{N}_0$.

It is proved in [68] that the mapping $A^* : h^* \to E^*$ is an *isomorphism* and $A^*[h] = E$.

Now let us introduce the operation of multiplication on the space of asymptotic distributions E^*, and thus for distributions from E.

Definition 14.1. The product of the asymptotic distributions f_1^*, f_2^* from E^* is defined as the asymptotic distribution

$$f^*(x) = f_1^*(x) \circ f_2^*(x) = A^*\Big[f_1^*(x, \varepsilon) f_2^*(x, \varepsilon)\Big] \in E^*, \qquad (14.2.5)$$

where $f_k^*(x, \varepsilon) = A^{*-1} f_k^*(x) \in h^*$ is a regularization of the asymptotic distribution f_k^*, $k = 1, 2$.

Since h^* is an associative and commutative algebra of functions and $A^* : h^* \to E^*$ is an isomorphism, the multiplication (14.2.5) is associative and commutative. The multiplication (14.2.5) turns the space of asymptotic distributions (the space of vector-valued distributions) E^* into an algebra.

Example 14.2.1. (see [68], [218], [220])

(a) $(x + i0)^{-m} \circ (x - i0)^{-n} = A^*\Big[(x + i\varepsilon)^{-m}(x - i\varepsilon)^{-n}\Big]$

$$= (-1)^{m-1} \left(\frac{i}{2}\right)^{m+n-2} \pi \sum_{s=0}^{m+n-2} B_s^+ \frac{i^s}{s!}\delta^{(s)}(x)\varepsilon^{-m-n+s+1}$$

$$+ \frac{(-1)^{n-1}i\pi}{(m+n-1)!}\Big((-1)^m + 2^{-m-n+2}B_{m+n-1}^+\Big)\delta^{(m+n-1)}(x) + P(x^{-m-n})$$

$$+ (-1)^{m-1} \left(\frac{i}{2}\right)^{m+n-1} \sum_{s=m+n}^{\infty} i^s\Big((-1)^s(x + i0)^{-s-1}B_s^+$$

$$- (x - i0)^{-s-1}B_s^-\Big)\varepsilon^{-m-n+s+1},$$

where $m, n = 1, 2, \ldots$; $B_s^+ = B_s(m, n)$, $B_s^- = B_s(n, m)$, and the coefficients $B_s(m, n)$ have the form

$$B_s(m, n) = \begin{cases} \sum_{k=0}^{s} C_{m+n-k-2}^{n-1} C_s^k (-2)^k, & 0 \le s \le m - 1, \\ \sum_{k=0}^{m-1} C_{m+n-k-2}^{n-1} C_s^k (-2)^k, & s \ge m. \end{cases}$$

Here the asymptotic distribution $(x + i0)^{-m} \circ (x - i0)^{-n} \in E^*$ is represented as the formal series (14.2.2).

(b) $\quad x \circ \delta(x) = A^* \left[\frac{1}{2} \left((x + i\varepsilon) + (x - i\varepsilon) \right) \cdot \frac{-1}{2\pi i} \left(\frac{1}{x + i\varepsilon} - \frac{1}{x - i\varepsilon} \right) \right]$

$$= \sum_{k=0}^{\infty} (-1)^k \left\{ \frac{1}{\pi} P \left(\frac{1}{x^{2k+1}} \right) \varepsilon^{2k+1} + \frac{1}{(2k+1)!} \delta^{(2k+1)}(x) \varepsilon^{2k+2} \right\}.$$

Here the asymptotic distribution $0 \ne x \circ \delta(x) \in E^*$ is represented as the formal series (14.2.2). We recall that in the framework of the classical Schwartz theory we have $x \circ \delta(x) = 0$.

(c) $\quad (\delta(x))^2 = A^* \left[\left(\frac{-1}{2\pi i} \left(\frac{1}{x + i\varepsilon} - \frac{1}{x - i\varepsilon} \right) \right)^2 \right]$

$$= \frac{\delta(x)}{2\pi \varepsilon} + \frac{1}{2\pi^2} \sum_{k=1}^{\infty} (-1)^k \left\{ \frac{\pi}{(2k)!} \delta^{(2k)}(x) \varepsilon^{2k-1} + P \left(\frac{1}{x^{2k+2}} \right) \varepsilon^{2k} \right\}.$$

Here we consider the asymptotic distribution $(\delta(x))^2 \in\in E^*$ as the formal series (14.2.2). On the other hand, this asymptotic distribution can be represented as the vector distribution (14.2.4):

$$(\delta(x))^2 = (\ldots, f_k(x), f_{k+1}(x), \ldots),$$

where $f_k(x) = 0$, $k < -1$, $f_{-1}(x) = \frac{1}{2\pi} \delta(x)$, $f_{2k-1}(x) = \frac{(-1)^k}{2\pi (2k)!} \delta^{(2k)}(x)$, $f_{2k}(x) = \frac{(-1)^k}{2\pi^2} P(x^{-2k-2})$, $k = 1, 2, \ldots$. For all test functions $\varphi_*(x) \in \mathcal{D}_*$ we have

$$\left\langle (\delta(x))^2, \varphi_*(x) \right\rangle = \frac{1}{2\pi} \left\langle \delta(x), \varphi_1(x) \right\rangle + \frac{1}{2\pi^2} \sum_{k=1}^{\infty} (-1)^k$$

$$\times \left\{ \frac{\pi}{(2k)!} \left\langle \delta^{(2k)}(x), \varphi_{-2k+1}(x) \right\rangle + \left\langle P \left(\frac{1}{x^{2k+2}} \right), \varphi_{-2k}(x) \right\rangle \right\},$$

where the last sum is finite.

Constructing approximations of *distributional products* in the form of weak asymptotic series (14.2.1) *has proved to be very useful* in the theory of discontinuous solutions to nonlinear equations. This formalism was used to develop the *weak asymptotics method* (for a summary of this method see [70]). The *dynamics of propagation and interaction* of nonlinear waves (infinitely narrow δ-solitons, shocks, δ-shocks) have been studied in [68]–[73], [220] and [224] using this method.

14.3. Construction of the p-adic Colombeau–Egorov algebra

14.3.1. Regularization of distributions

We call $\omega_k \in \mathcal{D}(\mathbb{Q}_p^n)$ a δ-*sequence* if there exists $N \in \mathbb{Z}$ such that $\omega_k(x) = \delta_k(x)$, $k \geq N$, $x \in \mathbb{Q}_p^n$, where $\delta_k(x)$ is a canonical δ-sequence (4.4.6), $k = 1, 2, \ldots$. It is clear that $\omega_k \to \delta$, $k \to \infty$ in $\mathcal{D}'(\mathbb{Q}_p^n)$.

In view of formula (4.7.8) from Proposition 4.7.4, the *regularization* of the distribution $f \in \mathcal{D}'(\mathbb{Q}_p^n)$ is the sequence

$$f_k(x) \stackrel{def}{=} f(x, p^{-k}) = \Big(f(x) * \omega_k(x)\Big)\Delta_k(x), \quad x \in \mathbb{Q}_p^n, \qquad (14.3.1)$$

where $*$ is a convolution, where $\Delta_k(x)$ is the canonical 1-sequence (4.4.5), $k = 1, 2, \ldots$. It is clear that $f_k \in \mathcal{D}(\mathbb{Q}_p^n)$.

Lemma 14.3.1. *Let f_k be the regularization of a distribution $f \in \mathcal{D}'(\mathbb{Q}_p^n)$. Then $f_k \to f$, $k \to \infty$ in $\mathcal{D}'(\mathbb{Q}_p^n)$. Moreover, for any test function $\varphi \in \mathcal{D}(\mathbb{Q}_p^n)$ we have*

$$\langle f_k, \varphi \rangle = \langle f, \varphi \rangle, \quad \forall k \geq -l,$$

where $l = l(\varphi)$ is the parameter of constancy of the function φ.

Proof. It is clear that

$$\langle f_k, \varphi \rangle = \langle f, \psi_k \rangle, \quad \psi_k(x) = \int p^{nk} \Omega(p^k |\xi|_p) \varphi(x - \xi) d\xi, \qquad (14.3.2)$$

$x \in \mathbb{Q}_p^n$. If $l = l(\varphi)$ is a parameter of constancy of the test function $\varphi \in \mathcal{D}(\mathbb{Q}_p^n)$ then we have

$$\psi_k(x) = \int_{\mathbb{Q}_p^n} p^{nk} \Omega(p^k |\xi|_p) \varphi(x - \xi) d\xi$$

$$= \varphi(x) p^{nk} \int_{B_{-k}^n} d\xi = \varphi(x), \qquad (14.3.3)$$

for all $k \geq -l$, $x \in \mathbb{Q}_p^n$.

Consequently, (14.3.2) and (14.3.3) imply $\langle f_k, \varphi \rangle = \langle f, \varphi \rangle$, for all $k \geq -l$. Thus for any function $\varphi \in \mathcal{D}(\mathbb{Q}_p^n)$ we have $\lim_{k \to \infty} \langle f_k, \varphi \rangle = \langle f, \varphi \rangle$. Moreover, $f_k = f$ for any compact $K \subset \mathbb{Q}_p^n$ and for sufficiently large k. $\qquad \square$

It is easy to prove the following lemma.

Lemma 14.3.2. *If $f \in C(\mathbb{Q}_p^n)$ then we have*

$$f_k(x) = f(x) * \delta_k(x) \to f(x), \quad k \to \infty,$$

pointwise for all x in any compact $K \subset \mathbb{Q}_p^n$.

14.3.2. The basic construction

Consider the set $\mathcal{P}(\mathbb{Q}_p^n)$ of all sequences $\{f_k : k \in \mathbb{N}\}$, where $f_k \in \mathcal{D}(\mathbb{Q}_p^n)$. Note that in $\mathcal{P}(\mathbb{Q}_p^n)$ there are constant sequences of the form $\{f_k(x) = f(x) : k \in \mathbb{N}\}$, where f are arbitrary functions from $\mathcal{D}(\mathbb{Q}_p^n)$. On $\mathcal{P}(\mathbb{Q}_p^n)$ we introduce an algebra structure, defining the operations componentwise:

$$\left(f_1 + f_2\right)_k(x) = f_{1,k}(x) + f_{2,k}(x),$$
$$\left(f_1 f_2\right)_k(x) = f_{1,k}(x) f_{2,k}(x), \quad x \in \mathbb{Q}_p^n,$$

where $\{f_{j,k}(x) : k \in \mathbb{N}\} \in \mathcal{P}(\mathbb{Q}_p^n)$, $j = 1, 2$. Let $\mathcal{N}_p(\mathbb{Q}_p^n)$ be the subalgebra of elements $\{f_k(x) : k \in \mathbb{N}\} \in \mathcal{P}(\mathbb{Q}_p^n)$ such that for any compact $K \subset \mathbb{Q}_p^n$ there exists $N \in \mathbb{N}$ such that $f_k(x) = 0$ for all $k \geq N$, $x \in K$. Clearly, $\mathcal{N}_p(\mathbb{Q}_p^n)$ is an ideal in the algebra $\mathcal{P}(\mathbb{Q}_p^n)$.

Now we introduce a *p-adic Colombeau-type algebra*.

Definition 14.2. The quotient algebra

$$\mathcal{G}_p(\mathbb{Q}_p^n) = \mathcal{P}(\mathbb{Q}_p^n)/\mathcal{N}_p(\mathbb{Q}_p^n)$$

will be called the *p-adic Colombeau–Egorov algebra*. The elements $\tilde{f}$ of this algebra will be called the Colombeau–Egorov *generalized functions*. The equivalence class of sequences which defines the element $\tilde{f}(x)$ will be denoted by $\tilde{f}(x) = [f_k(x)]$.

The algebra $\mathcal{G}_p(\mathbb{Q}_p^n)$ of generalized functions is associative and commutative.

Theorem 14.3.3. *There is a linear embedding*

$$\mathcal{D}'(\mathbb{Q}_p^n) \subset \mathcal{G}_p(\mathbb{Q}_p^n),$$

i.e., any distribution on $\mathbb{Q}_p^n$ is a generalized function on $\mathbb{Q}_p^n$. This embedding $\mathcal{D}(\mathbb{Q}_p^n) \subset \mathcal{G}_p(\mathbb{Q}_p^n)$ is a homomorphism of algebras and the space $\mathcal{D}(\mathbb{Q}_p^n)$ is a subalgebra of the algebra $\mathcal{G}_p(\mathbb{Q}_p^n)$.

Proof. Let us embed the distribution $f \in \mathcal{D}'(\mathbb{Q}_p^n)$ into $\mathcal{G}_p(\mathbb{Q}_p^n)$ by the mapping

$$\mathcal{D}'(\mathbb{Q}_p^n) \ni f(x) \mapsto \tilde{f}(x) = \{f_k(x)\} + \mathcal{N}_p(\mathbb{Q}_p^n) \in \mathcal{G}_p(\mathbb{Q}_p^n), \quad x \in \mathbb{Q}_p^n,$$

where f_k is the regularization of the distribution f given by (14.3.1).

Let us prove that $\{f_k(\cdot)\} \in \mathcal{N}_p(\mathbb{Q}_p^n)$ if and only if $f(x) = 0$. Indeed, if $f(x) = 0$ then $f_k(x) = 0$ for all k, and $\{f_k(x)\} \in \mathcal{N}_p(\mathbb{Q}_p^n)$. Suppose that $\{f_k(\cdot)\} \in \mathcal{N}_p(\mathbb{Q}_p^n)$. Hence according to (14.3.2) and (14.3.3),

$$\langle f_k, \varphi \rangle = \langle f, \varphi \rangle = 0,$$

for all $\varphi \in \mathcal{D}(\mathbb{Q}_p^n)$ and sufficiently big k, $k \geq -l$, i.e., $f(x) = 0$.

Thus any distribution $f \in \mathcal{D}'(\mathbb{Q}_p^n)$ determines a unique sequence $\{f_k : k \in \mathbb{N}\}$, which determines a unique generalized function $\tilde{f} \in \mathcal{G}_p(\mathbb{Q}_p^n)$.

Since $\mathcal{P}(\mathbb{Q}_p^n)$ contains the constant sequences $\{f_k\}$, $f_k = f \in \mathcal{D}(\mathbb{Q}_p^n)$, from Lemma 14.3.2 and the above arguments, we find that the mapping

$$\mathcal{D}(\mathbb{Q}_p^n) \ni f(x) \mapsto \tilde{f}(x) = \{f_k(x) = f(x)\} + \mathcal{N}_p(\mathbb{Q}_p^n) \in \mathcal{G}_p(\mathbb{Q}_p^n)$$

is a homomorphism of algebras.   □

Remark 14.1. In the usual real case Egorov theory [86], the embedding $\mathcal{D}' \subset \mathcal{G}$ *is not* injective, and hence any distribution $f \in \mathcal{D}'(\mathbb{Q}_p^n)$ determines *infinitely many* generalized functions belonging to the Colombeau algebra. For the p-adic theory the embedding $\mathcal{D}'(\mathbb{Q}_p^n)$ into $\mathcal{G}_p(\mathbb{Q}_p^n)$ is injective as in the real case of the Colombeau theory.

14.4. Properties of Colombeau–Egorov generalized functions

14.4.1. Function of generalized functions

Let $g(u_1, \ldots, u_m)$ be a continuous function, i.e. $g \in C(\mathbb{R}^m)$, and let $\tilde{f}_1, \ldots, \tilde{f}_m$ be generalized functions from $\mathcal{G}_p(\mathbb{Q}_p^n)$. We define the generalized function $g(\tilde{f}_1, \ldots, \tilde{f}_m)$ by the sequence $\{g_k : k \in \mathbb{N}\}$, where $g_k(x) = g(f_{1,k}(x), \ldots, f_{m,k}(x))\Delta_k(x)$, and $\Delta_k(x)$ is the canonical 1-sequence (4.4.5), $k = 1, 2, \ldots$. It is clear that $g(\tilde{f}_1, \ldots, \tilde{f}_m) \in \mathcal{G}_p(\mathbb{Q}_p^n)$.

14.4.2. Associated distribution

We say that the generalized function $\tilde{f} = [f_k] \in \mathcal{G}_p(\mathbb{Q}_p^n)$ *is associated with the distribution* $g \in \mathcal{D}'(\mathbb{Q}_p^n)$, if we have $\lim_{k\to\infty}\langle f_k, \varphi \rangle = \langle g, \varphi \rangle$, for all $\varphi \in \mathcal{D}(\mathbb{Q}_p^n)$. We shall write this fact using the notation $\tilde{f} \approx g$.

Example 14.4.1. According to (4.4.6), the generalized function $\delta^2(x) \in \mathcal{G}_p(\mathbb{Q}_p)$, $x \in \mathbb{Q}_p$ is defined by the sequence $\left\{p^{2k}\Omega^2(p^k|x|_p)\right\}$ for sufficiently large k. It is clear that $\delta^2 \notin \mathcal{D}'(\mathbb{Q}_p)$. Moreover, since $\lim_{k \to \infty} \langle p^{2k} \Omega^2 (p^k |x|_p), \varphi \rangle$ does not exist, $\delta^2(x)$ is not associated with any distribution.

Example 14.4.2. The following relation (which is also true in the usual Colombeau theory) holds

$$\left(1 + \left(\delta(x)\right)^n\right)^{1/n} \approx 1 + \delta(x), \quad x \in \mathbb{Q}_p.$$

Indeed, consider the integral

$$J_k = \int_{\mathbb{Q}_p} \left\{ \left(1 + \left(\delta_k(x)\right)^n\right)^{1/n} - 1 - \delta_k(x) \right\} \varphi(x) \, dx$$

$$= \int_{\mathbb{Q}_p} \left\{ \left(1 + p^{nk} \Omega^n (p^k |x|_p)\right)^{1/n} - 1 - p^k \Omega(p^k |x|_p) \right\} \varphi(x) \, dx,$$

where $\delta_k(x)$ is a regularization of the δ-function, $\varphi \in \mathcal{D}(\mathbb{Q}_p)$. Using formulas (3.3.1) and (3.2.6), we have

$$J_k = \left\{ \left(1 + p^{nk}\right)^{1/n} - 1 - p^k \right\} \varphi(0) \int_{B_{-k}} dx$$

$$= \varphi(0) \left\{ \left(1 + p^{-nk}\right)^{1/n} - 1 - p^{-k} \right\}$$

for sufficiently large k. Thus $\lim_{k \to \infty} J_k = 0$.

14.4.3. The point value

We are going to define the Colombeau–Egorov *generalized numbers*. Let $\overline{\mathbb{C}} = \mathbb{C} \cup \infty$ be the compactification of $\mathbb{C}$. Consider the set $\mathcal{F}$ of all sequences $\{c_k \in \overline{\mathbb{C}} : k \in \mathbb{N}\}$ and the subset $\mathcal{I} \subset \mathcal{F}$ of all sequences such that there exists $N \in \mathbb{N}$ such that $c_k = 0$ for all $k \geq N$. The equivalence classes

$$\widetilde{\mathcal{C}} = \mathcal{F}/\mathcal{I}$$

are called *Colombeau–Egorov generalized numbers*. It is clear that $\widetilde{\mathcal{C}}$ is an algebra. The equivalence class of sequences which defines the generalized number $\widetilde{c}$ will be denoted by $\widetilde{c} = [c_k]$. Since the set $\widetilde{\mathcal{C}}$ contains the constant sequences $\widetilde{c} = \{c_k = c\}$, where $c \in \overline{\mathbb{C}}, k \in \mathbb{N}$, the embedding $\overline{\mathbb{C}} \subset \widetilde{\mathcal{C}}$ holds.

Definition 14.3. Let $\widetilde{f} = [f_k] \in \mathcal{G}_p(\mathbb{Q}_p^n)$ be a generalized function. We define the *value* $\widetilde{f}(x_0)$ *at a point* $x_0 \in \mathbb{Q}_p^n$ as the generalized number from $\widetilde{\mathcal{C}}$, determined by the sequence $\left\{ f_k(x_0) : k \in \mathbb{N} \right\} \in \mathcal{F}$.

This definition is correct. Indeed, by the definitions of the ideals $\mathcal{N}_p(\mathbb{Q}_p^n)$ and $\mathcal{I}$, if $[f_k(x)] \in \mathcal{N}_p(\mathbb{Q}_p^n)$ then for any element of the class $f_k(x) \in [f_k(x)]$ we have $f_k(x_0) \in \mathcal{I}$.

Let $\tilde{f} = [f_k]$, $\tilde{f} \in \mathcal{D}(\mathbb{Q}_p^n) \subset \mathcal{G}_p(\mathbb{Q}_p^n)$. According to Lemma 14.3.2 and Theorem 14.3.3, the "classical" value $\tilde{f}(x_0)$ at a point $x_0 \in \mathbb{Q}_p^n$ coincides with the generalized value at a point $\tilde{f}(x_0)$ in the sense of Definition 14.3.

Remark 14.2. In the standard theories over the reals of Colombeau, Egorov, and Livchak, generalized functions are *not uniquely* determined by their point values [66], [86], [97, 1.2.4], [194], [196]. For example, consider the generalized function $f(x) = x \cdot \delta(x) \in \mathcal{G}$. Since the Dirac δ-function is defined as the equivalence class

$$\delta(x) = [\delta(\psi_\varepsilon, x)], \qquad \delta(\psi_\varepsilon, x) = \frac{1}{\varepsilon^n} \psi\left(\frac{x}{\varepsilon}\right),$$

then $f(x) = [f(\psi_\varepsilon, x)]$, where $f(\psi_\varepsilon, x) = x \frac{1}{\varepsilon^n} \psi\left(\frac{x}{\varepsilon}\right)$. It is clear that $f(x) \neq 0$ as a generalized function. On the other hand, it is easy to show ([66], [194]) that at any point $x \in \mathbb{R}^n$ we have the point value $f(x) = 0$. Indeed, $f(\psi_\varepsilon, 0) = 0$, and for any $x_0 \neq 0$ $f(\psi_\varepsilon, x_0) \to 0$ faster than any power of ε (see the definition of the Colombeau ideal $\mathcal{N}(\Omega)$ in Chapter 14, Section 14.2).

For lack of the point value characterization, in [196, Remark 2.11.], [97, 1.2.4.] a new type of characterization of elements of the real Colombeau and Egorov algebras was introduced. The basic idea of this characterization is to introduce *generalized points* which are analogues of *nonstandard numbers* (see, e.g., [7]). In particular, in the framework of the Egorov algebra on the set

$$\Omega_c = \{x_k \in \Omega : \exists K \subset\subset \Omega \, \exists N \in \mathbb{N} : x_k \in K, \, \forall k \geq N\}$$

an equivalence relation was introduced by $x_k \sim y_k$ if for some $N \in \mathbb{N}$ $x_k = y_k$ for all $k \geq N$. Then $\Omega_{c0} = \Omega_c / \sim$ is the set of compactly supported points of Ω_c. In the above mentioned papers the following theorem was proved: $f = 0$ in $\widehat{\mathcal{G}}(\Omega)$ iff $f(z) = 0$ in $\widetilde{\mathcal{C}}$ for all $z \in \Omega_{c0}$.

In the p-adic case generalized functions also are *not uniquely* determined by their point values [183]. In the paper [183], following an idea due to M. Kunzinger and M. Oberguggenberger [196], a *generalized point value characterization* of the p-adic Colombeau–Egorov algebra is given.

Example 14.4.3. Let $n = 1$.

(i) For point values of the δ-function we obtain the following result: if $x_0 = 0$ then $\delta(x_0) = \tilde{c}$, and if $x_0 \neq 0$ then $\delta(x_0) = 0$, where $\tilde{c} \in \widetilde{\mathcal{C}}$ is the generalized number defined by the sequence $\{c_k = p^k \Omega(0) = p^k : k \in \mathbb{N}\}$.

(ii) Consider the generalized function $\tilde{f}(x) = |x|_p$, $\tilde{f} \in \mathcal{D}'(\mathbb{Q}_p) \subset \mathcal{G}_p(\mathbb{Q}_p)$. In the classical sense, we have $|0|_p = 0$. Since the generalized function $\tilde{f}$ is defined by the sequence $\left(|x|_p\right)_k = p^k \int |x - \xi|_p \Omega(p^k|\xi|_p)\, d\xi$, the value $\tilde{f}(0)$ is a generalized number determined by the number sequence

$$\left(|0|_p\right)_k = p^k \int |\xi|_p \Omega(p^k|\xi|_p)\, d\xi = p^k \int_{B_{-k}} |\xi|_p\, d\xi = \frac{p^{-k+1}}{1+p}.$$

It is clear that $\left(|0|_p\right)_k \notin \mathcal{I}$, and, consequently, this sequence determines a generalized number $\tilde{f}(0) \in \tilde{C}$ which is not equal to zero. Thus for generalized functions $\tilde{f} \notin \mathcal{D}(\mathbb{Q}_p^n)$ the classical value *may not coincide* with the point value in the sense of generalized functions.

14.4.4. The Fourier transform

Let $\tilde{f}(x) = \left[f_k(x)\right]$ be a generalized function from $\mathcal{G}_p(\mathbb{Q}_p^n)$. Its Fourier transform is the generalized function $F[\tilde{f}](\xi) = \left[F[f_k(x)](\xi)\right]$. This definition describes $F[\tilde{f}](\xi)$ uniquely, since $F[f_k(x)](\xi) \in \mathcal{D}(\mathbb{Q}_p^n)$, and if $f_k(x) \in \mathcal{N}_p(\mathbb{Q}_p^n)$ then there is $N \in \mathbb{N}$ such that $F[f_k(x)](\xi) = 0$ for all $k \geq N$, i.e., $F[\mathcal{N}_p(\mathbb{Q}_p^n)] \subset \mathcal{N}_p(\mathbb{Q}_p^n)$.

14.4.5. Convolution

The convolution $\tilde{f} * \tilde{g}$ of the generalized functions $\tilde{f} = \left[f_k\right]$, $\tilde{g} = \left[g_k\right]$ is defined as the generalized function

$$\tilde{h}(x) = \left(\tilde{f} * \tilde{g}\right)(x) = \left[h_k(x)\right], \quad x \in \mathbb{Q}_p^n, \qquad (14.4.1)$$

where

$$h_k(x) = f_k(x) * g_k(x) = \int f_k(\xi) g_k(x - \xi)\, d\xi.$$

Definition (14.4.1) determines the convolution $\tilde{f} * \tilde{g}$ uniquely. Indeed, here $h_k(x) = \int f_k(\xi) g_k(x - \xi)\, d\xi \in \mathcal{D}(\mathbb{Q}_p^n)$, and if $f_k(x) \in \mathcal{N}_p(\mathbb{Q}_p^n)$ or $g_k(x) \in \mathcal{N}_p(\mathbb{Q}_p^n)$ then $h_k(x) = \int f_k(\xi) g_k(x - \xi)\, d\xi \in \mathcal{N}_p(\mathbb{Q}_p^n)$.

The convolution (14.4.1) turns $\mathcal{G}_p(\mathbb{Q}_p^n)$ into an *associative convolution algebra*.

It is clear that

$$F[\tilde{f} * \tilde{g}] = F[\tilde{f}]F[\tilde{g}], \qquad \tilde{f}, \tilde{g} \in \mathcal{G}_p(\mathbb{Q}_p^n).$$

Example 14.4.4. Since according to (4.8.10), $\delta_k(x) = F[\Delta_k(\xi)](x)$ and $\delta = \left[\delta_k\right]$, $1 = \left[\Delta_k\right]$, then

$$1 * 1 = F[\delta^2].$$

14.5. Fractional operators in the Colombeau–Egorov algebra

It is known that the space of test functions $\mathcal{D}(\mathbb{Q}_p^n)$ is not invariant under the fractional operators $D_\times^\alpha$, $\alpha \in \mathbb{C}^n$, and D^α, $\alpha \in \mathbb{C}$. Indeed, since both functions $|\xi_j|_p^\alpha$ and $|\xi_j|_p^\alpha F[\varphi](\xi)$, $\xi \in \mathbb{Q}^n$, are not locally constant on $\mathbb{Q}_p^n$, in general $D_{x_j}^\alpha \varphi \notin \mathcal{D}(\mathbb{Q}_p^n)$ for $\varphi \in \mathcal{D}(\mathbb{Q}_p^n)$. Since, according to Lemma 4.8.3, we have $F[\varphi](\xi) \in \mathcal{D}_{-l}^{-N}(\mathbb{Q}_p^n)$ for $\varphi \in \mathcal{D}_N^l(\mathbb{Q}_p^n)$ then $|\xi_j|_p^\alpha F[\varphi](\xi) \in \mathcal{D}_{-l}^{-k}(\mathbb{Q}_p^n)$ for all $\xi \in \mathbb{Q}_p^n \setminus B_{-k}$, $B_{j;-k} = \{\xi \in \mathbb{Q} : |\xi_j|_p \le p^{-k}\}$, $k > N$, and

$$\int_{\mathbb{Q}_p^n \setminus B_{j;-k}} |\xi_j|_p^l F[\varphi](\xi)\chi_p(-\xi x)\, d\xi \in \mathcal{D}_k^l(\mathbb{Q}_p^n), \quad k > N, \quad j = 1, \ldots, n.$$

Taking into account the above fact, we introduce the fractional operators on generalized functions $\tilde{f} \in \mathcal{G}_p(\mathbb{Q}_p^n)$, using some ideas from Chapter 7.

Let us define the fractional operators (studied in Chapter 9, Section 9.2) in the Colombeau–Egorov algebra.

Definition 14.4. Let $\tilde{f} = [f_k]$ be a generalized function from $\mathcal{G}_p(\mathbb{Q}_p^n)$. The Vladimirov operator of $\tilde{f}$ is the generalized function $D_\times^\alpha \tilde{f}$, $\alpha \in \mathbb{C}^n$, defined by the sequence $\{D_{(\times,k)}^\alpha f_k : k \in \mathbb{N}\}$, where the operator $D_{(\times,k)}^\alpha$ is defined by

$$\left(D_{(\times,k)}^\alpha f_k\right)(x) = F^{-1}[|\xi|_p^\alpha \Upsilon_{(\times,-k)}(\xi) F[f_k](\xi)], \quad \alpha \in \mathbb{C}^n,$$

and $\Upsilon_{(\times,-k)}(\xi) = \prod_{j=1}^n (1 - \Delta_{-k}(\xi_j))$ is the characteristic function of the set $\{\xi \in \mathbb{Q}_p^n : |\xi_j|_p > p^{-k}, j = 1, \ldots, n\}$, $\Delta_{-k}(\xi_j) = \Omega(p^k |\xi_j|_p)$, $j = 1, \ldots, n$. Here $|\xi|_p^\alpha = |\xi_1|_p^{\alpha_1} \cdots |\xi_n|_p^{\alpha_n}$.

Theorem 14.5.1. (a) *If $\tilde{f} \in \mathcal{G}_p(\mathbb{Q}_p^n)$ then $D_\times^\alpha \tilde{f}$ is a uniquely determined generalized function, i.e., Definition 14.4 is correct.*

(b) *The family of operators $\{D_\times^\alpha : \alpha \in \mathbb{C}^n\}$ on the space of generalized function $\mathcal{G}_p(\mathbb{Q}_p^n)$ forms an n-parametric abelian group: if $\tilde{f} \in \mathcal{G}_p(\mathbb{Q}_p^n)$ then*

$$D_\times^\alpha D_\times^\beta \tilde{f} = D_\times^\beta D_\times^\alpha \tilde{f} = D_\times^{\alpha+\beta} \tilde{f},$$
$$D_\times^\alpha D_\times^{-\alpha} \tilde{f} = \tilde{f}, \quad \alpha, \beta \in \mathbb{C}^n. \tag{14.5.1}$$

Proof. (a) Let $\tilde{f} = [f_k] \in \mathcal{G}_p(\mathbb{Q}_p^n)$. Since $F[f_k] \in \mathcal{D}(\mathbb{Q}_p^n)$ for $f_k \in \mathcal{D}(\mathbb{Q}_p^n)$, then for all $\alpha \in \mathbb{C}^n$ we have $\Upsilon_{(\times,-k)}(\xi) F[f_k](\xi) = 0$, $|\xi|_p^\alpha \Upsilon_{(\times,-k)}(\xi) F[f_k](\xi) = 0$ for $\xi_j = 0$, $j = 1, 2, \ldots, n$. Thus,

$$\Upsilon_{(\times,-k)}(\xi) F[f_k](\xi), \quad |\xi|_p^\alpha \Upsilon_{(\times,-k)}(\xi) F[f_k](\xi) \in \Psi_\times(\mathbb{Q}_p^n),$$

and, consequently, $\left(D_{(\times,k)}^\alpha f_k\right) \in \Phi_\times(\mathbb{Q}_p^n) \subset \mathcal{D}(\mathbb{Q}_p^n)$, where the spaces $\Psi_\times(\mathbb{Q}_p^n)$ and $\Phi_\times(\mathbb{Q}_p^n)$ are defined in Chapter 7, Section 7.3.1 (here $\Phi_\times(\mathbb{Q}_p^n)$ is the Lizorkin space of test functions of the first kind). If $f_k \in \mathcal{N}_p(\mathbb{Q}_p^n)$ then there is $N \in \mathbb{N}$ such that $F[f_k(x)](\xi) = 0$ for all $k \geq N$, and $\left(D^{\alpha}_{(\times,k)} f_k \right)(x) = 0$ for all $k \geq N$, i.e., $D^{\alpha}_{(\times,k)} \mathcal{N}_p(\mathbb{Q}^n_p) \subset \mathcal{N}_p(\mathbb{Q}^n_p)$. Thus, $D^{\alpha}_{\times} \tilde{f}$ is a uniquely determined generalized function, i.e., Definition 14.4 is correct.

(b) Since $\left(D^{\alpha}_{(\times,k)} f_k \right)(x) \in \Phi_{\times}(\mathbb{Q}^n_p)$, and according to Lemma 9.2.2, the Lizorkin space $\Phi_{\times}(\mathbb{Q}^n_p)$ is invariant under the Vladimirov fractional operator, we have

$$\left(D^{\beta}_{(\times,k)} D^{\alpha}_{(\times,k)} f_k \right)(x)$$
$$= F^{-1}[|\xi|^{\beta}_p \Upsilon_{(\times,-k)}(\xi) |\xi_j|^{\alpha}_p \Upsilon_{(\times,-k)}(\xi) F[f_k](\xi)], \quad \alpha, \beta \in \mathbb{C}^n.$$

Taking into account that $\left(\Upsilon_{(\times,-k)}(\xi) \right)^2 = \Upsilon_{(\times,-k)}(\xi)$, we deduce that

$$\left(D^{\beta}_{(\times,k)} D^{\alpha}_{(\times,k)} f_k \right)(x) = F^{-1}[|\xi|^{\alpha+\beta}_p \Upsilon_{(\times,-k)}(\xi) F[f_k](\xi)] = \left(D^{\alpha+\beta}_{(\times,k)} f_k \right)(x).$$

for any $\alpha, \beta \in \mathbb{C}^n$. Thus (14.5.1) holds. $\qquad \square$

Definition 14.5. Let $\tilde{f} = \left[f_k \right]$ be a generalized function from $\mathcal{G}_p(\mathbb{Q}^n_p)$. The Taibleson operator of $\tilde{f}$ is the generalized function $D^{\alpha} \tilde{f}$, $\alpha \in \mathbb{C}$, defined by the sequence $\{ D^{\alpha}_{(k)} f_k : k \in \mathbb{N} \}$, where the operator $D^{\alpha}_{(k)}$ is defined as

$$\left(D^{\alpha}_{(k)} f_k \right)(x) = F^{-1}[|\xi_j|^{\alpha}_p \Upsilon_{(-k)}(\xi) F[f_k](\xi)], \quad \alpha \in \mathbb{C}^n,$$

and $\Upsilon_{(-k)}(\xi) = 1 - \Delta_{-k}(\xi)$ is the characteristic function of the set $\{\xi \in \mathbb{Q}^n_p : |\xi|_p > p^{-k}\}$, $\Delta_{-k}(\xi) = \Omega(p^k |\xi|_p)$. Here $|\xi|^{\alpha}_p$ is defined by (7.2.1).

Theorem 14.5.2. (a) *If $\tilde{f} \in \mathcal{G}_p(\mathbb{Q}^n_p)$ then $D^{\alpha} \tilde{f}$ is a uniquely determined generalized function, i.e., Definition 14.5 is correct.*

(b) *The family of operators $\{ D^{\alpha} : \alpha \in \mathbb{C} \}$ on the space of generalized functions $\mathcal{G}_p(\mathbb{Q}^n_p)$ forms a one-parametric abelian group: if $\tilde{f} \in \mathcal{G}_p(\mathbb{Q}^n_p)$ then*

$$D^{\alpha} D^{\beta} \tilde{f} = D^{\beta} D^{\alpha} \tilde{f} = D^{\alpha+\beta} \tilde{f},$$
$$D^{\alpha} D^{-\alpha} \tilde{f} = \tilde{f}, \quad \alpha, \beta \in \mathbb{C}. \tag{14.5.2}$$

Proof. (a) Let $\tilde{f} = \left[f_k \right] \in \mathcal{G}_p(\mathbb{Q}^n_p)$. Since $F[f_k] \in \mathcal{D}(\mathbb{Q}^n_p)$ for $f_k \in \mathcal{D}(\mathbb{Q}^n_p)$, then for all $\alpha \in \mathbb{C}$ we have that $\Upsilon_{(-k)}(\xi) F[f_k](\xi) = 0$ and $|\xi|^{\alpha}_p \Upsilon_{(\times,-k)}(\xi) F[f_k](\xi) = 0$ for $\xi_j = 0$, $j = 1, 2, \ldots, n$. Thus,

$$\Upsilon_{(-k)}(\xi) F[f_k](\xi), \quad |\xi|^{\alpha}_p \Upsilon_{(-k)}(\xi) F[f_k](\xi) \in \Psi(\mathbb{Q}^n_p),$$

and, consequently, $\left(D^{\alpha}_{(k)} f_k \right)(x) \in \Phi(\mathbb{Q}^n_p) \subset \mathcal{D}(\mathbb{Q}^n_p)$, where the spaces $\Psi(\mathbb{Q}^n_p)$ and $\Phi(\mathbb{Q}^n_p)$ are defined in Chapter 7, Section 7.3.2 (here $\Phi(\mathbb{Q}^n_p)$ is the Lizorkin space of test functions of the second kind). It is easy to see that $D^{\alpha}_{(k)} \mathcal{N}_p(\mathbb{Q}^n_p) \subset \mathcal{N}_p(\mathbb{Q}^n_p)$. Thus, $D^{\alpha} \tilde{f}$ is a uniquely determined generalized function, i.e., Definition 14.5 is correct.

(b) Since $\left(D_{(k)}^{\alpha} f_k\right)(x) \in \Phi(\mathbb{Q}_p^n)$, and according to Lemma 9.2.5, the Lizorkin space $\Phi(\mathbb{Q}_p^n)$ is invariant under the Taibleson fractional operator, we have

$$\left(D_{(k)}^{\beta} D_{(k)}^{\alpha} f_k\right)(x) = F^{-1}[|\xi|_p^{\beta} \Upsilon_{(-k)}(\xi)|\xi|_p^{\alpha} \Upsilon_{(-k)}(\xi) F[f_k](\xi)], \quad \alpha, \beta \in \mathbb{C}.$$

Taking into account that $\left(\Upsilon_{(-k)}(\xi)\right)^2 = \Upsilon_{(-k)}(\xi)$, we deduce that

$$\left(D_{(k)}^{\beta} D_{(k)}^{\alpha} f_k\right)(x) = F^{-1}[|\xi|_p^{\alpha+\beta} \Upsilon_{(-k)}(\xi) F[f_k](\xi)] = \left(D_{(k)}^{\alpha+\beta} f_k\right)(x)$$

for any $\alpha, \beta \in \mathbb{C}$. Thus (14.5.2) holds. $\square$

Remark 14.3. In the general case, the fractional operators $D_{\times}^{\alpha}$, $\alpha \in \mathbb{C}^n$, and D^{α}, $\alpha \in \mathbb{C}$ are well defined on a distribution $f \in \mathcal{D}'(\mathbb{Q}_p^n)$ only as generalized functions $D_{\times}^{\alpha} f$ and $D^{\alpha} f$ from $\mathcal{G}_p(\mathbb{Q}_p^n)$. It is obvious that if $D_{\times}^{\alpha} f$ (or $D^{\alpha} f$) exists for $f \in \mathcal{D}'(\mathbb{Q}_p^n)$ as a distribution and $f = \tilde{f} \in \mathcal{G}_p(\mathbb{Q}_p^n)$ then $D_{\times}^{\alpha} f$ (or $D^{\alpha} f$) coincides with the generalized function $D_{\times}^{\alpha} \tilde{f}$ (or $D^{\alpha} \tilde{f}$).

Definitions 14.4, 14.5 and formula (4.9.8) imply that for $\tilde{f}, \tilde{g} \in \mathcal{G}_p(\mathbb{Q}_p^n)$ we have

$$D_{\times}^{\alpha}\left(\tilde{f} * \tilde{g}\right) = \left(D_{\times}^{\alpha} \tilde{f}\right) * \tilde{g} = \tilde{f} * \left(D_{\times}^{\alpha} \tilde{g}\right), \quad \alpha, \beta \in \mathbb{C}^n;$$

and

$$D^{\alpha}\left(\tilde{f} * \tilde{g}\right) = \left(D^{\alpha} \tilde{f}\right) * \tilde{g} = \tilde{f} * \left(D^{\alpha} \tilde{g}\right), \quad \alpha, \beta \in \mathbb{C}.$$

14.6. The algebra $\mathcal{A}^*$ of p-adic asymptotic distributions

14.6.1. Weak asymptotics of regularizations of distributions

Using formula (14.3.1), we construct the regularizations of the p-adic QAHDs introduced in Chapter 6. The regularization of the δ-function is given by (4.4.6). According to (14.3.1) and (6.3.2), the regularization of the distribution $\pi_{\alpha}(x) \log_p^m |x|_p$, $m \in \mathbb{N}_0$ is given by the formula:

$$\pi_{\alpha;m\ k}(x) = \pi_{\alpha}(x) \log_p^m |x|_p * \delta_k(x)$$

$$= \left\langle \pi_{\alpha}(\xi) \log_p^m |\xi|_p, p^k \Omega(p^k |x - \xi|_p) \right\rangle = \sum_{j=1}^{3} \pi_{\alpha;m\ k}^{(j)}(x), \quad (14.6.1)$$

where

$$
\pi^{(1)}_{\alpha;m\ k}(x) = p^k \int_{B_0} |\xi|_p^{\alpha-1} \pi_1(\xi) \log_p^m |\xi|_p
$$

$$
\times \Big(\Omega(p^k|x - \xi|_p) - \Omega(p^k|x|_p) \Big) d\xi,
$$

$$
\pi^{(2)}_{\alpha;m\ k}(x) = p^k \int_{\mathbb{Q}_p \backslash B_0} |\xi|_p^{\alpha-1} \pi_1(\xi) \log_p^m |\xi|_p \Omega(p^k|x - \xi|_p) d\xi, \qquad (14.6.2)
$$

$$
\pi^{(3)}_{\alpha;m\ k}(x) = p^k \Omega(p^k|x|_p) \int_{B_0} |\xi|_p^{\alpha-1} \pi_1(\xi) \log_p^m |\xi|_p d\xi.
$$

Here $\pi_{\alpha\ k}(x) = \pi_{\alpha;0\ k}(x)$. In view of (14.3.1) and (6.3.5), the regularization of the distribution $P\big(|x|_p^{-1} \log_p^m |x|_p\big)$, $m \in \mathbb{N}_0$, is given by the formula:

$$
P_{m\ k}(x) = P\big(|x|_p^{-1} \log_p^m |x|_p\big) * \delta_k(x)
$$

$$
= \langle P\big(|x|_p^{-1} \log_p^m |x|_p\big), p^k \Omega(p^k|x - \xi|_p) \rangle = \sum_{j=1}^{2} P^{(j)}_{m\ k}(x). \qquad (14.6.3)
$$

If we set $\alpha = 0$ and $\pi_1(\xi) = 1$ then $P^{(1)}_{m\ k}(x)$, $P^{(2)}_{m\ k}(x)$ are defined by formula (14.6.2). We set $P_k(x) = P_{0\ k}(x)$.

Let $\mathcal{A}$ be the linear span of the set of QAHDs and let

$$
\mathcal{H} = \{ f_k(x) = f(x, p^{-k}), k \in \mathbb{N} : f \in \mathcal{A} \}
$$

be the space of regularizations of distributions from $\mathcal{A}$. It is clear that there is a one-to-one correspondence between the elements of the spaces $\mathcal{H}$ and $\mathcal{A}$.

Let us introduce an *associative and commutative* algebra $\mathcal{H}^*$ generated by the space $\mathcal{H}$. The elements $f_k^*(x) = f^*(x, p^{-k})$ of this algebra are finite sums of finite products of elements from $\mathcal{H}$. In order to construct the *p-adic algebra of asymptotic distributions* we need to construct *asymptotic expansions in the weak sense (weak asymptotics)* of elements of the algebra $\mathcal{H}^*$, as $k \to \infty$, i.e., the asymptotic expansions of $\langle f_k^*, \varphi \rangle$, $k \to \infty$, where $f_k^* \in \mathcal{H}^*$, $\varphi \in \mathcal{D}(\mathbb{Q}_p)$.

Now we state the basic theorem whose proof we relegate to Appendix C (due to its technical character involving a large amount of calculations). This theorem forms the basis of our construction of the algebra of *asymptotic distributions*.

Theorem 14.6.1. *Let $f_k^* \in \mathcal{H}^*$. Then for any test function $\varphi \in \mathcal{D}(\mathbb{Q}_p)$ there exists a unique representation in the form of a weak asymptotics*

$$
\langle f_k^*, \varphi \rangle = \langle f_0, \varphi \rangle + \varphi(0) \sum_{m=1}^{M} \sum_{n=0}^{N} c_{mn} p^{-\lambda_m k} \log_p^{N-n}(p^{-k}) + o\big(p^{-\lambda_M k}\big),
$$

$$
k \to \infty, \qquad (14.6.4)
$$

where $f_0(x)$ is a distribution from the space $\mathcal{A}$, $\lambda_m \in \mathbb{R}$ is an increasing sequence, $M \in \mathbb{N}$, $N \in \mathbb{N}_0$, c_{mn} are constants (which depend on f_k^*). In particular, if $f_k^* \in \mathcal{H} \subset \mathcal{H}^*$ then in the representation (14.6.4) $c_{mn} = 0$ for all m, n. Moreover, there exists a stabilization parameter $s(\varphi)$ depending on φ, such that for all $k > s(\varphi)$ the remainder $o(p^{-\lambda_M k})$ disappears and the asymptotic expansion (14.6.4) turns into an exact equality (cf. Lemma 14.3.1). We will write the weak asymptotics (14.6.4) as

$$f_k^*(x) = f_0(x) + \delta(x) \sum_{m=1}^{M} \sum_{n=0}^{N} c_{mn} p^{-\lambda_m k} \log_p^{N-n}(p^{-k}), \quad k \to \infty. \quad (14.6.5)$$

Remark 14.4. In our construction

$$p^{-\lambda_m k} \log_p^{N-n}(p^{-k}) = p^{-\lambda_m k}(-k)^{N-n}, \quad k \to \infty,$$

$n = 0, 1, \ldots, N, m = 1, 2, \ldots, M$, is an asymptotic sequence which is an analog of the asymptotic sequence $\varepsilon^{\alpha_m} \log^n \varepsilon$, $\varepsilon \to +0$, arising in the construction of the real associative algebra of *asymptotic distributions* [68], [149], [150], [220]; the role of the small parameter ε is played by p^{-k} (see Section 14.2.2).

Remark 14.5. For $f_k^* \in \mathcal{H}^*$ we can construct a weak asymptotics of the type (14.6.5) from the superposition $g(f_k^*(x))$, $k \to \infty$, where $g(u) \in C^0(\mathbb{R})$. For example, if $g(0) = 0$ then

$$g(\delta_k(x)) = \frac{g(p^k)}{p^k} \delta(x), \quad k \to \infty. \quad (14.6.6)$$

Indeed, for all $\varphi \in \mathcal{D}(\mathbb{Q}_p)$,

$$\langle g(\delta_k(\cdot)), \varphi(\cdot) \rangle = \int g(p^k \Omega(p^k |x|_p)) \varphi(x) \, dx$$

$$= g(p^k) \int_{B_{-k}} \varphi(x) \, dx = g(p^k) p^{-k} \varphi(0), \quad k \to \infty.$$

14.6.2. Asymptotic distributions

Let A^* be the linear mapping which associates each element $f_k^* \in \mathcal{H}^*$ with the unique (in view of Theorem 14.6.1) vector-valued distribution:

$$f^*(x) = \big(f_{(\lambda_m, n)}(x)\big), \quad \lambda_m \in \mathbb{R}, \quad m, n \in \mathbb{Z}_+, \quad x \in \mathbb{Q}_p, \quad (14.6.7)$$

where $\lambda_0 = 0$, $\lambda_m \in \mathbb{R}$ is an increasing sequence; and the components of the vector $f_{(\lambda_0, N)}(x) = f_0(x) \in \mathcal{A}$ and $f_{(\lambda_m, n)}(x) = c_{mn} \delta(x)$ are corresponding coefficients of the asymptotics (14.6.5); $m = 1, 2, \ldots, M$, $n = 0, 1, \ldots, N$, $M \in \mathbb{N}$, $N \in \mathbb{N}_0$; c_{mn} and M, N are constants. According to Theorem 14.6.1,

A^* associates to each element $f_k^* \in \mathcal{H} \subset \mathcal{H}^*$ the unique vector-valued distribution (14.6.7) whose components $f_{(\lambda_m, n)}$ are equal to zero except for $f_{(0,N)}(x) = \lim_{k \to \infty} f_k^* \in \mathcal{A}$. According to (14.6.5), any vector-valued distribution $f^*(x)$ is finite from the left and from the right. Theorem 14.6.1 implies that A^* is an isomorphism.

The image $f^*(x) = A^*[f_k^*(x)]$, $f_k^* \in \mathcal{H}^*$ is called a *p-adic asymptotic distribution*. By

$$\mathcal{A}^* = A^*[\mathcal{H}^*] = \{f^* = A^*[f_k^*(\cdot)] : f_k^* \in \mathcal{H}^*\}$$

we denote the *linear space of all asymptotic distributions*. In view of Theorem 14.6.1, each distribution $f(x)$ from the space $\mathcal{A}$ can be identified with the p-adic vector-valued distribution $f^*(x) = \big(f_{(\lambda_m, n)}(x)\big)$, where $f_{0,N}(x) = f(x)$, while all the other components are equal to zero, i.e., there is an embedding $\mathcal{A} \subset \mathcal{A}^*$. Let $\mathcal{D}_*$ be the linear space of all test vector-valued functions

$$\varphi_*(x) = \big(\varphi_{(\beta_m, n)}(x)\big), \quad \varphi_{(\beta_m, n)} \in \mathcal{D}(\mathbb{Q}_p),$$

where $\beta_m \in \mathbb{R}$ is an increasing sequence, $m, n \in \mathbb{N}_0$. Then a *vector-valued distribution* $f^* \in \mathcal{A}^*$ is a linear continuous functional over the space of test functions $\mathcal{D}_*$. For all $\varphi_* \in \mathcal{D}_*$ we set

$$\langle f^*, \varphi_* \rangle \stackrel{def}{=} \sum_{m=0}^{\infty} \sum_{n=0}^{\infty} \langle f_{(\alpha_m, n)}, \varphi_{(-\alpha_m, -n)} \rangle.$$

Here the sums contain only a finite number of terms. Thus the space of asymptotic distributions $\mathcal{A}^*$ is a linear space of vector-valued Schwartz distributions.

In view of the representation (14.6.5), any asymptotic distribution $f^* \in \mathcal{A}^*$ can be identified with the sum

$$f^*(x) = f_0(x) + \delta(x) \sum_{m=1}^{M} \sum_{n=0}^{N} c_{mn} p^{-\lambda_m k} \log_p^{N-n}(p^{-k}), \qquad (14.6.8)$$

where k is sufficiently large (we write it as $k \to \infty$), $f_0 \in \mathcal{A}$, $\lambda_m \in \mathbb{R}$ is an increasing sequence, M, N, c_{mn} are constants depending on f^*. In view of the stabilization property of the representation (14.6.4), this identification is natural, since the right-hand side of (14.6.8) is the final value of the corresponding sequence $f_k^* \in \mathcal{H}^*$.

It is easy to show that if f is QAHD of degree $\pi_\alpha(x)$ and order m and $f_k(x) = f(x, p^{-k})$ is its regularization (14.3.1) then for $t \in \mathbb{Q}_p^\times$

$$f(tx, |t|_p p^{-k}) = \pi_\alpha(t) f(x, p^{-k}) + \sum_{j=1}^{m} \pi_\alpha(t) \log_p^j |t|_p f_{m-j}(x, p^{-k}),$$

$$(14.6.9)$$

where $f_{m-j}(x, p^{-k}) = \langle f_{m-j}(\xi), p^k \Omega(p^k |x - \xi|_p) \rangle$ is the regularization of QAHD f_{m-j} of degree $\pi_\alpha(x)$ and order $m - j$, $j = 1, 2, \ldots, m$. Thus $f_k(x) = f(x, p^{-k})$ is a quasi associated homogeneous function of degree $\pi_\alpha(x)$ and order m with respect to the variables (x, p^{-k}). In view of (14.6.9) and Definition 6.2, we introduce the following definition.

Definition 14.6. A weak asymptotic distribution $f^* \in \mathcal{A}^*$ is called *quasi associated homogeneous* of degree $\pi_\alpha(x)$ and order $m \in \mathbb{N}_0$ if the right-hand side of the representation (14.6.8) is a quasi associated homogeneous function of degree $\pi_\alpha(x)$ and order m with respect to the variables (x, p^{-k}).

Definition 14.7. We say that the asymptotic distribution (vector-valued distribution)

$$h^*(\cdot) = f^*(\cdot) \circ g^*(\cdot) = A^* \big[f_k^*(\cdot) g_k^*(\cdot) \big] \in \mathcal{A}^* \qquad (14.6.10)$$

is the *product* of the asymptotic distributions (vector-valued distributions) f^* and g^* from $\mathcal{A}^*$ (or distributions from $\mathcal{A}$). Here $f_k^*(\cdot) = A^{*-1} f^*(\cdot) \in \mathcal{H}^*$ and $g_k^*(\cdot) = A^{*-1} g^*(\cdot) \in \mathcal{H}^*$ are regularizations of f^* and g^*, respectively.

Since $\mathcal{H}^*$ is an associative and commutative algebra of functions, and $A^* : \mathcal{H}^* \to \mathcal{A}^*$ is an isomorphism, the multiplication (14.6.10) is *associative and commutative*.

It is clear that if $f^* \in \mathcal{A}^*$ is an associated homogeneous asymptotic distribution of degree $\pi_\alpha(x)$ and order m, and $g^* \in \mathcal{A}^*$ is an associated homogeneous asymptotic distribution of degree $\pi_\beta(x)$ and order n (in the sense of Definition 14.6) then their product $f^*(\cdot) g^*(\cdot) \in \mathcal{A}^*$ is an associated homogeneous asymptotic distribution of degree $\pi_\alpha(x) \pi_\beta(x)$ and order $m + n$.

Example 14.6.1. Using Propositions C.1–C.9, we can calculate the following products of distributions.

(i) $\delta^m(x) = \delta(x) p^{k(m-1)}$, $k \to \infty$;

(ii) $\delta^m(x) \circ \left(P\left(\frac{1}{|x|_p} \right) \right)^n = (-1)^n \delta(x) p^{(m+n-1)k} \log_p^n \left(p^{-k} \right) \left(1 - p^{-1} \right)^n$,

as $k \to \infty$;

(iii) If $\alpha, \beta \neq \alpha_j = \frac{2\pi i}{\ln p} j$, $m(\alpha - 1) + n(\beta - 1) + 1 = \frac{2\pi i}{\ln p} j$, $j \in \mathbb{Z}$ and $\pi_1(x) \equiv 1$ then

$$\pi_\alpha^m(x) \circ \pi_\beta^n(x) = P\left(\frac{1}{|x|_p} \right) + I_0^m(\alpha) I_0^n(\beta) \delta(x) - (1 - p^{-1}) \log_p \left(p^{-k} \right) \delta(x),$$

as $k \to \infty$, where $m, n \in \mathbb{N}$.

Here the asymptotic distributions (products of distributions) $\delta^m(x)$, $\delta^m(x) \circ \left(P\left(\frac{1}{|x|_p}\right)\right)^n$, and $\pi_\alpha^m(x) \circ \pi_\beta^n(x)$ belong to the algebra $\mathcal{A}^*$ and are represented in the form (14.6.8). These products can be represented as vector distributions (14.6.7).

According to Definition 14.6, the product (ii) is an associated homogeneous asymptotic distribution of degree $|x|_p^{-m-n}$ and order n; the products (i) and (iii) are homogeneous asymptotic distributions of degrees $|x|_p^{-m}$ and $|x|_p^{m(\alpha-1)+n(\beta-1)} \pi_1^{m+n}(x)$, respectively.

(iv) Using formula (14.6.6), we define the *singular superposition*

$$g\big(\delta(x)\big) = g(p^k)p^{-k}\delta(x), \quad k \to \infty \quad \in \mathcal{A}^*,$$

where $g(u) \in C^0(\mathbb{R})$. Using asymptotics similar to (14.6.6), we can define a singular superposition $g\big(f^*(x)\big)$, where $f^* \in \mathcal{A}^*$ is an asymptotic distribution, $g(u)$ is a smooth function.

Note that the right-hand sides of the above products can be represented as vector-valued distributions.

We define the Fourier transform of the *asymptotic distribution* (*vector-valued distribution*) (14.6.8) from $\mathcal{A}^*$ by the formula

$$g(\xi) = F[f^*](\xi) = F[f_0](\xi) + \sum_{m=1}^{M} \sum_{n=0}^{N} c_{mn} p^{-\lambda_m k} \log_p^{N-n}(p^{-k}), \quad k \to \infty.$$

$$(14.6.11)$$

According to Theorem 6.4.1, $F : \mathcal{A}^* \to \mathcal{A}^*$.

14.6.3. $\mathcal{A}^*$ as a convolution algebra.

The *generalized convolution* of asymptotic distributions $g_1^*(\xi)$, $g_2^*(\xi)$ from $\mathcal{A}^*$ (or distributions from $\mathcal{A}$) is defined as the asymptotic (vector-valued) distribution

$$\begin{aligned} g^*(\xi) &= g_1^*(\xi) \star g_2^*(\xi) \\ &= F^{-1}\big[F[g_1^*(\xi)](x) \circ F[g_1^*(\xi)](x) \big] \in \mathcal{A}^*. \end{aligned} \quad (14.6.12)$$

Since the algebra $\mathcal{A}^*$ is closed under multiplication (14.6.10) and the Fourier transform (according to Theorem 6.4.1), $\mathcal{A}^*$ is also an associative and commutative convolution algebra. Note that in the general case the associativity does not hold for the ordinary p-adic distributional convolution [241, IX.1.].

Example 14.6.2. Examples 14.6.1 and 6.4.1 imply:

(i) $1 * 1 = p^k, \quad k \to \infty;$

(ii) $1 * \left(\dfrac{p-1}{p \ln p} \ln |\xi|_p + \dfrac{1}{p} \right) = p^k \log_p \left(p^{-k} \right) \left(1 - p^{-1} \right), \quad k \to \infty.$

14.7. $\mathcal{A}^*$ as a subalgebra of the Colombeau–Egorov algebra

According to Sections 14.3.1 and 14.6.1, we have

$$\mathcal{H} \subset \mathcal{H}^* \subset \mathcal{P}(\mathbb{Q}_p), \qquad \mathcal{H}^* \cap \mathcal{N}_p(\mathbb{Q}_p) = 0.$$

In view of the latter relations it is clear that the mappings

$$\mathcal{H} \ni f_k(x) \mapsto f_k(x) + \mathcal{N}_p(\mathbb{Q}_p) \in \mathcal{G}_p(\mathbb{Q}_p),$$
$$\mathcal{H}^* \ni f_k^*(x) \mapsto f_k^*(x) + \mathcal{N}_p(\mathbb{Q}_p) \in \mathcal{G}_p(\mathbb{Q}_p),$$
$$(14.7.1)$$

are one-to-one.

Taking into account (14.7.1) and the fact that $A^* : \mathcal{H}^* \to \mathcal{A}^*$ is an isomorphism, we can identify the product

$$\left(f_k^*(x) + \mathcal{N}_p(\mathbb{Q}_p) \right) \left(g_k^*(x) + \mathcal{N}_p(\mathbb{Q}_p) \right) \in \mathcal{G}_p(\mathbb{Q}_p)$$

with the *asymptotic distribution* $A^* \left[f_k^*(x) g_k^*(x) \right] \in \mathcal{A}^*$. Therefore, by (14.7.1) we can define the mapping $\mathcal{A}^* \to \mathcal{G}_p(\mathbb{Q}_p)$ which is an algebra monomorphism. Thus the product (14.6.10) in the algebra $\mathcal{A}^*$ coincides with the corresponding product in the algebra $\mathcal{G}_p(\mathbb{Q}_p)$.

Appendix A

The theory of associated and quasi associated homogeneous real distributions

A.1. Introduction

In this appendix we construct and study *associated homogeneous distributions* (*AHDs*) and *quasi associated homogeneous distributions* (*QAHDs*) for the real case. These results are based on the paper [223]. The results of this appendix are used in Chapter 6 to develop the theory of p-adic associated and quasi associated homogeneous distributions.

The concept of AHD was first introduced and studied in the book [95, Ch.I, §4.1.] (see Definitions A.2 and A.3 by analogy with the notion of the *associated eigenvector* (A.2.2)). Later the concept of an AHD was introduced in the paper [232, Ch.X, 8.] by Definition A.4, and in the books [87, (2.6.19)], [88, (2.110)] by Definition A.5. In the book [95, Ch.I, §4] and in the paper [232, Ch.X, 8.] a theorem was given (without proof), in which all AHDs were described (see Proposition A.2.1). In Section A.2.2 we discuss and analyse Definitions A.3, A.4, A.5, (A.2.9) of an *AHD* and show that they are *self-contradictory* for $k \geq 2$. Moreover, these definitions come into conflict with Proposition A.2.1. According to Section A.2.2, there exist *only* AHDs of order $k = 0$, i.e., *homogeneous distributions* (*HDs*) (given by Definition A.1) and of order $k = 1$ (given by Definition (A.2.3) or Definition A.2). Thus one can see that the concept of an *AHD* requires a special study.

In Section A.3, we study the symmetry of the class of distributions mentioned in Proposition A.2.1 under the action of dilatation operators U_a, $a > 0$. We prove that the dilatation operator U_a acts in this class (and, consequently, in the space $\mathcal{AH}_0(\mathbb{R})$) according to formulas (A.3.7) and (A.3.11) but not according to Definition A.3.

According to the above result, a *direct transfer* of the notion of *associated eigenvector* to the case of distributions *is impossible for $k \geq 2$*. This is connected with the fact that any HD is an eigenfunction of *all* dilatation operators

285

$U_a f(x) = f(ax)$ (for *all* $a > 0$), while for $k \geq 2$ *no* distribution $x_{\pm}^{\lambda} \log^k x_{\pm}$, $P\left(x_{\pm}^{-n} \log^{k-1} x_{\pm}\right)$ is an AHD for *all* the dilatation operators.

In Section A.4.1, in view of the results of Section A.3, a natural generalization of Definition A.3 is given. Namely, we introduce Definition A.8 of a *distribution of degree λ and of order k* by the relation:

$$U_a f_k(x) = f_k(ax) = a^{\lambda} f_k(x) + \sum_{r=1}^{k} a^{\lambda} \log^r a f_{k-r}(x), \ \forall \, a > 0, \quad (A.1.1)$$

where $f_{\lambda;k-r}(x)$ is a *distribution of degree λ and of order $k - r$*, $r = 1, 2, \ldots, k$; $k = 0, 1, 2, \ldots$. Here for $k = 0$ we suppose that the sum in the right-hand side of (A.1.1) is empty. By Theorem A.4.2 the linear span $\mathcal{AH}_1(\mathbb{R})$ of all *distributions of degree λ and of order k* ($\lambda \in \mathbb{C}$, $k = 0, 1, 2, \ldots$) is described. Moreover, it is proved that $\mathcal{AH}_1(\mathbb{R}) = \mathcal{AH}_0(\mathbb{R})$.

In Section A.4.2, taking into account Definition A.8, we give a natural generalization of the notion of an *associated eigenvector* (A.2.2) and introduce Definition A.9 of a *quasi associated homogeneous distribution (QAHD)* of degree λ and order k by the relation:

$$U_a f_k(x) = f_k(ax) = a^{\lambda} f_k(x) + \sum_{r=1}^{k} h_r(a) f_{k-r}(x), \ \forall a > 0,$$

where $f_{k-r}(x)$ is a QAHD of order $k - r$, $h_r(a)$ is a differentiable function, $r = 1, 2, \ldots, k$; $k = 0, 1, 2, \ldots$. (For $k = 0$ we suppose that the sum in the right-hand side of the latter relation is empty.) Here the dilatation operator U_a acts as a discrete convolution

$$U_a f_k(x) = f_k(ax) = \big(f(x) * h(a)\big)_k$$

of a sequence $f(x) = \{f_0(x), f_1(x), f_2, \ldots\}$ and the sequence $h(a) = \{h_0(a), h_1(a), h_2(a), \ldots\}$, where $h_0(a) = a^{\lambda}$.

Thus the QAHD of order k is reproduced by the dilatation operator U_a (for all $a > 0$) up to a linear combination of QAHDs of orders $k - 1, k - 2, \ldots, 0$ (see (A.4.7)).

According to the main Theorem A.4.3, in order to introduce a QAHD of degree λ and order k we can use Definition A.10 instead of Definition A.9, i.e., the relation (A.1.1).

Remark A.1. Comparing Definition A.10, i.e., the relation (A.1.1), and Definition A.7 (see Grudzinski's book [99, Definition 2.28.]), we can see that *quasi associated homogeneous distributions* are a special case of *almost quasihomogeneous distributions*. The motivation for the concept of *almost quasihomogeneous distributions* in [99] is not connected with the problem of *associated* *eigenvectors* (A.2.2). In [99, Ch. 2, (c)] *almost quasihomogeneous distributions* appear as zero order Laurent coefficients of certain meromorphic functions of *quasihomogeneous distributions*.

From Theorems A.4.2 and A.4.3, the linear span $\mathcal{AH}(\mathbb{R})$ of all QAHDs coincides with the Gel'fand–Shilov class of distributions

$$\mathcal{AH}_0(\mathbb{R}) = \text{span}\{x_\pm^\lambda \log^k x_\pm, \ P(x_\pm^{-n} \log^{m-1} x_\pm) :$$
$$\lambda \neq -1, -2, \dots, -n, \dots; \quad n, m \in \mathbb{N}, \ k \in \{0\} \cup \mathbb{N}\},$$

introduced in [95, Ch.I, §4.] as the class of AHDs (see Proposition A.2.1). From Definitions A.1, A.2 and A.10, the classes of QAHDs of orders $k = 0$ and $k = 1$ coincide with those of HDs and AHDs of order $k = 1$, respectively.

In Section A.5, multidimensional QAHDs are introduced. By Theorem A.5.2 it is proved that f_k is a QAHD of order k, $k \geq 1$, if and only if it satisfies the Euler type system of differential equations (A.5.2). The Euler type system (A.5.2) can be rewritten as (A.5.14). This result generalizes the well-known classical statement for homogeneous distributions (see Theorem A.5.1). It is a special case of Proposition 2.31 from Grudzinski's book [99].

In Section A.6, a mathematical description of the Fourier transform of QAHDs is given for the multidimensional case. In Section A.7, Γ-functions of a new type generated by QAHDs are defined. In particular, these Γ-functions are calculated and their properties derived for $k = 1$.

Let us note that our results can be used in the theory of noncommutative residues for the algebra of pseudo-differential operators with log-polyhomogeneous symbols [175]. These symbols are represented as sums of *QAHDs*. Some areas for applications can also be found in the book [99].

A.2. Definitions of associated homogeneous distributions and their analysis

A.2.1. Associated homogeneous distributions

The concept of the one-dimensional *associated homogeneous distribution* (*AHD*) was first introduced and studied in the book [95, Ch.I, §4.1.]. Later, *AHDs* were studied in the books [87, (2.6.19)] and [88, (2.110)]. Let us repeat the main ideas and definitions of these books (we shall discuss in A.2.2 some problems which arise with these definitions). Define the dilatation operator on the space $\mathcal{D}'(\mathbb{R})$ by the formula $U_a f(x) = f(ax)$, $a > 0$. The definition of a *homogeneous distribution* (*HD*) is as follows.

Definition A.1. ([95, Ch.I, §3.11., (1)], [232, Ch.X, 8.], [106, 3.2.]) A distribution $f_0 \in \mathcal{D}'(\mathbb{R})$ is said to be *homogeneous* of degree λ if for any $a > 0$ and $\varphi \in \mathcal{D}(\mathbb{R})$ we have

$$\left\langle f_0(x), \varphi\left(\frac{x}{a}\right) \right\rangle = a^{\lambda+1} \langle f_0(x), \varphi(x) \rangle,$$

i.e.,

$$U_a f_0(x) = f_0(ax) = a^\lambda f_0(x). \qquad (A.2.1)$$

Thus a HD of degree λ is an eigenfunction of *any* dilatation operator U_a, $a > 0$ with the eigenvalue a^λ, where $\lambda \in \mathbb{C}$, and $\mathbb{C}$ is the set of complex numbers.

It is well known that in addition to an eigenfunction belonging to a given eigenvalue, a linear transformation will ordinarily also have so-called *associated functions* of various orders [95, Ch.I, §4.1.]. The functions $f_1, f_2, \ldots, f_k, \ldots$ are said to be *associated with the eigenfunction* f_0 of the transformation U if

$$\begin{aligned} U f_0 &= c f_0, \\ U f_k &= c f_k + d f_{k-1}, \quad k = 1, 2, \ldots, \end{aligned} \qquad (A.2.2)$$

where c, d are constants. Consequently, U reproduces an *associated function* of kth order up to some *associated function* of $(k-1)$th order.

(Def 1) In view of these facts, in the book [95, Ch.I, §4.1.], by analogy with Definition (A.2.2), the following definition is introduced: a function f_1 is said to be *associated homogeneous* of order 1 and of degree λ if for any $a > 0$

$$f_1(ax) = a^\lambda f_1(x) + h(a) f_0(x), \qquad (A.2.3)$$

where f_0 is a homogeneous function of degree λ. Here, in view of (A.2.1) and (A.2.2), $c = a^\lambda$. Next, in [95, Ch.I, §4.1.] it is proved that up to a constant factor

$$h(a) = a^\lambda \log a. \qquad (A.2.4)$$

Thus, by setting in the relation (A.2.3) $c = a^\lambda$ and $d = h(a) = a^\lambda \log a$, Definition (A.2.3) takes the following form:

Definition A.2. ([95, Ch.I, §4.1., (1), (2)]) A distribution $f_1 \in \mathcal{D}'(\mathbb{R})$ is called *associated homogeneous* (AHD) of order 1 and of degree λ if for any $a > 0$ and $\varphi \in \mathcal{D}(\mathbb{R})$

$$\left\langle f_1, \varphi\left(\frac{x}{a}\right) \right\rangle = a^{\lambda+1} \langle f_1, \varphi \rangle + a^{\lambda+1} \log a \langle f_0, \varphi \rangle,$$

i.e.,

$$U_a f_1(x) = f_1(ax) = a^\lambda f_1(x) + a^\lambda \log a f_0(x),$$

where f_0 is a homogeneous distribution of degree λ.

It is clear that the class of AHDs of order $k = 0$ coincides with the class of HDs.

Finally, according to (A.2.1), (A.2.2) and (A.2.4), using $c = a^\lambda$ and $d = h(a) = a^\lambda \log a$, the following definition is introduced.

Definition A.3. ([95, Ch.I, §4.1., (3)]) A distribution $f_k \in \mathcal{D}'(\mathbb{R})$ is called an *AHD* of order k, $k = 2, 3, \ldots$, and of degree λ if for any $a > 0$ and $\varphi \in \mathcal{D}(\mathbb{R})$

$$\left\langle f_k, \varphi\left(\frac{x}{a}\right)\right\rangle = a^{\lambda+1}\langle f_k, \varphi\rangle + a^{\lambda+1}\log a\langle f_{k-1}, \varphi\rangle, \qquad (A.2.5)$$

where f_{k-1} is an *AHD* of order $k - 1$ and of degree λ.

In the book [95, Ch.I, §4] (see also the paper [232, Ch.X, 8.]) the following proposition is stated (without proof).

Proposition A.2.1. *Any AHD of order k and of degree λ is a linear combination of the following linearly independent AHDs of order k, $k = 1, 2, \ldots$, and of degree λ:*
(a) $x_\pm^\lambda \log^k x_\pm$ *for $\lambda \neq -1, -2, \ldots$;*
(b) $P\left(x_\pm^{-n} \log^{k-1} x_\pm\right)$ *for $\lambda = -1, -2, \ldots$;*
(c) $(x \pm i0)^\lambda \log^k(x \pm i0)$ *for all λ.*

The above distributions are defined in Section A.3.

Denote by $\mathcal{AH}_0(\mathbb{R})$ the linear span of distributions mentioned in Proposition A.2.1.

(Def 2) In contradiction to Definition A.3 (from the book [95]), in the paper [232, Ch.X, 8.] based on the book [95], the following definition is used.

Definition A.4. ([232, Ch.X, 8.]) A distribution $f_k \in \mathcal{D}'(\mathbb{R})$ is called an *AHD* of order k, $k = 2, 3, \ldots$, and of degree λ if for any $a > 0$

$$f_k(ax) = a^\lambda f_k(x) + a^\lambda \log^k a f_{k-1}(x),$$

where f_{k-1} is an *AHD* of order $k - 1$ and of degree λ.

Here an analog of relation (A.2.5) is used, where on the right-hand side of (A.2.5) the term $\log a$ is replaced by $\log^k a$.

In [232, Ch.X, 8.] Proposition A.2.1 is also given (without proof).

(Def 3) In the books [87] and [88], according to (A.2.1), (A.2.2), the concept of an *AHD* is defined recursively.

Definition A.5. ([87, (2.6.19)], [88, (2.110)]) An *AHD* $f_k \in \mathcal{D}'(\mathbb{R})$ of order k and of degree λ is such that for any $a > 0$

$$f_k(ax) = a^\lambda f_k(x) + a^\lambda e(a) f_{k-1}(x), \qquad (A.2.6)$$

where f_{k-1} is an *associated homogeneous distribution* of order $k - 1$ and of degree λ, and $e(a)$ is some function.

Next, in these books *it is stated* that formula (A.2.6) (i.e., formula [87, (2.6.19)], [88, (2.110)]) *implies* the relation

$$e(ab) = e(a) + e(b), \qquad (A.2.7)$$

i.e.,

$$e(a) = K \log a \qquad (A.2.8)$$

for some constant K, which can be absorbed in f_{k-1} [87, p. 67], [88, p. 76]. Finally, the authors of these books conclude that in view of (A.2.6)–(A.2.8) one can define an *AHD* of order $k - 1$ and of degree λ by the equality

$$f_k(ax) = a^\lambda f_k(x) + a^\lambda \log a f_{k-1}(x), \quad \forall a > 0, \qquad (A.2.9)$$

where f_{k-1} is an *associated homogeneous distribution* of order $k - 1$ and of degree λ.

Thus according to the above arguments from the books [87], [88], Definition (A.2.9) ([87, (2.6.19)], [88, (2.110)]) coincides with Definition A.3 ([95, Ch.I, §4.1., (3)]).

It remains to note that in the book [106], the concept of an AHD is not discussed. It is only stated that for the distribution $P(x_+^{-n})$ "the homogeneity is partly lost". However, according to Definition A.2 and Proposition A.2.1 this distribution is an AHD of order 1 and of degree $-n$, i.e., has a special symmetry.

(Def 4) We would like to give special attention to the excellent book by Olaf von Grudzinski [99]. In this book some problems related to *quasihomogeneous distributions* are studied. The definition of a *quasihomogeneous distribution* (see Definitions 1.13., 2.1. in [99]) is the following.

Definition A.6. A distribution $f \in \mathcal{D}'(\mathbb{R}^n)$ is said to be *quasihomogeneous* of degree λ and of type $p = (p_1, \ldots, p_n) \in \mathbb{R}^n \setminus \{0\}$ if

$$f(M_a x) = a^\lambda f(x), \quad \forall a > 0, \qquad (A.2.10)$$

where $M_a x = (a^{p_1} x_1, \ldots, a^{p_n} x_n)$.

Homogeneous distributions correspond to $p = (1, \ldots, 1)$.

In the book [99] the concept of *almost quasihomogeneous distributions* is also introduced (see Definitions 1.32., 2.28. in [99]) and the problems related to these distributions are studied. The definition of an *almost quasihomogeneous distribution* is as follows:

Definition A.7. A distribution $f_k \in \mathcal{D}'(\mathbb{R}^n)$ is called *almost quasihomogeneous* of degree λ (and of type $p = (p_1, \ldots, p_n) \in \mathbb{R}^n \setminus \{0\}$) and of order $\leq k$ if there exist distributions $f_{k-1}, \ldots, f_1, f_0$ such that

$$f_k(M_a x) = a^\lambda f_k(x) + \sum_{r=1}^{k} a^\lambda \omega_r(a) f_{k-r}(x), \quad \forall a > 0, \quad (A.2.11)$$

where $\omega_r(a) = \frac{1}{r!} \log^r a$. (According to [99], every distribution f_j, $j = 0, 1, \ldots, k-1$ appearing in (A.2.11) is unique; it is called *jth order deficiency* of f_k.)

In the book by Grudzinski more general definitions than Definitions A.6 and A.7 are also presented.

In the book [99] the relationship between *almost quasihomogeneous* and *associated homogeneous* distributions is not considered.

A.2.2. Analysis of definitions of AHD

Unfortunately, Definitions A.3, A.4, A.5, and (A.2.9) of an *associated homogeneous distribution (AHD)* fore-quoted in Section A.2.1 are not quite consistent and have many inaccuracies. Now we discuss and analyse these definitions of an *AHD* and show that they are *self-contradictory* for $k \geq 2$. Moreover, these definitions come into conflict with Proposition A.2.1. According to our analysis, there exist *only* AHDs of order $k = 0$, i.e., HDs (given by Definition A.1), and of order $k = 1$ (given by Definition (A.2.3) or Definition A.2).

Comments on (Def 1–2). (i) For example, according to Proposition A.2.1, $\log^2(x_\pm)$ is an AHD of order 2 and of degree 0. Nevertheless, for all $a > 0$ we have

$$\log^2(a x_\pm) = \log^2 x_\pm + 2 \log a \log x_\pm + \log^2 a.$$

which contradicts Definitions A.3 and A.4.

(ii) In Section A.3, relations (A.3.7) and (A.3.11) imply that in compliance with Proposition A.2.1, $x_\pm^\lambda \log x_\pm$ and $P(x_\pm^{-n})$ are AHDs of order $k = 1$ and of degree λ and $-n$, respectively (in the sense of Definition A.2). However, for $k \geq 2$, relations (A.3.7) and (A.3.11) imply that $x_\pm^\lambda \log^k x_\pm$ and $P(x_\pm^{-n} \log^{k-1} x_\pm)$

are not AHDs of order k (in the sense of the above Definition A.3 or Definition A.4) which contradicts Proposition A.2.1.

(iii) We also point out that using Definition A.3 for an AHD of degree λ and of order k we get contradictions to certain results relating to *distributional quasi-asymptotics*. Indeed, if we temporarily assume that an AHD of degree λ and of order k is defined by Definition A.3, in view of (A.2.5), we have the asymptotic formulas:

$$f_k(ax) = a^\lambda f_k(x) + a^\lambda \log a f_{k-1}(x), \qquad a \to \infty,$$

$$f_k\left(\frac{x}{a}\right) = a^{-\lambda} f_k(x) - a^{-\lambda} \log a f_{k-1}(x), \qquad a \to \infty.$$

Here the coefficients of the *leading terms* of both asymptotics $f_{k-1}(x)$ and $-f_{k-1}(x)$ are AHDs of degree λ and of order $k-1$.

In view of the above asymptotics, and according to [80], [240, Ch.I, Sec. 3.3., Sec. 3.4.] (see also Chapter 12, Section 12.3), the distribution f_k has the *distributional quasi-asymptotics* $f_{k-1}(x)$ at infinity with respect to the *automodel* function $a^\lambda \log a$, and the distributional *quasi-asymptotics* $-f_{k-1}(x)$ at zero with respect to the *automodel* function $a^{-\lambda} \log^k a$:

$$f_k(x) \overset{\mathcal{D}'}{\sim} f_{k-1}(x), \qquad x \to \infty \quad (a^\lambda \log a),$$
$$f_k(x) \overset{\mathcal{D}'}{\sim} -f_{k-1}(x), \qquad x \to 0 \quad (a^{-\lambda} \log a). \tag{A.2.12}$$

Here both distributional quasi-asymptotics are AHDs of degree λ and of order $k-1$ (in the sense of Definition A.3), $k \geq 2$. However, according to [80], [240, Ch.I, Sec. 3.3., Sec. 3.4.], any distributional quasi-asymptotics is a *homogeneous distribution*. Thus we have a contradiction.

Remark A.2. Let $f_k \in \mathcal{AH}_0(\mathbb{R})$ be a QAHD of degree λ and of order k, $k \geq 1$. In view of Definition A.10 (see (A.1.1)), we have the asymptotic formulas:

$$f_k(ax) = a^\lambda f_k(x) + \sum_{r=1}^k a^\lambda \log^r a f_{k-r}(x), \qquad a \to \infty,$$

$$f_k\left(\frac{x}{a}\right) = a^{-\lambda} f_k(x) + \sum_{r=1}^k (-1)^r a^{-\lambda} \log^r a f_{k-r}(x), \qquad a \to \infty. \tag{A.2.13}$$

Here the coefficients at the *leading terms* of both asymptotics are homogeneous distributions f_0 and $(-1)^k f_0$ of degree λ.

According to [80], [240, Ch.I, Sec. 3.3., Sec. 3.4.] and formulas (A.2.13), the distribution f_k has the *distributional quasi-asymptotics* $f_0(x)$ at infinity with respect to an *automodel* function $a^\lambda \log^k a$, and the distributional *quasi-asymptotics* $(-1)^k f_0(x)$ at zero with respect to an *automodel* function

$a^{-\lambda} \log^k a$:

$$f_k(x) \overset{\mathcal{D}'}{\sim} f_0(x), \qquad\qquad x \to \infty \quad (a^{\lambda} \log^k a),$$
$$f_k(x) \overset{\mathcal{D}'}{\sim} (-1)^k f_0(x), \qquad x \to 0 \quad (a^{-\lambda} \log^k a). \qquad (A.2.14)$$

In contrast to (A.2.12), both the distributional quasi-asymptotics (A.2.14) are *homogeneous distributions*. This *is in compliance* with the corresponding result from [80], [240, Ch.I, Sec. 3.3., Sec. 3.4.]: a distributional quasi-asymptotics is a *homogeneous distribution*. Thus our Definition A.10, unlike Definition A.3, implies the "correct" results relating to distributional quasi-asymptotics.

(iv) Let us make an attempt "to preserve" Definition (A.3) by some minor technical modifications.

By analogy with relation (A.2.3) we will seek a function $h_1(a)$ such that if $f_2(x)$ is an AHD of order 2 and of degree λ then for any $a > 0$

$$U_a f_2(x) = f_2(ax) = a^{\lambda} f_2(x) + h_1(a) f_1(x), \qquad (A.2.15)$$

where $f_1(x)$ is an AHD of order 1 and of degree λ.

Similarly to [95, Ch.I, §4.1.], using (A.2.15) and Definition A.2, we obtain

$$\begin{aligned} f_2(abx) &= (ab)^{\lambda} f_2(x) + h_1(ab) f_1(x) = a^{\lambda} f_2(bx) + h_1(a) f_1(bx) \\ &= a^{\lambda}\big(b^{\lambda} f_2(x) + h_1(b) f_1(x)\big) + h_1(a)\big(b^{\lambda} f_1(x) + b^{\lambda} \log b \widetilde{f_0}(x)\big) \\ &= (ab)^{\lambda} f_2(x) + \Big(a^{\lambda} h_1(b) + b^{\lambda} h_1(a)\Big) f_1(x) + h_1(a) b^{\lambda} \log b \widetilde{f_0}(x), \end{aligned}$$

where $\widetilde{f_0}(x)$ is a HD of degree λ. Then for all $a, b > 0$:

$$\big(h_1(ab) - a^{\lambda} h_1(b) + b^{\lambda} h_1(a)\big) f_1(x) - h_1(a) b^{\lambda} \log b \widetilde{f_0}(x) = 0.$$

It is easy to prove that a HD of degree λ and an AHD of order 1 and of degree λ are linearly independent (see Lemma A.4.1 below). Consequently, there are two possibilities. If $h_1(a) \equiv 0$ then, according to (A.2.15), $f_2(x)$ is a HD of degree λ. If $\widetilde{f_0}(x) \equiv 0$ then $h_1(ab) = a^{\lambda} h_1(b) + b^{\lambda} h_1(a)$, $h_1(1) = 0$. As mentioned above, the latter equation has the solution (A.2.4), and, consequently, $f_2(x)$ is an AHD of order 1 and of degree λ.

Thus even for $k = 2$ it is impossible to preserve relation (A.2.2) for all dilatation operators $U_a f(x) = f(ax)$, $a > 0$. Consequently, *it is impossible* to construct an AHD of order $k \geq 2$ defined by the relation (A.2.2) with the coefficients $c = a^{\lambda}$ and $d = h(a) = a^{\lambda} \log a$.

Remark A.3. Definitions A.2 and A.3 are given in compliance with the book [95, Ch.I, §4.1., (3)]. Thus, in the case of Definition A.2 (which defines an

AHD of order 1) one can clearly see that the distribution f_0 does not depend on a. In the case of Definition A.3 (which defines an AHD of order k for $k \geq 2$), it is not clear what the authors of the book [95, Ch.I, §4.1., (3)] mean concerning the independence of f_{k-1} from a. However, it is impossible "to preserve" the definition [95, Ch.I, §4.1.] even if we suppose that a distribution f_{k-1} may depend on the variable a.

Indeed, if we suppose that in Definition A.3 f_{k-1} may depend on a, we will need to define AHD of degree λ and of order $k \geq 2$ by the following relation

$$f_k(ax) = a^\lambda f_k(x) + e(a) f_{k-1}(x, a), \quad \forall\, a > 0,$$

where $f_{k-1}(x, a)$ is an AHD (with respect to x) of degree λ and of order $k - 1$. In this case, due to the dependence of the distribution f_{k-1} on a, it is *impossible to determine* a function $e(a)$ uniquely.

Thus, unfortunately it seems impossible to find elements satisfying Definition A.3 (from [95, Ch.I, §4.1., (3)] as well as Definition A.4 (from [232, Ch.X, 8.]) and, consequently, the recursive step for $k = 2$ is impossible.

Comments on (Def 3). Let us prove that formula (A.2.6) (i.e., formula [87, (2.6.19)], [88, (2.110)]) *does not imply* relation (A.2.7) for any $k \geq 2$, and, consequently, Definition (A.2.9) does not follow from Definitions A.5. Indeed, in view of (A.2.6), for any $a, b > 0$ we have

$$f_k(abx) = (ab)^\lambda f_k(x) + (ab)^\lambda e(ab) f_{k-1}(x) = a^\lambda f_k(bx) + a^\lambda e(a) f_{k-1}(bx),$$
$$(A.2.16)$$

and

$$f_k(bx) = b^\lambda f_k(x) + b^\lambda e(b) f_{k-1}(x),$$
$$f_{k-1}(bx) = b^\lambda f_{k-1}(x) + b^\lambda e(b) f_{k-2}(x),$$
$$(A.2.17)$$

where f_{k-1} and f_{k-2} are AHDs of degree λ and of order $k - 1$ and $k - 2$, respectively, $e(a)$ is some function. By substituting relations (A.2.18) into (A.2.16), we obtain

$$(ab)^\lambda f_k(x) + (ab)^\lambda e(ab) f_{k-1}(x)$$
$$= a^\lambda \big(b^\lambda f_k(x) + b^\lambda e(b) f_{k-1}(x)\big) + a^\lambda e(a)\big(b^\lambda f_{k-1}(x) + b^\lambda e(b) f_{k-2}(x)\big).$$

Thus we have for all $a, b > 0$

$$\big(e(ab) - e(a) - e(b)\big) f_{k-1}(x) - e(a)e(b) f_{k-2}(x) = 0, \qquad (A.2.18)$$

$k = 1, 2, \ldots$. Here we set $f_{-1}(x) = 0$.

It is clear that in contrast to the above cited statement from [87] and [88], relation (A.2.18) is equivalent to (A.2.7) only if $f_{k-2}(x) = 0$, i.e., $k = 1$.

Indeed, setting $k = 1$, one can calculate that $e(a) = K \log a$, i.e., (A.2.9) holds for $k = 1$.

Let $k = 2$. In this case, using (A.2.18) and (A.2.8), we obtain

$$\left(\log ab - \log a - \log b \right) f_1(x) - \log a \log b f_0(x) = 0,$$

i.e., $f_0(x) \equiv 0$, which means that f_1 is a *homogeneous*, but is not an *associated homogeneous* distribution. Consequently, we have a contradiction.

Conclusion. According to the above result, an AHD of order k is reproduced by the dilatation operator U_a (for all $a > 0$) up to an AHD of order $k - 1$ *only* if $k = 1$. Thus *direct transfer* of the notion of an *associated eigenvector* to the case of distributions *is impossible for $k \geq 2$.*

Consequently, there exist *only* AHDs of order $k = 0$, i.e., HDs (given by Definition A.1) and of order $k = 1$ (given by Definition (A.2.3) or Definition A.2). Definition A.3 (from [95, Ch.I, §4.1., (3)]), which defines AHDs of order $k \geq 2$ contains, unfortunately, no nontrivial elements.

This is related to the fact that any HD is an eigenfunction of *all* dilatation operators $U_a f(x) = f(ax)$ (for *all* $a > 0$), while for $k \geq 2$ *no* distribution $x_\pm^\lambda \log^k x_\pm$, $P\left(x_\pm^{-n} \log^{k-1} x_\pm \right)$ is an AHD for *all* dilatation operators.

A.3. Symmetry of the class of distributions $\mathcal{AH}_0(\mathbb{R})$

The distributions mentioned in Proposition A.2.1 (which are sometimes called "pseudo-functions") are defined as regularizations of slowly divergent integrals. For all $\varphi \in \mathcal{D}(\mathbb{R})$ and for Re $\lambda > -1$ we set

$$\left\langle x_+^\lambda \log^k x_+, \varphi(x) \right\rangle \overset{def}{=} \int_0^\infty x^\lambda \log^k x \varphi(x) \, dx. \tag{A.3.1}$$

For Re $\lambda > -n - 1$, $\lambda \neq -1, -2, \ldots, -n$, according to the book [95, Ch.I, §4.2., (2), (6)], we have

$$\left\langle x_+^\lambda \log^k x_+, \varphi(x) \right\rangle = \int_0^1 x^\lambda \log^k x \left(\varphi(x) - \sum_{j=0}^{n-1} \frac{x^j}{j!} \varphi^{(j)}(0) \right) dx$$

$$+ \int_1^\infty x^\lambda \log^k x \varphi(x) \, dx + \sum_{j=0}^{n-1} \frac{(-1)^k k!}{j!(\lambda + j + 1)^{k+1}} \varphi^{(j)}(0). \tag{A.3.2}$$

The latter formula gives an analytical continuation of relation (A.3.1).

The distribution $P\left(x_+^{-n} \log^k x_+ \right)$ is the *principal value of the function* $x_+^{-n} \log^k x_+$ (it is not the value of the distribution $x_+^\lambda \log^k x_+$ at the point

$\lambda = -n$) (e.g., see [95, Ch.I, §4.2.]). According to [95, Ch.I, §4.2., (4), (7)], we have

$$\left\langle P\left(x_+^{-n} \log^k x_+\right), \varphi(x) \right\rangle \overset{def}{=} \int_0^\infty x^{-n} \log^k x$$

$$\times \left(\varphi(x) - \sum_{j=0}^{n-2} \frac{x^j}{j!} \varphi^{(j)}(0) - \frac{x^{n-1}}{(n-1)!} \varphi^{(n-1)}(0) H(1-x) \right) dx \qquad \text{(A.3.3)}$$

where $H(x)$ is the Heaviside function.

The other distributions mentioned in Proposition A.2.1 are defined in the following way:

$$\left\langle x_-^\lambda \log^k x_-, \varphi(x) \right\rangle \overset{def}{=} \left\langle x_+^\lambda \log^k x_+, \varphi(-x) \right\rangle,$$

$$\left\langle P\left(x_-^{-n} \log^k x_-\right), \varphi(x) \right\rangle \overset{def}{=} \left\langle P\left(x_+^{-n} \log^k x_+\right), \varphi(-x) \right\rangle, \qquad \text{(A.3.4)}$$

for all $\varphi \in \mathcal{D}(\mathbb{R})$.

The distributions $(x \pm i0)^\lambda \log^k(x \pm i0)$ are represented as linear combinations of the distributions $x_\pm^\lambda \log^k x_\pm$ and $P\left(x_\pm^{-n} \log^k x_\pm\right)$ [95, Ch.I, §4.5.]. In particular, for all λ [95, Ch.I, §3.6.]

$$(x \pm i0)^\lambda = x_+^\lambda + e^{\pm i\pi\lambda} x_-^\lambda, \quad \lambda \neq -n,$$

$$(x \pm i0)^{-n} = P\left(x^{-n}\right) \mp \frac{i\pi(-1)^{n-1}\delta^{n-1}(x)}{(n-1)!}, \quad n \in \mathbb{N}, \qquad \text{(A.3.5)}$$

where the distribution $P\left(x^{-n}\right) = P\left(x_+^{-n}\right) + (-1)^n P\left(x_-^{-n}\right)$ is called the principal value of the function x^{-n}. This distribution is homogeneous of degree $-n$. For $\lambda \neq -1, -2, \ldots$, the distribution $(x \pm i0)^\lambda \log^k(x \pm i0)$ can be obtained by differentiating the first relation in (A.3.5) with respect to λ.

Let us see how distributions from the class $\mathcal{AH}_0(\mathbb{R})$ (mentioned above in Proposition A.2.1) are transformed by the dilatation operators $U_a, a > 0$.

1. For Re $\lambda > -1$, $k \in \mathbb{N}$, and for all $\varphi \in \mathcal{D}(\mathbb{R})$, $a > 0$, definition (A.3.1) implies

$$\left\langle x_+^\lambda \log^k x_+, \varphi\left(\frac{x}{a}\right) \right\rangle = a^{\lambda+1} \int_0^\infty \xi^\lambda \log^k(a\xi)\varphi(\xi)\,d\xi$$

$$= a^{\lambda+1} \sum_{j=0}^k \log^j a C_k^j \int_0^\infty \xi^\lambda \log^{k-j} \xi \varphi(\xi)\,d\xi$$

$$= a^{\lambda+1} \left\langle x_+^\lambda \log^k x_+, \varphi(x) \right\rangle$$

$$+ \sum_{j=1}^k a^{\lambda+1} \log^j a \left\langle f_{\lambda;k-j}(x), \varphi(x) \right\rangle, \qquad \text{(A.3.6)}$$

where $f_{\lambda;k-j}(x) = C_k^j x_+^\lambda \log^{k-j} x_+$, C_k^j are the binomial coefficients, $j = 1, 2, \ldots, k$. For all $\lambda \neq -1, -2, \ldots$ we define (A.3.6) by means of analytic continuation. Thus

$$\left\langle x_+^\lambda \log^k x_+, \varphi\left(\frac{x}{a}\right)\right\rangle = a^{\lambda+1}\langle x_+^\lambda \log^k x_+, \varphi(x)\rangle \sum_{j=1}^k a^{\lambda+1} \log^j a\langle f_{\lambda;k}(\cdot), \varphi(\cdot)\rangle,$$

(A.3.7)

for all $\lambda \neq -1, -2, \ldots$.

2. For $k \in \mathbb{N}$ and for all $\varphi \in \mathcal{D}(\mathbb{R})$ definition (A.3.3) implies the following relations.

(a) $0 < a < 1$:

$$\left\langle P\left(x_+^{-n} \log^k x_+\right), \varphi\left(\frac{x}{a}\right)\right\rangle$$

$$= \int_0^1 x^{-n} \log^k x\left(\varphi(x/a) - \sum_{j=0}^{n-1} \frac{(x/a)^j}{j!} \varphi^{(j)}(0)\right) dx$$

$$+ \int_1^\infty x^{-n} \log^k x\left(\varphi(x/a) - \sum_{j=0}^{n-2} \frac{(x/a)^j}{j!} \varphi^{(j)}(0)\right) dx$$

$$= a^{-n+1}\left\{\int_0^{1/a} \xi^{-n} \log^k(a\xi)\left(\varphi(\xi) - \sum_{j=0}^{n-1} \frac{\xi^j}{j!} \varphi^{(j)}(0)\right) d\xi\right.$$

$$\left. + \int_{1/a}^\infty \xi^{-n} \log^k(a\xi)\left(\varphi(\xi) - \sum_{j=0}^{n-2} \frac{\xi^j}{j!} \varphi^{(j)}(0)\right) d\xi\right\}$$

$$= a^{-n+1}\left\{\int_0^1 \xi^{-n} \log^k(a\xi)\left(\varphi(\xi) - \sum_{j=0}^{n-1} \frac{\xi^j}{j!} \varphi^{(j)}(0)\right) d\xi\right.$$

$$\left. + \int_1^\infty \xi^{-n} \log^k(a\xi)\left(\varphi(\xi) - \sum_{j=0}^{n-2} \frac{\xi^j}{j!} \varphi^{(j)}(0)\right) d\xi - \frac{\varphi^{(n-1)}(0)}{(n-1)!} I_1\right\}$$

$$= a^{-n+1}\left\{\sum_{r=0}^k \log^r aC_k^r \langle P\left(x_+^{-n} \log^{k-r} x_+\right), \varphi(x)\rangle - \frac{\varphi^{(n-1)}(0)}{(n-1)!} I_1\right\},$$

(A.3.8)

where

$$I_1 = \int_1^{1/a} \frac{\log^k(a\xi)}{\xi} d\xi = \sum_{r=0}^k \log^r aC_k^r \int_1^{1/a} \frac{\log^{k-r} \xi}{\xi} d\xi$$

$$= \log^{k+1} a \sum_{r=0}^k C_k^r \frac{(-1)^{k+1-r}}{k+1-r} = -\frac{1}{k+1} \log^{k+1} a.$$

(A.3.9)

(b) $a = 1$:

$$\left\langle P\left(x_+^{-n} \log^k x_+\right), \varphi\left(\frac{x}{a}\right)\right\rangle = \left\langle P\left(x_+^{-n} \log^k x_+\right), \varphi(x)\right\rangle.$$

(c) $a > 1$:

$$\left\langle P\left(x_+^{-n} \log^k x_+\right), \varphi\left(\frac{x}{a}\right)\right\rangle$$

$$= a^{-n+1}\left\{ \int_0^{1/a} \xi^{-n} \log^k(a\xi)\left(\varphi(\xi) - \sum_{j=0}^{n-1} \frac{\xi^j}{j!}\varphi^{(j)}(0)\right) d\xi \right.$$

$$\left. + \int_{1/a}^\infty \xi^{-n} \log^k(a\xi)\left(\varphi(\xi) - \sum_{j=0}^{n-2} \frac{\xi^j}{j!}\varphi^{(j)}(0)\right) d\xi \right\}$$

$$= a^{-n+1}\left\{ \sum_{r=0}^k \log^r a C_k^r \langle P\left(x_+^{-n} \log^{k-r} x_+\right), \varphi(x)\rangle + \frac{\varphi^{(n-1)}(0)}{(n-1)!} I_2 \right\},$$

$$\text{(A.3.10)}$$

where

$$I_2 = \int_{1/a}^1 \frac{\log^k(a\xi)}{\xi} d\xi = -I_1 = \frac{1}{k+1} \log^{k+1} a.$$

Thus, (A.3.8)–(A.3.10) imply

$$\left\langle P\left(x_+^{-n} \log^k x_+\right), \varphi\left(\frac{x}{a}\right)\right\rangle = a^{-n+1}\left\langle P\left(x_+^{-n} \log^k x_+\right), \varphi(x)\right\rangle$$

$$+ \sum_{r=1}^{k+1} a^{-n+1} \log^r a\langle f_{-n;k+1-r}(x), \varphi(x)\rangle$$

$$\text{(A.3.11)}$$

where

$$f_{-n;0}(x) = \frac{(-1)^{n-1}}{(k+1)(n-1)!}\delta^{(n-1)}(x),$$

$$f_{-n;k+1-r}(x) = C_k^r P\left(x_+^{-n} \log^{k-r} x_+\right), \quad r = 1, 2, \dots, k.$$

For the distributions $x_-^\lambda \log^k x_-$, $P\left(x_-^{-n} \log^k x_-\right)$ relations of the type (A.3.7) and (A.3.11) can be obtained from (A.3.4).

A.4. Real quasi associated homogeneous distributions

A.4.1. A class of distributions $\mathcal{AH}_1(\mathbb{R})$

In Section A.3 (see also Section A.2.2), it is recognized that the dilatation operator U_a for all $a > 0$ does not reproduce a distribution of order k from

$\mathcal{AH}_0(\mathbb{R})$ with accuracy up to a distribution of order $(k - 1)$ from $\mathcal{AH}_0(\mathbb{R})$. It follows from the results of Section A.3 that the dilatation operator U_a acts in $\mathcal{AH}_0(\mathbb{R})$ by formulas (A.3.7) and (A.3.11). Now, by analogy with the transformation laws (A.3.7) and (A.3.11), we introduce the following definition.

Definition A.8. A distribution $f_{\lambda;k} \in \mathcal{D}'(\mathbb{R})$ is said to have *a degree λ and order k*, $k = 0, 1, 2, \ldots$, if for any $a > 0$ and $\varphi \in \mathcal{D}(\mathbb{R})$

$$\left\langle f_{\lambda;k}(x), \varphi\left(\frac{x}{a}\right) \right\rangle = a^{\lambda+1}\langle f_{\lambda;k}(x), \varphi(x) \rangle + \sum_{r=1}^{k} a^{\lambda+1} \log^r a \langle f_{\lambda;k-r}(x), \varphi(x) \rangle,$$

(A.4.1)

i.e.,

$$U_a f_{\lambda;k}(x) = f_{\lambda;k}(ax) = a^{\lambda} f_{\lambda;k}(x) + \sum_{r=1}^{k} a^{\lambda} \log^r a f_{\lambda;k-r}(x), \quad \text{(A.4.2)}$$

where $f_{\lambda;k-r}(x)$ is a *distribution of degree λ and of order $k - r$, $r = 1, 2, \ldots, k$.* Here for $k = 0$ we suppose that the sums in the right-hand sides of (A.4.1) and (A.4.2) are empty.

Let us denote by $\mathcal{AH}_1(\mathbb{R})$ the linear span of all *distributions $f_{\lambda;k} \in \mathcal{D}'(\mathbb{R})$ of order k and degree λ, $\lambda \in \mathbb{C}$, $k = 0, 1, 2, \ldots$,* defined by Definition A.8.

In view of Definitions A.1, A.2 and A.8, a HD of degree λ is a *distribution of order $k = 0$ and degree λ*, while an AHD of order 1 and degree λ is a *distribution of order $k = 1$ and degree λ*. According to (A.3.7) and (A.3.11), $x_{\pm}^{\lambda} \log^k x_{\pm}$, and $P\left(x_{\pm}^{-n} \log^{k-1} x_{\pm}\right)$ are *distributions of order k and of degree λ and $-n$*, respectively. Thus $\mathcal{AH}_0(\mathbb{R}) \subset \mathcal{AH}_1(\mathbb{R})$.

Remark A.4. The sum of a distribution of degree λ and of order k (from $\mathcal{AH}_1(\mathbb{R})$) and a distribution of degree λ and of order $r \leq k - 1$ (from $\mathcal{AH}_1(\mathbb{R})$) is a distribution of degree λ and of order k (from $\mathcal{AH}_1(\mathbb{R})$).

Lemma A.4.1. *Distributions from $\mathcal{AH}_1(\mathbb{R})$ of different degrees and orders are linearly independent.*

Proof. This lemma is proved in the same way as the analogous result for linearly independent homogeneous distributions [95, §3.11., 4.].

Suppose that

$$c_1 f^1(x) + \cdots + c_m f^m(x) = 0,$$

where $f^s \in \mathcal{AH}_1(\mathbb{R})$ is a distribution of degree λ and of order k_s, such that all λ_s or k_s, $s = 1, 2, \dots, m$ are different. Then, by Definition A.8, for all $a > 0$ and $\varphi \in \mathcal{D}(\mathbb{R})$ we have:

$$c_1 a^{\lambda_1} \left(\langle f^1(x), \varphi(x) \rangle + \sum_{r=1}^{k_1} \log^r a \langle f^1_{k_1-r}(x), \varphi(x) \rangle \right)$$

$$+ \cdots + c_m a^{\lambda_m} \left(\langle f^m(x), \varphi(x) \rangle + \sum_{r=1}^{k_m} \log^r a \langle f^m_{k_m-r}(x), \varphi(x) \rangle \right) = 0,$$

where $f^s_{k_s-r}(x) \in \mathcal{AH}_1(\mathbb{R})$ is a distribution of degree λ and of order $(k_s - r)$, $r = 1, 2, \dots, k_s$, $s = 1, 2, \dots, m$.

If all λ_s are different, by choosing different values a, it is easy to see that $c_s \equiv 0$, $s = 1, 2, \dots, m$.

If, for example, $\lambda_1 = \lambda_2$ and $k_1 > k_2$, then for all $a > 0$ and $\varphi(x) \in \mathcal{D}(\mathbb{R})$ we have

$$c_1 \left(\langle f^1(x), \varphi(x) \rangle + \sum_{r=1}^{k_1} \log^r a \langle f^1_{k_1-r}(x), \varphi(x) \rangle \right)$$

$$+ c_2 \left(\langle f^2(x), \varphi(x) \rangle + \sum_{r=1}^{k_2} \log^r a \langle f^2_{k_2-r}(x), \varphi(x) \rangle \right) = 0.$$

The latter relation implies that $c_1 f^1_{k_1-r}(x) = 0$, $r = k_2 + 1, \dots, k_1$, and, consequently, $c_1 \equiv 0$. Thus, $c_2 \equiv 0$. $\square$

Theorem A.4.2. *Each distribution $f \in \mathcal{AH}_1(\mathbb{R})$ of degree λ and order $k \in \mathbb{N}$ (up to a distribution of order $\leq k - 1$) can be represented as a linear combination of linearly independent distributions:*

(a) $x_{\pm}^{\lambda} \log^k x_{\pm}$, if $\lambda \notin -\mathbb{N}$;

(b) $P\left(x_{\pm}^{-n} \log^{k-1} x_{\pm}\right)$, if $\lambda \in -\mathbb{N}$, where $P\left(x_{\pm}^{-n} \log^{k-1} x_{\pm}\right)$ is the principal value of the function $x_{\pm}^{-n} \log^{k-1} x_{\pm}$.

Thus the class of distributions $\mathcal{AH}_1(\mathbb{R})$ coincides with the class $\mathcal{AH}_0(\mathbb{R})$ from Proposition A.2.1.

Proof. We prove this theorem by induction. (a) Let us consider the case $\lambda \neq -1, -2, \dots$. According to Definitions A.2 and A.8, a distribution $f_1 \in \mathcal{AH}_1(\mathbb{R})$ of degree λ and order $k = 1$ is an AHD of degree λ and order $k = 1$, and for all $a > 0$ satisfies the equation

$$f_1(ax) = a^{\lambda} f_1(x) + a^{\lambda} \log a f_0(x), \tag{A.4.3}$$

where $f_0(x)$ is a HD of degree λ. In view of Theorem [95, Ch.I, §3.11.], $f_0(x) = A_1 x_+^\lambda + A_2 x_-^\lambda$, where A_1, A_2 are constants. If we differentiate (A.4.3) with respect to a and set $a = 1$, we obtain the differential equation

$$xf_1'(x) = \lambda f_1(x) + A_1 x_+^\lambda + A_2 x_-^\lambda. \tag{A.4.4}$$

For $\pm x > 0$ this equation can be integrated in the ordinary sense.

Thus, for $x > 0$ equation (A.4.4) coincides with the equation

$$xf_1'(x) = \lambda f_1(x) + A_1 x^\lambda.$$

Integrating this equation, we obtain $f_1(x) = A_1 x_+^\lambda \log x_+ + B_1 x_+^\lambda$, where B_1 is a constant. Similarly, we can prove that $f_1(x) = A_2 x_-^\lambda \log x_- + B_2 x_-^\lambda$ for $x < 0$. Thus the distribution $g(x) = f_1(x) - A_1 x_+^\lambda - A_2 x_-^\lambda - B_1 x_+^\lambda - B_2 x_-^\lambda$ satisfies equation (A.4.4) being concentrated at the point $x = 0$. Therefore, $g(x) = \sum_{m=0}^{M} C_m \delta^{(m)}(x)$, where $C_1, \ldots, C_M$ are constants. However, since $\delta^{(m)}(x)$ is a HD of degree $-m - 1$, $g(x) = 0$ by Lemma A.4.1. Thus

$$f_1(x) = A_1 x_+^\lambda \log x_+ + A_2 x_-^\lambda \log x_- + B_1 x_+^\lambda + B_2 x_-^\lambda.$$

Consequently, up to the distribution $B_1 x_+^\lambda + B_2 x_-^\lambda \in \mathcal{AH}_1(\mathbb{R})$ of order 0, we have $f_1(x) = A_1 x_+^\lambda \log x_+ + A_2 x_-^\lambda \log x_-$.

Let us assume that the distribution $f_{k-1}(x) \in \mathcal{AH}_1(\mathbb{R})$ of degree λ and order $(k - 1)$ is represented as the linear combination

$$f_{k-1}(x) = \sum_{j=0}^{k-1} \left(A_{1j} x_+^\lambda \log^j x_+ + A_{2j} x_-^\lambda \log^j x_- \right). \tag{A.4.5}$$

A distribution $f_k \in \mathcal{AH}_1(\mathbb{R})$ of degree λ and of order $k \geq 2$ satisfies (A.4.2) for all $a > 0$. Differentiating this equation with respect to a and setting $a = 1$, we obtain

$$xf_k'(x) = \lambda f_k(x) + f_{k-1}(x). \tag{A.4.6}$$

Taking (A.4.5) into account and integrating (A.4.6) for $x \neq 0$, we calculate

$$f_k(x) = \sum_{j=0}^{k-1} \left(\frac{A_{1j}}{j+1} x_+^\lambda \log^{j+1} x_+ + \frac{A_{2j}}{j+1} x_-^\lambda \log^{j+1} x_- \right) + B_1 x_+^\lambda + B_2 x_-^\lambda,$$

where B_1, B_2 are constants. By repeating the above reasoning we obtain

$$f_k(x) = \frac{A_{1\,k-1}}{k} x_+^\lambda \log^k x_+ + \frac{A_{2\,k-1}}{k} x_-^\lambda \log^k x_-$$

up to distributions of degree λ and of order $\leq k - 1$.

By induction, the statement (a) is proved.

The statement (b), when $\lambda = -n$, $n \in \mathbb{N}$, can be proved similarly. $\quad\square$

A.4.2. QAHDs

Taking relations (A.3.7) and (A.3.11) into account, by analogy with (A.2.3), we can naturally generalize the notion of an *associated eigenvector* (A.2.2).

Definition A.9. A distribution $f_k \in \mathcal{D}'(\mathbb{R})$ is said to be *quasi associated homogeneous* of degree λ and of order k, $k = 0, 1, 2, 3, \ldots$ if for any $a > 0$ and $\varphi \in \mathcal{D}(\mathbb{R})$

$$\left\langle f_k(x), \varphi\left(\frac{x}{a}\right)\right\rangle = a^{\lambda+1}\langle f_k(x), \varphi(x)\rangle + \sum_{r=1}^{k} h_r(a)\langle f_{k-r}(x), \varphi(x)\rangle,$$

i.e.,

$$U_a f_k(x) = f_k(ax) = a^\lambda f_k(x) + \sum_{r=1}^{k} h_r(a) f_{k-r}(x), \tag{A.4.7}$$

where $f_{k-r}(x)$ is a QAHD of degree λ and of order $k - r$, $h_r(a)$ is a differentiable function, $r = 1, 2, \ldots, k$. Here for $k = 0$ we suppose that sums on the right-hand sides of the above relations are empty.

Let us denote by $\mathcal{AH}(\mathbb{R})$ the linear span of all *QAHDs* of order k and degree λ, $\lambda \in \mathbb{C}$, $k = 0, 1, 2, \ldots$, defined by Definition A.9. In view of Definition A.8, $\mathcal{AH}_1(\mathbb{R}) \subset \mathcal{AH}(\mathbb{R})$.

Theorem A.4.3. *Any QAHD $f_k(x)$ of degree λ and of order k, $k = 0, 1, \ldots$ (see Definition A.9) is a distribution of degree λ and of order k (from $\mathcal{AH}_1(\mathbb{R})$) (see Definition A.8 and Theorem A.4.2), i.e., $f_k(x)$ satisfies relation (A.4.2).*

Thus $\mathcal{AH}(\mathbb{R}) = \mathcal{AH}_1(\mathbb{R})$, and, consequently, the class $\mathcal{AH}(\mathbb{R})$ coincides with the Gel'fand–Shilov class $\mathcal{AH}_0(\mathbb{R})$ from Proposition A.2.1.

Proof. We prove this theorem by induction.

1. For $k = 1$ this theorem is proved in [95, Ch.I, §4.1.] (see also Section A.2.1).

2. If $k = 2$, according to Definition A.9, for a QAHD $f_2(x)$ of degree λ and of order $k = 2$ we have

$$f_2(ax) = a^\lambda f_2(x) + h_1(a) f_1(x) + h_2(a) f_0(x), \quad \forall a > 0, \tag{A.4.8}$$

where $f_1(x)$ is an AHD of degree λ and of order $k = 1$, $f_0(x)$ is a HD of degree λ, and $h_1(a)$, $h_2(a)$ are the desired functions.

Taking into account that $f_1(bx) = b^\lambda f_1(x) + b^\lambda \log b f_0^{(1)}(x)$, where $f_0^{(1)}(x)$ is a HD of degree λ, in view of (A.4.8) and Definition A.2, we obtain for all $a, b > 0$:

$$
\begin{aligned}
f_2(abx) &= (ab)^\lambda f_2(x) + h_1(ab) f_1(x) + h_2(ab) f_0(x) \\
&= a^\lambda f_2(bx) + h_1(a) f_1(bx) + h_2(a) f_0(bx) \\
&= a^\lambda \left(b^\lambda f_2(x) + h_1(b) f_1(x) + h_2(b) f_0(x) \right) \\
&\quad + h_1(a) \left(b^\lambda f_1(x) + b^\lambda \log b f_0^{(1)}(x) \right) + h_2(a) b^\lambda f_0(x) \\
&= (ab)^\lambda f_2(x) + \left(a^\lambda h_1(b) + b^\lambda h_1(a) \right) f_1(x) \\
&\quad + \left(a^\lambda h_2(b) + h_2(a) b^\lambda \right) f_0(x) + h_1(a) b^\lambda \log b f_0^{(1)}(x).
\end{aligned}
$$

Obviously, this implies that for all $a, b > 0$

$$
\begin{aligned}
&\left(h_1(ab) - a^\lambda h_1(b) - b^\lambda h_1(a) \right) f_1(x) \\
&+ \left(h_2(ab) - a^\lambda h_2(b) - b^\lambda h_2(a) \right) f_0(x) - h_1(a) b^\lambda \log b f_0^{(1)}(x) = 0.
\end{aligned}
$$

$$(A.4.9)$$

According to [95, Ch.I, §3.11.], there are two *linearly independent* HDs of degree λ, such that each HD is their linear combination. Thus there are two possibilities: either $f_0^{(1)}(x)$ and $f_0(x)$ are linearly independent HDs, or $f_0^{(1)}(x) = C f_0(x)$, where C is a constant.

Thus, in the first case, since from Lemma A.4.1 any HD and AHD of order 1 are linearly independent, relation (A.4.9) implies $h_1(a) = 0$ and $h_2(ab) = a^\lambda h_2(b) + b^\lambda h_2(a)$. The solution of the latter equation constructed in [95, Ch.I, §4.1.] (see also Section A.2.1) is given by (A.2.4), i.e., $h_2(a) = a^\lambda \log a$. Thus, relation (A.4.8), Definition A.2, and Theorem A.4.2 imply that $f_2(x)$ is an AHD of order 1. Consequently, we obtain a *trivial solution*.

In the second case, in view of Lemma A.4.1, $f_0(x)$ and $f_1(x)$ are linearly independent, and, consequently, relation (A.4.9) implies the following system of functional equations:

$$
\begin{aligned}
h_1(ab) &= a^\lambda h_1(b) + b^\lambda h_1(a), \\
h_2(ab) &= a^\lambda h_2(b) + h_2(a) b^\lambda + C h_1(a) b^\lambda \log b, \quad \forall \, a, b > 0,
\end{aligned}
$$

$$(A.4.10)$$

where $h_1(1) = 0$, $h_2(1) = 0$. According to [95, Ch.I, §4.1.] (see also (A.2.4)), $h_1(a) = a^\lambda \log a$. Then the second equation in (A.4.10) implies

$$h_2(ab) = h_2(a)b^\lambda + a^\lambda h_2(b) + C(ab)^\lambda \log a \log b \qquad (A.4.11)$$

and, consequently, the function $\widetilde{h}_2(a) = \frac{h_2(a)}{a^\lambda}$ satisfies the equation

$$\widetilde{h}_2(ab) = \widetilde{h}_2(a) + \widetilde{h}_2(b) + C \log a \log b, \quad \forall\, a, b > 0. \qquad (A.4.12)$$

Making the change of variables $\psi_2(z) = \widetilde{h}_2(e^z)$, where $\psi_2(0) = 0$ and $a = e^\xi$, $b = e^\eta$, we can see that (A.4.12) can be rewritten as

$$\psi_2(\xi + \eta) = \psi_2(\xi) + \psi_2(\eta) + C\xi\eta, \quad \forall\, \xi, \eta. \qquad (A.4.13)$$

We will seek a solution of equation (A.4.13) in the class of differentiable functions. Differentiating the relation (A.4.13) with respect to η, we obtain for all ξ, η

$$\psi_2'(\xi + \eta) = \psi_2'(\eta) + C\xi, \qquad \psi_2(0) = 0.$$

Setting $\eta = 0$, we have the differential equation

$$\psi_2'(\xi) = \psi_2'(0) + C\xi, \quad \psi_2(0) = 0,$$

whose solution has the form

$$\psi_2(\xi) = \psi_2'(0)\xi + \frac{C}{2}\xi^2.$$

Since $a = e^\xi$, then $\widetilde{h}_2(a) = A_2 \log a + \frac{C}{2} \log^2 a$ and

$$h_2(a) = A_2 a^\lambda \log a + \frac{C}{2} a^\lambda \log^2 a, \qquad (A.4.14)$$

where $A_2 = \widetilde{h}_2'(1) = h_2'(1)$ is a constant.

By substituting the functions $h_1(a)$, $h_2(a)$ given by (A.2.4) and (A.4.14) into (A.4.8), we obtain

$$f_2(ax) = a^\lambda f_2(x) + a^\lambda \log a f_1(x) + \left(A_2 a^\lambda \log a + \frac{C}{2} a^\lambda \log^2 a \right) f_0(x).$$

The latter relation can be rewritten in the desired form (A.4.2):

$$f_2(ax) = a^\lambda f_2(x) + a^\lambda \log a \, \widetilde{f}_1(x) + a^\lambda \log^2 a \, \widetilde{f}_0(x), \quad \forall\, a > 0,$$

$$(A.4.15)$$

where $\widetilde{f}_1(x) = f_1(x) + A_2 f_0(x)$ is an AHD of degree λ and of order $k = 1$, $\widetilde{f}_0(x) = \frac{C}{2} f_0(x)$ is a HD. Thus $f_2(x)$ is a distribution of degree λ and of order 2 (in the sense of Definition A.8), and, according to Theorem A.4.2, $f_2(x) \in \mathcal{AH}_0(\mathbb{R})$.

3. Let $f_k(x)$ be a QAHD of degree λ and of order k. Let us assume that any QAHD $f_j(x)$, $j = 0, 1, \ldots, k-1$ is a distribution of degree λ and of order j (in the sense of Definition A.8). Then, according to Theorem A.4.2, $f_j(x) \in \mathcal{AH}_0(\mathbb{R})$, and the relation (A.4.2) holds. Thus, in view of our assumption, (A.4.7) and (A.4.2) imply for all $a, b > 0$:

$$
\begin{aligned}
f_k(abx) &= (ab)^\lambda f_k(x) + \sum_{r=1}^{k} h_r(ab) f_{k-r}(x) = a^\lambda f_k(bx) + \sum_{r=1}^{k} h_r(a) f_{k-r}(bx) \\
&= a^\lambda \Big(b^\lambda f_k(x) + \sum_{r=1}^{k} h_r(b) f_{k-r}(x) \Big) \\
&\quad + \sum_{r=1}^{k-1} h_r(a) \Big(b^\lambda f_{k-r}(x) + \sum_{j=1}^{k-r} b^\lambda \log^j b f_{k-r-j}^{(k-r)}(x) \Big) + h_k(a) b^\lambda f_0(x) \\
&= (ab)^\lambda f_k(x) + \sum_{r=1}^{k} \Big(a^\lambda h_r(b) + b^\lambda h_r(a) \Big) f_{k-r}(x) \\
&\quad + \sum_{r=1}^{k-1} \sum_{j=1}^{k-r} h_r(a) b^\lambda \log^j b f_{k-r-j}^{(k-r)}(x),
\end{aligned}
$$

where $f_{k-r-j}^{(k-r)}(x)$ is a distribution of degree λ and of order $k - r - j$ (in the sense of Definition A.8) which belongs to $\mathcal{AH}_0(\mathbb{R})$, $r = 1, \ldots, k-1$, $j = 1, \ldots, k-r$. By changing the order of summation, we can easily see that for all $a, b > 0$:

$$
\begin{aligned}
\sum_{r=1}^{k} h_r(ab) f_{k-r}(x) &= \sum_{r=1}^{k} \Big(a^\lambda h_r(b) + b^\lambda h_r(a) \Big) f_{k-r}(x) \\
&\quad + \sum_{r=2}^{k} \sum_{j=1}^{r-1} h_{r-j}(a) b^\lambda \log^j b f_{k-r}^{(k-r+j)}(x). \quad \text{(A.4.16)}
\end{aligned}
$$

Since, in view of Lemma A.4.1, the distribution $f_{k-1} \in \mathcal{AH}_1(\mathbb{R})_0$ of order $k - 1$ and the distributions $f_{k-r}, f_{k-r}^{(k-r+j)} \in \mathcal{AH}_1(\mathbb{R})_0$ of order $k - r$, are linearly independent, $r = 2, \ldots, k$, $j = 1, \ldots, r-1$, the relation (A.4.16) implies that for all $a, b > 0$

$$h_1(ab) = a^\lambda h_1(b) + b^\lambda h_1(a),$$

$$h_2(ab) f_{k-2} = \left(a^\lambda h_2(b) + b^\lambda h_2(a) \right) f_{k-2} + h_1(a) b^\lambda \log b f_{k-2}^{(k-1)},$$

$$h_3(ab) f_{k-3} = \left(a^\lambda h_3(b) + b^\lambda h_3(a) \right) f_{k-3} + \sum_{j=1}^{2} h_{3-j}(a) b^\lambda \log^j b f_{k-3}^{(k-3+j)},$$

$$\dots\dots\dots\dots\dots\dots\dots\dots\dots\dots\dots\dots\dots\dots\dots\dots\dots\dots\dots$$

$$h_k(ab) f_0 = \left(a^\lambda h_k(b) + b^\lambda h_k(a) \right) f_0 + \sum_{j=1}^{k-1} h_{k-j}(a) b^\lambda \log^j b f_0^{(j)}.$$

Taking into account that the function $\frac{h_j(ab)}{(ab)^\lambda} - \frac{h_j(a)}{(a)^\lambda} - \frac{h_j(b)}{(b)^\lambda}$ is symmetric in a and b, it is easy to see that the latter system has a *nontrivial* solution only if $f_{k-r}^{(k-r+j)}(x) = C_{k-r}^{(k-r+j)} f_{k-r}(x)$, where $C_{k-r}^{(k-r+j)}$ are constants, $r = 2, 3, \dots, k$, $j = 1, 2, \dots, r - 1$. Thus in view of Lemma A.4.1, we obtain the following system of functional equations:

$$h_1(ab) = a^\lambda h_1(b) + b^\lambda h_1(a),$$

$$h_2(ab) = a^\lambda h_2(b) + b^\lambda h_2(a) + C_{k-2}^{(k-1)} h_1(a) b^\lambda \log b,$$

$$h_3(ab) = a^\lambda h_3(b) + b^\lambda h_3(a) + \sum_{j=1}^{2} C_{k-3}^{(k-3+j)} h_{3-j}(a) b^\lambda \log^j b, \quad \text{(A.4.17)}$$

$$\dots\dots\dots\dots\dots\dots\dots\dots\dots\dots\dots\dots\dots\dots\dots\dots\dots$$

$$h_k(ab) = a^\lambda h_k(b) + b^\lambda h_k(a) + \sum_{j=1}^{k-1} C_0^{(j)} h_{k-j}(a) b^\lambda \log^j b.$$

Consequently, the functions $\widetilde{h}_j(a) = \frac{h_j(a)}{a^\lambda}$ satisfy the system of equations

$$\widetilde{h}_1(ab) = \widetilde{h}_1(b) + \widetilde{h}_1(a),$$

$$\widetilde{h}_2(ab) = \widetilde{h}_2(b) + \widetilde{h}_2(a) + C_{k-2}^{(k-1)} \widetilde{h}_1(a) \log b,$$

$$\widetilde{h}_3(ab) = \widetilde{h}_3(b) + \widetilde{h}_3(a) + \sum_{j=1}^{2} C_{k-3}^{(k-3+j)} \widetilde{h}_{3-j}(a) \log^j b, \quad \text{(A.4.18)}$$

$$\dots\dots\dots\dots\dots\dots\dots\dots\dots\dots\dots\dots\dots\dots\dots\dots\dots$$

$$\widetilde{h}_k(ab) = \widetilde{h}_k(b) + \widetilde{h}_k(a) + \sum_{j=1}^{k-1} C_0^{(j)} \widetilde{h}_{k-j}(a) \log^j b,$$

where $\widetilde{h}_j(1) = 0$, $j = 1, 2, \dots, k$.

By changing variables $\psi_j(z) = \widetilde{h}_j(e^z)$, where $\psi_j(0) = 0$, $j = 1, 2, \dots, k$ and $a = e^\xi$, $b = e^\eta$, the system (A.4.18) can be rewritten as

$$\psi_1(\xi + \eta) = \psi_1(\xi) + \psi_1(\eta),$$

$$\psi_2(\xi + \eta) = \psi_2(\xi) + \psi_2(\eta) + C_{k-2}^{(k-1)} \psi_1(\xi) \eta,$$

$$\psi_3(\xi + \eta) = \psi_3(\xi) + \psi_3(\eta) + \sum_{j=1}^{2} C_{k-3}^{(k-3+j)} \psi_{3-j}(\xi) \eta^j, \quad \text{(A.4.19)}$$

$$\dots\dots\dots\dots\dots\dots\dots\dots\dots\dots\dots\dots\dots\dots\dots\dots\dots$$

$$\psi_k(\xi + \eta) = \psi_k(\xi) + \psi_k(\eta) + \sum_{j=1}^{k-1} C_0^{(j)} \psi_{k-j}(\xi) \eta^j.$$

Differentiating relations (A.4.19) with respect to η and setting $\eta = 0$, we obtain the following system of differential equations:

$$\psi_1'(\xi) = \psi_1'(0),$$
$$\psi_2'(\xi) = \psi_2'(0) + C_{k-2}^{(k-1)}\psi_1(\xi),$$
$$\psi_3'(\xi) = \psi_3'(0) + C_{k-1}^{(k-3+j)}\psi_2(\xi),$$

$$\dots\dots\dots\dots\dots\dots\dots\dots$$

$$\psi_k'(\xi) = \psi_k'(0) + C_0^{(1)}\psi_{k-1}(\xi),$$

where $\psi_j(0) = 0$, $j = 1, 2, \dots, k$. (A.4.20)

By successive integration, it is easy to see that the solution of the system (A.4.20) has the form

$$\psi_r(\xi) = \sum_{j=1}^{r} A_r^j \xi^j,$$

where A_r^j are constants which can be calculated, $r = 1, 2, \dots, k$, $j = 1, 2, \dots, r$.

Since $a = e^\xi$, $\psi_j(z) = \tilde{h}_j(e^z)$, then $\tilde{h}_r(a) = \sum_{j=1}^{r} A_r^j \log^j a$ and

$$h_r(a) = a^\lambda \sum_{j=1}^{r} A_r^j \log^j a,$$ (A.4.21)

where A_r^j are constants, $r = 1, 2, \dots, k$, $j = 1, 2, \dots, r$.

By substituting the functions (A.4.21) into (A.4.7), the latter relation can be rewritten in the form (A.4.7):

$$f_k(ax) = a^\lambda f_k(x) + a^\lambda \sum_{r=1}^{k} \sum_{j=1}^{r} A_r^j \log^j a f_{k-r}(x)$$

$$= a^\lambda f_k(x) + \sum_{r=1}^{k} a^\lambda \log^r a \tilde{f}_{k-r}(x),$$ (A.4.22)

where, according to our assumption, $\tilde{f}_{k-r}(x) = \sum_{j\geq r}^{k} A_j^r f_{k-j}(x) \in \mathcal{AH}_0(\mathbb{R})$, $r = 1, 2, \dots, k$. Moreover, in view of Remark A.4, $\tilde{f}_{k-r}(x)$ is a distribution of degree λ and of order $k - r$ (in the sense of Definition A.8), $r = 1, 2, \dots, k$. Consequently, $f_k(x)$ satisfies the relation (A.4.2).

By the induction axiom, $\mathcal{AH}(\mathbb{R}) = \mathcal{AH}_1(\mathbb{R})$.

According to Theorems A.4.2, the class $\mathcal{AH}(\mathbb{R})$ of QAHDs coincides with the class $\mathcal{AH}_0(\mathbb{R})$.

Thus the theorem is proved. $\square$

A.4.3. Resume

In [95, Ch.I, §4.1.] it was proved that in order to introduce an AHD of order $k = 1$, one can use Definition A.2 instead of Definition (A.2.3). Similarly, according to Theorem A.4.3, in order to introduce a QAHD, instead of Definition A.9 one can use the following definition (in fact, Definition A.8).

Definition A.10. The distribution $f_k \in \mathcal{D}'(\mathbb{R})$ is called a *QAHD* of degree λ and of order k, $k = 0, 1, 2, \ldots$, if for any $a > 0$ and $\varphi \in \mathcal{D}(\mathbb{R})$

$$\left\langle f_k(x), \varphi\left(\frac{x}{a}\right) \right\rangle = a^{\lambda+1} \langle f_k(x), \varphi(x) \rangle + \sum_{r=1}^{k} a^{\lambda+1} \log^r a \langle f_{k-r}(x), \varphi(x) \rangle,$$

(A.4.23)

i.e.,

$$f_k(ax) = a^\lambda f_k(x) + \sum_{r=1}^{k} a^\lambda \log^r a f_{k-r}(x),$$

(A.4.24)

where $f_{k-r}(x)$ is an AHD of degree λ and of order $k - r$, $r = 1, 2, \ldots, k$. Here for $k = 0$ we suppose that the sums on the right-hand sides of (A.4.23), (A.4.24) are empty.

Thus instead of the term *"distribution of degree λ and of order k"* one can use the term *"QAHD of degree λ and of order k"*.

According to Remark A.4, the sum of a QAHD of degree λ and of order k and a QAHD of degree λ and of order $r \leq k - 1$ is a QAHD of degree λ and of order k.

A.5. Real multidimensional quasi associated homogeneous distributions

Definition A.11. (see [95, Ch.III, §3.1., (1)]) A distribution $f_0(x) = f_0(x_1, \ldots, x_n)$ from $\mathcal{D}'(\mathbb{R}^n)$ is said to be *homogeneous* of degree λ if for any $a > 0$ and $\varphi \in \mathcal{D}(\mathbb{R}^n)$

$$\left\langle f_0, \varphi\left(\frac{x_1}{a}, \ldots, \frac{x_n}{a}\right) \right\rangle = a^{\lambda+n} \langle f_0, \varphi(x_1, \ldots, x_n) \rangle,$$

i.e.,

$$f_0(ax_1, \ldots, ax_n) = a^\lambda f_0(x_1, \ldots, x_n).$$

Let us recall a well-known theorem.

Theorem A.5.1. (*see* [95, Ch.III, §3.1.]) *A distribution* $f_0(x)$ *is homogeneous of degree* λ *if and only if it satisfies the Euler equation*

$$\sum_{j=1}^{n} x_j \frac{\partial f_0}{\partial x_j} = \lambda f_0,$$

Now we introduce a multidimensional analog of Definition A.10, and prove a multidimensional analog of Theorem A.5.1.

Definition A.12. We say that a distribution $f_k \in \mathcal{D}'(\mathbb{R}^n)$ is a *QAHD* of degree λ and of order k, $k = 0, 1, 2, \ldots$, if for any $a > 0$ we have

$$f_k(ax) = f_k(ax_1, \ldots, ax_n) = a^\lambda f_k(x) + \sum_{r=1}^{k} a^\lambda \log^r a f_{k-r}(x), \quad \text{(A.5.1)}$$

where $f_{k-r}(x)$ is a QAHD of degree λ and of order $k - r$, $r = 1, 2, \ldots, k$. (Here we suppose that for $k = 0$ the sum on the right-hand side of (A.4.2) is empty.)

The characterization of a QAHD is given by the following theorem.

Theorem A.5.2. $f_k(x)$ *is a QAHD of degree* λ *and of order* k, $k \geq 1$, *if and only if it satisfies the Euler type system of equations, i.e., there exist distributions* $f_{k-1}, \ldots, f_0$ *such that*

$$\begin{aligned}
\sum_{j=1}^{n} x_j \frac{\partial f_0}{\partial x_j} &= \lambda f_0, \\
\sum_{j=1}^{n} x_j \frac{\partial f_r}{\partial x_j} &= \lambda f_r + f_{r-1}, \quad r = 1, \ldots k,
\end{aligned} \quad \text{(A.5.2)}$$

i.e., for all $\varphi \in \mathcal{D}(\mathbb{R}^n)$

$$-\left\langle f_0, \sum_{j=1}^{n} x_j \frac{\partial \varphi}{\partial x_j} \right\rangle = (\lambda + n)\langle f_0, \varphi \rangle,$$

$$-\left\langle f_r, \sum_{j=1}^{n} x_j \frac{\partial \varphi}{\partial x_j} \right\rangle = (\lambda + n)\langle f_r, \varphi \rangle + \langle f_{r-1}, \varphi \rangle, \quad r = 1, \ldots, k.$$

Proof. Let $f_k \in \mathcal{D}'(\mathbb{R}^n)$ be a QAHD of degree λ and of order k. According to Definition A.12, there are distributions f_j, $j = 0, 1, 2, \ldots, k - 1$ and $f_{k-s-r}^{(k-s)}$, $s = 0, 1, 2, \ldots, k - 2, r = 2, \ldots, k - s$ such that

$$f_k(ax_1, \ldots, ax_n) = a^\lambda f_k(x) + a^\lambda \log a f_{k-1}(x)$$
$$+ \textstyle\sum_{r=2}^k a^\lambda \log^r a f_{k-r}^{(k)}(x),$$
$$f_{k-1}(ax_1, \ldots, ax_n) = a^\lambda f_{k-1}(x) + a^\lambda \log a f_{k-2}(x)$$
$$+ \textstyle\sum_{r=2}^{k-1} a^\lambda \log^r a f_{k-1-r}^{(k-1)}(x),$$
$$f_{k-2}(ax_1, \ldots, ax_n) = a^\lambda f_{k-2}(x) + a^\lambda \log a f_{k-3}(x) \qquad \text{(A.5.3)}$$
$$+ \textstyle\sum_{r=2}^{k-2} a^\lambda \log^r a f_{k-2-r}^{(k-2)}(x),$$

$$\cdots\cdots\cdots\cdots\cdots\cdots\cdots\cdots\cdots\cdots\cdots$$

$$f_1(ax_1, \ldots, ax_n) = a^\lambda f_1(x) + a^\lambda \log a f_0(x),$$
$$f_0(ax_1, \ldots, ax_n) = a^\lambda f_0(x).$$

Differentiating (A.5.3) with respect to a and setting $a = 1$, we derive the system (A.5.2).

Conversely, let $f_k \in \mathcal{D}'(\mathbb{R}^n)$ be a distribution satisfying the system (A.5.2), i.e., there are distributions $f_j \in \mathcal{D}'(\mathbb{R}^n)$, $j = 0, 1, 2, \ldots, k - 1$ such that the system (A.5.2) holds. We prove by induction that f_k is a QAHD of degree λ and of order k.

For $k = 0$ this statement follows from Theorem A.5.1.

If $k = 1$ then the system of equations

$$\sum_{j=1}^n x_j \frac{\partial f_1}{\partial x_j} = \lambda f_1 + f_0,$$
$$\sum_{j=1}^n x_j \frac{\partial f_0}{\partial x_j} = \lambda f_0, \qquad \text{(A.5.4)}$$

holds. Here, in view of Theorem A.5.1, the second equation in (A.5.4) implies that f_0 is a HD.

Consider the function

$$g_1(a) = f_1(ax_1, \ldots, ax_n) - a^\lambda f_1(x) - a^\lambda \log a f_0(x).$$

It is clear that $g_1(1) = 0$. By differentiation we have

$$g_1'(a) = \sum_{j=1}^{n} x_j \frac{\partial f_1}{\partial x_j}(ax_1, \ldots, ax_n) - \lambda a^{\lambda-1} f_1(x) - (\lambda a^{\lambda-1} \log a + a^{\lambda-1}) f_0(x)$$

(A.5.5)

Applying the first relation in (A.5.4) to the arguments $ax_1, \ldots, ax_n$ we find that

$$\sum_{j=1}^{n} x_j \frac{\partial f_1}{\partial x_j}(ax_1, \ldots, ax_n) = \frac{\lambda}{a} f_1(ax_1, \ldots, ax_n) + \frac{1}{a} f_0(ax_1, \ldots, ax_n).$$

(A.5.6)

Substituting (A.5.6) into (A.5.5) and taking into account the relation $\frac{1}{a} f_0(ax_1, \ldots, ax_n) = a^{\lambda-1} f_0(x_1, \ldots, x_n)$, we find that $g_1(a)$ satisfies the differential equation with the initial data

$$g_1'(a) = \frac{\lambda}{a} g_1(a), \qquad g_1(1) = 0. \tag{A.5.7}$$

Obviously, its solution is $g_1(a) = 0$. Thus $g_1(a) = f_1(ax_1, \ldots, ax_n) - a^{\lambda} f_1(x) - a^{\lambda} \log a f_0(x) = 0$, i.e., $f_1(x)$ is an AHD of order $k = 1$, i.e., a QAHD of order $k = 1$.

Let us assume that for $k - 1$ the theorem holds, i.e., if f_{k-1} satisfies all the equations in (A.5.2) except for the first one, then f_{k-1} is a QAHD of degree λ and of order $k - 1$.

Now, we suppose that there exist distributions $f_{k-1}, \ldots, f_0$ such that (A.5.2) holds. Note that in view of our assumption f_{k-1} is a QAHD of order $k - 1$.

Consider the function

$$g_k(a) = f_k(ax_1, \ldots, ax_n) - a^{\lambda} f_k(x) - a^{\lambda} \log a f_{k-1}(x). \tag{A.5.8}$$

It is clear that $g_k(1) = 0$. By differentiation we have

$$g_k'(a) = \sum_{j=1}^{n} x_j \frac{\partial f_k}{\partial x_j}(ax_1, \ldots, ax_n) - \lambda a^{\lambda-1} f_k(x)$$

$$- (\lambda a^{\lambda-1} \log a + a^{\lambda-1}) f_{k-1}(x). \tag{A.5.9}$$

Applying the first relation in (A.5.2) to the arguments $ax_1, \ldots, ax_n$ we find that

$$\sum_{j=1}^{n} x_j \frac{\partial f_k}{\partial x_j}(ax_1, \ldots, ax_n) = \frac{\lambda}{a} f_k(ax_1, \ldots, ax_n) + \frac{1}{a} f_{k-1}(ax_1, \ldots, ax_n).$$

(A.5.10)

Substituting (A.5.10) into (A.5.9) and taking into account that according to our assumption, f_{k-1} is a QAHD of order $k - 1$, i.e.,

$$f_{k-1}(ax_1, \ldots, ax_n) = a^\lambda f_{k-1}(x) + \sum_{r=1}^{k-1} a^\lambda \log^r a f_{k-1-r}^{(k-1)}(x),$$

where $f_{k-1-r}^{(k-1)}(x)$ is a QAHD of order $k - 1 - r$, $r = 1, 2, \ldots, k - 1$, we find that $g_k(a)$ satisfies the linear differential equation

$$g_k'(a) = \frac{\lambda}{a} g_k(a) + \sum_{r=1}^{k-1} a^{\lambda-1} \log^r a f_{k-1-r}^{(k-1)}(x), \qquad g_1(1) = 0. \quad \text{(A.5.11)}$$

Now it is easy to see that its general solution has the form

$$g_k(a) = \sum_{r=1}^{k-1} a^\lambda \log^{r+1} a \frac{f_{k-1-r}^{(k-1)}(x)}{r+1} + a^\lambda C(x),$$

where $C(x)$ is a distribution. Taking into account that $g_1(1) = 0$, we get $C(x) = 0$. Thus

$$g_k(a) = \sum_{r=1}^{k-1} a^\lambda \log^{r+1} a \frac{f_{k-1-r}^{(k-1)}(x)}{r+1}. \quad \text{(A.5.12)}$$

By substituting (A.5.12) into (A.5.8), we find

$$f_k(ax_1, \ldots, ax_n) = a^\lambda f_k(x) - a^\lambda \log a f_{k-1}(x) + \sum_{r=2}^{k} a^\lambda \log^r a \frac{f_{k-r}^{(k-1)}(x)}{r},$$

$$\text{(A.5.13)}$$

where by our assumption f_{k-1} is a QAHD of order $k - 1$, and, consequently, $f_{k-r}^{(k-1)}(x)$ is a QAHD of order $k - r$, $r = 2, \ldots, k$. Thus, in view of Definition A.12, f_k is a QAHD of order k.

By the induction axiom, the theorem is proved. $\qquad \square$

It is easy to see that Theorem A.5.2 can be reformulated in the following way:

Theorem A.5.3. *$f_k(x)$ is a QAHD of degree λ and of order k, $k \geq 1$, if and only if*

$$\left(\sum_{j=1}^{n} x_j \frac{\partial}{\partial x_j} - \lambda \right)^{k+1} f_k(x) = 0. \quad \text{(A.5.14)}$$

In this form, Theorem A.5.2 is a special case of Proposition 2.31 presented in Grudzinski's book [99].

A.6. The Fourier transform of real quasi associated homogeneous distributions

The Fourier transform of $\varphi \in \mathcal{D}(\mathbb{R}^n)$ is defined by

$$F[\varphi](\xi) = \int_{\mathbb{R}^n} \varphi(x)e^{i\xi \cdot x}\, d^n x, \quad \xi \in \mathbb{R}^n,$$

where $\xi \cdot x$ is the scalar product of vectors x and ξ. We define the Fourier transform $F[f]$ of a distribution [95, Ch.II]

$$\langle F[f], \varphi \rangle = \langle f, F[\varphi] \rangle, \quad \forall \varphi \in \mathcal{D}(\mathbb{R}^n).$$

Let $f \in \mathcal{D}'(\mathbb{R}^n)$. If $a \neq 0$ is a constant then

$$F[f(ax)](\xi) = F[f(ax_1, \ldots, ax_n)](\xi) = |a|^{-n} F[f(x)]\left(\frac{\xi}{a}\right). \quad (A.6.1)$$

Theorem A.6.1. *If $f \in \mathcal{D}'(\mathbb{R}^n)$ is a QAHD of degree λ and of order k, then its Fourier transform $F[f]$ is a QAHD of degree $-\lambda - 1$ and of order k, $k = 0, 1, 2, \ldots$.*

Proof. We prove this theorem by induction.

If $k = 0$ then using (A.6.1) and Definition A.11, we have for all $a > 0$

$$F\big[f(x)\big](a\xi) = a^{-n} F\Big[f\Big(\frac{x}{a}\Big)\Big](\xi) = a^{-\lambda - n} F\big[f(x)\big](\xi), \quad (A.6.2)$$

i.e., $F[f(x)](\xi)$ is a HD of degree $-\lambda - n$.

Let $k = 1$. Using (A.6.1) and Definition A.12, we obtain for all $a > 0$

$$\begin{aligned}
F\big[f(x)\big](a\xi) &= a^{-n} F\Big[f\Big(\frac{x}{a}\Big)\Big](\xi) \\
&= a^{-\lambda - n} F\big[f(x)\big](\xi) - a^{-\lambda - n} \log a F\big[f_0(x)\big](\xi),
\end{aligned}$$

where f_0 is a HD of degree λ. In view of (A.6.2), $F[f_0](\xi)$ is a HD of degree $-\lambda - n$, hence, according to Definition A.12, $F[f(x)](\xi)$ is an AHD of degree $-\lambda - n$ and of order $k = 1$, i.e., a QAHD of degree $-\lambda - n$ and of order $k = 1$.

Let f be a QAHD of degree λ and order k, $k = 2, 3, \ldots$. By using (A.6.1) and Definition A.12, for all $a > 0$ we have

$$F[f(x)](a\xi) = a^{-n}F\left[f\left(\frac{x}{a}\right)\right](\xi)$$

$$= a^{-\lambda-n}F[f(x)](\xi) + \sum_{r=1}^{k}(-1)^r a^{-\lambda-n}\log^r a F[f_{k-r}(x)](\xi),$$

where $f_{k-r}(x)$ is a QAHD of degree λ and order $k - r$, $r = 1, 2, \ldots, k$.

Suppose that the theorem holds for QAHDs of degree λ and order $k = 1, 2 \ldots, k - 1$. Hence, by induction, the latter relation implies that $F[f](\xi)$ is a QAHD of degree λ and of order k.

The theorem is thus proved.

Taking into account Theorem A.4.2 and Remark A.4, we can prove the above theorem at once by calculating the Fourier transform of the distributions $P\left(x_{\pm}^{-n}\log^{k-1}x_{\pm}\right)$ and $x_{\pm}^{\lambda}\log^k x_{\pm}$, where $\lambda \neq -1, -2, \ldots$. $\qquad\square$

Thus $F[\mathcal{AH}_0(\mathbb{R})] = \mathcal{AH}_0(\mathbb{R})$.

A.7. New type of real Γ-functions

Consider the Fourier transform of the homogeneous distribution x_+^{λ}, $\lambda \neq -1, -2, \ldots$, which according to Theorem A.6.1 is represented as

$$F[x_+^{\lambda}](\xi) = C(\xi + i0)^{-\lambda-1}, \qquad (A.7.1)$$

where C is a constant, and the distribution $(x \pm i0)^{\lambda}$ is given by (A.3.5). Setting $\xi = i$, we can calculate that

$$C = i^{\lambda+1}\int_0^{\infty}x^{\lambda}e^{-x}\,dx = i^{\lambda+1}\Gamma(\lambda+1). \qquad (A.7.2)$$

Thus the factor of proportionality in (A.7.1) is (up to $i^{\lambda+1}$) the Γ-function, $\Gamma(\lambda+1) = \int_0^{\infty}x^{\lambda}e^{-x}\,dx$.

In view of Theorem A.6.1 and Remark A.4, for $\lambda \neq -1, -2, \ldots$ we have

$$F[x_+^{\lambda}\log^k x_+](\xi) = \sum_{j=0}^{k}A_{k-j}(\xi+i0)^{-\lambda-1}\log^{k-j}(\xi+i0), \quad (A.7.3)$$

and for $\lambda = -n$, $n \in \mathbb{N}$,

$$F[P(x_+^{-n}\log^{k-1}x_+)](\xi) = \sum_{j=0}^{k}B_{k-j}\xi^{n-1}\log^{k-j}(\xi+i0), \qquad (A.7.4)$$

where A_j, B_j are constants, $j = 1, \dots, k$. Here $(\xi + i0)^{-\lambda-1} \log^{k-j}(\xi + i0)$ and $\xi^{n-1} \log^{k-j}(\xi + i0)$ are QAHDs of order $k - j$ and of degree $-\lambda - 1$ and $n - 1$, respectively (we set $(\xi + i0)^{n-1} \equiv \xi^{n-1}$, $n \in \mathbb{N}$).

Similarly to (A.7.2), the factors

$$\Gamma_j(\lambda + 1; k) = i^{-\lambda-1} \log^j i \, A_j \qquad (A.7.5)$$

and

$$\Gamma_j(-n + 1; k) = i^{n-1} \log^j i \, B_j \qquad (A.7.6)$$

in the relations (A.7.3) and (A.7.4) are said to be *Γ-functions of the type* $(j; k, \lambda)$ and $(j; k, -n)$, respectively, where $\lambda \neq -n$, $n \in \mathbb{N}$, $j = 0, 1, \dots, k$.

By successively substituting $\xi = i, 2i, \dots, (k + 1)i$ into (A.7.3) and (A.7.4), we obtain the linear systems of equations for $A_0, \dots, A_k$ and $B_0, \dots, B_k$. Solving these systems, we can calculate the *associated homogeneous Γ-functions* $\Gamma_j(\lambda + 1; k)$ and $\Gamma_j(-n + 1; k)$, respectively.

Now we calculate the Γ-functions for $k = 1$.

Let $\lambda \neq -1, -2, \dots$. According to [95, Ch.II, §2.4., (1)],

$$F\big[x_+^\lambda \log x_+\big](\xi) = -i^{\lambda+1}\Gamma(\lambda + 1)(\xi + i0)^{-\lambda-1} \log(\xi + i0)$$
$$+ i^{\lambda+1}\Big(\Gamma'(\lambda + 1) + i\frac{\pi}{2}\Gamma(\lambda + 1)\Big)(\xi + i0)^{-\lambda-1}.$$

This relation and (A.7.5) imply that

$$\begin{aligned}
\Gamma_1(\lambda + 1; 1) &= -i\tfrac{\pi}{2}\,\Gamma(\lambda + 1), \\
\Gamma_0(\lambda + 1; 1) &= \Gamma'(\lambda + 1) + i\tfrac{\pi}{2}\Gamma(\lambda + 1).
\end{aligned} \qquad (A.7.7)$$

Let $\lambda = -1, -2, \dots$. According to [95, Ch.II, §2.4., (14)],

$$F\big[x_+^{-n}\big](\xi) = -a_{-1}^{(n)}\xi^{n-1} \log(\xi + i0) + a_0^{(n)}\xi^{n-1},$$

where

$$\begin{aligned}
a_{-1}^{(n)} &= \tfrac{i^{n+1}}{(n-1)!}, \\
a_0^{(n)} &= \tfrac{i^{n+1}}{(n-1)!}\Big(1 + \tfrac{1}{2} + \dots + \tfrac{1}{n-1} + \Gamma'(1) + i\tfrac{\pi}{2}\Big).
\end{aligned}$$

Thus, in view of (A.7.6), we have

$$\begin{aligned}
\Gamma_1(-n + 1; 1) &= -i\tfrac{\pi}{2}\tfrac{(-1)^n}{(n-1)!}, \\
\Gamma_0(-n + 1; 1) &= \tfrac{(-1)^n}{(n-1)!}\Big(1 + \tfrac{1}{2} + \dots + \tfrac{1}{n-1} + \Gamma'(1) + i\tfrac{\pi}{2}\Big).
\end{aligned} \qquad (A.7.8)$$

Formulas (A.7.7) and (A.7.8) can be derived directly as well.

According to (A.7.7) and (A.7.8), we have

$$\Gamma_1(\lambda + 1; 1) = \lambda\Gamma_1(\lambda; 1),$$

$$\Gamma_0(\lambda + 1; 1) = \lambda\Gamma_0(\lambda; 1) + \Gamma(\lambda),$$

and

$$\Gamma_1(-n + 1; 1) = (-n)\Gamma_1(-n; 1),$$

$$\Gamma_0(-n + 1; 1) = (-n)\Gamma_0(-n; 1) - \frac{(-1)^n}{n!},$$

where $\mathrm{res}_{\lambda=-n}\Gamma(\lambda) = \frac{(-1)^n}{n!}$.

In the same way, we can calculate the Γ-functions of the types $\Gamma_j(\lambda + 1; k)$ and $\Gamma_j(-n + 1; k)$ and study their properties.

Appendix B

Two identities

Let us recall that the Bernoulli numbers $\mathbf{B}_\gamma$ are defined by the following recurrence relation:

$$\mathbf{B}_0 = 1, \qquad \sum_{r=0}^{\gamma-1} C_\gamma^r \mathbf{B}_r = 0. \tag{B.1}$$

In particular, $\mathbf{B}_1 = -\frac{1}{2}$, $\mathbf{B}_{2j-1} = 0$, $j = 2, 3, \ldots$, $\mathbf{B}_2 = \frac{1}{6}$, $\mathbf{B}_4 = -\frac{1}{30}$.

Now we present the well-known relation, which we will use repeatedly in our construction.

Proposition B.1. (*see* [167])

$$
\begin{aligned}
\mathbf{S}_s(\gamma_0) = \sum_{\gamma=1}^{\gamma_0} \gamma^s &= \frac{1}{s+1} \sum_{r=0}^{s} C_{s+1}^r \mathbf{B}_r \gamma_0^{s+1-r} + \gamma_0^s \\
&= \frac{1}{s+1} \left(\gamma_0^{s+1} - C_{s+1}^1 \mathbf{B}_1 \gamma_0^s + C_{s+1}^2 \mathbf{B}_2 \gamma_0^{s-1} + \cdots + C_{s+1}^s \mathbf{B}_s \gamma_0 \right),
\end{aligned}
\tag{B.2}
$$

where $\gamma_0 \geq 1$; C_{s+1}^r are the binomial coefficients; $r = 0, 1, \ldots, s, s = 0, 1, 2, \ldots$; $\mathbf{B}_\gamma$ are the Bernoulli numbers.

By formula (B.2) we can consider $\mathbf{S}_s(\gamma_0)$ as a polynomial with respect to γ_0 for which the following relation holds.

Lemma B.2. *If we consider* $\mathbf{S}_s(\gamma_0)$ *as a polynomial with respect to* γ_0 *then for* $\gamma_0 \leq -1$, $s = 0, 1, 2, \ldots$ *we have*

$$\sum_{\gamma=\gamma_0+1}^{0} \gamma^s = -\mathbf{S}_s(\gamma_0). \tag{B.3}$$

Proof. Let us rewrite the above sum by using the relation (B.2) as

$$\sum_{\gamma=\gamma_0+1}^{0} \gamma^s = (-1)^s \sum_{\gamma=1}^{-\gamma_0-1} \gamma^s$$

$$= -\frac{1}{s+1}\Big((\gamma_0+1)^{s+1} - (-1)C_{s+1}^1\mathbf{B}_1(\gamma_0+1)^s$$

$$+ C_{s+1}^2\mathbf{B}_2(\gamma_0+1)^{s-1} + \cdots + (-1)^s C_{s+1}^s\mathbf{B}_s(\gamma_0+1)\Big). \quad (B.4)$$

Using (B.1), it is easy to see that the coefficients of γ_0^{s+1}, γ_0^s, γ_0^{s-1} in the above sum are equal to 1, $C_{s+1}^1 + C_{s+1}^1\mathbf{B}_1 = -C_{s+1}^1\mathbf{B}_1$, $C_{s+1}^2 + C_{s+1}^1\mathbf{B}_1C_s^1 + C_{s+1}^2\mathbf{B}_2 = C_{s+1}^2\mathbf{B}_2$, respectively. Taking into account the relation (B.2) and the relation $\mathbf{B}_{2j-1} = 0$, $j = 2, 3, \ldots$, we obtain the coefficient of γ_0^{s-j}:

$$C_{s+1}^{s-j} + C_{s+1}^1\mathbf{B}_1C_s^{s-j} + C_{s+1}^2\mathbf{B}_2C_{s-1}^{s-j} - C_{s+1}^3\mathbf{B}_3C_{s-2}^{s-j}$$

$$+ \cdots + (-1)^j C_{s+1}^j\mathbf{B}_jC_{s-j}^{s+1-j} + (-1)^{j+1}C_{s+1}^{j+1}\mathbf{B}_{j+1}$$

$$= C_{s+1}^{j+1}\sum_{r=0}^{j} C_{j+1}^r\mathbf{B}_r + C_{s+1}^{j+1}\mathbf{B}_{j+1} = C_{s+1}^{j+1}\mathbf{B}_{j+1},$$

$j = 2, 3, \ldots, s - 1$. The coefficient of γ_0^0 is equal to:

$$\sum_{r=0}^{s} C_{s+1}^r\mathbf{B}_r = 0.$$

$\square$

Appendix C

Proof of a theorem on weak asymptotic expansions

Now we prove the main Theorem 14.6.1, which gives weak asymptotic expansions for the products of regularizations of p-adic QAHDs. The proof of this theorem is reduced to the construction of the weak asymptotics of the elements of the algebra $\mathcal{H}^*$:

$$\delta_k^m(x) \Big(P_{a_1\ k}(x) \Big)^{n_1} \cdots \Big(P_{a_r\ k}(x) \Big)^{n_r} \cdot$$
$$\cdot \Big(\pi_{\alpha_1;b_1\ k}(x) \Big)^{m_1} \cdots \Big(\pi_{\alpha_s;b_s\ k}(x) \Big)^{m_s}, \quad k \to \infty, \qquad \text{(C.1)}$$

where $m, n_j, m_k \in \mathbb{N}_0$, $j = 1, 2, \ldots, r$, $k = 1, 2, \ldots, s$. The above-mentioned weak asymptotics are constructed by using Propositions C.1–C.9, which we shall prove below.

Proposition C.1. *In the weak sense we have*

$$\Big(\delta_k(x) \Big)^s = \delta(x) p^{k(s-1)}, \quad k \to \infty, \quad s = 1, 2, \ldots. \qquad \text{(C.2)}$$

Proof. Using (3.3.1) and the regularization of the delta-function (4.4.6), we obtain

$$\Big\langle \big(\delta_k(x)\big)^s, \varphi(x) \Big\rangle = p^{ks} \int \Omega^s(p^k|x|_p)\varphi(x)\,dx$$
$$= p^{ks} \int_{B_{-k}} \varphi(x)\,dx = p^{ks}\varphi(0) \int_{B_{-k}} dx = p^{k(s-1)}\varphi(0),$$
$$k \to \infty,$$

for all $\varphi \in \mathcal{D}(\mathbb{Q}_p)$ and all sufficiently large k. $\qquad \square$

Proposition C.2. *If $\alpha \neq \alpha_j = \frac{2\pi i}{\ln p} j$, $j \in \mathbb{Z}$, then the following asymptotics holds in the weak sense:*

$$\delta_k^s(x)\pi_{\alpha;m\ k}^n(x) = \delta(x)p^{(-n(\alpha-1)+s-1)k}\Big(\sum_{j=0}^{m} C_m^j I_0(\alpha; m-j)\log_p^j p^{-k}\Big)^n,$$

$$k \to \infty. \quad \text{(C.3)}$$

Proof. For $m \in \mathbb{N}_0$, $s, n \in \mathbb{N}$ let us calculate the following asymptotics

$$J_k = \langle \delta_k^s(x)\pi_{\alpha;m\ k}^n(x), \varphi(x)\rangle, \quad k \to \infty,$$

where $\alpha \neq \alpha_j = \frac{2\pi i}{\ln p} j$, $j \in \mathbb{Z}$, and, by (14.6.1), $\pi_{\alpha;m\ k}(x) = \pi_{\alpha;m\ k}^{(1)}(x) + \pi_{\alpha;m\ k}^{(2)}(x) + \pi_{\alpha;m\ k}^{(3)}(x)$.

First, using (14.6.1) and (14.6.2), we calculate the asymptotics

$$J_{2k} = \Big\langle \delta_k^s(x)\Big(\pi_{\alpha;m\ k}^{(2)}(x)\Big)^r, \varphi(x)\Big\rangle = p^{(s+n)k} \int \Omega^s(p^k|x|_p)$$

$$\times \Big\{ \int_{\mathbb{Q}_p \backslash B_0} |\xi|_p^{\alpha-1}\pi_1(\xi)\log_p^m |\xi|_p \Omega(p^k|x - \xi|_p)\,d\xi \Big\}^n \varphi(x)\,dx$$

$$= p^{(s+n)k} \int_{B_{-k}} \Big\{ \int_{\mathbb{Q}_p \backslash B_0} |\xi|_p^{\alpha-1}\pi_1(\xi)\log_p^m |\xi|_p \Omega(p^k|x - \xi|_p)\,d\xi \Big\}^n \varphi(x)\,dx.$$

where $r = 1, 2, \ldots$. Since $|x|_p \leq \frac{1}{p^k}$ and $\xi \in \mathbb{Q}_p \backslash B_0 = \{|\xi|_p > 1\}$ then $|x - \xi|_p > \frac{1}{p^k}$. Otherwise, we would have

$$|\xi|_p = |x - (x - \xi)|_p \leq \max(|x|_p, |x - \xi|_p) = \frac{1}{p^k}.$$

Thus $\Omega(p^k|x - \xi|_p) = 0$, i.e., $J_{2k} = 0$ for all sufficiently large k.

Hence, using the binomial formula, we find

$$J_k = \Big\langle \delta_k^s(x)\Big(\pi_{\alpha;m\ k}^{(1)}(x) + \pi_{\alpha;m\ k}^{(3)}(x)\Big)^n, \varphi(x)\Big\rangle = p^{(s+n)k} \int \Omega^s(p^k|x|_p)$$

$$\times \Big\{ \int_{B_0} |\xi|_p^{\alpha-1}\pi_1(\xi)\log_p^m |\xi|_p \Big(\Omega(p^k|x - \xi|_p) - \Omega(p^k|x|_p)\Big)\,d\xi$$

$$+ \Omega(p^k|x|_p) \int_{B_0} |\xi|_p^{\alpha-1}\pi_1(\xi)\log_p^m |\xi|_p\,d\xi \Big\}^n \varphi(x)\,dx$$

$$= p^{(s+n)k} \int_{B_{-k}} \left\{ \int_{B_0} |\xi|_p^{\alpha-1} \pi_1(\xi) \log_p^m |\xi|_p \Big(\Omega(p^k|x - \xi|_p) - 1 \Big) d\xi \right.$$

$$\left. + \int_{B_0} |\xi|_p^{\alpha-1} \pi_1(\xi) \log_p^m |\xi|_p \, d\xi \right\}^n \varphi(x) \, dx$$

$$= p^{(s+n-1)k} \varphi(0) \left\{ \int_{B_0} |\xi|_p^{\alpha-1} \pi_1(\xi) \log_p^m |\xi|_p \Omega(p^k|\xi|_p) \, d\xi \right\}^n$$

$$= p^{(s+n-1)k} \varphi(0) \left\{ \int_{B_{-k}} |\xi|_p^{\alpha-1} \pi_1(\xi) \log_p^m |\xi|_p \, d\xi \right\}^n. \tag{C.4}$$

Making the change of variables $\xi \to p^{-k}\xi$ and taking into account the relation $\pi_1(p^k\xi) = \pi_1(\xi)$ and formulas (6.2.3) and (6.3.3), we obtain

$$I_{-k}(\alpha; m) = \int_{B_{-k}} |\xi|_p^{\alpha-1} \pi_1(\xi) \log_p^m |\xi|_p \, d\xi$$

$$= p^{-k\alpha} \int_{B_0} |\xi|_p^{\alpha-1} \pi_1(\xi) \log_p^m \left(p^{-k} |\xi|_p \right) d\xi$$

$$= p^{-k\alpha} \sum_{j=0}^m C_m^j (-k)^j I_0(\alpha; m - j), \tag{C.5}$$

where the integral $I_0(\alpha; m)$ is defined by (6.3.3).

Using (C.5), we have from (C.4) that

$$J_k = \varphi(0) p^{(s+n-n\alpha-1)k} \left(\sum_{j=0}^m C_m^j (-k)^j I_0(\alpha; m - j) \right)^n, \quad k \to \infty.$$

$\square$

Proposition C.3. *The following asymptotics holds in the weak sense:*

$$\delta_k^m(x) \Big(P_k(x) \Big)^n = (-1)^n \delta(x) p^{(m+n-1)k} \log_p^n \left(p^{-k} \right) \left(1 - p^{-1} \right)^n, \quad k \to \infty, \tag{C.6}$$

where $m \in \mathbb{N}$, $n \in \mathbb{N}_0$.

Proof. Using calculations similar to those carried out in Proposition C.2, it is easy to show that for any test function φ and sufficiently large k:

$$\left\langle \delta_k^m(x)\Big(P_k(x)\Big)^n, \varphi(x)\right\rangle$$

$$= p^{(m+n-1)k}\varphi(0)\left(\int_{B_0} |\xi|_p^{-1}\Big(\Omega(p^k|\xi|_p) - 1\Big)\,d\xi\right)^n$$

$$= p^{(m+n-1)k}\varphi(0)\left(\sum_{\gamma=-k+1}^{0}\int_{S_\gamma} |\xi|_p^{-1}\,d\xi\right)^n$$

$$= p^{(m+n-1)k}\log_p^n\left(p^{-k}\right)\Big(1 - p^{-1}\Big)^n(-1)^n\varphi(0).$$

$\square$

Repeating calculations similar to those in Propositions C.1–C.3, we can prove the following

Proposition C.4. *We have the following asymptotics in the weak sense:*

$$\delta_k^m(x)\Big(\pi_{\alpha_1;a_1\,k}(x)\Big)^{m_1}\cdots\Big(\pi_{\alpha_s;a_s\,k}(x)\Big)^{m_s}\Big(P_{b_1\,k}(x)\Big)^{n_1}\cdots\Big(P_{b_r\,k}(x)\Big)^{n_r}$$

$$= \delta(x)p^{\gamma k}R_N\left(\log_p p^{-k}\right), \quad k\to\infty,$$

where $R_N(z)$ is a polynomial of degree $N = \sum_{j=1}^s m_j a_j + \sum_{j=1}^r n_j$ with the computable coefficients $\gamma = m - 1 - \sum_{j=1}^s m_j(\alpha_j - 1) + \sum_{j=1}^r n_j$.

We also have the following.

Proposition C.5. *Suppose that $\alpha \neq \alpha_j = \frac{2\pi i}{\ln p}j$, $j \in \mathbb{Z}$, $n \in \mathbb{N}$. Then we have in the weak sense:*

(a) if $n(\alpha - 1) + 1 \neq \alpha_j$ and $\pi_1(x) \not\equiv 1$ then

$$\pi_{\alpha;m\,k}^n(x) = |x|_p^{n(\alpha-1)}\pi_1^n(x)\log_p^{nm}|x|_p$$

$$+ p^{-(n(\alpha-1)+1)k}\left\{\left(\sum_{j=0}^m C_m^j \log_p^j\left(p^{-k}\right)I_0(\alpha; m - j)\right)^n\right.$$

$$\left.- \sum_{j=0}^{nm} C_{nm}^j \log_p^j\left(p^{-k}\right)I_0(n(\alpha - 1) + 1; nm - j)\right\}\delta(x), \quad \text{(C.7)}$$

as $k \to \infty$;

(b) *if $n(\alpha - 1) + 1 = \alpha_j$ and $\pi_1(x) \equiv 1$ then*

$$\pi_{\alpha;m\ k}^n(x) = P\left(\frac{\log_p^{nm} |x|_p}{|x|_p}\right) + \left\{\left(\sum_{j=0}^{m} C_m^j \log_p^j \left(p^{-k}\right) I_0(\alpha; m - j)\right)^n\right.$$

$$\left. -(1 - p^{-1}) S_{nm}\left(\log_p(p^{-k})\right)\right\} \delta(x), \quad k \to \infty. \quad (C.8)$$

In particular, the following assertion holds for $m = 0$:
(a) *if $n(\alpha - 1) + 1 \neq \alpha_j$ and $\pi_1(x) \not\equiv 1$ then*

$$\pi_{\alpha\ k}^n(x) = |x|_p^{n(\alpha-1)} \pi_1^n(x)$$

$$+ p^{-(n(\alpha-1)+1)k}\left\{\left(I_0(\alpha)\right)^n - I_0(n(\alpha - 1) + 1)\right\} \delta(x), \quad (C.9)$$

as $k \to \infty$;
(b) *if $n(\alpha - 1) + 1 = \alpha_j$ and $\pi_1(x) \equiv 1$ then*

$$\pi_{\alpha;\ k}^n(x) = P\left(\frac{1}{|x|_p}\right) + \left\{\left(I_0(\alpha)\right)^n - (1 - p^{-1}) \log_p\left(p^{-k}\right)\right\} \delta(x), \quad (C.10)$$

as $k \to \infty$.

Here the integrals $I_0(\alpha)$ and $I_0(\alpha; m - j)$ are defined by (6.2.3) and (6.3.3), respectively, the function $S_s(r)$ is defined by (B.2).

Proof. Let us calculate the asymptotics of $J_k = \langle \pi_{\alpha;m\ k}^n, \varphi \rangle, n \in \mathbb{N}$, represented in the form of the sum

$$J_k = \int_{B_0} \pi_{\alpha;m\ k}^n(x)\left(\varphi(x) - \varphi(0)\right) dx$$

$$+ \int_{\mathbb{Q}_p \setminus B_0} \pi_{\alpha;m\ k}^n(x)\varphi(x)\, dx + \varphi(0) \int_{B_0} \pi_{\alpha;m\ k}^n(x)\, dx.$$

Since $\pi_\alpha(x) \log_p^m |x|_p$ is a continuous function for $x \neq 0$, in view of Lemma 14.3.2, for $x \neq 0$ we have

$$\pi_{\alpha;m\ k}^n(x) \to \left(\pi_\alpha(x) \log_p^m |x|_p\right)^n = |x|_p^{n(\alpha-1)} \pi_1^n(x) \log_p^{nm} |x|_p, \quad k \to \infty,$$

in the sense of pointwise convergence. Let $J_{1k} = \int_{B_0} \pi_{\alpha;m\ k}^n(x)\left(\varphi(x) - \varphi(0)\right) dx$, $J_{2k} = \int_{\mathbb{Q}_p \setminus B_0} \pi_{\alpha;m\ k}^n(x)\varphi(x)\, dx$. Denote by $l(\varphi)$ the parameter of constancy of the function φ. Taking into account that $\varphi(x) - \varphi(0) = 0$ for all $x \in B_{l(\varphi)} \subset B_0$, according to Lemma 14.3.1, for any φ and for sufficiently large

k we have

$$
\begin{aligned}
J_{1k} &= \int_{B_0} \left(\pi_\alpha(x)\log_p^m |x|_p\right)^n \left(\varphi(x) - \varphi(0)\right) dx, \\
J_{2k} &= \int_{\mathbb{Q}_p \setminus B_0} \left(\pi_\alpha(x)\log_p^m |x|_p\right)^n \varphi(x)\, dx.
\end{aligned}
\tag{C.11}
$$

Now we calculate the asymptotics of the integral

$$
J_{3k} = \int_{B_0} \pi_{\alpha;m\ k}^n(x)\, dx = \int_{B_0} \left(\sum_{j=1}^{3} \pi_{\alpha;m\ k}^{(j)}(x)\right)^n dx = J_{3,1k} + J_{3,2k},
$$

where, in view of (14.6.2),

$$
\begin{aligned}
J_{3,1k} &= \int_{B_0} \left(\pi_{\alpha;m\ k}^{(1)}(x) + \pi_{\alpha;m\ k}^{(2)}(x)\right)^n dx, \\
J_{3,2k} &= \sum_{r=0}^{n-1} C_n^r \int_{B_0} \left(\pi_{\alpha;m\ k}^{(1)}(x)\right)^r \left(\pi_{\alpha;m\ k}^{(3)}(x)\right)^{n-r} dx.
\end{aligned}
\tag{C.12}
$$

By calculations analogous to those used for (C.4), we derive from (C.12), (14.6.2), (6.3.3) and (C.5)

$$
\begin{aligned}
J_{3,2k} &= \sum_{r=0}^{n-1} C_n^r \int_{B_0} \left(\pi_{\alpha;m\ k}^{(1)}(x)\right)^r \left(\pi_{\alpha;m\ k}^{(3)}(x)\right)^{n-r} dx = p^{nk} \sum_{r=0}^{n-1} C_n^r \\
&\quad \times \int_{B_0} \left(\int_{B_0} |\xi|_p^{\alpha-1}\pi_1(\xi)\log_p^m |\xi|_p \left(\Omega(p^k|x-\xi|_p) - \Omega(p^k|x|_p)\right) d\xi\right)^r \\
&\quad \times \Omega^{n-r}(p^k|x|_p)\left(\int_{B_0} |\xi|_p^{\alpha-1}\pi_1(\xi)\log_p^m |\xi|_p\, d\xi\right)^{n-r} dx \\
&= p^{(n-1)k} \sum_{r=0}^{n-1} C_n^r \left(\int_{B_0} |\xi|_p^{\alpha-1}\pi_1(\xi)\log_p^m |\xi|_p \left(\Omega(p^k|\xi|_p) - 1\right) d\xi\right)^r \\
&\quad \times \left(\int_{B_0} |\xi|_p^{\alpha-1}\pi_1(\xi)\log_p^m |\xi|_p\, d\xi\right)^{n-r} \\
&= p^{(n-1)k} \sum_{r=0}^{n-1} C_n^r \left(\int_{B_{-k}} |\xi|_p^{\alpha-1}\pi_1(\xi)\log_p^m |\xi|_p\, d\xi\right. \\
&\quad \left. - \int_{B_0} |\xi|_p^{\alpha-1}\pi_1(\xi)\log_p^m |\xi|_p\, d\xi\right)^r \left(\int_{B_0} |\xi|_p^{\alpha-1}\pi_1(\xi)\log_p^m |\xi|_p\, d\xi\right)^{n-r} \\
&= p^{(n-1)k} \sum_{r=0}^{n-1} C_n^r \left\{ p^{-k\alpha} \sum_{j=0}^{m} C_m^j(-k)^j I_0(\alpha; m-j) - I_0(\alpha; m)\right\}^r
\end{aligned}
$$

$$\times \left(I_0(\alpha;m)\right)^{n-r} = p^{-(n(\alpha-1)+1)k} \left\{ \left(\sum_{j=0}^{m} C_m^j(-k)^j I_0(\alpha;m-j)\right)^n \right.$$

$$\left. - \left(\sum_{j=0}^{m} C_m^j(-k)^j I_0(\alpha;m-j) - p^{k\alpha} I_0(\alpha;m)\right)^n \right\}, \quad k \to \infty.$$

$$(C.13)$$

Next, we construct the asymptotics of the integral $J_{3,1k}$. It is clear (see (14.6.2)) that for $|x|_p \le 1$ and for sufficiently large k

$$\pi_{\alpha;m\ k}^{(2)}(x) = p^k \int_{\mathbb{Q}_p \setminus B_0} |\xi|_p^{\alpha-1} \pi_1(\xi) \log_p^m |\xi|_p \Omega(p^k|x-\xi|_p)\,d\xi = 0.$$

Indeed, since $|x|_p \le 1$ and $\xi \in \mathbb{Q}_p \setminus B_0 = \{|\xi|_p > 1\}$ then $|x-\xi|_p > \frac{1}{p^k}$. Otherwise, supposing $|x-\xi|_p \le \frac{1}{p^k}$, we would have $|\xi|_p = |x-(x-\xi)|_p \le \max(|x|_p, |x-\xi|_p) = 1$, which is a contradiction. Thus $\Omega(p^k|x-\xi|_p) = 0$ and $\pi_{\alpha;m\ k}^{(2)}(x) = 0$. Thus we have

$$J_{3,1k} = \int_{B_0} \left(\pi_{\alpha;m\ k}^{(1)}(x) + \pi_{\alpha;m\ k}^{(2)}(x)\right)^n dx = \int_{B_0} \left(\pi_{\alpha;m\ k}^{(1)}(x)\right)^n dx$$

$$= \int_{B_{-k}} \left(\pi_{\alpha;m\ k}^{(1)}(x)\right)^n dx + \int_{B_0 \setminus B_{-k}} \left(\pi_{\alpha;m\ k}^{(1)}(x)\right)^n dx, \quad (C.14)$$

as $k \to \infty$, where, in view of (14.6.2),

$$\pi_{\alpha;m\ k}^{(1)}(x) = p^k \int_{B_0} |\xi|_p^{\alpha-1} \pi_1(\xi) \log_p^m |\xi|_p \left(\Omega(p^k|x-\xi|_p) - \Omega(p^k|x|_p)\right) d\xi.$$

Now we calculate the first integral in (C.14). Let $|x|_p \le p^{-k}$. Hence, taking into account the relation (C.5), we obtain

$$\pi_{\alpha;m\ k}^{(1)}(x) = -p^k \int_{p^{-k} < |\xi|_p \le 1} |\xi|_p^{\alpha-1} \pi_1(\xi) \log_p^m |\xi|_p\,d\xi$$

$$= -p^k \sum_{\gamma=-k+1}^{0} \int_{S_\gamma} |\xi|_p^{\alpha-1} \pi_1(\xi) \log_p^m |\xi|_p\,d\xi$$

$$= -p^k \left(I_0(\alpha;m) - I_{-k}(\alpha;m)\right)$$

$$= -p^k \left(I_0(\alpha;m) - p^{-k\alpha} \sum_{j=0}^{m} C_m^j(-k)^j I_0(\alpha;m-j)\right), \quad (C.15)$$

where the integrals $I_0(\alpha;s)$ and $I_{-k}(\alpha;s)$ are determined by the formulas (6.3.3) and (C.5), respectively, $\alpha \ne \alpha_j = \frac{2\pi i}{\ln p} j$, $j \in \mathbb{Z}$. Here the latter expression was obtained by the analytical continuation with respect to α. Using (C.15), (3.3.1) and (3.2.6), we write the first integral in (C.14) in the form

$$\int_{B_{-k}} \left(\pi^{(1)}_{\alpha;m \ k}(x)\right)^n dx = (-1)^n p^{(n-1)k} \Big(I_0(\alpha; m)$$

$$-p^{-k\alpha} \sum_{j=0}^{m} C_m^j (-k)^j I_0(\alpha; m-j)\Big)^n, \quad k \to \infty.$$

(C.16)

Next, we calculate the second integral in (C.14). Let $x \in B_0 \setminus B_{-k}$. It is clear that $\Omega(p^k|x-\xi|_p) = 0$. Otherwise, we would have $|x-\xi|_p \le \frac{1}{p^k}$ and $|x|_p = |\xi + (x-\xi)|_p \le \max(|\xi|_p, |x-\xi|_p) = \frac{1}{p^k}$. Thus, for $p^{-k} < |x|_p \le 1$ we have

$$\pi^{(1)}_{\alpha;m \ k}(x) = p^k \int_{p^{-k} < |\xi|_p \le 1} |\xi|_p^{\alpha-1} \pi_1(\xi) \log_p^m |\xi|_p \Omega(p^k|x-\xi|_p) d\xi. \quad (C.17)$$

Let $\pi_1(x) \not\equiv 1$. In view of (3.3.5), the relation (C.17) implies

$$\pi^{(1)}_{\alpha;m \ k}(x) = p^k \sum_{\gamma = \log_p |\xi| \in [-k+1,0]; \ \xi \in B_{-k}(x)} p^{\gamma(\alpha-1)} \gamma^m$$

$$\times \int_{S_\gamma} \pi_1(\xi)\varphi(x) d\xi = 0, \quad p^{-k} < |x|_p \le 1. \quad (C.18)$$

Let $\pi_1(x) \equiv 1$. Since the convolution is commutative, (C.17) can be rewritten as

$$\pi^{(1)}_{\alpha;m \ k}(x) = p^k \int_{\mathbb{Q}_p} |\xi|_p^{\alpha-1} \log_p^m |\xi|_p \Big(1 - \Delta_{-k}(\xi)\Big) \Delta_0(\xi) \Omega(p^k|x-\xi|_p) d\xi$$

$$= \Big(|x|_p^{\alpha-1} \log_p^m |x|_p \big(1 - \Delta_{-k}(x)\big) \Delta_0(x)\Big) * \delta_k(x),$$

$$= p^k \int_{\mathbb{Q}_p} |x-\xi|_p^{\alpha-1} \log_p^m |x-\xi|_p \Big(1 - \Delta_{-k}(x-\xi)\Big)$$

$$\times \Delta_0(x-\xi) \Omega(p^k|\xi|_p) d\xi, \quad p^{-k} < |x|_p \le 1,$$

where $\Big(1 - \Delta_{-k}(x)\Big)\Delta_0(x)$ is the characteristic function of the set $B_0 \setminus B_{-k}$. Analogously to (14.3.3), taking into account that for all $p^{-k} < |x|_p \le 1$ and $|\xi|_p \le p^{-k}$, we have $|x-\xi|_p = |x|_p$. We can see that the latter integral is equal to

$$\pi^{(1)}_{\alpha;m \ k}(x) = |x|_p^{\alpha-1} \log_p^m |\xi|_p \big(1 - \Delta_{-k}(x)\big)\Delta_0(x). \quad (C.19)$$

Substituting (C.19) into the second integral in (C.14), we obtain

$$\int_{B_0 \setminus B_{-k}} \left(\pi^{(1)}_{\alpha;m \ k}(x)\right)^n dx = I_1(n(\alpha-1)+1; nm), \quad (C.20)$$

where $I_1(n(\alpha - 1) + 1; nm) = \int_{p^{-k} < |x|_p \leq 1} |x|_p^{n(\alpha-1)} \log_p^{nm} |x|_p \, dx$.

If $n(\alpha - 1) + 1 \neq \alpha_j$, from (6.3.3) and (C.5), we have

$$I_1(n(\alpha - 1) + 1; nm) = \int_{p^{-k} < |x|_p \leq 1} |x|_p^{n(\alpha-1)} \pi_1^n(x) \log_p^{nm} |x|_p \, dx$$

$$= I_0(n(\alpha - 1) + 1; nm) - I_{-k}(n(\alpha - 1) + 1; nm)$$

$$= I_0(n(\alpha - 1) + 1; nm) - p^{-k(n(\alpha-1)+1)}$$

$$\times \sum_{j=0}^{nm} C_{nm}^j (-k)^j I_0(n(\alpha - 1) + 1; nm - j). \quad \text{(C.21)}$$

If $n(\alpha - 1) + 1 = \alpha_j$, taking into account that $p^{\gamma \alpha_j} = 1$, the equality (B.2), and Lemma B.2, we can obtain

$$I_1(n(\alpha - 1) + 1; nm) = \sum_{\gamma=-k+1}^{0} \int_{S_\gamma} |x|_p^{n(\alpha-1)} \pi_1^n(x) \log_p^{nm} |x|_p \, dx$$

$$= (1 - p^{-1}) \sum_{\gamma=-k+1}^{0} p^{\gamma(-1+\alpha_j)} p^\gamma \gamma^{nm}$$

$$= -(1 - p^{-1}) \mathbf{S}_{nm}(-k), \quad \text{(C.22)}$$

where the function $\mathbf{S}_s(r)$ is defined by (B.2).

Summarizing the above results and using the formulas (C.11), (C.13), (C.14), (C.16), (C.20), (C.21) and (C.22), we calculate

$$J_k = J_{1k} + J_{2k} + \varphi(0)(J_{3,1k} + J_{3,2k})$$

$$= \int_{B_0} |x|_p^{n(\alpha-1)} \pi_1^n(x) \log_p^{nm} |x|_p (\varphi(x) - \varphi(0)) \, dx$$

$$+ \int_{\mathbb{Q}_p \setminus B_0} |x|_p^{n(\alpha-1)} \pi_1^n(x) \log_p^{nm} |x|_p \varphi(x) \, dx + \varphi(0) C_k, \quad k \to \infty,$$

$$\text{(C.23)}$$

where if $n(\alpha - 1) + 1 \neq \alpha_j$ then

$$C_k = J_{3,1k} + J_{3,2k} = p^{-(n(\alpha-1)+1)k} \left(\sum_{j=0}^{m} C_m^j (-k)^j I_0(\alpha; m - j) \right)^n$$

$$+ I_0(n(\alpha - 1) + 1; nm) - p^{-k(n(\alpha-1)+1)}$$

$$\times \sum_{j=0}^{nm} C_{nm}^j (-k)^j I_0(n(\alpha - 1) + 1; nm - j); \quad \text{(C.24)}$$

and if $n(\alpha - 1) + 1 = \alpha_j$ and $\pi_1(x) \equiv 1$ then

$$
C_k = J_{3,1k} + J_{3,2k} = \left(\sum_{j=0}^{m} C_m^j (-k)^j I_0(\alpha; m - j) \right)^n
$$

$$
- (1 - p^{-1}) S_{nm}(-k). \tag{C.25}
$$

In view of the formulas (6.3.2), (6.3.3) and (6.3.5), the relations (C.23)–(C.25) can be rewritten in the weak sense as (C.7) and (C.8).   □

Analogously one can prove the following proposition.

Proposition C.6. (a) *If* $\alpha, \beta, m(\alpha - 1) + n(\beta - 1) + 1 \neq \alpha_j = \frac{2\pi i}{\ln p} j, \ j \in \mathbb{Z}$, *then the following asymptotics holds in the weak sense:*

$$
\pi_{\alpha}^{m}{}_k(x) \pi_{\beta}^{n}{}_k(x) = \pi_{m(\alpha-1)+n(\beta-1)+1}(x) \pi_1^{m+n-1}(x)
$$
$$
+ \delta(x) p^{-(m(\alpha-1)+n(\beta-1)+1)k} \Big(I_0^m(\alpha) I_0^n(\beta) - I_0(m(\alpha - 1)
$$
$$
+ n(\beta - 1) + 1) \Big),
$$

as $k \to \infty$.

(b) *If* $\alpha, \beta \neq \alpha_j = \frac{2\pi i}{\ln p} j, \ m(\alpha - 1) + n(\beta - 1) + 1 = \alpha_j = \frac{2\pi i}{\ln p} j$ *and* $\pi_1(x) \equiv 1$, *then the following asymptotics holds in the weak sense:*

$$
\pi_{\alpha}^{m}{}_k(x) \pi_{\beta}^{n}{}_k(x) = P\left(\frac{1}{|x|_p} \right) + \Big(I_0^m(\alpha) I_0^n(\beta) - (1 - p^{-1}) \log_p \left(p^{-k} \right) \Big) \delta(x),
$$

as $k \to \infty$.

Proposition C.7. *The following asymptotics holds in the weak sense:*

$$
\left(P_k(x) \right)^n = |x|_p^{-n} + p^{(n-1)k} \Big(\log_p^n \left(p^{-k} \right) \left(1 - p^{-1} \right)^n - I_0(-n + 1) \Big) \delta(x),
$$

as $k \to \infty, n = 2, 3, \ldots$.

Proof. To construct the asymptotics $J_k = \langle \left(P_k(x) \right)^n, \varphi(x) \rangle$ we use the same calculations as in the proof of Proposition C.5. Let us represent the latter expression as a sum

$$
J_k = J_{1k} + J_{2k} + \varphi(0) J_{3k},
$$

where

$$J_{1k} = \int_{B_0} \left(P_k(x)\right)^n \left(\varphi(x) - \varphi(0)\right) dx,$$

$$J_{2k} = \int_{\mathbb{Q}_p \setminus B_0} \left(P_k(x)\right)^n \varphi(x) dx,$$

$$J_{3k} = \int_{B_0} \left(P_k(x)\right)^n dx.$$

By repeating the proof of the relations (C.11) word for word for all φ and sufficiently large k, we have

$$J_{1k} = \int_{B_0} |x|_p^{-n} \left(\varphi(x) - \varphi(0)\right) dx,$$

$$J_{2k} = \int_{\mathbb{Q}_p \setminus B_0} |x|_p^{-n} \varphi(x) dx.$$

Next, representing J_{3k} as the sum

$$J_{3k} = \int_{B_0} \left(P_k(x)\right)^n dx = \int_{B_{-k}} \left(P_k^{(1)}(x)\right)^n dx + \int_{B_0 \setminus B_{-k}} \left(P_k^{(1)}(x)\right)^n dx,$$

where

$$P_k^{(1)}(x) = p^k \int_{B_0} |\xi|_p^{-1} \left(\Omega(p^k|x - \xi|_p) - \Omega(p^k|x|_p)\right) d\xi,$$

and repeating the calculations used in (C.16) and (C.20) almost word for word, we obtain

$$\int_{B_{-k}} \left(P_k^{(1)}(x)\right)^n dx = (-1)^n p^{(n-1)k} \left(\int_{p^{-k} < |\xi|_p \le 1} |\xi|_p^{-1} d\xi\right)^n$$

$$= (-1)^n p^{(n-1)k} \left(\sum_{\gamma=-k+1}^{0} \int_{S_\gamma} |\xi|_p^{-1} d\xi\right)^n$$

$$= (-1)^n p^{(n-1)k} k^n \left(1 - p^{-1}\right)^n$$

$$= \int_{B_0 \setminus B_{-k}} \left(P_k^{(1)}(x)\right)^n dx = \int_{p^{-k} < |x|_p \le 1} |x|_p^{-n} dx$$

$$= \sum_{\gamma=-k+1}^{0} \int_{S_\gamma} |x|_p^{-n} dx = \left(1 - p^{(n-1)k}\right) I_0(-n + 1).$$

To construct the asymptotics of J_k, we note that the following relation (analogous to (C.19)) holds:

$$\left(|x|_p^{-1}\left(1 - \Delta_{-k}(x)\right)\Delta_0(x)\right) * \delta_k(x) = |x|_p^{-1}\left(1 - \Delta_{-k}(x)\right)\Delta_0(x).$$

Thus,

$$J_k = \left\langle \left(P_k(x)\right)^n, \varphi(x)\right\rangle = J_{1k} + J_{3k} + \varphi(0)J_{3k}$$

$$= \int_{B_0} |x|_p^{-n}\left(\varphi(x) - \varphi(0)\right) dx + \int_{\mathbb{Q}_p \setminus B_0} |x|_p^{-n}\varphi(x)\, dx + \varphi(0)I_0(-n+1)$$

$$+ \varphi(0)p^{(n-1)k}\left((-k)^n\left(1-p^{-1}\right)^n - I_0(-n+1)\right), \quad k \to \infty.$$

$\square$

Analogously one can prove the following proposition.

Proposition C.8. (a) *If* α, $m(\alpha-1)-n+1 \neq \alpha_j = \frac{2\pi i}{\ln p} j$, $j \in \mathbb{Z}$, *then the following asymptotics holds in the weak sense:*

$$\pi_{\alpha\ k}^m(x)\left(P_k(x)\right)^n = \pi_{m(\alpha-1)-n+1}\pi_1^{m-1}(x)$$

$$+ p^{-(m(\alpha-1)-n+1)k}\left\{I_0^m(\alpha)\left(1-p^{-1}\right)^n \log_p^n\left(p^{-k}\right)\right.$$

$$\left.- I_0(m(\alpha-1)-n+1)\right\}\delta(x), \quad k \to \infty.$$

(b) *If* $\alpha \neq \alpha_j = \frac{2\pi i}{\ln p} j$, $m(\alpha-1)-n+1 = \alpha_j = \frac{2\pi i}{\ln p} j$ *and* $\pi_1(x) \equiv 1$, *then the following asymptotics holds in the weak sense:*

$$\pi_{\alpha\ k}^m(x)\left(P_k(x)\right)^n$$

$$= P\left(\frac{1}{|x|_p}\right) + \left\{I_0^m(\alpha)\left(1-p^{-1}\right)^n \log_p^n\left(p^{-k}\right) - \left(1-p^{-1}\right)\log_p\left(p^{-k}\right)\right\}\delta(x),$$

$$k \to \infty.$$

Repeating the calculations, analogous to those used in Propositions C.5–C.8, one can prove the following proposition.

Proposition C.9. *We have the following asymptotics in the weak sense:*

$$\left(\pi_{\alpha_1;a_1\ k}(x)\right)^{m_1}\cdots\left(\pi_{\alpha_s;a_s\ k}(x)\right)^{m_s}\left(P_{b_1\ k}(x)\right)^{n_1}\cdots\left(P_{b_r\ k}(x)\right)^{n_r}$$

$$= |x|_p^\beta \pi_1^m(x)\log_p^a |x|_p + \delta(x)p^{\gamma k}R_N\left(\log_p p^{-k}\right), \quad k \to \infty,$$

where $\beta = \sum_{j=1}^s m_j(\alpha_j-1) - \sum_{j=1}^r n_j$, $m = \sum_{j=1}^s m_j$, $a = \sum_{j=1}^s m_j a_j$, $\gamma = -1 - \sum_{j=1}^s m_j(\alpha_j-1) + \sum_{j=1}^r n_j$, $R_N(z)$ *is a polynomial of degree* $N = \sum_{j=1}^s m_j a_j + \sum_{j=1}^r n_j$ *with the computable coefficients.*

Appendix D

One "natural" way to introduce a measure on $\mathbb{Q}_p$

The following remark was communicated to us by A. V. Kosyak and follows A. Kosyak's book [161].

For simplicity we consider the case of the ring of p-adic integer numbers $\mathbb{Z}_p$ (see Section 1.7), but a similar procedure can be used to construct a product measure on $\mathbb{Q}_p$. Since $\mathbb{Z}_p$ can be realized as

$$\mathbb{Z}_p = \left\{ \sum_{n \in \mathbb{N}_0} a_n p^n : a_n \in \mathbb{F}_p, \ n \in \mathbb{N}_0 \right\},$$

where $\mathbb{F}_p = \mathbb{Z}/p\mathbb{Z}$ is the *finite field* with p elements, one can consider the space $\mathbb{Z}_p$ as the infinite product of the rings $\mathbb{F}_p$:

$$\mathbb{Z}_p = \prod_{n \in \mathbb{N}_0} (\mathbb{F}_p)_n, \qquad (\mathbb{F}_p)_n = \mathbb{F}_p.$$

Consequently, a measure on $\mathbb{Z}_p$ can be constructed, for example, as the infinite tensor product of the probability measures ν_n on $\mathbb{F}_p$:

$$\nu = \bigotimes_{n \in \mathbb{N}_0} \nu_n, \qquad \nu_n(\mathbb{F}_p) = 1, \qquad \sum_{k \in \mathbb{F}_p} \nu_n(k) = 1. \tag{D.1}$$

Remark D.1. One can obtain the Haar measure $h_{\mathbb{Z}_p}$ on $\mathbb{Z}_p$ as a particular case of (D.1), i.e., as the tensor product of the invariant measures on $\mathbb{F}_p$:

$$h_{\mathbb{Z}_p} = \bigotimes_{n \in \mathbb{N}_0} \nu_n^{inv},$$

where the invariant measures ν_n^{inv} on $\mathbb{F}_p$ are defined as

$$\nu_n^{inv}(k) = \frac{1}{p}, \qquad k \in \mathbb{F}_p, \quad n \in \mathbb{N}_0.$$

In [160] the following lemma is used:

331

Lemma D.1. *Two measures* $\mu = \bigotimes_{n \in \mathbb{N}_0} \mu_n$, $\nu = \bigotimes_{n \in \mathbb{N}_0} \nu_n$ *on* $\mathbb{Z}_p$ *are equivalent if and only if their* Hellinger integral

$$H(\mu, \nu) = \prod_{n \in \mathbb{N}_0} \int_{\mathbb{F}_p} \sqrt{\frac{d\mu_n}{d\nu_n}(k)} d\nu_n(k) = \prod_{n \in \mathbb{N}_0} \sum_{k \in \mathbb{F}_p} \sqrt{\mu_n(k)\nu_n(k)}$$

is positive, i.e., $H(\mu, \nu) > 0$. *Otherwise,* $\mu \perp \nu$.

The definition and the properties of the Helliger integral can be found in [168].

References

[1] S. Albeverio, R. Cianci and A. Yu. Khrennikov, *On the spectrum of the p-adic position operator*, J. Physics A: Math. and General, **30**, (1997), 881–889.

[2] S. Albeverio, R. Cianci and A. Yu. Khrennikov, *A representation of quantum field Hamiltonian in a p-adic Hilbert space*, (Russian), Teoret. Mat. Fiz., **112**, (1997), no. 3, 355–374; translation in Theoret. and Math. Phys., **112**, (1997), no. 3, 1081–1096.

[3] S. Albeverio, R. Cianci and A. Yu. Khrennikov, *On the Fourier transform and the spectral properties of the p-adic momentum and Schrödinger operators*, J. Physics A, Math. and General, **30**, (1997), 5767–5784.

[4] S. Albeverio, S. Evdokimov, M. Skopina, *p-adic multiresolution analysis and wavelet frames*. To appear in journal of Fourier Analysis and Applications (2010). Preprint at the url: http://arxiv.org/abs/0802.1079v1

[5] S. Albeverio, S. Evdokimov, M. Skopina, *p-adic multiresolution analysis*, (2008), Preprint at the url: http://arxiv.org/abs/0810.1147v1

[6] S. Albeverio, S. Evdokimov, M. Skopina, *p-adic non-orthogonal wavelet bases*, Proc. Steklov Inst. Math., **v. 265**, Moscow, 2009, 1–12.

[7] S. Albeverio, J. E. Fenstad, R. R. Høegh-Krohn, T. Lindstrøm, *Nonstandard methods in stochastic analysis and mathematical physics*, Academic Press, Orlando, 1986. Published in 2009, Dover Publications (Mineola, NY)

[8] S. Albeverio, Z. Haba, F. Russo, *Trivial solutions for a non-linear two-space-dimensional wave equation perturbed by space-time white noise*, Stochastics Stochastics Rep. **56**, (1996), no. 1–2, 127–160.

[9] S. Albeverio, F. Gesztesy, R. Høegh-Krohn, and H. Holden, *Solvable Models in Quantum Mechanics*. New York–Berlin–Heidelberg–London–Paris–Tokyo, Springer, 1988; 2nd ed. (with an appendix by P. Exner), AMS Chelsea Publishing, Providence, RI, 2005.

[10] S. Albeverio, E. I. Gordon, A. Yu. and Khrennikov, *Finite-dimensional approximations of operators in the Hilbert spaces of functions on locally compact abelian groups*, Acta Appl. Math., **64**, no. 1, (2000), 33–73.

[11] S. Albeverio, W. Karwowski, *A random walk on p-adics, the generator and its spectrum*, Stochastic Process Appl., **53**, (1994), 1–22.

[12] S. Albeverio, W. Karwowski, *Jump processes on leaves of multibranching trees*, J. Math. Phys., **49**, (2008), 093503–093523.

[13] S. Albeverio and W. Karwowski, *Diffusion on p-adic numbers in Gaussian random fields*, (Nagoya, 1990), 86–99, Ser. Probab. Statist. 1, World Sci. Publ., River Edge, N.J., 1991.

[14] S. Albeverio, A. Yu. Khrennikov, *Representation of the Weyl group in spaces of square integrable functions with respect to p-adic valued Gaussian distributions*, J. of Phys. A, **29**, (1996), 5515–5527.

[15] S. Albeverio, A. Yu. Khrennikov, *p-adic Hilbert space representation of quantum systems with an infinite number of degrees of freedom*, Int. J. of Modern Phys. B, **10**, (1996), 1665–1673.

[16] S. Albeverio, A. Yu. Khrennikov, V. M. Shelkovich, *Associated homogeneous p-adic distributions*, J. Math. An. Appl., **313**, (2006), 64–83.

[17] S. Albeverio, A. Yu. Khrennikov, V. M. Shelkovich, *Associated homogeneous p-adic generalized functions*, Dokl. Ross. Akad. Nauk, **393**, no. 3, (2003), 300–303. English transl. in Russian Doklady Mathematics., **68**, no. 3, (2003), 354–357.

[18] S. Albeverio, A. Yu. Khrennikov, and V. M. Shelkovich, *p-adic Colombeau–Egorov type theory of generalized functions*, Math. Nachr., **278**, no. 1–2, (2005), 3–16.

[19] S. Albeverio, A. Yu. Khrennikov, and V. M. Shelkovich, *Associative algebras of p-adic distributions*, Proc. Steklov Inst. Math., v. **245**, Moscow, 2004, 22–33.

[20] S. Albeverio, A. Yu. Khrennikov, and V. M. Shelkovich, *Nonlinear problems in p-adic analysis: associative algebras of p-adic distributions*, Izvestia Ross. Akademii Nauk, Seria Math., **69**, no. 2, 3–44; English transl. in Izvestiya: Mathematics, **69**, no. 2, 221–263.

[21] S. Albeverio, A. Yu. Khrennikov, V. M. Shelkovich, *Harmonic analysis in the p-adic Lizorkin spaces: fractional operators, pseudo-differential equations, p-adic wavelets, Tauberian theorems*, Journal of Fourier Analysis and Applications, Vol. 12, Issue 4, (2006), 393–425.

[22] S. Albeverio, A. Yu. Khrennikov, and V. M. Shelkovich, *Pseudo-differential operators in the p-adic Lizorkin space*, in: *p*-Adic Mathematical Physics. 2nd International Conference, Belgrade, Serbia and Montenegro, 15–21 September 2005, Eds: Branko Dragovich, Zoran Rakic, Melville, New York, 2006, AIP Conference Proceedings – March 29, 2006, Vol. 826, Issue 1, pp. 195–205.

[23] S. Albeverio, A. Yu. Khrennikov, and V. M. Shelkovich, *p-adic semi-linear evolutionary pseudo-differential equations in the Lizorkin space* Dokl. Ross. Akad. Nauk, **415**, no. 3, (2007), 295–299. English transl. in Russian Doklady Mathematics, **76**, no. 1, (2007), 539–543.

[24] S. Albeverio, A. Yu. Khrennikov, and B. Tirozzi, *p-adic neural networks*, Mathematical Models and Methods in Applied Sciences, **9** (9), (1999), 1417–1437.

[25] S. Albeverio, S. V. Kozyrev, *Multidimensional ultrametric pseudodifferential equations*, Proc. Steklov Inst. Math., v. **265**, Moscow, 2009, 13–29.

[26] S. Albeverio, S. V. Kozyrev, *Frames of p-adic wavelets and orbits of the affine group*, *p*-Adic Numbers, Ultrametric Analysis and Applications, **1**, no. 1, (2009), 18–33.

[27] S. Albeverio, P. Kurasov, *Singular perturbations of differential operators*, in: London Math. Soc. Lecture Note Ser. Vol. 271, Cambridge Univ. Press, Cambridge, 2000.

[28] S. Albeverio and P. Kurasov, *Pseudo-differential operators with point interactions*, Lett. Math. Phys. **41**, (1997), 79–92.

[29] S. Albeverio and S. Kuzhel, *Pseudo-Hermiticity and theory of singular perturbations*, Lett. Math. Phys., **67**, (2004), 223–238.

[30] S. Albeverio and S. Kuzhel, *One dimensional Schrödinger operators with $\mathcal{P}$-symmetric zero-range potentials*, J. Phys. A., **38**, (2005), 4975–4988.

[31] S. Albeverio, S. Kuzhel, and S. Torba, *p-adic Schrödinger-type operator with point interactions*, J. Math. Anal. Appl., **338**, (2008), 1267–1281.

[32] S. Albeverio and X. Zhao, *On the relation between different constructions of random walks on p-adics*, Markov Process. Related Fields **6**, (2000), 239–256.

[33] S. Albeverio and X. Zhao, *Measure-valued branching processes associated with random walks on p-adics*, Ann. Probab. **28**, (2000), 1680–1710.

[34] S. Albeverio and X. Zhao, *A decomposition theorem of Lévy processes on local fields*, J. Theoret. Probab. **14**, (2001), 1–19.

[35] D. Aldous, and S. N. Evans, *Dirichlet forms on totally disconnected spaces and bipartite Markov chains*, J. Theoret. Probab., **12**, no. 3, (1999), 839–857.

[36] V. Anashin and A. Yu. Khrennikov, *Applied algebraic dynamics*, De Gruyter, Berlin, New York, 2009.

[37] T. Ando and K. Nishio, *Positive selfadjoint extensions of positive symmetric operators*, Tohoku Math. J. 22 (1970) 65–75.

[38] I. Ya. Aref'eva, B. G. Dragovic, and I. V. Volovich, *On the adelic string amplitudes*, Phys. Lett. **B, 209**, no. 4, (1998), 445–450.

[39] I. Ya. Aref'eva, I. V. Volovich, *Strings, gravity and p-adic space-time*, Quantum Gravity. Proceedings of the Fourth Seminar on Quantum Gravity, held May 25–29, 1987, in Moscow, USSR. Edited by M. A. Markov, V. A. Berezin and V. P. Frolov. Published by World Scientific, Singapore, 1988, p. 409

[40] I. Ya. Aref'eva, B. Dragovich, P. H. Frampton and I. V. Volovich, *Wave function of the Universe and p-adic gravity*, Int. J. Mod. Phys. A, 6 (1991), 4341–2307.

[41] I. Ya. Aref'eva, P. H. Frampton, *Beyond Planck energy to nonarchimedean geometry*, Mod. Phys. Lett. A, 6 (1991), 313–316.

[42] V. Arnold, S. Gusein-Zade, A. Varchenko, *Singularities of Differentiable Maps*, Vol. II. Birkhäuser, Boston Basel Berlin, 1988.

[43] V. A. Avetisov, A. H. Bikulov, S. V. Kozyrev, *Application of p-adic analysis to models of spontaneous breaking of replica symmetry*, J. Phys. A: Math. Gen., **32**, no. 50, (1999), 8785–8791. http://xxx.lanl.gov/abs/cond-mat/9904360

[44] V. A. Avetisov, A. H. Bikulov, S. V. Kozyrev, and V. A. Osipov, *p-adic models of ultrametric diffusion constrained by hierarchical energy landscapes*, J. Phys. A: Math. Gen., **35**, no. 2, (2002), 177–189. http://xxx.lanl.gov/abs/cond-mat/0106506

[45] V. A. Avetisov, A. H. Bikulov, and V. A. Osipov, *p-adic description of characteristic relaxation in complex systems*, J. Phys. A: Math. Gen., **36**, no. 15, (2003), 4239–4246.

[46] T. Ya. Azizov and I. S. Iokhvidov, *Linear Operators in Spaces with Indefinite Metric*, Longman/Wiley, New York, 1989.

[47] G. Bachman, *Introduction to p-adic numbers and valuation theory*, Academic Press, New York and London, 1964.

[48] C. M. Bender, D. C. Brody, and H. F. Jones, *Must a Hamiltonian be Hermitian?* Amer. J. Phys., **71**, (2003), 1095–1102.

[49] C. M. Bender, B. Tan, *Calculation of the hidden symmetry operator for a $\mathcal{PT}$-symmetric square well*, J. Phys. A., **39**, (2006), 1945–1953.

[50] J. J. Benedetto, and R. L. Benedetto, *A wavelet theory for local fields and related groups*, The Journal of Geometric Analysis, **3**, (2004), 423–456.

[51] J. J. Benedetto, A. I. Zayed (eds.), *Sampling, wavelets, and tomography*, Birkhäuser Boston, Basel, Berlin, 2004.

[52] R. L. Benedetto, *Examples of wavelets for local fields*, Wavelets, Frames, and Operator Theory, (College Park, MD, 2003), Am. Math. Soc., Providence, RI, (2004), 27–47.

[53] F. A. Berezin, L. D. Faddeev, *Remark on the Schrödinger equation with singular potential*, (Russian) Dokl. Akad. Nauk SSSR, **137**, (1961), 1011–1014.

[54] H. A. Biagioni, *A nonlinear theory of generalized functions*, Springer–Verlag, Berlin–Heildelberg–New York, 1990.

[55] A. H. Bikulov, and I. V. Volovich, *p-adic Brownian motion*, Izvestia Ross. Akademii Nauk, Seria Math., **61** no. 3, 75–90; English transl. in Izvestiya: Mathematics, **61** no. 3, 537–552.

[56] C. de Boor, R. DeVore, A. Ron, *On construction of multivariate (pre) wavelets*, Constr. Approx., V. 9 (1993), 123–166.

[57] H. Bremermann, *Distributions, Complex Variables, and Fourier Transforms*, Addison-Wesley Publ. Comp, Reading, Massachusetts, 1965.

[58] Yu. A. Brychkov, *Asymptotical expansions of distributions*, I. Theor. Math. Phys. **5** (1970), no. 1, 98–109.

[59] Yu. A. Brychkov, *On asymptotical expansions of distributions*, Mathematical Notices **12** (1972), no. 2, 131–138.

[60] Yu. A. Brychkov, Yu. M. Shirokov, *On some limiting formulas for generalized functions*, Mathematical Notices **2** (1967), no. 1, 81–91.

[61] Yu. A. Brychkov, Yu. M. Shirokov, *On asymptotical behavior of Fourier transform*. Theor. Math. Phys. **4** (1970), no. 3, 301–309.

[62] F. Bruhat, *Distributions sur un groupe localement compact et applications à l'étude des représentations des groupes p-adiques*, Bull. Soc. math. France, **89**, (1961), 43–75.

[63] Nguyen Minh Chuong, Nguyen Van Co, *The Cauchy problem for a class of pseudodifferential equations over p-adic field*, Journal of Mathematical Analysis and Applications, **340** (2008), 629–645.

[64] R. Cianci, A. Yu. Khrennikov, *p-adic numbers and the renormalization of eigenfunctions in quantum mechanics*, Phys. Letters B, **328**, Issues 1–2, (1994) 109–112.

[65] R. Cianci and A. Yu. Khrennikov, *Energy levels corresponding to p-adic quantum states*, Dokl. Akad. Nauk, **342**, (1995), no. 5, 603–606.

[66] J. F. Colombeau, *Elementary introduction to new generalized functions*. North-Holland Mathematics Studies 113, North-Holland, Amsteradm 1985.

[67] V. G. Danilov, V. M. Shelkovich, *Generalized solutions of nonlinear differential equations and the Maslov algebras of distributions,* Integral Transformations and Special Functions, **6**, (1997), no. 1–4, 137–146.

[68] V. G. Danilov, V. P. Maslov, V. M. Shelkovich, *Algebra of singularities of singular solutions to first-order quasilinear strictly hyperbolic systems,* (Russian), Teoret. Mat. Fiz., **114**, (1998), no. 1, 3–55; translation in Theoret. and Math. Phys., **114**, (1998), no. 1, 1–42.

[69] V. G. Danilov and V. M. Shelkovich, *Propagation and interaction of shock waves of quasilinear equation,* Nonlinear Studies, v. 8, no. 1, (2001), 135–170.

[70] V. G. Danilov, G. A. Omel'yanov, V. M. Shelkovich, *Weak Asymptotics Method and Interaction of Nonlinear Waves,* in Mikhail Karasev (ed.), "Asymptotic Methods for Wave and Quantum Problems", Amer. Math. Soc. Transl., Ser. 2, **208**, 2003, 33–165.

[71] V. G. Danilov, V. M. Shelkovich, *Propagation and interaction of δ-shock waves to hyperbolic systems of conservation laws,* Dokl. Ross. Akad. Nauk, **394**, no. 1, (2004), 10–14. English transl. in Russian Doklady Mathematics., **69**, no. 1, 4–8.

[72] V. G. Danilov, V. M. Shelkovich, *Delta-shock wave type solution of hyperbolic systems of conservation laws,* Quarterly of Applied Mathematics, **63**, no. 3, (2005), 401–427.

[73] V. G. Danilov, V. M. Shelkovich, *Dynamics of propagation and interaction of delta-shock waves in conservation law systems,* Journal of Differential Equations, **211**, (2005), 333–381.

[74] V. Derkach, S. Hassi, and H. de Snoo, *Singular perturbations of self-adjoint operators,* Math. Phys. Anal. Geometry, **6**, (2003), 349–384.

[75] V. A. Derkach and M. M. Malamud, *Generalized resolvents and the boundary value problems for Hermitian operators with gaps,* J. Funct. Anal., **95**, (1991), 1–95.

[76] I. Daubechies, *Ten Lectures on wavelets,* CBMS-NSR Series in Appl. Math., SIAM, 1992.

[77] Branko G. Dragovic, *On signature change in p-adic space-times,* Mod. Phys. Lett., A6 (1991), 2301–2307.

[78] B. Dragovich, A. Yu. Khrennikov, S. V. Kozyrev, and I. V. Volovich, *On p-adic mathematical physics p*-Adic Numbers, Ultrametric Analysis and Applications, **1**, no. 1, (2009), 1–17.

[79] B. Dragovich, Lj. Nesic, *On p-adic numbers in gravity,* Balkan Physics Letters 6, (1998), 78–81.

[80] Yu. N. Drozzinov, B. I. Zavialov, *Quasi-asymptotics of generalized functions and Tauberian theorems in the complex domain,* Math. Sb., **102**, (1977), no. 3, 372–390. English transl. in Math. USSR Sb., **31**.

[81] Yu. N. Drozzinov, B. I. Zavialov, *Multidimensional Tauberian theorems for generalized functions with values in Banach spaces,* Math. Sb., **194**, (2003), no. 11, 17–64. English transl. in Sb. Math., **194**, (2003), no. 11–12, 1599–1646.

[82] S. D. Eidelman, and A. N. Kochubei, *Cauchy problem for fractional diffusion equations,* Journal of Differential Equations, **199**, (2004), 211–255.

[83] S. D. Eidelman, S. D. Ivasyshen, and A. N. Kochubei, *Analytic Methods in the Theory of Differential and Pseudo-Differential Equations of Parabolic Type,* Birkhäuser, Basel, 2004.

[84] A. Erdélyi, *Asymptotic representations of Fourier integrals and the method of stationary phase*, J. Soc. Indust. Appl. Math., **3**, (1955), 17–27.

[85] A. Erdélyi, *Asymptotic expansions of Fourier integrals involving logarithmic singularities*, J. Soc. Indust. Appl. Math., **4**, (1956), 38–47.

[86] Yu. V. Egorov, *On contribution to the theory generalized functions*, Uspehi Mat. Nauk, **45**, no. 5, 1990, 3–40. English transl. in Russian Math. Surveys, **45**, no. 5, (1990), 1–49.

[87] R. Estrada, R. P. Kanwal, *Asymptotic analysis: A distributional approach.* Birkhäuser Boston, Inc., Boston, MA, 1994.

[88] R. Estrada, R. P. Kanwal, *A distributional approach to asymptotics: Theory and applications.* Second edition. Birkhäuser Boston, Inc., Boston, 2002.

[89] S. N. Evans, *p-adic white noise, chaos expansions, and stochastic integration*, Probability measures on groups and related structures, XI, (Oberwolfach, 1994), World Sci. Publishing, River Edge, NJ, (1995), pp. 102–115.

[90] S.A. Evdokimov, and M.A. Skopina, *2-adic wavelet bases*, Proceedings of Institute of Mathematics and Mechanics of the Ural Branch of the Russian Academy of Sciences **v. 15**, no. 1, (2009), 135–146. (in Russian)

[91] Yu. A. Farkov, and V. Yu. Protasov, *Dyadic wavelets and refinable functions on a half-line*, Mat. Sbornik, **197**, no. 10, (2006), 129–160; English transl. in Sbornik: Mathematics, **197**, (2006), 1529–1558.

[92] M. V. Fedoryuk, *Asymptotics: Integrals and Series*, Moscow, Nauka (in Russian), 1987.

[93] S. Fischenko, E. Zelenov, *p-Adic Models of Turbulence* in: *p*-Adic Mathematical Physics. 2nd International Conference, Belgrade, Serbia and Montenegro, 15–21 September 2005, Eds: Branko Dragovich, Zoran Rakic, Melville, New York, 2006, AIP Conference Proceedings – March 29, 2006, Vol. 826, Issue 1, pp. 174–191.

[94] P. G. O. Freund, and E. Witten, *Adelic string amplitudes*, Phys. Lett. **B, 199**, (1987), 191–195.

[95] I. M. Gel'fand, and G. E. Shilov, *Generalized functions. vol 1: Properties and operations.* New York, Acad. Press, 1964.

[96] I. M. Gel'fand, M. I. Graev, and I. I. Piatetskii-Shapiro, *Generalized functions. vol 6: Representation theory and automorphic functions.* Nauka, Moscow, 1966. Translated from the Russian by K.A. Hirsch, Published in 1990, Academic Press (Boston).

[97] M. Grosser, M. Kunzinger, M. Oberguggenberger, R. Steinbauer, *Geometric theory of generalized functions with applications to general relativity.* Academic Publ., Dordrecht, 2001.

[98] F. Q. Gouvêa, *p-Adic Numbers, An Introduction*, Springer-Verlag, Berlin–Heidelberg–New York, second edition, 1997.

[99] Olaf von Grudzinski, *Quasihomogeneous distributions.* Amsterdam, New York: North-Holland, Elsevier Science Pub. Co., 1991.

[100] U. Gunther, F. Stefani, and M. Znojil, *MHD α^2-dynamo, Squire equation and $\mathcal{PT}$-symmetric interpolation between square well and harmonic oscillator*, J. Math. Phys., **46**, (2005), 063504–063526.

[101] A. Haar, *Zur Theorie der orthogonalen Funktionensysteme*, Math. Ann., **69**, (1910), 331–371.

[102] S. Hassi and S. Kuzhel, *On symmetries in the theory of finite rank singular perturbations*, Journal of Functional Analysis, **256**, (2009), 777–809.

[103] S. Hassi and S. Kuzhel, *On symmetries in the theory of singular perturbations*, Preprint University of Vaasa, 2006 (http://www.uwasa.fi/julkaisu/).

[104] D. B. Heifetz, *p-adic oscillatory integrals and wave front sets*, Pacific Journal of Mathematics, **116**, (1985), 285–305.

[105] E. Hewitt, K. A. Ross, *Abstract Harmonic Analysis*. v. I and II, Springer 1963.

[106] L. Hörmander, *The Analysis of Linear Partial Differential Operators*. vol 1: Distribution Theory and Fourier Analysis. Springer-Verlag, Berlin, Heidelberg, New York, Tokyo, 1983.

[107] J. Igusa, *A stationary phase formula for p-adic integrals and its applications*, In: C. L. Bajaj (ed.), "Algebraic Geometry and Its Applications", Springer, New York, 1994, pp. 175–194.

[108] V. K. Ivanov, *Asymptotical approximation to the product of generalized functions*, Izv. Vyssh. Uchebn. Zaved. Mat., no. 1, (1981), 19–26 (in Russian).

[109] D. S. Jones, *Generalized Functions*. McGraw-Hill, 1966.

[110] I. Karamata, *Sur un mode de croissance régulière des fonctions*, Matematica (Gluj), **4**, (1930), 38–53.

[111] G. Kaiser, *A friendly guide to wavelets*, Birkhäuser Boston, 1994.

[112] S. Katok, *p-adic analysis compared with real*, American Mathematical Society, 2007.

[113] B. L. Keyfitz, *Conservation laws, delta-shocks and singular shocks*, In: Grosser, M., Horman, G., Kunzinger, M., Oberguggenberger, M. (eds) Nonlinear theory of generalized functions. Boca Raton: Chapman and Hall/CRC press, (1999), 99–111.

[114] A. Yu. Khrennikov, *Fundamental solutions over the field of p-adic numbers*, St. Petersburg Math. J., **4**, no. 3, (1993), 613–628.

[115] A. Yu. Khrennikov, *p-adic valued distributions in mathematical physics*, Kluwer Academic Publ., Dordrecht, 1994.

[116] A. Yu. Khrennikov, *Non-archimedean analysis: quantum paradoxes, dynamical systems and biological models*, Kluwer Academic Publ., Dordrecht, 1997.

[117] A. Yu. Khrennikov, *Information dynamics in cognitive, psychological, social and anomalous phenomena*, Kluwer Academic Publ., Dordrecht, 2004.

[118] A. Yu. Khrennikov, *p-adic discrete dynamical systems and their applications in physics and cognitive science*, Russian Journal of Mathematical Physics, **11**, no. 1, (2004), 45–70.

[119] A. Yu. Khrennikov. *Mathematical methods of the non-archimedean physics*, Uspekhi Mat. Nauk, **45**:4(274), (1990), 79–110. English transl. in Russian Math. Surveys, **45**:4, (1990), 87–125.

[120] A. Yu. Khrennikov, *Quantum mechanics over Galois extensions of number fields*, Dokl. Akad. Nauk SSSR, Physics, **315**, (1990), no. 4, 860–864; translation in Soviet Physics Doklady, v. 35, (1990), 1032–1036.

[121] A. Yu. Khrennikov. *p-adic quantum mechanics with p-adic valued functions*, J. Math. Phys., **32**, (1991), 932–937.

[122] A. Yu. Khrennikov, *Real-non-Archimedean structure of space-time* (Russian), Teoret. Mat. Fiz., **86**, (1991), no. 2, 177–190; translation in Theoret. and Math. Phys., **86**, (1991), no. 2, 121–130.

[123] A. Yu. Khrennikov, *Description of experiments detecting p-adic statistics in quantum diffraction experiments*, Russian Doklady Mathematics, **58**, (1998), no. 3, 478–480.

[124] A. Yu. Khrennikov, *Ultrametric Hilbert space representation of quantum mechanics with a finite exactness*, Found. Physics, **26**, (1996), no. 8, 1033–1054.

[125] A. Yu. Khrennikov, *On probablity distributions on the field of p-adic numbers*, Theory of Probab. and Appl., **40**, (1995), no. 1, 189–192.

[126] A.Yu. Khrennikov, *p-adic probability and statistics*, Dokl. Akad. Nauk, **322**, (1992), 1075–1079.

[127] A. Yu. Khrennikov, *p-adic probability theory and its applications. The principle of statistical stabilization of frequencies*, Theor. and Math. Phys., **97**, (1993), no. 3, 348–363.

[128] A.Yu. Khrennikov and Huang Zhiyan, *Generalized functionals of p-adic white noise*, Dokl. Akad. Nauk, **344**, (1995), no. 1, 23–26.

[129] A. Yu. Khrennikov and Zhiyuan Huang, *A model for white noise analysis in p-adic number fields*, Acta Mathematica Scientia (China), **16**, (1996), no. 1, 1–14.

[130] A. Yu. Khrennikov, *Interpretations of probability*, De Gruyter, Berlin, New York, 2009, second edition.

[131] A. Yu. Khrennikov, and S. V. Kozyrev, *Wavelets on ultrametric spaces*, Applied and Computational Harmonic Analysis, **19**, (2005), 61–76.

[132] A. Yu. Khrennikov, and S. V. Kozyrev, *Pseudodifferential operators on ultrametric spaces and ultrametric wavelets*, Izvestia Ross. Akademii Nauk, Seria Math., **69**, no. 5, 133–148; English transl. in Izvestiya: Mathematics, **69**, no. 5, 989–1003.

[133] A. Yu. Khrennikov, and S. V. Kozyrev, *Localization in space for free particle in ultrametric quantum mechanics*, Dokl. Ross. Akad. Nauk, **411** no. 3, (2006), 319–322. English transl. in Russian Doklady Mathematics, **74**, no. 3, (2006), 906–911.

[134] A.Yu. Khrennikov, S.V. Kozyrev, *Wavelets and the Cauchy problem for the Schrödinger equation on analytic ultrametric space*, P.344–350, in: Proceedings of the 2nd Conference on Mathematical Modelling of Wave Phenomena 2005, 14–19 August 2005, Växjö, Sweden, eds. B. Nilsson, L. Fishman, AIP Conference Proceedings, Vol. 834, Melville, New York, 2006.

[135] A. Yu. Khrennikov, S. V. Kozyrev, *Ultrametric random field*, Infinite Dimensional Analysis, Quantum Probability and Related Topics, Vol. 9, No. 2 (2006), P.199–213.

[136] A. Yu. Khrennikov, S. V. Kozyrev, *Replica symmetry breaking related to a general ultrametric space I: replica matrices and functionals*, Physica A: Statistical Mechanics and its Applications, **359**, (2006), 222–240.

[137] A. Yu. Khrennikov, S. V. Kozyrev, *Replica symmetry breaking related to a general ultrametric space II: RSB solutions and the $n \rightarrow 0$ limit*, Physica A: Statistical Mechanics and its Applications, **359**, (2006), 241–266.

[138] A. Yu. Khrennikov, S. V. Kozyrev, *Replica symmetry breaking related to a general ultrametric space III: The case of general measure*, Physica A: Statistical Mechanics and its Applications, **378**, no. 2, (2007), 283–298.
http://arxiv.org/abs/cond-mat/0603694

[139] A. Yu. Khrennikov and M. Nilsson, *p-Adic Deterministic and Random Dynamics*, Kluwer, Dordrecht, 2004.

[140] A. Yu. Khrennikov, and V. M. Shelkovich, *Tauberian theorems for p-adic distributions*, Integral Transforms and Special Functions, **17**, no. 02–03, (2006), 141–147.

[141] A. Yu. Khrennikov, and V. M. Shelkovich, *Distributional asymptotics and p-adic Tauberian and Shannon-Kotelnikov theorems*, Asymptotical Analysis, **46** (2), (2006), 163–187.

[142] A. Yu. Khrennikov, and V. M. Shelkovich, *p-adic multidimensional wavelets and their application to p-adic pseudo-differential operators*, (2006), Preprint at the url: http://arxiv.org/abs/math-ph/0612049

[143] A. Yu. Khrennikov, V. M. Shelkovich, *Non-Haar p-adic wavelets and their application to pseudo-differential operators and equations*, Applied and Computational Harmonic Analysis, (2010), 1–23.

[144] A. Yu. Khrennikov, V. M. Shelkovich, *Non-Haar p-adic wavelets and pseudo-differential operators*, Dokl. Ross. Akad. Nauk, **418**, no. 2, (2008), 167–170. English transl. in Russian Doklady Mathematics, **77**, no. 1, (2008), 42–45.

[145] A. Yu. Khrennikov, V. M. Shelkovich, *Asymptotical behavior of one class of p-adic singular Fourier integrals*, J. Math. An. Appl., **350**, Issue 1, (2009), 170–183.

[146] A. Yu. Khrennikov, V. M. Shelkovich, *An infinite family of p-adic non-Haar wavelet bases and pseudo-differential operators*, p-Adic Numbers, Ultrametric Analysis and Applications, **1**, no. 3, (2009), 204–216.

[147] A.Yu. Khrennikov, V.M. Shelkovich, M. Skopina, *p-adic orthogonal wavelet bases*, p-Adic Numbers, Ultrametric Analysis and Applications, **1**, no. 2, (2009), 145–156.

[148] A.Yu. Khrennikov, V.M. Shelkovich, M. Skopina, *p-adic refinable functions and MRA-based wavelets*, Journal of Approximation Theory, **161**, (2009), 226–238. (See also as the Preprint at the url: http://arxiv.org/abs/0711.2820)

[149] A. Yu. Khrennikov, V. M. Shelkovich, and O. G. Smolyanov, *Multiplicative structures in the linear space of vector-valued distributions*, Dokl. Ross. Akad. Nauk, **383**, no. 1, (2002) 28–31; English transl. in Russian Doklady Mathematics., **65**, no. 2, (2002), 169–172.

[150] A. Yu. Khrennikov, V. M. Shelkovich, and O. G. Smolyanov, *Locally convex spaces of vector-valued distributions with multiplicative structures*, Infinite-Dimensional Analysis, Quantum Probability and Related Topics, **5**, no. 4, (2002), 1–20.

[151] A. Yu. Khrennikov, V. M. Shelkovich, and O. G. Smolyanov, *An associative algebra of vector-valued distributions and singular solutions of nonlinear equations*, in Mathematical modelling in physics, engineering and cognitive sciences. v. 7. Proceedings of the conference "Mathematical Modelling of Wave Phenomena", November 2002. Edited by B. Nilsson and L. Fishman, Växjö University Press, 2004, 191–205.

[152] N. Koblitz, *p-adic numbers, p-adic analysis, and zeta-functions*. Second edition. Graduate Texts in Mathematics, 58. Springer-Verlag, New York, 1984.

[153] A. N. Kochubei, *Parabolic equations over the field of p-adic numbers*, Izv. Akad. Nauk SSSR Ser. Mat., **55**, no. 6 (1991), 1312–1330. English transl. in Math. USSR Izv., **39**, (1992), 1263–1280

[154] A. N. Kochubei, *Schrödinger-type operator over the p-adic number field*, Teoret. Mat. Fiz., **86**, (1991), no. 3, 323–333; translation in Theoret. and Math. Phys., **86**, (1991), no. 3, 221–228.

[155] A. N. Kochubei, *The differentiation operator on subset of the field of p-adic numbers*, Russ. Acad. Sci. Izv. Math. **41**, (1993), 289–305.

[156] A. N. Kochubei, *Additive and multiplicative fractional differentiations over the field of p-adic numbers*, in: W .H. Schikhov et al. (Eds.), *p*-adic Functional Analysis, Lecture Notes in Pure Appl. Math., vol. 192, Marcel Dekker, New York, 1997, pp.275–280.

[157] A. N. Kochubei, *Pseudo-differential equations and stochastics over non-archimedean fields*, Marcel Dekker. Inc. New York, Basel, 2001.

[158] A. N. Kochubei, *Fundamental solutions of pseudo-differential equations associated with p-adic quadratic forms*, Russ. Acad. Sci. Izv. Math. **62**, (1998), 1169–1188.

[159] A. N. Kochubei, *A non-Archimedean wave equation*, Pacific Journal of Mathematics, **235**, (2008), no. 2, 245–261.

[160] J. Korevaar, *Tauberian theory. A century of developments*. Grundlehren der Mathematischen Wissenschaften [Fundamental Principles of Mathematical Sciences], 329. Springer-Verlag, Berlin, 2004.

[161] A. V. Kosyak, *Regular, quasiregular and induced representations of infinite-dimensional groups*, 450 p. (in preparation).

[162] S. V. Kozyrev, *Methods and applications of ultrametric and p-adic analysis: from wavelet theory to biophysics*, Sovrem. Probl. Mat. Issue 12. Steklov Inst. Math., Moscow, 2008. http://www.mi.ras.ru/spm/pdf/012.pdf.

[163] S. V. Kozyrev, *Wavelet analysis as a p-adic spectral analysis*, Izvestia Ross. Akademii Nauk, Seria Math., **66** no. 2, 149–158; English transl. in Izvestiya: Mathematics, **66** no. 2, 367–376.

[164] S. V. Kozyrev, *p-adic pseudodifferential operators: methods and applications*, Proc. Steklov Inst. Math., **v. 245**, Moscow, 2004, 154–165.

[165] S. V. Kozyrev, *p-adic pseudodifferential operators and p-adic wavelets*, Theor. Math. Physics, **138**, no. 3, 1–42; English transl. in Theoretical and Mathematical Physics, **138**, no. 3, 322–332.

[166] S. V. Kozyrev, *Towards ultrametric theory of turbulence*, Teoret. Mat. Fiz., **157**, no. 3, (2008), 413–424; translation in Theoretical and Mathematical Physics, **157**, no. 3, (2008), 1713–1722.

[167] V. A. Kudryavzev, *Summation of the powers to the positive integers and the Bernoulli numbers*, M.-L., ONTI NKTP USSR, 1936. (In Russian)

[168] H. H. Kuo, *Gaussian measures in Banach spaces*, in: Lecture Notes Mathematics, **463**, Springer, Berlin, 1975.

[169] S. Kuzhel, and S. Torba, *p-adic fractional differential operator with point interactions*, Meth. Funct. Anal. Topology, **13**, (2007), no. 2, 169–180.

[170] W. C. Lang, *Wavelet analysis on the Cantor dyadic group*, Houston J. Math., **24**, (1998), 533–544.

[171] W. C. Lang, *Orthogonal wavelets on the Cantor dyadic group*, SIAM J. Math. Anal., **27**, (1996), 305–312.

[172] D. Laugwitz, *Anwendungen unendlicher kleiner Zahlen: I. Zur Theorie der Distributionen*, J. reine und angew. Math., **207** (1961), no. 1–2, 53–70.

[173] D. Laugwitz, *Anwendungen unendlicher kleiner Zahlen: II. Ein Zugang zur Operatorenrechnung von Mikusinski*, J. reine und angew. Math. **208** (1961), no. 1–2, 22–34.

[174] E. Yu. Lerner and M. D. Missarov, *Scalar models of p-adic quantum field theory and hierarchical models*, Theoretical and Mathematical Physics, **78**, (1989), 177–184. Translated from Teoreticheskaya i Matematicheskaya Fizika, **78**, no. 2, (1989), 248–257.

[175] M. Lesch, *On the noncommutative residue for pseudodifferential operators with log-polyhomogeneous symbols*, **17**, (1999), 151–187.

[176] Li Bang-He, *Non-standard analysis and multiplication of distributions*, Acta scientia sinica, (1978), v. 21, no. 5, 561–585.

[177] Ya. B. Livchak, *On limiting distributions*, Mat. Zametki (Mathematical Notes) **VI** (1968), no. 3, Ural'sky University, Sverdlovsk, 29–37 (in Russian).

[178] Ya. B. Livchak, *On the definition of generalized function*, Mat. Zametki (Mathematical Notes) **VI** (1968), no. 3, Ural'sky University, Sverdlovsk, 38–44 (in Russian).

[179] Ya. B. Livchak, *To the theory of generalized functions*. Trudy Rizhskogo Algebr. Seminar (Proceedings of the Riga Algebraic Seminar), Izd. Rizhsk. Gos. Univ., Riga 1969, 98–164 (In Russian).

[180] P. I. Lizorkin, *Generalized Liouville differentiation and the functional spaces $L_p{}^r(E_n)$. Imbedding theorems*, (Russian) Mat. Sb. (N.S.), **60** (102), (1963), 325–353.

[181] P. I. Lizorkin, *Generalized Liouville differentiation and the multiplier method in the theory of imbeddings of classes of differentiable functions*, (Russian) Trudy Mat. Inst. Steklov., **105**, (1969), 89–167.

[182] P. I. Lizorkin, *Operators connected with fractional differentiation, and classes of differentiable functions*, (Russian) Studies in the theory of differentiable functions of several variables and its applications, IV. Trudy Mat. Inst. Steklov. **117**, (1972), 212–243.

[183] E. Mayerhofer, *On the characterrization of p-adic Colombeau–Egorov generalized functions by their point values*, Mathematische Nachrichten, **280**, no. 11, (2007), 1297–1301.

[184] S. Mallat, *Multiresolution representation and wavelets*, Ph. D. Thesis, University of Pennsylvania, Philadelphia, PA, 1988.

[185] S. Mallat, *An efficient image representation for multiscale analysis*, In: Proc. of Machine Vision Conference, Lake Taho. 1987.

[186] Y. Meyer, *Ondelettes and fonctions splines*, Séminaire EDP. Paris. December 1986.

[187] Y. Meyer, *Principe d'incertitude, bases hilbertiennes et algèbres d'opérateurs*, Séminaire N. Bourbaki, 1985–1986, exp. no. 662, p. 209–223.

[188] A. Monna, *Analyse non-Archimedienne*, Springer-Verlag, New York, 1970.

[189] A. Mostafazadeh, *Pseudo-Hermiticity versus PT-symmetry: the necessary condition for the reality of the spectrum of a non-Hermitian Hamiltonian*, J. Math. Phys. **43**, (2002), 205–214.

[190] A. Mostafazadeh, *Pseudo-Hermiticity and generalized PT and CPT-symmetries*, J. Math. Phys., **44**, (2003), 979–989.

[191] D. Nikolić-Despotović, S. Pilipović, *The quasiasymptotic expansion at the origin. Abelian-type results for Laplace and Stieltjes transform*, Math. Nachr. **174**, (1995), 231–238.

[192] I. Novikov, V. Protassov, and M. Skopina, *Wavelet Theory*. Moscow: Fizmatlit, 2005. (In Russian)

[193] M. Oberguggenberger, *Generalized functions in nonlinear models – a survey*, Nonlinear Analysis, **47** (2001), 5029–5040.

[194] M. Oberguggenberger, *Multiplication of distributions and applications to partial differential equations*. Longman, Harlow, UK, 1992.

[195] M. Oberguggenberger and F. Russo, *Nonlinear SPDEs: Colombeau solutions and pathwise limits*. In: L. Decreusefond, J. Gjerde, B. Oksendal, and A. S. Üstünel (eds.), *Stochastic Analysis and Related Topics*, VI, Birkhaeuser, Boston, 1998, pp. 319–332.

[196] M. Oberguggenberger, M. Kunzinger, *Characterization of Colombeau generalized functions by their pointvalues*, Math. Nachr., **203**, (1999), 147–157.

[197] M. Oberguggenberger, S. Pilipovic, D. Scarpalezos, *Local properties of Colombeau generalized functions*, Math. Nachr., **256**, 2003, 88–99.

[198] G. Parisi, *p-adic functional integral*, Mod. Phys. Lett., **A**, **4**, (1988), 369–374.

[199] G. Parisi, N. Sourlas, *p-adic numbers and replica symmetry breaking*, The European Physical Journal B, **14**, (2000), 535–542. http://arxiv.org/abs/cond-mat/9906095

[200] F. Pham, *Singularités des processus de diffusion multiple*, Annales de l'Institut Henry Poincaré, Section A., Physique Théorique, Vol. VI, no. 2, 1967, 89–204.

[201] A. Posilicano, *A Krein-like formula for singular perturbations of self-adjoint operators and applications*, J. Funct. Anal., **183**, (2001), 109–147.

[202] E. M. Radyno, and Ya. V. Radyno, *Distributions and mnemofunctions on adeles. The Fourier transform*. Proc. Steklov Inst. Math., **v. 245**, no. 2, Moscow, 2004, 215–227.

[203] H. Reiter, J. D. Stegeman, *Classical Harmonic Analysis and Locally Compact Groups*, Oxford University Press; Second Edition, 2000.

[204] A. M. Robert, *A course in p-adic analysis*, Graduate Texts in Math., vol. 198, Springer-Verlag, New York, 2000.

[205] J. J. Rodríguez-Vega and W. A. Zuniga-Galindo, *Taibleson operators, p-adic parabolic equations and ultrametric diffusion*, Pacific Journal of Mathematics, **237**, (2008), no. 2, 327–347.

[206] B. G. Rubin, *Fractional integrals and potentials*, Addison Wesley Longman Limited, Edinburgh Gate, Harlow, Essex, 1996.

[207] W. Rudin, *Real and complex analysis*, McGraw-Hill International Editions: Mathematics Series, 1987.

[208] S. G. Samko, *Test functions that vanish on a given set, and division by a function*, (Russian) Mat. Zametki, **21**, no. 5, (1977), 677–689; English transl. Mathematical Notes, **21**, no. 5, (1977), 379–386.

[209] S. G. Samko, *Density in $L_p(R^n)$ of spaces Φ_V of Lizorkin type*, (Russian) Mat. Zametki, **31**, no. 6, (1982), 855–865; English transl. Mathematical Notes, **31**, no. 6, (1982), 432–437.

[210] S. G. Samko, *Hypersingular integrals and their applications*. Taylor & Francis, London, 2002.

[211] S. G. Samko, A. A. Kilbas, and O. I. Marichev, *Fractional integrals and derivatives and some of their applications*. Minsk, Nauka i Tekhnika, 1987 (in Russian); English translation: Fractional integrals and derivatives. Theory and applications, Gordon and Breach, London, 1993.

[212] A. M. Savchuk, A. A. Shkalikov, *Sturm–Liouville operators with distribution potentials*, Trans. Mosc. Math. Soc. 2003, 143–192, (2003); translation from Tr. Mosk. Mat. O.-va **64** (2003), 159–212.

[213] H. Schaefer, *Topological vector spaces*. New York–London, 1966.

[214] W. H. Schikhof, *Ultrametric calculus. An introduction to p-adic analysis*. Cambridge University Press, Cambridge, 1984.

[215] C. Schmieden, D. Laugwitz, *Eine Erweiterung der Infinitesimalrechnung*, Math. Z. **69** (1958), 1–39.

[216] E. Seneta, *Regularly varying functions*, Lecture Notes in Math., vol. 508, Springer-Verlag, Berlin, 1976.

[217] V. M. Shelkovich, *An associative algebra of distributions and multipliers*, Dokl. Akad. Nauk SSSR, **314**, (1990), no. 1, 159–164; translation in Soviet Math. Dokl., v. 42, 1991, 409–414.

[218] V. M. Shelkovich, A. P. Yuzhakov, *The structure of a class of asymptotic distributions of V. K. Ivanov*, (Russian) Izv. Vyssh. Uchebn. Zaved. Mat., (1991), 70–73; translation in Soviet Math. (Iz. VUZ), **35**, (1991), 70–72.

[219] V. M. Shelkovich, *The Colombeau algebra and an algebra of asymptotical distributions*, Proceedings of the International Conference on Generalized Functions (ICGF 2000), eds. A. Delcroix, M. Hasler, J.-A. Marti, V. Valmorin, University of French West Indies. Cambridge Scientific Publishers Ltd., 2004, 317–328.

[220] V. M. Shelkovich, *Associative and commutative distribution algebra with multipliers, and generalized solutions of nonlinear equations*, Mat. Zametki, **57**, no. 5, (1995),765–783 (in Russian); English translation in Mathematical Notices, **57**, no. 5, (1995), 536–549.

[221] V. M. Shelkovich, *Colombeau generalized functions: a theory based on harmonic regularizations*, Mat. Zametki, **63**, no. 2, (1998), 313–316 (in Russian); English transl. in Mathematical Notices, **63**, no. 2, (1998), 313–316.

[222] V. M. Shelkovich, *New versions of the Colombeau algebras*, Mathematische Nachrichten, **278**, no. 11, 2005, 1–23.

[223] V. M. Shelkovich, *Associated and quasi associated homogeneous distributions (generalized functions)*, J. Math. An. Appl., **338**, (2008), 48–70.

[224] V. M. Shelkovich, *Delta-shock waves of a class of hyperbolic systems of conservation laws*, in A. Abramian, S. Vakulenko, V. Volpert (Eds.), "Patterns and Waves", Akadem Print, St. Petersburg, 2003, 155–168.

[225] V. M. Shelkovich, M. Skopina, *p-adic Haar multiresolution analysis and pseudo-differential operators*, Journal of Fourier Analysis and Applications, Vol. 15, Issue 3, (2009), 366–393.

[226] Yu. M. Shirokov, *On microcovariance and microcausality*. Nucl. Phys. **46** (1963), 617–638.

[227] B. Simon, *Trace Ideals and Their Applications*, 2nd ed. AMS Publishing, Providence, RI, Vol. 120, 2005.

[228] V. A. Smirnov, *Renormalization in p-adic quantum mechanics*, Modern Phys. Lett. A **6**(15), (1991), 1421–1427.

[229] M. H. Taibleson, *Harmonic analysis on n-dimensional vector spaces over local fields. I. Basic results on fractional integration*, Math. Annalen, **176**, (1968), 191–207.

[230] M. H. Taibleson, *Fourier analysis on local fields*. Princeton University Press, Princeton, 1975.

[231] T. Tanaka, *General aspects of $\mathcal{PT}$-symmetric and $\mathcal{P}$-self-adjoint quantum theory in a Krein space*, J. Phys. A., **39**, (2006), 14175–14203.

[232] N. Ya. Vilenkin, *Generalized functions*, In the book: Funktsional'nyi analiz. (Russian) [Functional analysis] Edited by S. G. Krein. Second edition, revised and augmented. Mathematical Reference Library. Izdat. "Nauka", Moscow, 1972, Ch.X.

[233] V. S. Vladimirov, *The Freund–Witten Adelic Formulas for Veneziano and Virasoro–Shapiro Amplitudes*, Usp. Mat. Nauk, **48**, (1993), no. 6, 3–38 (in Russian); English translation in Russ. Math. Surv., **48**, (1993), no. 6, 1–39.

[234] V. S. Vladimirov, *Adelic Formulas for Gamma and Beta Functions of Completions of Algebraic Number Fields and Their Applications to String Amplitudes*, Izv. Ross. Akad. Nauk, Ser. Mat., **60**, (1996), no. 1, 63–86 (in Russian); English translation in Izvestiya: Mathematics, **60**, (1996), no. 1, 67–90.

[235] V. S. Vladimirov, *Adelic Formulas for Gamma- and Beta-Functions in Algebraic Number Fields*, in p-Adic Functional Analysis, New York: Dekker, 1997, pp. 383–395. (Lect. Notes Pure Appl. Math., vol. 192).

[236] V. S. Vladimirov, *Generalized Functions in Mathematical Physics*, MIR, Moscow, 1979 (Translated from Russian).

[237] V. S. Vladimirov, *On the spectrum of some pseudo-differential operators over p-adic number field*, Algebra and analysis, **2**, no. 6, 1990, 107–124.

[238] V. S. Vladimirov, *p-adic analysis and p-adic quantum mechanics*, Ann. of the NY Ac. Sci.: Symposium in Frontiers of Math., 1988.

[239] V. S. Vladimirov, *Generalized functions over p-adic number field*, Russian Math. Surveys, **43**, no. 5, 19–64; translated from Uspekhi Mat. Nauk., **43**, no. 5, (1988), 17–53.

[240] V. S. Vladimirov, Yu. N. Drozzinov, and B. I. Zavyalov, *Tauberian theorems for generalized functions*, Kluwer Academic Publishers, Dordrecht–Boston–London, 1988.

[241] V. S. Vladimirov, I. V. Volovich, and E. I. Zelenov, *p-adic analysis and mathematical physics*. World Scientific Publishing, River Edge, NJ, 1994.

[242] V. S. Vladimirov and I. V. Volovich, *p-adic quantum mechanics*, Commun. Math. Phys., **123**, (1989), 659–676.

[243] V. S. Vladimirov and I.V. Volovich, *p-adic Schrödinger-type equation*, Lett. Math. Phys., **18**, (1989), 43–53.

[244] V. S. Vladimirov, and T. M. Zapuzhak, *Adelic formulas for string amplitudes in fields of algebraic numbers*, Lett. Math. Phys., **37**, (1996), 232–242.

[245] I. V. Volovich, *p-adic space-time and the string theory*, Theor. Math. Phys., **71**, no. 3, (1987), 337–340.

[246] I. V. Volovich, *p-adic string*, Class. Quant. Grav., **4**, (1987), L83–L87.

[247] A. Weil, *L'intégration dans les groupes topologiques et ses applications*, 2ᵉ ed., Hermann, Paris, 1953.

[248] A. L. Yakymiv, *Probabilistic applications of Tauberian theorems*, Modern Probability and Statistics, VSP, Utrecht, 2005.

[249] K. Yasuda, *Additive processes on local fields*, J. Math. Sci. Univ. Tokyo, **3**, (1996), 629–654.

[250] K. Yasuda, *On infinitely divisible distributions on locally compact abelian groups*, J. Theoret. Probab. **13**, (2000), 635–657.

[251] K. Yasuda, *Extension of measures to infinite-dimensional spaces over p-adic field*, Osaka J. Math., **37**, (2000), 967–985.

[252] A. I. Zaslavsky, *Multidimensional analogue of Erdélyi lemma and the Radon transform*, In E. Quinot, M. Cheney, P. Kuchment (Eds.), Tomography, Impedance Imaging, and Integral Geometry. Lecture in Applied Mathematics, Vol. 30, 1994.

[253] A. I. Zayed, *Advances in Shannon's Sampling Theory*. CRC Press, Boca Raton, 1993.

[254] E. I. Zelenov, *p-adic path integrals*, Journal Math. Phys., **32**, no. 1, (1991), 147–152.

[255] M. Znojil, *Experiments in PT-symmetric quantum mechanics. Pseudo-Hermitian Hamiltonians in quantum physics*, Czechoslovak J. Phys., **54**, (2004), 151–156.

[256] W. A. Zuniga-Galindo, *Pseudo-differential equations connected with p-adic forms and local zeta functions*, Bull. Austral. Math. Soc., **70**, no. 1, (2004), 73–86.

[257] W. A. Zuniga-Galindo, *Fundamental solutions of pseudo-differential operators over p-adic fields*, Rend. Sem. Mat. Univ. Padova, **109**, (2003), 241–245.

[258] W. A. Zuniga-Galindo, *p-adic oscillatory integrals and Newton polyhedra*, Rev. Acad. Colombiana Cienc. Exact. Fis. Natur. **28**, (2004), no. 106, 95–99.

Index